Lecture Notes in Computer Science

Lecture Notes in Artificial Intelligence 14829

Founding Editor

Jörg Siekmann

Series Editors

Randy Goebel, *University of Alberta, Edmonton, Canada*
Wolfgang Wahlster, *DFKI, Berlin, Germany*
Zhi-Hua Zhou, *Nanjing University, Nanjing, China*

The series Lecture Notes in Artificial Intelligence (LNAI) was established in 1988 as a topical subseries of LNCS devoted to artificial intelligence.

The series publishes state-of-the-art research results at a high level. As with the LNCS mother series, the mission of the series is to serve the international R & D community by providing an invaluable service, mainly focused on the publication of conference and workshop proceedings and postproceedings.

Andrew M. Olney · Irene-Angelica Chounta ·
Zitao Liu · Olga C. Santos · Ig Ibert Bittencourt
Editors

Artificial Intelligence in Education

25th International Conference, AIED 2024
Recife, Brazil, July 8–12, 2024
Proceedings, Part I

 Springer

Editors
Andrew M. Olney ⓘ
University of Memphis
Memphis, TN, USA

Irene-Angelica Chounta ⓘ
University of Duisburg-Essen
Duisburg, Germany

Zitao Liu ⓘ
Jinan University
Guangzhou, China

Olga C. Santos ⓘ
UNED
Madrid, Spain

Ig Ibert Bittencourt ⓘ
Universidade Federal de Alagoas
Maceio, Brazil

ISSN 0302-9743   ISSN 1611-3349 (electronic)
Lecture Notes in Artificial Intelligence
ISBN 978-3-031-64301-9   ISBN 978-3-031-64302-6 (eBook)
https://doi.org/10.1007/978-3-031-64302-6

LNCS Sublibrary: SL7 – Artificial Intelligence

Preface

The 25th International Conference on Artificial Intelligence in Education (AIED 2024) was hosted by Centro de Estudos e Sistemas Avançados do Recife (CESAR), Brazil from July 8 to July 12, 2024. It was set up in a face-to-face format but included an option for an online audience. AIED 2024 was the next in a longstanding series of annual international conferences for the presentation of high-quality research on intelligent systems and the cognitive sciences for the improvement and advancement of education. Note that AIED is ranked A in CORE (top 16% of all 783 ranked venues), the well-known ranking of computer science conferences. The AIED conferences are organized by the prestigious International Artificial Intelligence in Education Society, a global association of researchers and academics, which has already celebrated its 30th anniversary, and aims to advance the science and engineering of intelligent human-technology ecosystems that support learning by promoting rigorous research and development of interactive and adaptive learning environments for learners of all ages across all domains.

The theme for the AIED 2024 conference was "AIED for a World in Transition". The conference aimed to explore how AI can be used to enhance the learning experiences of students and teachers alike when disruptive technologies are turning education upside down. Rapid advances in Artificial Intelligence (AI) have created opportunities not only for personalized and immersive experiences but also for ad hoc learning by engaging with cutting-edge technology continually, extending classroom borders, from engaging in real-time conversations with large language models (LLMs) to creating expressive artifacts such as digital images with generative AI or physically interacting with the environment for a more embodied learning. As a result, we now need new approaches and measurements to harness this potential and ensure that we can safely and responsibly cope with a world in transition. The conference seeks to stimulate discussion of how AI can shape education for all sectors, how to advance the science and engineering of AI-assisted learning systems, and how to promote broad adoption.

AIED 2024 attracted broad participation. We received 334 submissions for the main program, of which 280 were submitted as full papers, and 54 were submitted as short papers. Of the full paper submissions, 49 were accepted as full papers, and another 27 were accepted as short papers. The acceptance rate for full papers and short papers together was 23%. These accepted contributions are published in the Springer proceedings volumes LNAI 14829 and 14830.

The submissions underwent a rigorous double-masked peer-review process aimed to reduce evaluation bias as much as possible. The first step of the review process was done by the program chairs, who verified that all papers were appropriate for the conference and properly anonymized. Program committee members were asked to declare conflicts of interest. After the initial revision, the program committee members were invited to bid on the anonymized papers that were not in conflict according to their declared conflicts of interest. With this information, the program chairs made the review assignment, which consisted of three regular members to review each paper plus a senior member to provide a meta-review. The management of the review process (i.e., bidding, assignment, discussion, and meta-review) was done with the EasyChair platform, which was configured so that reviewers of the same paper were anonymous to each other. A subset of the program committee members were not included in the initial assignment but were asked to be ready to do reviews that were not submitted on time (i.e., the emergency review period). To avoid a situation where program committee members would be involved in too many submissions, we balanced review assignments and then rebalanced them during the emergency review period.

As a result, each submission was reviewed anonymously by at least three Program Committee (PC) members and then a discussion was led by a Senior Program Committee (SPC) member. PC and SPC members were selected based on their authorship in previous AIED conferences, their experience as reviewers in previous AIED editions, their h-index as calculated by Google Scholar, and their previous positions in organizing and reviewing related conferences. Therefore all members were active researchers in the field, and SPC members were particularly accomplished on these metrics. SPC members served as meta-reviewers whose role was to seek consensus to reach the final decision about acceptance and to provide the corresponding meta-review. They were also asked to check and highlight any possible biases or inappropriate reviews. Decisions to accept/reject were taken by the program chairs. For borderline cases, the contents of the paper were read in detail before reaching the final decision. In summary, we are confident that the review process assured a fair and equal evaluation of the submissions received without any bias, as far as we are aware.

Beyond paper presentations, the conference included a Doctoral Consortium Track, Late-Breaking Results, a Workshops and Tutorials Track, and an Industry, Innovation and Practitioner Track. There was a WideAIED track, which was established in 2023, where opportunities and challenges of AI in education were discussed with a global perspective and with contributions coming also from areas of the world that are currently under-represented in AIED. Additionally, a BlueSky special track was included with contributions that reflect upon the progress of AIED so far and envision what is to come in the future. The submissions for all these tracks underwent a rigorous peer review process. Each submission was reviewed by at least three members of the AIED community, assigned by the corresponding track organizers who then took the final decision about acceptance.

The participants of the conference had the opportunity to attend three keynote talks: "Navigating Strategic Challenges in Education in the Post-Pandemic AI Era" by Blaženka Divjak, "Navigating the Evolution: The Rising Tide of Large Language Models for AI and Education" by Peter Clark, and "Artificial Intelligence in Education and Public Policy: A Case from Brazil" by Seiji Isotani. These contributions are published in the Springer proceedings volumes CCIS 2150 and 2151.

The conference also included a Panel with experts in the field and the opportunity for the participants to present a demonstration of their AIED system in a specific session of Interactive Events. A selection of the systems presented is included as showcases on the web page of the IAIED Society[1]. Finally, there was a session with presentations

---

[1] https://iaied.org/showcase.

of papers published in the International Journal of Artificial Intelligence in Education[2], the journal of the IAIED Society indexed in the main databases, and a session with the best papers from conferences of the International Alliance to Advance Learning in the Digital Era (IAALDE)[3], an alliance of research societies that focus on advances in computer-supported learning, to which the IAIED Society belongs.

For making AIED 2024 possible, we thank the AIED 2024 Organizing Committee, the hundreds of Program Committee members, the Senior Program Committee members, and the AIED proceedings chairs Paraskevi Topali and Rafael D. Araújo. In addition, we would like to thank the Executive Committee of the IAIED Society for their advice during the conference preparation, and specifically two of the working groups, the Conference Steering Committee, and the Diversity and Inclusion working group. They all gave their time and expertise generously and helped with shaping a stimulating AIED 2024 conference. We are extremely grateful to everyone!

July 2024

Andrew M. Olney
Irene-Angelica Chounta
Zitao Liu
Olga C. Santos
Ig Ibert Bittencourt

---

[2] https://link.springer.com/journal/40593.
[3] https://alliancelss.com/.

Organization

Conference General Co-chairs

Olga C. Santos — UNED, Spain
Ig Ibert Bittencourt — Universidade Federal de Alagoas, Brazil

Program Co-chairs

Andrew M. Olney — University of Memphis, USA
Irene-Angelica Chounta — University of Duisburg-Essen, Germany
Zitao Liu — Jinan University, China

Doctoral Consortium Co-chairs

Yu Lu — Beijing Normal University, China
Elaine Harada T. Oliveira — Universidade Federal do Amazonas, Brazil
Vanda Luengo — Sorbonne Université, France

Workshop and Tutorials Co-chairs

Cristian Cechinel — Federal University of Santa Catarina, Brazil
Carrie Demmans Epp — University of Alberta, Canada

Interactive Events Co-chairs

Leonardo B. Marques — Federal University of Alagoas, Brazil
Ben Nye — University of Southern California, USA
Rwitajit Majumdar — Kumamoto University, Japan

Industry, Innovation and Practitioner Co-chairs

Diego Dermeval   Federal University of Alagoas, Brazil
Richard Tong   IEEE Artificial Intelligence Standards
                                    Committee, USA
Sreecharan Sankaranarayanan   Amazon, USA
Insa Reichow   German Research Center for Artificial
                                    Intelligence, Germany

Posters and Late Breaking Results Co-chairs

Marie-Luce Bourguet   Queen Mary University of London, UK
Qianru Liang   Jinan University, China
Jingyun Wang   Durham University, UK

Panel Co-chairs

Julita Vassileva   University of Saskatchewan, Canada
Alexandra Cristea   Durham University, UK

Blue Sky Co-chairs

Ryan S. Baker   University of Pennsylvania, USA
Benedict du Boulay   University of Sussex, UK
Mirko Marras   University of Cagliari, Italy

WideAIED Co-chairs

Isabel Hilliger   Pontificia Universidad Católica, Chile
Marco Temperini   Sapienza University of Rome, Italy
Ifeoma Adaji   University of British Columbia, Canada
Maomi Ueno   University of Electro-Communications, Japan

Local Organising Co-chairs

Rafael Ferreira Mello   Universidade Federal Rural de Pernambuco,
Brazil
Taciana Pontual   Universidade Federal Rural de Pernambuco,
Brazil

AIED Mentoring Fellowship Co-chairs

Amruth N. Kumar   Ramapo College of New Jersey, USA
Vania Dimitrova   University of Leeds, UK

Diversity and Inclusion Co-chairs

Rod Roscoe   Arizona State University, USA
Kaska Porayska-Pomsta   University College London, UK

Virtual Experiences Co-chairs

Guanliang Chen   Monash University, Australia
Teng Guo   Jinan University, China
Eduardo A. Oliveira   University of Melbourne, Australia

Publicity Co-chairs

Son T. H. Pham   Nha Viet Institute, USA
Pham Duc Tho   Hung Vuong University, Vietnam
Miguel Portaz   UNED, Spain

Volunteer Co-chairs

Isabela Gasparini   Santa Catarina State University, Brazil
Lele Sha   Monash University, Australia

Proceedings Co-chairs

Paraskevi Topali   Radboud University, Netherlands
Rafael D. Araújo   Federal University of Uberlândia, Brazil

Awards Co-chairs

Ning Wang   University of Southern California, USA
Beverly Woolf   University of Massachusetts, USA

Sponsorship Chair

Tanci Simões Gomes   CESAR, Brazil

Scholarship Chair

Patrícia Tedesco   Universidade Federal de Pernambuco, Brazil

Steering Committee

Noboru Matsuda   North Carolina State University, USA
Eva Millan   Universidad de Málaga, Spain
Sergey Sosnovsky   Utrecht University, Netherlands
Ido Roll   Israel Institute of Technology, Israel
Maria Mercedes T. Rodrigo   Ateneo de Manila University, Philippines

Senior Program Committee Members

Laura Allen   University of Minnesota, USA
Ryan Baker   University of Pennsylvania, USA
Gautam Biswas   Vanderbilt University, USA
Nigel Bosch   University of Illinois at Urbana-Champaign, USA
Jesus G. Boticario   UNED, Spain
Bert Bredeweg   University of Amsterdam, Netherlands
Christopher Brooks   University of Michigan, USA
Guanliang Chen   Monash University, Australia

Ruth Cobos	Universidad Autónoma de Madrid, Spain
Cesar A. Collazos	Universidad del Cauca, Colombia
Mutlu Cukurova	University College London, UK
Maria Cutumisu	McGill University, Canada
Tenzin Doleck	Simon Fraser University, Canada
Ralph Ewerth	Leibniz Universität Hannover, Germany
Dragan Gasevic	Monash University, Australia
Isabela Gasparini	UDESC, Brazil
Sébastien George	Le Mans Université, France
Alex Sandro Gomes	Universidade Federal de Pernambuco, Brazil
Jason Harley	McGill University, Canada
Peter Hastings	DePaul University, USA
Bastiaan Heeren	Open University, Netherlands
Neil Heffernan	Worcester Polytechnic Institute, USA
Tomoo Inoue	University of Tsukuba, Japan
Seiji Isotani	Harvard University, USA
Johan Jeuring	Utrecht University, Netherlands
Srecko Joksimovic	University of South Australia, Australia
Simon Knight	UTS, Australia
Kenneth Koedinger	Carnegie Mellon University, USA
Irena Koprinska	University of Sydney, Australia
Vitomir Kovanovic	University of South Australia, Australia
Amruth N. Kumar	Ramapo College of New Jersey, USA
H. Chad Lane	University of Illinois at Urbana-Champaign, USA
James Lester	North Carolina State University, USA
Fuhua Lin	Athabasca University, Canada
Roberto Martinez-Maldonado	Monash University, Australia
Riichiro Mizoguchi	Japan Advanced Institute of Science and Technology, Japan
Kasia Muldner	Carleton University, Canada
Tomohiro Nagashima	Saarland University, Germany
Roger Nkambou	Université du Québec à Montréal, Canada
Radek Pelánek	Masaryk University Brno, Czechia
Elvira Popescu	University of Craiova, Romania
Kaska Porayska-Pomsta	University College London, UK
Maria Mercedes T. Rodrigo	Ateneo de Manila University, Philippines
Jonathan Rowe	North Carolina State University, USA
Demetrios Sampson	Curtin University, Australia
Mohammed Saqr	University of Eastern Finland, Finland
Filippo Sciarrone	Universitas Mercatorum, Italy
Sergey Sosnovsky	Utrecht University, Netherlands
Michelle Taub	University of Central Florida, USA

Craig Thompson	University of British Columbia, Canada
Stefan Trausan-Matu	University Politehnica of Bucharest, Romania
Maomi Ueno	University of Electro-Communications, Japan
Rosa Vicari	Universidade Federal do Rio Grande do Sul, Brazil
Vincent Wade	Trinity College Dublin, Ireland
Alistair Willis	Open University, UK
Diego Zapata-Rivera	Educational Testing Service, USA
Gustavo Zurita	Universidad de Chile, Chile

Program Committee Members

Mark Abdelshiheed	University of Colorado Boulder, USA
Kamil Akhuseyinoglu	University of Pittsburgh, USA
Carlos Alario-Hoyos	Universidad Carlos III de Madrid, Spain
Laia Albó	Universitat Pompeu Fabra, Spain
Giora Alexandron	Weizmann Institute of Science, Israel
Isaac Alpizar Chacon	Utrecht University, Netherlands
Ioannis Anastasopoulos	University of California, Berkeley, USA
Antonio R. Anaya	Universidad Nacional de Educación a Distancia, Spain
Tracy Arner	Arizona State University, USA
Burcu Arslan	Educational Testing Service, USA
Juan I. Asensio-Pérez	Universidad de Valladolid, Spain
Michelle Banawan	Asian Institute of Management, Philippines
Abhinava Barthakur	University of South Australia, Australia
Beata Beigman Klebanov	Educational Testing Service, USA
Brian Belland	Pennsylvania State University, USA
Francisco Bellas	Universidade da Coruna, Spain
Emmanuel Blanchard	Le Mans Université, France
Nathaniel Blanchard	Colorado State University, USA
Maria Bolsinova	Tilburg University, Netherlands
Miguel L. Bote-Lorenzo	Universidad de Valladolid, Spain
Anthony F. Botelho	University of Florida, USA
François Bouchet	Sorbonne Université - LIP6, France
Marie-Luce Bourguet	Queen Mary University of London, UK
Rex Bringula	University of the East, Philippines
Julien Broisin	Université Toulouse III - Paul Sabatier, France
Armelle Brun	LORIA - Université de Lorraine, France
Minghao Cai	University of Alberta, Canada
Dan Carpenter	North Carolina State University, USA

R. McKell Carter	University of Colorado Boulder, USA
Paulo Carvalho	Carnegie Mellon University, USA
Alberto Casas	UNED, Spain
Teresa Cerratto-Pargman	Stockholm University, Sweden
Cs Chai	Chinese University of Hong Kong, China
Jiahao Chen	TAL, China
Penghe Chen	Beijing Normal University, China
Thomas K. F. Chiu	Chinese University of Hong Kong, China
Heeryung Choi	Massachusetts Institute of Technology, USA
Chih-Yueh Chou	元智大工程系, Taiwan
Jody Clarke-Midura	Utah State University, USA
Keith Cochran	DePaul University, USA
Maria de Los Angeles Constantino González	Tecnológico de Monterrey Campus Laguna, Mexico
Evandro Costa	Federal University of Alagoas, Brazil
Alexandra Cristea	Durham University, UK
Jeffrey Cross	Tokyo Institute of Technology, Japan
Mihai Dascalu	University Politehnica of Bucharest, Romania
Jeanine DeFalco	University of New Haven, USA
Carrie Demmans Epp	University of Alberta, Canada
Vanessa Dennen	Florida State University, USA
Michel Desmarais	Polytechnique Montréal, Canada
M. Ali Akber Dewan	Athabasca University, Canada
Yannis Dimitriadis	University of Valladolid, Spain
Konomu Dobashi	Aichi University, Japan
Fabiano Dorça	Universidade Federal de Uberlandia, Brazil
Mohsen Dorodchi	University of North Carolina Charlotte, USA
Benedict du Boulay	University of Sussex, UK
Cristina Dumdumaya	University of Southeastern Philippines, Philippines
Nicholas Duran	Arizona State University, USA
Kareem Edouard	Drexel University, USA
Fahmid Morshed Fahid	North Carolina State University, USA
Xiuyi Fan	Nanyang Technological University, Singapore
Alexandra Farazouli	Stockholm University, Sweden
Effat Farhana	Vanderbilt University, USA
Mingyu Feng	WestEd, USA
Márcia Fernandes	Federal University of Uberlândia, Brazil
Brendan Flanagan	Kyoto University, Japan
Carol Forsyth	Educational Testing Service, USA
Reva Freedman	Northern Illinois University, USA

Mirko Marras	University of Cagliari, Italy
Alejandra Martínez-Monés	Universidad de Valladolid, Spain
Goran Martinovic	J.J. Strossmayer University of Osijek, Croatia
Jeffrey Matayoshi	McGraw Hill ALEKS, USA
Diego Matos	Federal University of Alagoas, Brazil
Noboru Matsuda	North Carolina State University, USA
Guilherme Medeiros Machado	ECE Paris, France
Christoph Meinel	Hasso Plattner Institute, Germany
Roberto Angel Melendez Armenta	Universidad Veracruzana, Mexico
Agathe Merceron	Berlin State University of Applied Sciences, Germany
Eva Millan	Universidad de Málaga, Spain
Marcelo Milrad	Linnaeus University, Sweden
Wookhee Min	North Carolina State University, USA
Tsunenori Mine	Kyushu University, Japan
Sein Minn	Inria, France
Péricles Miranda	UFPE, Brazil
Tsegaye Misikir Tashu	University of Groningen, Netherlands
Tanja Mitrovic	University of Canterbury, New Zealand
Phaedra Mohammed	University of the West Indies, Trinidad and Tobago
Mukesh Mohania	IIIT Delhi, India
Bradford Mott	North Carolina State University, USA
Ana Mouta	USAL, Portugal
Chrystalla Mouza	University of Illinois at Urbana-Champaign, USA
Pedro J. Muñoz-Merino	Universidad Carlos III de Madrid, Spain
Takashi Nagai	Institute of Technologists, Japan
Tricia Ngoon	Carnegie Mellon University, USA
Huy Nguyen	Carnegie Mellon University, USA
Narges Norouzi	UC Berkeley, USA
Dr. Nasheen Nur	Florida Institute of Technology, USA
Benjamin Nye	University of Southern California, USA
Xavier Ochoa	New York University, USA
Jaclyn Ocumpaugh	University of Pennsylvania, USA
Elaine H. T. Oliveira	Universidade Federal do Amazonas, Brazil
Jennifer Olsen	University of San Diego, USA
Ranilson Paiva	Universidade Federal de Alagoas, Brazil
Viktoria Pammer-Schindler	Graz University of Technology, Austria
Luc Paquette	University of Illinois at Urbana-Champaign, USA
Abelardo Pardo	University of Adelaide, Australia
Zach Pardos	University of California, Berkeley, USA
Rebecca Passonneau	Pennsylvania State University, USA

Rumana Pathan	Indian Institute of Technology Bombay, India
Mar Perez-Sanagustin	Université Paul Sabatier Toulouse III, France
Yang Pian	Beijing Normal University, China
Niels Pinkwart	Humboldt-Universität zu Berlin, Germany
Eduard Pogorskiy	Open Files Ltd., UK
Oleksandra Poquet	Technical University of Munich, Germany
Ethan Prihar	École polytechnique fédérale de Lausanne, Switzerland
David Pynadath	University of Southern California, USA
Yuvaraj R.	Anna University, India
Mladen Rakovic	Monash University, Australia
Sowmya Ramachandran	Stottler Henke Associates Inc., USA
Steven Ritter	Carnegie Learning, Inc., USA
Luiz Rodrigues	Federal University of Alagoas, Brazil
José Raúl Romero	University of Cordoba, Spain
Carolyn Rose	Carnegie Mellon University, USA
Rinat B. Rosenberg-Kima	Technion - Israel Institute of Technology, Israel
José A. Ruipérez Valiente	University of Murcia, Spain
Sreecharan Sankaranarayanan	Carnegie Mellon University, USA
Eileen Scanlon	Open University, UK
Robin Schmucker	Carnegie Mellon University, USA
Bertrand Schneider	Harvard University, USA
Kazuhisa Seta	Osaka Metropolitan University, Japan
Lele Sha	Monash University, Australia
Tasmia Shahriar	North Carolina State University, USA
Kshitij Sharma	Norwegian University of Science and Technology, Norway
Bruce Sherin	Northwestern University, USA
Atsushi Shimada	Kyushu University, Japan
Valerie Shute	FSU, USA
Aditi Singh	Cleveland State University, USA
Sean Siqueira	Federal University of the State of Rio de Janeiro (UNIRIO), Brazil
Andy Smith	North Carolina State University, USA
Nancy Songer	University of Utah, USA
John Stamper	Carnegie Mellon University, USA
Christian M. Stracke	University of Bonn, Germany
Abhijit Suresh	University of Colorado Boulder, USA
Vinitra Swamy	EPFL, Switzerland
Paula Toledo Palomino	Sao Paulo State College of Technology (FATEC) - Matão, Brazil
Maya Usher	Technion, Israel

International Artificial Intelligence in Education Society

Management Board

President

Secretary/Treasurer

Vania Dimitrova   University of Leeds, UK

Journal Editors

Vincent Aleven   Carnegie Mellon University, USA
Judy Kay   University of Sydney, Australia

Finance Chair

Benedict du Boulay   University of Sussex, UK

Membership Chair

Benjamin D. Nye   University of Southern California, USA

IAIED Officers

Yancy Vance Paredes   North Carolina State University, USA
Son T. H. Pham   Nha Viet Institute, USA
Miguel Portaz   UNED, Spain

Executive Committee

Akihiro Kashihara   University of Electro-Communications, Japan
Amruth Kumar   Ramapo College of New Jersey, USA
Christothea Herodotou   Open University, UK
Jeanine A. Defalco   CCDC-STTC, USA
Judith Masthoff   Utrecht University, Netherlands
Maria Mercedes T. Rodrigo   Ateneo de Manila University, Philippines
Ning Wang   University of Southern California, USA
Olga Santos   UNED, Spain
Rawad Hammad   University of East London, UK
Zitao Liu   Jinan University, China
Bruce M. McLaren   Carnegie Mellon University, USA
Cristina Conati   University of British Columbia, Canada
Diego Zapata-Rivera   Educational Testing Service, Princeton, USA
Erin Walker   University of Pittsburgh, USA
Seiji Isotani   University of São Paulo, Brazil
Tanja Mitrovic   University of Canterbury, New Zealand

Noboru Matsuda	North Carolina State University, USA
Min Chi	North Carolina State University, USA
Alexandra I. Cristea	Durham University, UK
Neil Heffernan	Worcester Polytechnic Institute, USA
Andrew M. Olney	University of Memphis, USA
Irene-Angelica Chounta	University of Duisburg-Essen, Germany
Beverly Park Woolf	University of Massachusetts, USA
Ig Ibert Bittencourt	Universidade Federal de Alagoas, Brazil

Contents – Part I

Contents – Part II

Short Papers

Full Papers

From Learning Actions to Dynamics: Characterizing Students' Individual Temporal Behavior with Sequence Analysis

Esteban Villalobos[(⊠)] [iD], Mar Pérez-Sanagustín[iD], and Julien Broisin[iD]

IRIT, Université Toulouse III - Paul Sabatier, Toulouse, France
{esteban.villalobos,mar.perez-sanagustin,julien.broisin}@irit.fr

Abstract. Researchers recognize the pivotal role of temporal analysis in unraveling learning processes from learners' trace data. However, most temporal analysis methods focus on clustering similar trajectories, and very few offer metrics to characterize the learners' temporal learning behavior. This study draws upon a set of sequence indicators that characterize students' temporal behavior dynamics. These metrics, focusing on sequence length, diversity, and complexity, provide insights into students' learning engagement with the course. Applied to a dataset of 91 students collected from a Blended Learning course, we studied these metrics in relation to the learners' final grades and their self-regulatory profiles. Additionally, we assessed the causal effects of introducing a Learning Analytics Dashboard, an intervention aimed at fostering self-regulated learning, on students' temporal behavior and final grades by applying Inverse Probability Weighting (IPW). The results show that these metrics serve (1) to characterize individual students' dynamics through the course and (2) to provide information about the effects of the intervention. In particular, we show that students who interacted with the dashboard had significantly more complex behavior, but no difference in their final course grades. We also revealed that engagement measured as consistent activity might be a stronger predictor than specific learning strategies in an order. This research contributes with a methodological approach for extracting metrics that enrich traditional methods, emphasizing the unique dynamics of individual students, and paving the way for future interventions.

Keywords: Sequence Analysis · Temporal Analysis · Learning Analytics · Inverse Probability Weighting

1 Introduction

In recent years, digital learning environments have provided vast data, leading Learning Analytics and Educational Data Mining communities to explore analytical methods for interpreting learning from trace data, often combining it with

A. M. Olney et al. (Eds.): AIED 2024, LNAI 14829, pp. 3–17, 2024.
https://doi.org/10.1007/978-3-031-64302-6_1

learners' self-reported measures [12,19]. These methods yield insights into general learning behaviors but often overlook the temporal dynamics of learning, primarily focusing on transition frequencies and probabilities [18].

Acknowledging that learning is an inherently dynamic and evolving process [9], there has been a paradigm shift towards interpreting learning from a temporal and sequential perspective [9,10,19]. Sequence Analysis (SA) has emerged as a robust methodology in this regard, providing a nuanced observation and understanding of learning tasks as temporal and sequential phenomena [23]. SA offers a unique lens to contextualize events, explore relationships, and delve into the effects of sequences over time. Recent works on this line have started investigating innovative approaches to extract meaningful metrics and indicators to elucidate individual learners' sequential patterns [2,10,16,22]. However, measures that reflect changes in study behavior are uncommon, despite their importance to understanding students' temporal behavior [16]. Some papers have recently appeared in this line [10,22]. These papers draw upon methods employed in the social sciences [17] for extracting indicators that profile learners' individual behavioral dynamics. However, research employing SA methods for evaluating the impact of educational interventions remains limited [2]. Scholars advocate for additional examples that utilize SA to analyze the effects of specific interventions [23].

In this paper, we contribute to expanding the current landscape of SA in education by presenting the results of analyzing an intervention in a Blended Learning (BL) course introducing a dashboard-based solution for supporting learners' Self-regulated Learning (SRL). Specifically, we apply SA methods proposed by Ritschard [17] to extract sequence indicators characterizing students' temporal behavior throughout the course. Then, we apply Inverse Probability Weighting (IPW) to analyze causal relationships between students' dashboard adoption, their behavior, and final grades by controlling for students' self-reported SRL proficiency, prior achievements (GPA), and course behavior before the intervention. The main aim of this study is to understand **how sequential indicators can inform students' behavior along the course, and to assess the impact of the intervention.**

1.1   Characterizing the Nature of Sequences

SA methods visualize sequences as consecutive states (e.g., AAABBCC), defining 'states' as specific conditions or stages within the sequence [17]. Researchers often represent sequences as a succession of 'spells' in different states. A spell is the successions of states with the same value. For example, AAABCC is represented in spells as A/3-B/1-C/2, where A/3 indicates a spell of length 2 in A.

Ritschard [17] presented a study of different metrics to measure persons' individual nature of sequences. The metrics are classified as Basic indicators (counts of states/spells), Diversity (the heterogeneity of visited states/spells), and Complexity (unpredictability of state/spell arrangement). Basic indicators encompass aspects like counts of visited states, transitions, and spells. Diversity metrics focus on the distribution of states and spells in the sequence. An example would be the normalized entropy of the distribution of visited states, which was used by Fussel et al. [4] to compare the diversity in life-course events between different countries. Complexity metrics refer to the instability or unpredictability of state arrangement in the sequence, and are prevalent in characterizing careers and family-life trajectories [1,3]. Crucially, these metrics are state-agnostic, focusing on the sequence's dynamics rather than the specific actions involved. Consequently, sequences like ABAB and CDCD would yield the same indicators, despite having no states in common. This underscores the focus on the process's temporal dynamics rather than the individual states.

From a more conceptual perspective, a recent work by Molenaar and Wise [13] proposes a framework for characterizing learners' temporal behavior. They distinguish between two essential temporal-related concepts: the passage of time and order in time. While the passage of time considers time as a continuous stream of events, the order focuses on the temporal organization of sequences of events. Within the passage of time, they introduce four metrics—position, duration, frequency, and rate—applicable across various statistical methodologies. Analysis of order in time, however, requires more advanced techniques. This aspect explores consistency, recurrent and non-recurrent change, and irregular change within the sequencing of events. This prior work establishes the theoretical foundation for defining and analyzing sequences from educational data but does provide few metrics to characterize them. We utilize SA methods for extracting students' temporal behavioral metrics and use this conceptual framework to interpret them.

1.2 Sequence Analysis in Education

SA in educational settings commonly relies on log data, such as clicks on a web platform or problem-answering attempts, to define its sequences' states [23]. Torre et al. [23] highlights that one of the most challenging parts of running SA analysis is to transform this raw data into analyzable sequences that reflect learners' interactions and cognitive processes. One common approach is to define learning actions based on interactions with the Learning Management System (LMS), such as opening a PDF, pausing a video, or editing code [11,20,25]. Another approach involves mapping these interactions to higher-level meta-cognitive actions, typically associated with SRL strategies such as planning, reviewing, or practicing, thereby aligning practical interactions with theoretical learning constructs [2,16]. Additionally, certain studies adopt a multi-step approach, creating session-based clusters and subsequently combining these to form comprehensive course-level sequences [20,25].

Traditionally, SA methods have been employed to cluster similar behavior across learners using pairwise dissimilarities [11,20,25]. These studies typically employ a variation of the Optimal Matching distance, which quantifies the number of insertions, deletions and substitutions required to transform one sequence into another. However, while this method efficiently groups analogous behaviors, it can potentially overlook the nuanced individual characteristics of each sequence [17]. This clustering often acts as a 'black box,' obscuring the rich, chronological sequence of students' actions, and hindering a clear interpretation of individual sequential behavior.

Current research initiatives investigate innovative approaches to extract meaningful metrics and indicators to elucidate individual learners' sequential patterns [10, 16, 20, 22] to address these methodological limitations. For example, Saqr et al. [20] used diversity metrics to analyze the diversity of course strategies students use during a higher education program, while Tan et al. [22] used complexity metrics to characterize students' procrastination behavior in an online course. Despite these advances, research in this area remains relatively scarce, underscoring the necessity for novel techniques to interpret individual learners' dynamics [16].

Our study aims to contribute to this gap by showing how sequence indicators can be used both to assess the impact of an intervention and to analyze their connection to students' personal characteristics. Specifically, and based on this prior work, we address the following research questions research questions:

RQ1 - How do sequence indicators (basic, diversity, and complexity) complement existing approaches to profile learners' based on their personal characteristics (SRL proficiency and prior achievements)?

RQ2 - How can sequence indicators inform on the effectiveness of an intervention aimed at supporting SRL strategies for influencing students' behavioral dynamics?

2   Methodology

2.1   Context

The research was conducted within a second-year course on databases offered in a French technical university's Degree in Management of Enterprises program. The course comprised 119 higher students, but only 91 consented to using their data for research purposes. The course followed a BL approach, using the Moodle Platform as the LMS. Students engaged in 1.5 h of in-person instruction once a week, alongside online activities and at-home projects, which were anticipated to require 1–2 hours of weekly commitment.

The course lasted 15 weeks and was structured into three periods with different instructional designs. The first period (weeks 1–5) used a Flipped Classroom (FC) approach, where students engaged with theoretical resources before in-class sessions and completed a self-evaluation questionnaire beforehand. The objective was to acquaint students with key concepts in Database modeling. In-class activities involved individual problem-solving, group discussions, and presentations. The second period (weeks 6–11) shifted towards a design for fostering individual and independent work. Students participated in in-class problem-solving sessions. These sessions were complemented with exercises and instructional videos that the students practiced at home. The aim was to develop proficiency in SQL database management and queries using PHPMyAdmin. In the final period (weeks 12–15), students were invited to conduct a hands-on project involving group collaboration to create a database for a library.

Table 1. Library of actions in the LMS

Action	Description
On-time	Accessing a resource, the corresponding week it was planned for
Catching-up	Accessing a resource planned for at least a week ago for the first time
Reviewing (Recent)	Revisiting a resource from $\leq$ 2 weeks ago
Reviewing (Old)	Revisiting a resource from $>$ 2 weeks ago
Evaluation	Interaction with the course quizzes
Idle	A break of at least 30 min of inactivity

On week 6, the teacher introduced the NoteMyProgress (NMP) tool, a Learning Analytics Dashboard (LAD) that aims to support students' SRL [14]. NMP provides visualizations and functionalities for supporting, specifically, Time Management (TM) and Strategic planning (SP). The visualizations include metrics such as the frequency of the usage of different resources along the course and the duration of the students' sessions, as well as a view for comparing students' individual behavior with their peers.

2.2 Data Sources

Various data collection techniques were used: (1) self-reported measures about students' SRL proficiency, (2) trace data generated from students' interactions with the course, and (3) metadata associated with the course. To measure students' SRL, we administered the MSLQ questionnaire [15]. The instrument was translated into French with good reliability (Cronbach's alpha=0.93). For this study, we only considered data about the construct's Cognitive Strategy Use and Self-regulation dimensions to analyze the possible connections between the sequence indicators and students' SRL proficiency. The trace data was obtained from the Moodle platform log files. They collect information about students' interaction with the course resources uploaded by the teacher: quizzes, assignments, videos, and reading material. The metadata associated to the course was provided by the teacher, indicating when the different activities took place and the instructional design of each period in the course.

Please, refer to [25] for more details about the collected data. In this paper, we re-used the data in [25] to enrich our analysis and analyze whether SA could help us get better insights into individual learners' behavioral dynamics. All data collection instruments and anonymized datasets are available in OSF[1].

2.3 Analytical Process

Data Pre-processing. The first step consisted of defining learning actions from the trace data. Taking previous studies as a basis [10,20], we first defined some low-level actions and then we used a Mixture Hidden Markov Model (MHMM) [7] to find clusters of similar behavior across students. Details about each of these phases are described in what follows.

---

[1] https://osf.io/daejr/.

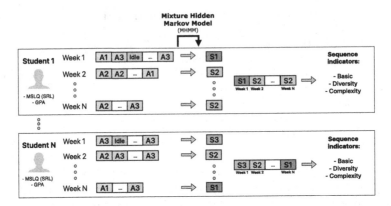

Fig. 1. Pipeline of the analytical framework to extract sequence indicators from learners' trace data. Students' self-regulated learning skills were measured using the MSLQ [15], while their GPA was collected as a measure of prior achievements. **A** stands for Action (e.g., On-time or Catching-up). **S** represents a 'weekly strategy' (i.e., cluster of weekly behavior).

Defining Learning Actions. Table 1 summarizes the actions defined from trace data. We defined actions that take advantage of the weekly nature of the course. All resources were meant to be seen by the students the week they were posted in the LMS. As such, we defined the 'On-time' action as accessing a resource the week it was planned for; 'Catching-up' as accessing a resource planned for at least a week ago for the first time; 'Evaluation' as interacting with a quiz; and 'Reviewing' as revisiting an already seen resource. The last one, 'Reviewing', was divided into two actions, differentiated on whether that resource is 'Recent' (less than or equal to two weeks) or 'Old' (more than two weeks). Finally, we computed an 'Idle' action representing at least 30 min of inactivity. This was meant to separate actions from different sessions, as it is common practice in LA [20,25].

Clustering Weekly Behavior. We concatenated the actions performed by a student during a week to create sequences of weekly behavior (see Fig. 1). These sequences were then clustered using the MHMM model, a model that has been widely used to cluster sequences in social science [7] and has recently been used to cluster students' trace data [20]. This model assumes that the students' sequences of learning actions are generated by a hidden (latent) Markov chain process. The MHMM then assumes that there are different sub-populations (clusters), each with its own Markov process. Consequently, each sequence was associated with a cluster, determined by the highest likelihood of it being generated by that cluster's process. This method allows for organizing students' weekly behavior into clusters, which we will refer to as 'weekly strategies'.

We estimated the MHMM with different numbers of clusters and states, and the best model was selected based on the lowest BIC values. The clustering

Table 2. Sequence indicators proposed by Ritschard [17] used in this study.

	State	Spells
Basic	Number of non-missing states, i.e., weeks in which the student was active *Code: Basic_State*	Proportion of transitions, i.e., proportion of times a student changed their weekly behavior *Code: Basic_Spell*
Diversity	Normalized entropy of weekly strategies *Code: Diversity_State*	Spell length standard deviation, including 0-length spells *Code: Diversity_Spells*
Complexity	Complexity index of weekly strategies *Code: Cplx_States*	Normalized Turbulence, including 0-length spells *Code: Cplx_Spells*

process was conducted using the R package seqHMM [7]. These clusters were then used to create course-level sequences, where each state is a cluster as computed by the MHMM.

Indicators Characterizing Sequence Behavior. All indicators defined in this work are based on the concept of state and spell used by [17]. A state corresponds to a weekly strategy as computed by the MHMM model (see Sect. 2.3). A spell is a group of consecutive weekly strategies.

We define sequences as successions of states taken from a finite alphabet A, i.e., the set of all possible states. Building upon the work by [17] (See Sect. 1.1), we selected two metrics of each type (basic, diversity, and complexity), one focusing on states and the other on spells. We made this distinction to obtain a richer perspective on the sequential behavior of the students.

Basic Indicators. While several count-based indicators can characterize a sequence, we based our analysis on the two basic ones proposed by [17]. These are the number of non-missing states in the sequence (*Basic_State*), and the proportion of transitions $tp = tn/(\ell - 1)$, where tn is the number of transitions to different states and ℓ is the length of the sequence (*Basic_Spell*, ℓ_d).

Indicators for Diversity. The diversity of a sequence refers to the diversity in the distribution of states or spell lengths. Ritschard [17] proposes two measures for sequence diversity.

A common indicator for diversity in the states is the longitudinal entropy. This measure calculates the distribution of the sequence's state according to the entropy measure, $h(X)$, proposed by Shannon [21]. In particular, we computed the normalized longitudinal entropy of the visited states (*Diver_States*). This metric is minimized at 0 when all states are equal and is maximized at 1 for a uniform distribution across all states in the alphabet.

Regarding the diversity of spells, we computed the standard deviation of spell lengths (*Diver_Spells*, s_d^*). As suggested by [17], we included non-visited states as spells of length zero in the standard deviation. This is to avoid counter-intuitive results obtained when not all spells are visited. The variance s_d^{*2} of the spell duration is zero only when all spells have the same length and all states are visited.

Table 3. Example of metrics following ordering in time classifications proposed by Molenaar and Wise [13]. Metrics are computed assuming an alphabet $A = \{1, 2, 3, 4, 5, 6\}$.

	Example Sequence	Basic Spell Proportion of Transitions	Diversity State Entropy	Spell STD of spell length	Complexity State Complexity index	Spell Turbulence
Consistency	111111111111	0.00	0.000	4.472	0.000	0.000
Recurrent regular	123123123123	1.00	0.613	0.400	0.738	0.897
Non-recurrent regular	111122223333	0.18	0.613	2.000	0.334	0.299
Irregular change	541256424532	1.00	0.927	0.000	0.963	0.970

Indicators for Complexity. The complexity of a sequence refers to the instability or unpredictability of state arrangement in the sequence. It involves multiple aspects, and complexity increases with, for example, the transitions, the number of visited states, and the unpredictability of the time spent in each state.

Gabadinho et al. [6] define the complexity index (*Cplx_States*, $C(x)$) as the geometric mean between the longitudinal entropy of a sequence and the proportion of transitions tp. The formula for complexity is then

$$C(x) = \sqrt{tp\frac{h(X)}{h_{max}}},$$

where h_{max} is the highest entropy achievable given alphabet. This metric is bounded between 0 and 1, where the sequence with a higher proportion of transitions gets a higher complexity index among two sequences with the same longitudinal entropy.

The turbulence index (*Cplx_Spells*), originally proposed by Elzinga and Liefbroer [3], is a composite index based on the variance of spell duration s_d^{*2} and the number $\phi(x)$ of subsequences that can be extracted from the DSS. As suggested by Ritschard [17], we used their corrected version of the turbulence, which includes the 0-length spells to prevent counter-intuitive results. The formula for the turbulence is then

$$T^*(x) = \log_2\left(\phi(x)\frac{s_{d,\,max}^{*2}(x) + 1}{s_d^{*2}(x) + 1}\right)$$

The maximum turbulence is achieved when creating a distinct sequence using all the states in the alphabet. For example, for an alphabet $A = \{a, b, c\}$, the state sequence of length 7 with the maximum turbulence is *abcabca*.

For illustrative purposes, in Table 3, we apply these metrics to the four sequencing examples originally classified by Molenaar and Wise [13] as consistency, recurrent and non-recurrent change, and irregular change. Consistency exhibits low entropy and complexity, signifying predictability. Conversely, the consistent behavior results in a high standard deviation for spell lengths. Recurrent change displays a moderate entropy and high complexity and turbulence, given the high number of transitions. Non-recurrent change involves a mix of elements with moderate entropy and complexity, suggesting a structured yet less predictable pattern. Irregular change introduces abrupt, unpredictable shifts with high complexity and low turbulence, indicating a highly disorderly and unstable sequence.

We computed the sequence indicators for students' weekly activity using the TraMineR package [5]. The only indicator that considers the weeks with no activities was Basic_States (i.e., active weeks). All other indicators are computed, ignoring the weeks where no activity was recorded.

(RQ1) Profiling Students' Behavioral Dynamics. To extract metrics for profiling students' behavioral dynamics, we first characterized students' SRL proficiency using the Cognitive Strategy Use and Self-Regulation dimensions of the MSLQ and their prior academic achievement through their GPA before the course. To analyze how self-reported measures and prior achievement influence course behavior, we computed the Spearman correlation and the confidence intervals using the Fisher transformation. We then adjusted for multiple comparisons using the Holm-Bonferroni method. All statistical analyses were obtained using the Pingouin Python library [24].

(RQ2) Assessing the Impact of an Intervention. To study the effect of the NMP dashboard on students' behavior, we used Inverse Probability Weighting (IPW). This technique allows to control for different covariates that may be related to the exposure to the intervention and the outcome variable. This is particularly useful when a controlled experiment is not possible [8]. For our analysis, we controlled for students' self-reported SRL proficiency, prior achievement, and the sequence indicators of the first 5 weeks of the course for characterizing students' behavioral dynamics prior to the intervention. The IPW consists of using these covariates to predict the probability of exposure to the intervention, also called the propensity score. This score is then used to balance samples by giving more importance to the samples with similar propensity scores across the treated and untreated populations. In our study, exposure is defined as having accessed the NMP dashboard at least two times. We will refer to those students accessing the dashboard as the 'treated group' and the 'untreated group' to those who did not. We then used the IPW to measure the differences between students' individual indicators after the intervention (using week seven onwards) and their final grades. We then adjusted for multiple comparisons using the Holm-Bonferroni method.

3   Results

For the data pre-processing, we assigned students' weekly behavior to different weekly strategies (clusters). We found 3 distinct weekly strategies, each one with a Markov model with 7 hidden states:

- *Weekly Strategy 1* (n = 374): Students mostly focus on jumping between **answering quizzes (evaluation) and reviewing recent material**. This behavior is usually split across several sessions during the week.

Table 4. Significant correlations between sequence indicators and their personal characteristics (Cognitive Strategy Use, Self-regulation and GPA). Only significant results after the Holm-Bonferroni adjustment are shown. The only indicator with significant correlations was Basic_States, i.e., the number of active weeks.

Sequence Indicators	Personal Characteristics	Spearman's Correlation	CI95%	p-value
Basic_States *(i.e., active weeks)*	Self-Regulation	0.426**	[0.23, 0.59]	0.0012
	GPA	0.389**	[0.19, 0.56]	0.0057

- *Weekly Strategy 2* (n = 396): Students mostly focus on **looking at material on time and reviewing old material**. It is also common for students to start their week by catching up on recent material.
- *Weekly Strategy 3* (n = 303): Students have a more **diverse behavior**, where they alternate between evaluations, catching up, and reviewing recent and old resources.

Using these weekly strategies, we created sequences for students' course behavior. These sequences allow us to characterize the changes in students' weekly strategies, which could represent different ways of adapting to the changes in the course. In the following sections, we show how sequence indicators complement traditional approaches to profiling learners (through self-reported metrics and GPA), and we analyze the effect of the intervention by analyzing their changes over time.

3.1 RQ1 - Differences Across SRL Proficiency and Prior Achievements

Overall, there were few correlations between students' sequence indicators throughout the course and their personal characteristics (Cognitive Strategy Use, Self-regulation, and GPA). The only indicator with significant correlations was the number of active weeks with students' self-regulation and prior achievements (see Table 4). Prior studies characterizing students' dynamics, which focused on finding groups of students with similar entropy of behavior, have also failed to find significant correlations with students' self-reported measures. [16]. Our study further expands these results by including more aspects of students' temporal dynamics while still showing no relation to students' self-reported measures. This suggests that students' self-reported measures may not be a good indicator of learners' variability in behavior. Additionally, we did not find any significant correlations between students' sequence indicators and their final grades.

3.2 RQ2 - Assessing the Impact of an Intervention

We used the IPW method to assess the average treatment effect of interacting with the NMP tool. We used the IPW to control for variables that could affect both the students' participation in the intervention and the outcomes, such as

Table 5. Average Treatment Effect (ATE) of the NMP intervention on difference variables, p-values have been adjusted using Holm-Bonferroni. * $p<0.05$, ** $p<0.01$, *** $p<0.001$'

Indicator Code	Indicator	ATE	STD Error	p-value
Basic_States	Active weeks	0.682*	0.258	0.039
Basic_Spells	Proportion of transitions	0.102**	0.028	0.003
Diver_States	Normalized Entropy	0.085*	0.032	0.039
Diver_Spells	Spell standard deviation	-0.135	0.070	0.115
Cplx_States	Complexity index	0.099**	0.027	0.003
Cplx_Spells	Normalized Turbulence	0.104**	0.028	0.002
	Final Grade	0.099	0.491	0.841

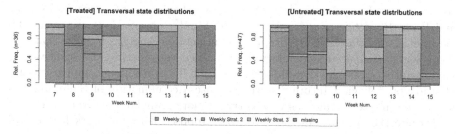

Fig. 2. State distribution of weekly strategies after the intervention on week 6 for the treated group (left), who interacted with the NMP tool, and the untreated group (right), who did not interact with the NMP tool.

self-reported measures, GPA and behavior prior to the intervention. The results show that the intervention significantly changed students' learning dynamics, but not their final grade in the course (see Table 5). In particular, students who used the tool (treated group) had significantly more complex behavior than those who did not (untreated group), meaning they adapted their strategies more often and chose a wider variety of strategies. Figure 2 shows the transversal state distributions for each group. This graph indicates that the treated group showed more varied weekly behavior than their counterparts after the intervention in week 6.

4   Discussion and Conclusion

This study contributes with a set of metrics that characterize students' learning sequences from trace data, focusing on three dimensions: (1) its basic components, (2) its diversity, and (3) its complexity. These metrics provide insights into diverse aspects of students' actions over the course and shed light on *how* their temporal behavior unfolded rather than exactly *what* they did. For example, two students may have similar behavior complexity while having different weekly strategies. This differs from prior approaches that cluster students based on optimal matching distances, which look at students doing similar things at similar points in time [11, 20, 25], which ignores the individual differences.

Our results show that there are few significant relations between students' sequence indicators and their personal characteristics (self-reported SRL and GPA) **(RQ1)**. In particular, only the number of active weeks was significantly related to self-regulation and GPA, while no diversity or complexity indicators presented correlations. This may suggest that personal characteristics collected before the start of the course may not be good predictors of the diversity or complexity of the student's behavior during the course.

However, when looking at the learners' temporal behavior using these same indicators after the intervention **(RQ2)**, we were able to show that students who interacted with the tool showed more complex behavior. It is important to note that these insights could not be appreciated using the clustering methods commonly used in prior literature [25]. This behavior could fit what Molenaar and Wise [13] define as 'irregular change'. Similar observations were made by Dever et al. [2], who used Recurrence Quantification Analysis to show that students who participated in an intervention showed less repetition in cognitive and metacognitive strategies than those who did not. To gain a more complete picture of students' behavior, future research can use a combination of these analyses with clustering techniques, as well as trying more granular time-frames [16].

Despite the potential of the indicators proposed here, our work has several limitations that should be considered. First, since we used a dataset from a real exploratory study, our sample was small. As a consequence, and even though we controlled for potential covariates using IPW, there may still be unobserved variables that influenced the results, such as demographic variables, motivational factors, and course interactions outside the LMS that we could not capture. This is particularly relevant in BL, where face-to-face classes play a significant part of the course dynamics. Additionally, we could not control for students who did not consent for the data to be used in this study. Other metrics based on engagement, such as the ones proposed and explored in the review by Wang and Mousavi [26], could also be considered in future work against the metrics proposed in this paper. Despite these limitations, we argue that our study still contributes to the field by illustrating how sequence indicators could help better understand how an intervention changes students' learning dynamics. Future research could consider gathering data from face-to-face classes and students' motivation throughout the course.

In conclusion, the metrics presented expand current research on the interplay of analytical methods and learning sciences by providing an analytical framework to conceptualize learning processes considering the learners' temporal dynamics. The results also offer a more detailed comprehension of how students navigate BL environments and the relationship between their engagement and the course instructional design. Furthermore, this study lays the groundwork for future educational tools and strategies, potentially guiding the creation of more transparent and effective learning support systems.

Acknowledgment. This paper has been funded by the ANR LASER (156322). The authors acknowledge PROF-XXI, which is an Erasmus+ Capacity Building in the Field of Higher Education project funded by the European Commission (609767-EPP-1-2019-1-ES-EPPKA2-CBHE-JP). This publication reflects the views only of the authors and funders cannot be held responsible for any use which may be made of the information contained therein.

References

1. Christensen, R.C.: Elite professionals in transnational tax governance. Global Netw. **21**(2), 265–293 (2021). https://doi.org/10.1111/glob.12269
2. Dever, D.A., Sonnenfeld, N.A., Wiedbusch, M.D., Schmorrow, S.G., Amon, M.J., Azevedo, R.: A complex systems approach to analyzing pedagogical agents' scaffolding of self-regulated learning within an intelligent tutoring system. Metacognition Learn. (2023). https://doi.org/10.1007/s11409-023-09346-x
3. Elzinga, C.H., Liefbroer, A.C.: De-standardization of family-life trajectories of young adults: a cross-national comparison using sequence analysis: déstandardisation des trajectoires de vie familiale des jeunes adultes: comparaison entre pays par analyse séquentielle. Eur. J. Population **23**(3–4), 225–250 (2007). https://doi.org/10.1007/s10680-007-9133-7
4. Fussell, E., Gauthier, A.H., Evans, A.: Heterogeneity in the Transition to Adulthood: The Cases of Australia, Canada, and the United States: Hétérogénéité dans le passage à l'âge adulte: Cas de l'Australie, du Canada et des Etats-Unis. Eur. J. Population **23**(3–4), 389–414 (2007). https://doi.org/10.1007/s10680-007-9136-4
5. Gabadinho, A., Ritschard, G., Müller, N.S., Studer, M.: Analyzing and visualizing state sequences in R with **TraMineR**. J. Stat. Soft. **40**(4) (2011). https://doi.org/10.18637/jss.v040.i04
6. Gabadinho, A., Ritschard, G., Studer, M., Müller, N.: Indice de Complexité Pour Le Tri et La Comparaison de Séquences Catégorielles (2010)
7. Helske, S., Helske, J.: Mixture hidden markov models for sequence data: the **seqHMM** package in R. J. Stat. Soft. **88**(3) (2019). https://doi.org/10.18637/jss.v088.i03
8. Karimi-Haghighi, M., Castillo, C., Hernández-Leo, D.: A causal inference study on the effects of first year workload on the dropout rate of undergraduates. In: Rodrigo, M.M., Matsuda, N., Cristea, A.I., Dimitrova, V. (eds.) Artificial Intelligence in Education, vol. 13355, pp. 15–27. Springer International Publishing, Cham (2022). https://doi.org/10.1007/978-3-031-11644-5_2
9. Knight, S., Friend Wise, A., Chen, B.: Time for change: why learning analytics needs temporal analysis. Learn. Analyt. **4**(3) (2017). https://doi.org/10.18608/jla.2017.43.2
10. López-Pernas, S., Saqr, M.: How the dynamics of engagement explain the momentum of achievement and the inertia of disengagement: a complex systems theory approach. Comput. Hum. Behav. **153**, 108126 (2024). https://doi.org/10.1016/j.chb.2023.108126
11. Matcha, W., Gašević, D., Ahmad Uzir, N., Jovanović, J., Pardo, A., Maldonado-Mahauad, J., Pérez-Sanagustín, M.: Detection of learning strategies: a comparison of process, sequence and network analytic approaches. In: Scheffel, M., Broisin, J., Pammer-Schindler, V., Ioannou, A., Schneider, J. (eds.) Transforming Learning with Meaningful Technologies, vol. 11722, pp. 525–540. Springer International Publishing, Cham (2019). https://doi.org/10.1007/978-3-030-29736-7_39

12. Matcha, W., Gašević, D., Uzir, N.A., Jovanović, J., Pardo, A.: Analytics of Learning Strategies: Associations with Academic Performance and Feedback. In: Proceedings of the 9th International Conference on Learning Analytics & Knowledge, pp. 461–470. ACM, Tempe AZ USA (Mar 2019). https://doi.org/10.1145/3303772.3303787

13. Molenaar, I., Wise, A.F.: Temporal aspects of learning analytics - grounding analyses in concepts of time. In: Handbook of Learning Analytics, p. 11 (2022)

14. Pérez-Sanagustín, M., Pérez-Álvarez, R., Maldonado-Mahauad, J., Villalobos, E., Sanza, C.: Designing a Moodle Plugin for Promoting Learners' Self-regulated Learning in Blended Learning. In: Hilliger, I., Muñoz-Merino, P.J., Laet, T.D., Ortega-Arranz, A., Farrell, T. (eds.) 17th European Conference on Technology Enhanced Learning (EC-TEL 2022). LNCS, vol. 13450, pp. 324–339. Springer International Publishing, Toulouse, France (Sep 2022). https://doi.org/10.1007/978-3-031-16290-9_24

15. Pintrich, P.R., Groot, E.V.D.: Motivational and Self-Regulated Learning Components of Classroom Academic Performance, p. 8 (1990)

16. Poquet, O., Jovanovic, J., Pardo, A.: Student profiles of change in a university course: a complex dynamical systems perspective. In: LAK23: 13th International Learning Analytics and Knowledge Conference, pp. 197–207. ACM, Arlington TX USA (Mar 2023). https://doi.org/10.1145/3576050.3576077

17. Ritschard, G.: Measuring the nature of individual sequences. Sociological Methods Res., 00491241211036156 (Sep 2021). https://doi.org/10.1177/00491241211036156

18. Saint, J., Fan, Y., Gašević, D., Pardo, A.: Temporally-focused analytics of self-regulated learning: a systematic review of literature. Comput. Educ. Artifi. Intell. **3**, 100060 (2022). https://doi.org/10.1016/j.caeai.2022.100060

19. Saint, J., Gašević, D., Matcha, W., Uzir, N.A., Pardo, A.: Combining analytic methods to unlock sequential and temporal patterns of self-regulated learning. In: Proceedings of the Tenth International Conference on Learning Analytics & Knowledge, pp. 402–411. ACM, Frankfurt Germany (Mar 2020). https://doi.org/10.1145/3375462.3375487

20. Saqr, M., López-Pernas, S., Jovanović, J., Gašević, D.: Intense, turbulent, or wallowing in the mire: a longitudinal study of cross-course online tactics, strategies, and trajectories. Internet Higher Educ. **57**, 100902 (2023). https://doi.org/10.1016/j.iheduc.2022.100902

21. Shannon, C.E.: A mathematical theory of communication. Bell Syst. Tech. J. **27**(3), 379–423 (1948). https://doi.org/10.1002/j.1538-7305.1948.tb01338.x

22. Tan, T.K., Samavedham, L.: The learning process matter: a sequence analysis perspective of examining procrastination using learning management system. Comput. Educ. Open **3**, 100112 (2022). https://doi.org/10.1016/j.caeo.2022.100112

23. Torre, M.V., Oertel, C., Specht, M.: The Sequence Matters in Learning - A Systematic Literature Review (2023). https://doi.org/10.1145/3636555.3636880

24. Vallat, R.: Pingouin: Statistics in Python. JOSS **3**(31), 1026 (2018). https://doi.org/10.21105/joss.01026

25. Villalobos, E., Pérez-Sanagustin, M., Sanza, C., Tricot, A., Broisin, J.: Supporting self-regulated learning in bl: exploring learners' tactics and strategies. In: Hilliger, I., Muñoz-Merino, P.J., De Laet, T., Ortega-Arranz, A., Farrell, T. (eds.) Educating for a New Future: Making Sense of Technology-Enhanced Learning Adoption, vol. 13450, pp. 407–420. Springer International Publishing, Cham (2022). https://doi.org/10.1007/978-3-031-16290-9_30

26. Wang, Q., Mousavi, A.: Which log variables significantly predict academic achievement? a systematic review and meta-analysis. Br. J. Edu. Technol. **54**(1), 142–191 (2023). https://doi.org/10.1111/bjet.13282

Explainable Automatic Grading
with Neural Additive Models

Aubrey Condor$^{(\boxtimes)}$ and Zachary Pardos

University of California Berkeley, Berkeley, CA 94720, USA
aubrey_condor@berkeley.edu

Abstract. The use of automatic short answer grading (ASAG) models may help alleviate the time burden of grading while encouraging educators to frequently incorporate open-ended items in their curriculum. However, current state-of-the-art ASAG models are large neural networks (NN) often described as "black box", providing no explanation for which characteristics of an input are important for the produced output. This inexplicable nature can be frustrating to teachers and students when trying to interpret, or learn from an automatically-generated grade. To create a powerful yet intelligible ASAG model, we experiment with a type of model called a Neural Additive Model that combines the performance of a NN with the explainability of an additive model. We use a Knowledge Integration (KI) framework from the learning sciences to guide feature engineering to create inputs that reflect whether a student includes certain ideas in their response. We hypothesize that indicating the inclusion (or exclusion) of predefined ideas as features will be sufficient for the NAM to have good predictive power and interpretability, as this may guide a human scorer using a KI rubric. We compare the performance of the NAM with another explainable model, logistic regression, using the same features, and to a non-explainable neural model, DeBERTa, that does not require feature engineering.

Keywords: Explainable AI · Automatic Grading · ASAG · Neural Additive Models

1 Introduction

It has been shown that the use of open-ended (OE) items is beneficial for student learning due to the generation effect [3] or in combination with self-explanation [6]. However, assessing OE items is time consuming for teachers [16] and consequently, educators default to using multiple choice (MC) questions instead. Automatic short answer grading (ASAG) may alleviate this time burden while encouraging educators to frequently incorporate OE items in their curriculum.

Many of the best performing ASAG models include some variation of a deep neural network (NN) [15]. NNs are impressive predictors for high dimensional inputs like text embeddings, but predictive power tends to come at the cost of intelligibility. These models are often described as "black box" which means that users only have access to inputs and outputs, yet no information as to the process

A. M. Olney et al. (Eds.): AIED 2024, LNAI 14829, pp. 18–31, 2024.
https://doi.org/10.1007/978-3-031-64302-6_2

in-between. Further, unlike explainable models such as a logistic regression (LR), with NN ASAG models, we are not choosing which features the model may consider, and the model provides no explanation for which characteristics of an input are important for the produced output. Whereas educators typically use a scoring rubric to provide an explanation to stakeholders about which features of a given response substantiate the assigned score, and despite that attempts have been made to integrate scoring rubrics into ASAG models [8], NN ASAG models provide no justification for their predictions. The inexplicable nature of ASAG models can be frustrating to both teachers and students when trying to make sense of, or learn from an automated grade. Teachers are unable to monitor student understandings at a fine-grained level, and students may not productively learn without knowing why they received a certain grade. It is critical that researchers find more explainable models if ASAG systems are to be of greater practical use.

In an effort to create a powerful yet intelligible ASAG model, we experiment with a type of model that combines the performance of a NN with the explainability of an additive model called a Neural Additive Model (NAM) [1]. NAMs allow us to visually examine the contribution of each feature to the final predicted score for each response, similar to testing the significance of a regression coefficient. For this type of model, we must engineer features of a response as inputs to the model instead of allowing the model to create its own text features like typical Large Language Model (LLM) classifiers do. The research questions we seek to answer include, (1) can NAMs provide intelligible automatic grading such that stakeholders can understand which features of a response are important for its prediction, and (2) is the predictive performance of NAMs better than that of legacy explainable models like a LR and commensurate with that of an LLM classifier? This research is unique as NAMs have not yet been explored for educational applications, or more specifically for explainable automatic grading.

We demonstrate our NAM approach for ASAG with one item designed under a Knowledge Integration (KI) perspective from the learning sciences and corresponding rubric to guide our feature engineering. KI is a framework for strengthening science understanding that emphasizes incorporating new ideas and sorting out alternative perspectives with evidence [22]. We hypothesize that the inclusion (or exclusion) of predefined KI ideas as features will be sufficient for the NAM to have good predictive power, as this is precisely what would guide a human scorer using the KI rubric. We compare the performance of the NAM with another explainable model, a LR, using the same features as those used for the NAM, and to a non-explainable model, DeBERTa, an improved version of the popular BERT model [17]. The DeBERTa model does not utilize the engineered features that are used by the NAM or LR model, but creates its own features from the text. For a more extensive comparison of the performance of these three models, we extend our analysis to include results from five additional questions from an open-source data set that has been used extensively in ASAG research - the Automatic Student Assessment Prize (ASAP) Short Answer Scoring (SAS) data [28].

2   Related Work

In this section, we describe previous works relating to Explainable AI and more particularly Explainable ASAG, as well as applications of NAMs.

2.1   Explainable AI and ASAG

A recent review article outlines the increase in demand for explainable AI (XAI), especially for people who are affected by AI driven decisions [36]. They describe the rising popularity of NN models for their state-of-the-art performance, despite that their inference processes are not known or interpretable. Many researchers aim to increase the explainability of AI systems. Initiatives such as DARPA's "Explainable AI (XAI) program" [14] and the European Union's "General Data Protection Regulation" which demands citizen's rights to an explanation for decisions made by AI [13] are contributing to an increased demand for XAI.

Only recently are researchers beginning to think about XAI in terms of ASAG models. Schlippe et al. [31] surveyed over 70 educators about preferences for explainability in ASAG and learned that most prefer to see matches between student answers and exemplary answers to justify scores over other explainability methods like highlighting the importance of certain words. Poulton et al. [27] proposed an XAI framework using SHAP values to assess popular LLMs for ASAG such as BERT and RoBERTa. Zeng et al. [37] investigated whether autograding models align with human graders in terms of the important words they use when assigning a grade by conducting a randomized controlled trial to see if highlighting words deemed important by an autograder can assist human grading. Singh et al. [32] introduced Summarize and Score (SASC) to explain text modules', which map text to scalar values within LLMs, providing a natural language explanation of the module's selectivity. Finally, Tornqvist et al. [33] introduced ExASAG, an explainable framework for ASAG that generates natural language explanations for predictions, and Condor and Pardos train a reinforcement learning agent to alter students' short responses so that these alterations provide insight as to why a ASAG score was assigned [7].

2.2   Applications of Neural Additive Models

Although Neural Additive Models (NAMs) were introduced only a few years ago [1], there has been substantial interest in their use as the machine learning community has increased efforts to promote understandable models. Some researchers have proposed altered or improved versions of NAMs for specific applications. Mariotti et al. [24] explored the tension between interpretability and performance of NAMs and introduce a constrainable NAM (CNAM) which includes specifications for model regularization. The CNAM model is able to consider the performance-interpretability tradeoff during training, and is demonstrated to work well for both regression and classification tasks [24]. Luber et al. [23] introduced Structural NAMs (SNAMs) which incorporate neural splines to a typical NAM. SNAMs offer improved intelligibility over NAMs by enabling direct interpretation of estimated parameters, as well as quantification of parameter uncertainty and post-hoc analysisr. Bouchiat et al. [4] take a Bayesian perspective and develop a Laplace approximated NAM (LA-NAM) to enhance interpretability, and Jo and Kim [18] introduced NAMs for multivariate time series modeling called NAM nowcasting (NAM-NC). Others have used NAMs for applications in finance [5], mortality prediction [26], and survival analysis [34]. Importantly, we found no previous works that used NAMs for educational applications.

3   Background

In this section, we provide a description of the data, and a brief overview of NAMs, LR, and the DeBERTa model.

3.1   The Data

The KI item used for this project was collected during a previous research project at the Web-based Inquiry Science Environment (WISE) research center at the University of California, Berkeley consisting of OE science items designed for middle school students [30]. Students accessed the items via an online classroom system, and responses were scored with a Knowledge Integration (KI) rubric. For this project, we use one item from a unit about the physics of sound waves that engages students to refine their ideas about concepts such as wavelengths, frequency and pitch. For this item, students must distinguish how the pitch of sound made by tapping a full glass of water compares to the pitch made by tapping an empty glass. They are asked to explain why they think the pitch of the sound waves may be the same or different for the two glasses [30]. See Fig. 1 for a visual of the Soundwaves item. The KI data include 1,313 OE student responses that were carefully rated from 1 to 5 by subject matter experts. A detailed rubric was created by researchers and used for scoring which includes a description of the rating level, examples of correct/incorrect mechanisms and conclusions, and exemplar student responses that would fall into each category. 2 provides a sample of the KI scoring rubric.

   The ASAP data used to extend our model comparison is from a 2012 Kaggle competition sponsored by the Hewlett Foundation, and consists of almost 13,000 short answer responses to 10 science and English questions [28]. We used only the 5 science questions to match the domain of the KI data. Each of the five items have between 1300–1800 OE student responses and corresponding human-rated scores between 1 and 4 with 4 being the most correct score. Each item has an expert created scoring rubric, and we aggregate the 5 items for training, testing and the resulting model comparisons.

Arlene has two glass cups. She leaves one empty and fills the other one with water. She then uses a chopstick to gently strike each glass. What will she hear?

How does the pitch of the sound made by the tapped full glass compare to the pitch of the tapped empty glass? (For reference: A whistle makes a high-pitched sound and a bear makes a low-pitched growl.)

○ There is no change; the pitch stays the SAME.

○ The pitch of the tapped full glass is LOWER than the pitch of the tapped empty glass.

○ The pitch of the tapped full glass is HIGHER than the pitch of the tapped empty glass.

SAVE

Explain <u>why</u> you think the pitches of the sound waves generated by striking the two glasses will be the same or different.

Fig. 1. The Sound Waves item bundle

KI Score	Description	Category of responses	Examples
1	I don't know or off topic	Response is off topic or "I don't know" • I don't know • Irrelevant ...	• "Because there is water in the cup" • "Idk" ...
2	Irrelevant or incorrect	Irrelevant or incorrect mechanism • sound bounces • sound wave has less space to move • the material inside the glass makes the sound echo ...	• "One of them has water, which makes the sound bounce, the other has air, which doesn't." • "There is more room in the glass to make a sound then the water full one."
3	Partially correct	Correct mechanism only • affects speed of vibration or oscillation • wave has higher/lower frequency Correct conclusion only • sound (or pitch) will be lower (or deeper) • sound/pitch is different (sufficient for a 3) ...	• "One is thicker than the other." • "The sound travels through the water slower so it comes out lower while sound travels through the empty space quicker so it come out higher."
4	Emerging understanding	One set of correct mechanism + correct conclusion • Lower pitch is linked to lower frequency in full glass • Higher/lower density of material is linked to higher/lower pitch ...	• "The pitch is lower because the water density changes the sound wave since it has to travel through the water. The empty glass sound wave will be a higher pitch because it doesn't have to travel through the water, keeping its length and amplitude the same." ...
5	Full understanding	Two sets of correct mechanism + correct conclusion • Full glass has more mass, so it is harder to vibrate, creating a lower pitched sound • Glass with water has more density which creates the sound wave to travel faster and a lower pitch ...	• "The full glass has mass inside of it making it more dense. This would cause it to have a lower pitch/frequency than an empty glass." ...

Fig. 2. The KI Scoring Rubric

3.2   Neural Additive Models

Neural Additive Models (NAMs) impose a restriction on the structure of NNs in order to make the model interpretable. NAMs belong to a family of models called Generalized Additive Models (GAMs) which have the form:

$$g(E[y]) = +f_1(x_1) + f_2(x_2) + \ldots + f_k(x_k)$$

where $x = (x_1, x_2, \ldots, x_k)$ represents the input with K features and each f_i is a univariate shape function with $E[f_i] = 0$ [1]. NAMs learn a linear combination of jointly-trained NNs where each NN attends to a single input feature. They can approximate complex, high-dimensional functions and their predictions are easily interpretable. NAMs provide advantages over standard GAMs such as their superior scalability with the use of GPU/TPU hardware developed for NNs, differentiability, and visual interpretability [1].

3.3 Logistic Regression

Multinomial LR is a classification model that predicts probabilities of different outcomes for a categorical dependent variable. To generalize to a K-class setting, the model runs K-1 independent binary LR models where one outcome is chosen as a "pivot" and other K-1 outcomes are separately regressed against the pivot outcome. We use the Limited-memory Broyden-Fletcher-Goldfarb-Shannon (LBfGS) algorithm for optimization [12], and incorporate L2 regularization.

3.4 DeBERTa

DeBERTa (Decoding-enhanced BERT with disentangled attention) is a Large Language Model (LLM) that improves upon the popular BERT [10] model by using a different version of the standard attention mechanism [35] and a novel position encoding method. DeBERTa uses a disentangled attention' mechanism where each word is represented with two vectors - one to encode the word's context, and another to encode the word's relative position. Further, an enhanced masked decoder incorporates absolute positions to predict masked tokens during training. Finally, DeBERTa uses a unique adversarial training method to fine-tune the model's generalizability performance on downstream tasks. We use an improved yet smaller version of the original DeBERTa model called DeBERTaV3-base [17], consisting of 12 layers, a hidden size of 768, and 86 million parameters.

4   Methods

To fit a NAM, we utilize the python library, "nam" [19], created by the researchers who introduced NAMs for easy implementation of the model. We altered the package code to accommodate our multi-class classification setting and allow for our evaluation procedure and corresponding metrics. Further described in the "Evaluation" section below, we use a cross validation (CV) evaluation procedure for model comparison, so for each model class, we train a total of 10 models on a different sample of the data. We train each of the 10 NAMs for 120 epochs with a batch size of 64 and a learning rate of 0.002. An epoch is one complete pass of the training data through the model, batch size represents the number of samples used in a single forward and backward pass through the network, and the learning rate governs the pace at which the model updates its parameter estimates. Further, we use a regularization technique called dropout with a value of 0.15 to help avoid over fitting. Hyperparameters were chosen using a standard grid search where different combinations of values are tested on a held-out validation set, and those that result in the best performance are chosen. Each of the DeBERTa models were trained for six epochs using a batch size of eight, and a learning rate of 0.0007. The LR models were fit using the scikit-learn "LogisticRegression" python library.

4.1 Feature Engineering

To create features from student response texts for the NAM and LR models, we utilize KI phrases from the rubric that represent both correct, and incorrect mechanisms and conclusions. We chose to use KI ideas to create features because these are exactly what a human identifies to assign a grade to a student response within the KI framework. Because students often express their ideas using a variety of words (i.e., the idea that *"sound moves faster in air"* could be stated as *"sound travels faster through air"*), we scan the responses for n-grams that are similar to the chosen KI phrases. An n-gram is a series of n consecutive words within a string of text. For example, 2-grams of the phrase *"the dog ran"* would be *"the dog"* and *"dog ran"*. We use n-grams of size n = 1 through 5 for each response as students can also express the same ideas in a different number of words. Features for the ASAP science questions were created in the same manner (and use the same number of features for each item), although we provide the feature phrases only from the KI data for demonstration.

To calculate the similarity between a given feature phrase and an n-gram from a student response, we first embed both the n-gram and the phrase using sentence-BERT. Sentence-BERT utilizes siamese and triplet network structures to create semantically meaningful sentence embeddings [29]. We then calculate the cosine similarity of the KI phrase and the n-gram. Cosine similarity is a metric used to measure the similarity of text using the cosine of the angle between the vector embeddings of the two texts being compared. Once the KI phrase has been compared with all respective n-grams in a response, we use the largest cosine similarity found as the feature for that phrase. Thus the feature representation for each phrase provides a measure of whether or not some form of the phrase is included in the student response. We use 34 phrases and 28 key words from the KI Sound Wave Rubric (shown in Fig. 2) for a total of 62 features. The same number of phrases and words were chosen for each ASAP item from their corresponding scoring rubrics. The KI phrases are shown below, and the individual words were chosen from the phrases:

"I don't know", "idk", "sound bounces", "water blocks sound", "water mutes ringing", "sound moves more in air", "sound moves less in water", "sound echoes", "sound sinks in water", "sound moves fast in air", "frequency is height of wave", "pitch higher in full glass", "the density is different", "water is more dense", "water vibrates less", "affects vibration", "pitch is lower in water", "pitch is different", "higher frequency in air", "frequency is different", "sound moves faster in water", "water has more mass", "mass is different", "sound is denser in water", "sound is slower in water", "amplitude is number of waves", "pitch lower in empty glass", "air is less dense", "empty glass vibrates more", "vibration is different", "pitch is higher in air", "lower frequency in water", "sound moves slower in air", "empty glass less mass".

4.2 Evaluation

We use a standard ASAG evaluation metric in our results table: Quadratic Weighted Cohen's Kappa (QWK). QWK reports the agreeability between two scores beyond random chance - a more robust measure than accuracy. A 5×2 cross validation (CV) paired t-test was used to evaluate the statistical significance of the difference in models. The 5×2 CV paired t-test is based on five iterations of twofold cross-validation, and is presented in [11] as the recommended approximate statistical test for whether one machine learning model outperforms another because of it's more acceptable type I error, and stronger statistical power than other methods such as McNemar's test, or a paired t-test based on 10-fold CV. We used the QWK metric for the 5×2 CV paired t-tests (with 5 degrees of freedom). We perform a t-test comparing the QWK metric for the NAM to the LR model and the DeBERTa model. The null hypothesis for each t-test states that the QWK metric for the NAM is no different than the QWK metric for LR and DeBERTa. Further, we report the magnitude of the difference in performance of the NAM to LR and DeBERTa using the QWK metric from the CV model fits with a Cohen's D effect size. Cohen's D is essentially a difference of means, scaled by a standard deviation value [20]. A low value of Cohen's D which represents a small magnitude of difference is typically below 0.20, and a large value would be around 0.80 or higher. We use a pooled standard deviation to calculate Cohen's D, which represents a weighted combination of the standard deviations from each model.

5   Results

Comparisons of each ASAG model's performance are presented in Tables 1 and 2. In Table 1, we present results of the 5×2 CV paired t-test using the QWK metric for both the KI data and the aggregate ASAP data. We compare the NAM against both LR and DeBERTa. For the KI data, the NAM performs better than LR at a significance level of 0.05 (t = 2.1732, p = 0.0409), and the NAM performs slightly worse than the DeBERTa model but not statistically at a significance level of 0.05 (t = −0.8410, p = 0.2194). Further, for the ASAP data, the NAM does not perform significantly better than the LR at a significance level of 0.05 (t = 1.691, p = 0.0758) although the p-value is quite small. For the ASAP data, we also see that the NAM performs worse than the DeBERTa model but not at a statistically significant level of 0.05 (t = −1.741, p = 0.071), although similarly, the p-value is quite small. As shown by the CV average metrics in Table 2, on average, DeBERTa performs better than the NAM and the NAM performs better than the LR across both data sets. Additionally, in Table 1 we present the Cohen's D effect sizes for the difference in performance from the NAM to the two other models. The magnitude of difference between the NAM and both models is quite large for the KI data, with effect sizes greater than 1.0 for each comparison. For the ASAP data, the effect size is moderate when comparing the NAM and LR.

Table 1. 5 × 2 Cross Validation Paired t-test (5 df) and Cohen's D Effect Sizes

NAM versus:	t(5)	p-val	Cohen's D
KI data			
DeBERTa	−0.8410	0.2194	−1.111
Log Reg	2.1732	0.0409	1.003
ASAP data			
DeBERTa	−1.741	0.0711	−1.070
Log Reg	1.691	0.0758	0.222

A visual of feature importance from the NAM for the KI data is presented in Fig. 3. We show the top 40 most important features in order of increasing importance. We see that the feature importance for the words *"density"* and *"higher"* is much larger than that of most other words and phrases. The phrase *"the density is different"* also has a comparatively high importance. Further, in Fig. 4, we present NAM Shape functions for the top 8 important words and phrases. As each feature of the NAM is handled independently by a learned shape function, we can see how the model makes its predictions, or in our case ratings, by graphing the shape functions for each feature. The scores are averaged and centered so that the shape functions can be directly compared on the same scale. As our NAM is a multi-class classifier, we can visualize five shape functions for each feature representing each of the five classes. We choose to show only the shape function for the lowest rating class and the highest rating class to retain the most important information yet avoid a messy, undecipherable visual. However, Fig. 5 provides an example of shape plots with all five shape functions shown for reference. A negative y-axis score signals a low probability of a certain class, and a positive score represents a high probability. Noticeably, the shape functions are jagged-like with sharp jumps. [1] emphasizes the unique capability of NAMs to model highly "jumpy" 1-dimensional functions due to the use of an exponential-centered (ExU) non-linear function, whereas standard neural network models are bias towards smoothness with the use of Rectified Linear Unit (ReLU) functions. We could hypothesize that the true shape functions for the features in this project are more smooth, and in future work we could employ regularization techniques such as weight decay to provide smoother shape functions. On the same plot, we see the data density in the form of pink shaded bars. The darker the shade of pink, the more data there is in that region. In some areas with a really low data density, the model may not have had enough data to adequately learn the shape function.

Table 2. 5 × 2 Cross Validation QWK Averages

Model	KI data	ASAP data
DeBERTa	0.7475	0.7176
Log Reg	0.6929	0.6342
NAM	0.7174	0.6473

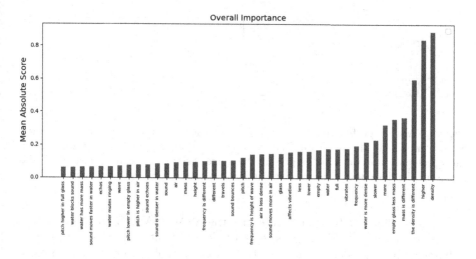

Fig. 3. NAM Mean Feature Importance

6 Discussion

The CV averages in Table 2 give us an idea for how well each model performs for ASAG. We are not surprised to see that the NAM provides more predictive power than the LR, but not as much as the DeBERTa model. The DeBERTa model uses the entire student response as input text and learns many hundreds of features for its predictions, whereas with the NAM and LR, we are limited to the particular features we choose and it would be an unreasonable task to engineer the same number of features as the DeBERTa model learns. Additionally, the results of the 5×2 CV paired t-test in Table 1 give evidence that the NAM outperforms the LR model - across both data sets at a significance level of 0.10 and more notably with the KI data at a significance level of 0.05. Importantly, the NAM and the LR model use the same exact features. Further, the Cohen's D effect size values show that the magnitude of difference in performance of the models is quite large for the KI data, and smaller but still notable for the ASAP data. Not only is the difference statistically significant, but it seems to be practically meaningful.

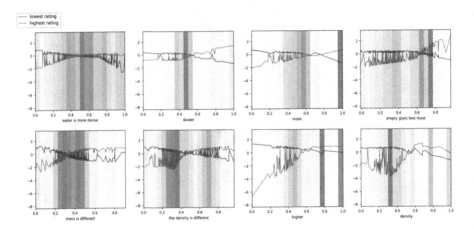

Fig. 4. NAM shape functions of the highest and lowest rating category for the top 8 most important phrases/words. The y-axes show the log odds of predicting a given rating category, and the x-axes represent the range of similarity scores. The pink shades represent the data density at varying similarity scores. (Color figure online)

Figure 3 allows us to visually interpret the ASAG NAM through overall feature importance, and Fig. 4 gives us more detail about how each feature contributes with NAM shape functions. Human grading of OE responses can often consist of identifying which ideas - correct or incorrect - are included in a student's response. The KI scoring rubric allowed us to identify *which* ideas (phrases and words) may help guide the NAM to correctly rate student responses, and the resulting NAM visualizations enable us to discern *how* these ideas contribute to an automated rating. The feature importance plot can help stakeholders identify which ideas are most indicative of the student's grade, and the NAM shape function gives a notion of how the directionality of the model's prediction changes for different values of different rating classes. For example, for the *"higher"* feature, the probability for the highest rating class (shown by the green line in Fig. 4) goes down significantly with decreasing cosine similarities. So, including the word *"higher"* in a response seems to correspond to higher scores. Additionally, with the density shading in the NAM shape function plots, we can infer the density of responses that include the given word or phrase. For example, the words *"more"*, and *"higher"* both have dense regions near high cosine similarities, so we can assume that many responses include these words or similar words. This can give an educator insight into which ideas students are more or less likely to write about.

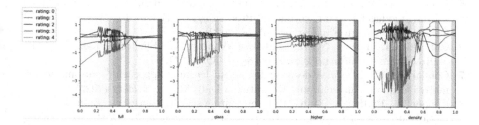

Fig. 5. An Example of NAM shape functions with **all** rating categories

Limitations of using a NAM for ASAG include that the performance of the model in terms of its match to human ratings is less than that of a LLM ASAG model. Also, engineering features for the NAM is more costly than preparing student responses for a LLM ASAG model, as the model will create features from the text itself. Further, we only show results for science questions so we cannot conclude that our results would generalize to other short answer items from different subject areas. Our results suggest that the NAM may be a sufficient alternative to legacy explainable models, like a LR. Not only did the NAM generally perform better than the LR with the same features, but the NAM provides easily interpretable visualizations of the model's prediction functions which can give an educator insights about their students' understanding of the item content. Many researchers in the field of Learning Analytics still use LR over NNs for classification when the advantages of intelligibility outweigh that of performance [2, 9, 21, 25]. The Learning Analytics community may benefit from investigating the use of NAMs for various classification tasks as well. In future work, we hope to experiment with using NAMs for ASAG with different question types from different domains, and perform qualitative interviews with educators to see if the NAM visualizations are understandable and helpful.

References

1. Agarwal, R., Frosst, N., Zhang, X., Caruana, R., Hinton, G.E.: Neural additive models: Interpretable machine learning with neural nets. arXiv preprint arXiv:2004.13912 (2020)

2. Alonso-Fernández, C., Martínez-Ortiz, I., Caballero, R., Freire, M., Fernández-Manjón, B.: Predicting students' knowledge after playing a serious game based on learning analytics data: A case study. J. Comput. Assist. Learn. **36**(3), 350–358 (2020)
3. Bertsch, S., Pesta, B.J., Wiscott, R., McDaniel, M.A.: The generation effect: a meta-analytic review. Mem. Cogn. **35**(2), 201–210 (2007)
4. Bouchiat, K., Immer, A., Yèche, H., Ràtsch, G., Fortuin, V.: Laplace-approximated neural additive models: improving interpretability with bayesian inference. arXiv preprint arXiv:2305.16905 (2023)
5. Chen, D., Ye, W.: Monotonic neural additive models: Pursuing regulated machine learning models for credit scoring. In: Proceedings of the Third ACM International Conference on AI in Finance, pp. 70–78 (2022)
6. Chi, M.T., De Leeuw, N., Chiu, M.H., LaVancher, C.: Eliciting self-explanations improves understanding. Cogn. Sci. **18**(3), 439–477 (1994)
7. Condor, A., Pardos, Z.: A deep reinforcement learning approach to automatic formative feedback. International Educational Data Mining Society (2022)
8. Condor, A., Pardos, Z., Linn, M.: Representing scoring rubrics as graphs for automatic short answer grading. In: Rodrigo, M.M., Matsuda, N., Cristea, A.I., Dimitrova, V. (eds.) AIED 2022. LNCS, vol. 13355, pp. 354–365. Springer, Cham (2022). https://doi.org/10.1007/978-3-031-11644-5_29
9. Deho, O.B., Zhan, C., Li, J., Liu, J., Liu, L., Le Duy, T.: How do the existing fairness metrics and unfairness mitigation algorithms contribute to ethical learning analytics? Br. J. Edu. Technol. **53**(4), 822–843 (2022)
10. Devlin, J., Chang, M.W., Lee, K., Toutanova, K.: BERT: pre-training of deep bidirectional transformers for language understanding. arXiv preprint arXiv:1810.04805 (2018)
11. Dietterich, T.G.: Approximate statistical tests for comparing supervised classification learning algorithms. Neural Comput. **10**(7), 1895–1923 (1998). https://doi.org/10.1162/089976698300017197
12. Fletcher, R.: Practical Methods of Optimization. Wiley, New York (2000)
13. Goodman, B., Flaxman, S.: European union regulations on algorithmic decision making and a "right to explanation". AI Mag. **38**(2), 781–796 (2017)
14. Gunning, D., Aha, D.: Darpa's explainable artificial intelligence (XAI) program. AI Mag. **40**(2), 44–58 (2019)
15. Haller, S., Aldea, A., Seifert, C., Strisciuglio, N.: Survey on automated short answer grading with deep learning: from word embeddings to transformers. arXiv preprint arXiv:2204.03503 (2022)
16. Hancock, C.L.: Implementing the assessment standards for school mathematics: enhancing mathematics learning with open-ended questions. Math. Teach. **88**(6), 496–499 (1995)
17. He, P., Gao, J., Chen, W.: DeBERTaV3: improving DeBERTa using ELECTRA-style pre-training with gradient-disentangled embedding sharing. arXiv preprint arXiv:2111.09543 (2021)
18. Jo, W., Kim, D.: Neural additive models for nowcasting. arXiv preprint arXiv:2205.10020 (2022)
19. Kayid, A., Frosst, N., Hinton, G.E.: Neural additive models library (2020)
20. Kelley, K., Preacher, K.J.: On effect size. Psychol. Methods **17**(2), 137 (2012)
21. Le, C.V., Pardos, Z.A., Meyer, S.D., Thorp, R.: Communication at scale in a MOOC using predictive engagement analytics. In: Penstein Rosé, C., et al. (eds.) AIED 2018. LNCS (LNAI), vol. 10947, pp. 239–252. Springer, Cham (2018). https://doi.org/10.1007/978-3-319-93843-1_18

22. Linn, M.C.: Designing the knowledge integration environment. Int. J. Sci. Educ. **22**(8), 781–796 (2000)
23. Luber, M., Thielmann, A., Säfken, B.: Structural neural additive models: enhanced interpretable machine learning. arXiv preprint arXiv:2302.09275 (2023)
24. Mariotti, E., Moral, J.M.A., Gatt, A.: Exploring the balance between interpretability and performance with carefully designed constrainable neural additive models. Inf. Fus. **99**, 101882 (2023)
25. Misiejuk, K., Wasson, B., Egelandsdal, K.: Using learning analytics to understand student perceptions of peer feedback. Comput. Hum. Behav. **117**, 106658 (2021)
26. Moslehi, S., Mahjub, H., Farhadian, M., Soltanian, A.R., Mamani, M.: Interpretable generalized neural additive models for mortality prediction of COVID-19 hospitalized patients in hamadan, iran. BMC Med. Res. Methodol. **22**(1), 339 (2022)
27. Poulton, A., Eliens, S.: Explaining transformer-based models for automatic short answer grading. In: Proceedings of the 5th International Conference on Digital Technology in Education, pp. 110–116 (2021)
28. Prize, A.S.A.: The Hewlett foundation: automated essay scoring (2019)
29. Reimers, N., Gurevych, I.: Sentence-BERT: sentence embeddings using Siamese BERT-networks. arXiv preprint arXiv:1908.10084 (2019)
30. Riordan, B., et al.: An empirical investigation of neural methods for content scoring of science explanations. In: Proceedings of the Fifteenth Workshop on Innovative Use of NLP for Building Educational Applications (2020)
31. Schlippe, T., Stierstorfer, Q., Koppel, M.t., Libbrecht, P.: Explainability in automatic short answer grading. In: Cheng, E.C.K., Wang, T., Schlippe, T., Beligiannis, G.N. (eds.) AIET 2022. LNCS, vol. 154, pp. 69–87. Springer, Singapore (2022). https://doi.org/10.1007/978-981-19-8040-4_5
32. Singh, C., et al.: Explaining black box text modules in natural language with language models. arXiv preprint arXiv:2305.09863 (2023)
33. Tornqvist, M., Mahamud, M., Guzman, E.M., Farazouli, A.: ExASAG: explainable framework for automatic short answer grading. In: Proceedings of the 18th Workshop on Innovative Use of NLP for Building Educational Applications (BEA 2023), pp. 361–371 (2023)
34. Utkin, L., Konstantinov, A.: An extension of the neural additive model for uncertainty explanation of machine learning survival models. In: Kravets, A.G., Bolshakov, A.A., Shcherbakov, M. (eds.) Cyber-Physical Systems: Intelligent Models and Algorithms, vol. 417, pp. 3–13. Springer, Cham (2022). https://doi.org/10.1007/978-3-030-95116-0_1
35. Vaswani, A., et al.: Attention is all you need. Adv. Neural Inf. Process. Syst. **30** (2017)
36. Xu, F., Uszkoreit, H., Du, Y., Fan, W., Zhao, D., Zhu, J.: Explainable AI: a brief survey on history, research areas, approaches and challenges. In: Tang, J., Kan, M.-Y., Zhao, D., Li, S., Zan, H. (eds.) NLPCC 2019. LNCS (LNAI), vol. 11839, pp. 563–574. Springer, Cham (2019). https://doi.org/10.1007/978-3-030-32236-6_51
37. Zeng, Z., Li, X., Gasevic, D., Chen, G.: Do deep neural nets display human-like attention in short answer scoring? In: Proceedings of the 2022 Conference of the North American Chapter of the Association for Computational Linguistics: Human Language Technologies, pp. 191–205 (2022)

PBChat: Enhance Student's Problem Behavior Diagnosis with Large Language Model

Penghe Chen[1,2], Zhilin Fan[2], Yu Lu[1,2(✉)], and Qi Xu[1]

[1] Advanced Innovation Center for Future Education, Beijing Normal University,
Beijing 100875, China
luyu@bnu.edu.cn
[2] School of Educational Technology, Faculty of Education, Beijing Normal
University, Beijing 100875, China

Abstract. Student's problem behaviors are undesirable behaviors encompass actions that deviate from established school standards, potentially impacting students' overall well-being and academic success significantly. Diagnosing these behaviors demands a multidisciplinary understanding, posing a challenge for conventional educators. Capitalizing on the advancements in Large Language Model (LLM) technology, we introduce this PBChat model, a specialized LLM designed for pinpointing problem behaviors. We articulate a theoretical framework for problem behavior diagnosis, laying the conceptual groundwork for PBChat. To train PBChat, we curate a multi-turn dialogue dataset based on annotated cases, and subsequently, fine-tune the ChatGLM2 base model using the QLoRA algorithm to build PBChat model. Experimental assessments gauge the performance of PBChat, with both automated and human evaluations revealing its efficacy in successfully diagnosing problem behaviors, surpassing the capabilities of general LLMs.

Keywords: Problem Behavior · Large Language Model ·
Parameter-Efficient Fine-Tuning

1 Introduction

Student's problem behaviors are undesirable behaviors that deviate from school standards, which can have profound implications on the overall well-being and academic success of students [22]. Most of those problem behaviors arise under the influence of undesirable internal and external factors [21], which lead to the unsatisfied psychological needs of individuals in the process of growing up, and then lead to deviant behaviors of primary and secondary school students with limited cognitive and behavioral abilities. To effectively tackle and mitigate the impact of these problem behaviors, problem behavior diagnosis aims to identify the underlying factors causing such behaviors through a systematic and comprehensive evaluation of behaviors.

A. M. Olney et al. (Eds.): AIED 2024, LNAI 14829, pp. 32–45, 2024.
https://doi.org/10.1007/978-3-031-64302-6_3

Historically, educators have predominantly relied on their accumulated experiences for the identification of student behavioral issues. However, given that the diagnosis of problem behaviors necessitates a nuanced understanding drawn from pedagogy, psychology, sociology, and physiology, teachers lacking such specialized expertise may encounter challenges in comprehensively analyzing students' behavioral concerns [25]. Moreover, the multifaceted nature of the underlying causes, encompassing factors like family dynamics and peer relationships, further complicates the diagnostic process. Consequently, an approach rooted solely in experiential insights may lack the systematic rigor required and be susceptible to inherent biases. Additionally, the temporal and energy constraints placed on teachers, who often need to cater to a broad student population, can impede the efficiency of the diagnosis process.

The emergence of Large Language Models (LLM) signifies a notable advancement in artificial intelligence, prompting our exploration of these challenges through LLM technology. LLM represents a sophisticated deep neural network, encompassing billions of parameters trained on extensive textual corpora [12]. Noteworthy for its ability to grasp contextual dependencies, infer meaning, and generate coherent and contextually relevant text across diverse linguistic domains, LLM stands as an apt choice for the diagnosis of student problem behaviors, which involve a sequence of natural language interactions and reasoning. Nevertheless, generic LLMs may encounter two specific challenges in supporting problem behavior diagnosis. Firstly, while excelling in responding to queries, LLMs exhibit limitations in formulating questions-a crucial aspect in the diagnostic process. Secondly, LLMs tend to furnish broad suggestions, often lacking in personalized recommendations [12].

In response to these challenges, we introduce the PBChat model, a specialized LLM designed for the targeted diagnosis of problematic behaviors. Our approach involves a fourfold strategy. Firstly, we establish a theoretical framework elucidating the factors and root causes of problem behaviors, providing a conceptual underpinning for our LLM. Secondly, we curate a multi-turn dialogue dataset derived from cases annotated in accordance with the aforementioned theoretical framework, laying the groundwork for the data-driven construction of our LLM. Thirdly, we craft the PBChat model by refining the ChatGLM2 [6] through the utilization of the QLoRA algorithm [5], aligning it with the proposed theoretical framework and the meticulously assembled dataset. Fourthly, we conduct a comprehensive evaluation encompassing both automated assessments and human evaluations to gauge the performance of PBChat.

2 Related Work

2.1 Problem Behavior Diagnosis

Student's Problem behaviors are undesirable behaviors that deviate from school standards, which can hinder the healthy physical and mental growth of students [22]. The Problem-Behavior Theory identifies three critical systems influencing behavior: the personality system, the perceived environment system, and the behavioral system [7]. Key environmental influences include parenting styles [19], family structure [8], socioeconomic status [10,17], and the quality of parent-child relationships [2,28], all of which are pivotal in shaping student behavior. Additionally, school dynamics, such as teacher-student relationships [25] and peer acceptance [4,26], are also predictive of behavior. Personality-wise, children with certain extreme traits are more prone to problematic behaviors in adolescence, underscoring the multifaceted nature of these issues [1]. Educational research has extensively explored various strategies for diagnosing student's problem behaviors. From the standpoint of influencing factors, enhancing students' relationships with their parents and guiding parental behavior have been shown to effectively mitigate such behaviors [13,15]. Moreover, fostering a positive school environment is instrumental in promoting desirable student conduct [20]. Methodologically, social-emotional learning (SEL) has emerged as a pivotal approach in bolstering adolescent resilience and fostering positive mental health [14], thereby aiding in character development and the reduction of problem behaviors [24]. Additionally, the School-Wide Positive Behavioral Interventions and Supports (SWPBIS) framework has proven effective in preventing problematic behaviors and intervening constructively [21]. All these research findings provides theoretical foundations to build the PBChat model.

2.2 Large Language Models and Educational Applications

LLM is a complicated deep neural network with billions of parameters trained with massive textual corpora [12]. LLM exhibits a remarkable capacity to capture contextual dependencies, infer meaning, and produce coherent and contextually relevant text across diverse linguistic domains. LLMs have been applied to various domains, in which the fine-tuning method is a common approach to building domain-specific LLM, which involves retraining a pre-trained model on a dataset specific to a particular task to better adapt the model to that task.

Researchers have also explored the usage of LLM to solve educational tasks. LLMs have been built to solve general educational tasks. For instance, EduChat [3] encompasses a wide range of educational scenarios and tasks like open-ended question answering, paper evaluation, etc. LLMs have also been developed for subject teaching and learning. For example, KWAIYIIMATH [9] performs well in processing mathematical reasoning problems in both Chinese and English. MAmmoTH [27] introduces a scalable and lightweight approach to adjust mathematical instructions, significantly improving the LLM's general mathematical reasoning capabilities. In the realm of children's emotional companionship, Qiaoban [23] facilitates effective emotion management and guidance for children through fine-tuning LLM with generic domain human-computer dialogue data, single-round instruction data, and children's emotional companionship dialogue data. All these explorations have motivated us to enhance student's problem behavior diagnosis with LLM technologies.

3   PBChat Model

In this work, we develop a specialized LLM, PBChat, aiming for the precise diagnosis of problem behaviors. The development and assessment of the PBChat model are delineated in four distinct phases, as depicted in Fig. 1. Firstly, we establish a theoretical framework for problem behavior by conducting a comprehensive review of existing literature on problem behavior diagnosis, thereby laying the foundational guidelines for PBChat. Subsequently, a conversational dataset for problem behavior diagnosis is constructed based on annotated cases, providing the requisite supervised dataset for the training of PBChat. In the third phase, we employ a parameter-efficient fine-tuning technique, QLoRA, to refine ChatGLM2 [6] and create PBChat. The final step involves the evaluation of PBChat using both automated and human-centric methods, which will be expounded upon and presented in Sect. 4.

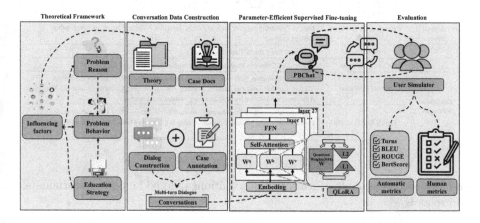

Fig. 1. System Architecture and Workflow for PBChat Model Building

3.1   Theoretical Framework

The development of a theoretical framework for student problem behaviors and an analysis of contributory factors are crucial for understanding, analyzing, and managing such behaviors effectively. Utilizing the Child Behavior Checklist (CBCL), we conducted an analysis to identify prevalent and impactful problem behaviors in educational settings, thereby establishing a categorization framework for these behaviors. As summarized in Table 1, we employed a broad to a specific approach to classifying the influences on problem behaviors, grounded in the principles of psychological management in primary and secondary education. This involved delineating both external environmental and internal individual factors contributing to student behavioral issues. Furthermore, drawing on Maslow's hierarchy of needs, we pinpointed the root causes of problem behaviors

Table 1. Theoretical Framework of Problem Behavior Diagnosis

Category		Subcategory
Problem Behaviors	Aggressive Behavior	Physical Aggression, Verbal Aggression, Relational Aggression
	Rule-Breaking Behavior	Non-Disturbing Discipline Violation in Class, Disturbing Discipline Violation in Class, Discipline Violation Outside Class
	delinquent behavior	Lying Behavior, Stealing Behavior, Immoral Behavior
	Social Withdrawal	Verbal Withdrawal, Behavioral Withdrawal, Psychological Withdrawal
	Emotional Problem	Depression, Anxiety
	Learning Problem	Learning-Ability Problem, Learning-Method Problem, Learning-Attitude Problem , Attention Problem
	Egocentricity	Self-Bragging, Stubbornness, Selfishness
	Special Problem	Addiction, Puppy Love, Extreme Behavior
Influencing factors	Family Background	Family Structure, Parenting Style, Family atmosphere, Education Background, Medical Condition, Delinquent Behaviors, Socioeconomic Status
	School Environment	Teacher Leadership, Peer Acceptance
	Sociocultural Factors	Mass Media, Cultural Custom
	Individual Factors	Gender, Grade, Health Condition, Social Group
Psychological Needs	Physiological Needs	Bad at mood control, Pathologic problem
	Safety Needs	Lack of Security
	Belongingness and Love Needs	Lack of Friendship Support, Lack of Family Affection, Frustrated Romantic Relationship, Lack of Attention
	Esteem Needs	Frustrated Self-Esteem, Lack of Confidence
	Cognition Needs	Mismatched Cognitive Need, Incorrect Perception, Lack of Proper Guidance
Education Strategies	School Education	Persuasive Education Method, Exemplary Demonstration Method, Emotional Edification Method, Conduct Guidance Method, Practical Training Method, Character Assessment Method, Appreciative Education Method, Learning Assistance Method
	Family Collaboration	Parenting Assistance Method, Home Learning Method, Parent-school Communication Method, Parent Volunteering Method, Decision-making Involvement Method, Community Collaboration Method

and proposed actionable educational interventions tailored for teachers, students, and schools.

3.2   Conversational Dataset Construction

Diagnosis fundamentally involves experts initiating dialogues by asking questions, a task at which general LLMs are notably less adept. To imbue an LLM with diagnostic capabilities, it is essential to fine-tune it using datasets reflective of such interactive questioning. Examination of expert diagnostic processes reveals four critical stages in a comprehensive diagnosis. Initially, experts inquire about behaviors to ascertain the type of problem behaviors exhibited by the student. Subsequently, upon identifying the problem behavior, they probe into factors influencing the student's psychological state and behavior. In the third stage, experts analyze these factors to uncover the underlying psychological needs driving the student's behavior, pinpointing them as the root causes. Finally, experts propose targeted intervention strategies tailored to assist the student in addressing these behaviors.

The diagnostic dialogue process is characterized by two key features. Firstly, it should mimic the conversational flow of a problem behavior diagnosis, engaging in a back-and-forth exchange akin to real-life expert consultations. Secondly, the dialogue must align with an established theoretical framework, incorporating a comprehensive analysis of various student-related factors. In the absence of such specialized datasets, we aim to create one derived from collected case studies. These cases, documented by seasoned educators, encapsulate their firsthand experiences in diagnosing and addressing students' problem behaviors, offering a wealth of practical insights and expertise. Each case provides a detailed account of the teacher's observations regarding a student's behavior, personality analysis, family background, and the teacher's deductions concerning the nature of the problem behaviors, underlying causes, and the pedagogical approaches applied. This wealth of information aligns with our theoretical framework. Therefore, our approach to constructing this dataset involves initially tagging the cases based on the theoretical framework and subsequently translating these annotations into a structured diagnostic dialogue format.

The initial step involves annotating each case in line with the designated theoretical framework, targeting six critical aspects: basic information inquiry, student problem behavior description, inquiry into influencing factors, analysis of behavioral performance, investigation of problem causes, and suggestion of educational strategies. The challenge lies in isolating text segments that articulate these aspects, framing this task as a text classification issue. In this study, a meta-learning-based approach [18] is employed to facilitate the automatic annotation of these cases.

The subsequent phase involves transforming the annotations into a diagnostic dialogue using a heuristic method. A diagnostic dialogue between an expert and a teacher usually comprises four distinct sections, as differentiated by colors in Table 2, we initiate by posing a question related to the student's behavior like *"What kinds of behaviors does the student normally exhibit?"*. Given the annotated behavior descriptions, we then craft a corresponding user response. Next, we formulate questions about various elements identified within the theoretical framework. For example, concerning the student's parenting style, a question like *"What is the student's parenting style in the family?"* may be posed, with the user's reply based on the case data. Elements not covered in the case data are omitted from the dialogue. Following this, we synthesize the type of problem behavior and its underlying reasons based on the behavior descriptions and factor information previously gathered. Lastly, we address the identified causes by suggesting intervention strategies documented in the case, thereby constructing a comprehensive diagnostic dialogue that mirrors the expert's approach while being grounded in the specific details of each case.

3.3   Parameter-Efficient Fine-Tuning

In this study, we adopt parameter-efficient fine-tuning techniques to refine our LLM for educational consulting tasks. Fine-tuning is a pivotal step in imbuing LLMs with the requisite dialogue competencies. While fine-tuning mirrors aspects of the initial pretraining phase, it uniquely focuses on leveraging domain-specific datasets to align the model's learning with task-specific requirements.

Table 2. One Example of Generated Dialogue

Turn	Role	Content
1	Expert	学生目前有什么问题行为表现吗？ What kinds of behaviors does the student normally exhibit?
2	Teacher	学生好像产生了习得性无助，表现在做作业不及时完成，而且字迹潦草。上课的时候目光呆滞，也不回答问题，学习成绩一直下滑。The student develops learned helplessness, which manifests itself in doing homework that is not completed in a timely manner and in scribbled handwriting. He stares blankly in class and does not answer questions, and his academic performance keeps slipping.
3	Expert	在家庭中父母对孩子的教养方式是怎样的呢？ What is the student's parenting style in the family?
4	Teacher	学生的爸爸忙着赚钱，无暇顾及学生的生活和学习，然后妈妈和爸爸关系不融洽，也很少关心这个学生。The student's father is too busy earning money to take care of the student's life and studies, and then the mom and dad don't have a good relationship and seldom care about the student.
		...
11	Expert	在学校中同学对孩子的同伴接纳类型是怎样的呢？ What is the type of peer acceptance of the child by peers at school?
12	Teacher	学生与同伴的相处一般，没有大的冲突，但是也没有很亲密的朋友。Students get along with their peers in general, no major conflicts, but no very close friends either.
13	Expert	根据上述情况，该学生的问题行为主要是 学习态度问题。 学生出现问题的主要原因是：缺乏自信、缺乏亲情关爱。 具体表现为，无论怎么努力，还是没有进步，然后就对学习失去了信心。父母经常吵架，都不管他。 Based on the above, the student's problematic behavior is primarily a **learning attitude problem**. The main reasons for the student's issue are: **Lack of self-confidence and Lack of affection and care**. This is evidenced by: This is manifested in the fact that no matter how hard he tries, there is still no progress, and then he loses confidence in learning. Parents often quarrel, both do not care about him. 我建议你， 及时鼓励进步：在一次课堂中，我特意把话题绕到了文明礼貌上面，然后表扬了他；教师课外帮扶：我决定要对他进行单独辅导，每次布置的作业，我都细心地跟他讲解；传授家庭教育知识：我尝试和他妈妈沟通，说学生最近学习有所进步，但是他需要家人更多的关怀。 I suggest you, **Timely Encouragement of Progress:** In one class, I purposely circled the topic to civility and then praised him; **Teacher extracurricular help:** I decided that I wanted to tutor him individually, and I carefully explained to him every time he was assigned homework; **Imparting knowledge about homeschooling:** I tried to communicate with his mom that the student has been making progress in his studies lately, but he needs more care from his family.

This process facilitates a harmonious integration of the model's inherent generalization abilities with the nuanced requirements of educational consulting, thus optimizing its performance. Specifically, we have selected ChatGLM2 [6] as the base model to build PBChat.

Fine-tuning aims to update model parameters, but traditional methods update all parameters, leading to high costs. Methods like LoRA [11] and QLoRA [5] are increasingly favored in the field due to their proficient utilization of existing data while minimizing the need for extensive computational resources. LoRA innovates by enhancing linear projections with an added factorized projection, maintaining the pre-trained model's weights while transforming the updated matrices into trainable, rank-decomposed matrices across the transformer architecture's layers. This approach significantly lowers the number of parameters needing training for new tasks. QLoRA extends LoRA's principles by integrating quantization, which reduces the numerical data's precision within the model, further cutting down on memory and computational needs. Specifically, QLoRA employs a 4-bit NormalFloat (NF4) data type for quantization and utilizes double quantization to further reduce memory requirements. As depicted in Fig. 2, for a given projection $XW = Y$, where $X \in R^{h \times b}$ and W

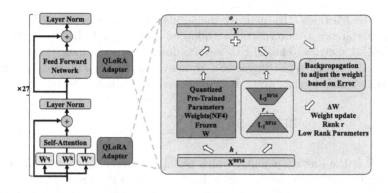

Fig. 2. Reparameterization Process of QLoRA

$\in R^{b \times o}$, the LoRA approach transforms the formula into

$$Y = XW + sXL_1L_2$$

where $L_1 \in R^{h \times r}$, $L_2 \in R^{r \times o}$, and s representing a scalar. Utilizing the QLoRA technique enables the updating of model parameters with a substantial reduction in memory usage, yet preserves the comprehensive effectiveness associated with 16-bit fine-tuning tasks.

4 Experiment

4.1 Experiment Setup

The primary objective of this study is to assess the efficacy of the proposed model in accurately diagnosing students' problem behaviors and recommending appropriate intervention strategies. For the experimental setup, the dataset is partitioned using a 9:1 ratio, where 90% is allocated for training the model, and the remaining 10% is used to evaluate its diagnostic capabilities on student behaviors. This evaluation also includes a comparative analysis with the baseline ChatGLM2 model to benchmark the performance of our proposed model. Furthermore, to gauge the effectiveness of the proposed model, we subject the training examples to the developed model to determine if it exhibits superior performance in diagnosing and addressing students' problem behaviors.

4.2 Evaluation Metrics

In this study, we utilize both automatic metrics and human metrics to assess the performance of the proposed model. The automatic metrics leverage automated natural language processing metrics, offering an objective and efficient means of assessment. Conversely, the human metrics involve expert reviewers manually examining the results, providing more in-depth but time-consuming and subjective analysis.

Automatic Metrics. To quantify the similarity between generated sentences and their reference counterparts, we adopted the BLEU-1, BLEU-2, BLEU-3, and BLEU-4 metrics. The BLEU metric, designed to evaluate the quality of translations, calculates the degree of n-gram overlap between machine-generated translations and reference texts. It is expressed mathematically as:

$$\text{BLEU} = \text{BP} \cdot \exp \left(\sum_{n=1}^{N} w_n \log p_n \right)$$

where BP is the brevity penalty, which penalizes overly short translations, w_n is the weight assigned to each n-gram level, and p_n is the precision of n-grams, reflecting the proportion of n-grams in the translated output that match the reference translations.

In addition to BLEU metrics, we utilize the ROUGE-1, ROUGE-2, and ROUGE-L metrics [16], which are specifically developed for the evaluation of automatic summarization and machine translation tasks. ROUGE distinguishes itself from BLEU by emphasizing recall, and assessing the extent of overlap between the content generated automatically and the reference summaries. The computation of its variant ROUGE-N is as follows:

$$\text{ROUGE} - \text{N} = \frac{\sum_{S \in \{ReferenceSummaries\}} \sum_{\text{gram}_n \in S} Count_{match}(gram_n)}{\sum_{S \in \{ReferenceSummaries\}} \sum_{\text{gram}_n \in S} Count(gram_n)}$$

were, gram_N represents the count of matching n-grams in the reference summary with the generated summary.

We further employ the BertScore metric to evaluate sentence similarity. BertScore accounts for contextual and semantic information, offering more accurate assessments of similarity, particularly for semantically similar but differently phrased sentences. The BertScore formula is as follows:

$$P = \frac{1}{|C|} \sum_{i \in C} \max_{j \in R} \cos(e_i^C, e_j^R), R = \frac{1}{|R|} \sum_{i \in R} \max_{j \in C} \cos(e_i^R, e_j^C), F = 2 \times \frac{P \times R}{P + R}$$

where C and R signify the words in the candidate and reference texts, respectively. e_j^C and e_i^R denote the embedding vectors for these words. The function "cos" represents the cosine similarity operation.

Human Metrics. The assessment of generated content is typically categorized into two principal dimensions: intrinsic evaluation and extrinsic evaluation. Intrinsic evaluation pertains to the scrutiny of the generated content's internal quality, encompassing aspects such as linguistic fluency, accuracy, and logical coherence. On the other hand, extrinsic evaluation emphasizes the appraisal of the dialog system's generated content in relation to its efficacy in fulfilling intended objectives. In this work, we have established a comprehensive framework to evaluate the diagnostic content generated by the PBChat model, which is summarized in Table 3.

This framework is meticulously designed to reflect the educational objectives and pedagogical value inherent in the PBChat through three dimensions. The educational value orientation dimension concerns the consistency in understanding the psychological determinants underlying students' mental health issues. The model functional application dimension concentrates on two key aspects: pertinency and rationality. Pertinency concerns whether acquired information during diagnosing is necessary for identifying causes of problem behaviors. Rationality measures whether the interventions suggested responses to these behaviors. The human-computer interaction guidance dimension gauge the accuracy and relevancy of generated dialogue content. Accuracy checks whether the diagnosing process is accurate and comprehensive, and relevancy measures whether the problem behaviors and corresponding causes are relevant.

Table 3. Human Metrics

Dimension	Aspect	Score	Description
Education Value Orientation	Consistency	0-5	Whether the analysis and summary of the main causes of the problems are consistent and correct with the content of the descriptions.
Model Function Application	Pertinency	0-5	Whether the information obtained during the counseling dialogue is comprehensive and pertinent.
	Rationality	0-5	Whether the recommended methods and specific operations of educational strategies are reasonable and effective.
Human-Computer Interaction Guidance	Accuracy	0-5	Whether the analysis and summary of students' problem manifestations are accurate and comprehensive.
	Relevancy	0-5	Whether the problem manifestations, the analysis of causes, and the educational countermeasures are relevant and focus.

5   Results Analysis

5.1   Results on Automatic Metrics

Table 4 provides a detailed comparison of performance metrics between the PBChat and ChatGLM2 models based on both training data and testing data. It can be seen that the PBChat model demonstrates a significant improvement in all evaluated metrics over the ChatGLM2 model. Especially for the metrics of B-3, B-4 and R-2, PBChat model obtain similar performance like B-1 and R-1, while ChatGLM2 has a significant performance decline. This indicates that proposed PBChat model showing a better content relevance and linguistic quality, demonstrating superior coherence and relevance in dialogue generation.

In addition, we also conduct a performance comparison of PBChat between the training data and testing data. The experimental results demonstrate that PBChat model performs a little better on training data than testing data. This is reasonable as PBChat is build with the training data and should have better performance. The performance on testing data implies the proposed PBChat has obtained the generalization ability in problem behavior diagnosis.

Table 4. Automatic Evaluation Results

Data	Model	Automatic							
		B-1	B-2	B-3	B-4	R-1	R-2	R-L	BertScore
Training Data	ChatGLM2	20.8	8.09	2.81	0.95	16.31	2.11	14.12	61.8
	PBChat	**66.4**	**61.7**	**59.5**	**57.8**	**65.0**	**56.6**	**61.9**	**85.6**
Testing Data	ChatGLM2	20.1	7.73	2.35	0.85	16.08	1.86	13.5	61.8
	PBChat	**38.1**	**28.5**	**24.1**	**20.8**	**28.7**	**12.4**	**25.7**	**70.9**

Table 5. Human Evaluation Result

Aspects	M±SD			One-way ANOVA		
	S1 (n = 409)	S2 (n = 409)	S3 (n = 409)	F	Sig.(2-tailed)	Post-hoc
Consistency	4.71±0.52	3.82±0.92	1.47±1.56	962.16**	.000	S1 > S2 > S3
Pertinency	4.63±0.70	4.36±0.73	0.16±0.62	5481.30**	.000	S1 > S2 > S3
Rationality	4.65±0.68	4.01±0.86	3.23±1.23	230.70**	.000	S1 > S2 > S3
Accuracy	4.88±0.39	4.38±0.80	1.66±1.55	1146.95**	.000	S1 > S2 > S3
Relevancy	4.67±0.58	3.86±0.74	2.60±1.04	688.62**	.000	S1 > S2 > S3

5.2   Results on Human Metrics

The investigation assesses three separate datasets: the original test dialogue data
(S1), dialogue content generated by the PBChat model (S2), and dialogue output from the ChatGLM2 model (S3). The evaluation is conducted according to
the five aspects proposed in the human metrics framework: consistency, pertinency, rationality, accuracy, and relevancy. ANOVA results indicate significant
disparities in the performance across these metrics for dialogues generated by the
three models (S1, S2, and S3) as outlined in Table 5. Subsequent post hoc multiple comparisons elucidated that the PBChat model markedly outperformed the
ChatGLM2 model in terms of consistency in summarizing reasons, relevancy in
gathering information pertinent to the counseling process, rationality in formulating parenting responses, accuracy in delineating problem manifestations, and
contextual consistency within the overall conversational content ($p < 0.01$). However, dialogues generated by the PBChat model still exhibit significant deviations
from the original test dialogue data across all five metrics ($p < 0.01$), particularly in the consistency of summarizing and describing the causative analysis
of problems ($M = 0.89$), and in the relevance of content relating to problem
manifestations, causative analysis, and parenting strategies ($M = 0.81$).

5.3   Exemplary Case of PBChat Diagnosis

An exemplary case of problem behavior diagnosis conducted by the PBChat
model is shown in Fig. 3. Initially, the system inquires and identifies behaviors
such as the student consistently avoiding eye contact and reluctance to communicate with others. It then gathers information on various factors that may
influence the student's development, tailored to the observed behaviors. This

我有一个学生行为相关的问题，你可以帮我解答吗? I have a question related to student behavior. Can you help me?

当然可以，我会尽力帮你解答。学生有哪些问题行为表现呢? Sure, I'll do my best to help you with your questions. What problem behaviors do students exhibit?

她总是低着头，一个人默默地坐在座位上。她有些抵触，看到我来到身边，她就有意避开，我跟她讲话，她也不理我。 She always sits quietly at her seat, head bowed, alone. Exhibiting some resistance, she deliberately avoids me whenever I approach. She remains unresponsive to my attempts at conversation.

这个学生的性别是什么? What is the gender of this student?

女生 Female Student

你能描述下这个学生的健康状况吗? Can you describe the health condition of this student?

她身体不好，从小就患有先天性心脏病。 She is not in good health and has been suffering from congenital heart disease since childhood.

那学生当前的所属群体状况呢? What is the current group status of the student?

她的父母在外地打工，平时就由外公外婆照顾。 Her parents work in other places and are usually taken care of by her grandparents

在家庭中父母对孩子的教育方式是怎样进行的呢? How do parents in the family carry out their parenting style with their children?

父母很忙，没有时间管她。 Her parents are busy. They don't have time for her.

周围的同学或同伴对该学生的接纳程度如何呢? How do classmates or peers accept this student?

在班级排座位时，很多家长都不愿意自己的孩子挨着她坐，还说怕被传染了。课余时间，同学们也不和她玩。 When arranging seats in the class, many parents do not want their children to sit next to her, fearing contagion. During extracurricular time, classmates also do not play with her.

综上发现学生的问题表现主要是：行为型退缩、言语型退缩
In summary, the student's main issues are: behavioral withdrawal, verbal withdrawal.
学生出现问题的主要原因是：缺乏自信、缺乏友情支持
The main reasons for the student's problems are: lack of confidence, lack of friendship support.
表现如下：她知道我关心她，但也不愿意跟我多说话。她看起来很孤独，没有朋友，自信心不足。
Manifested as follows: She knows I care about her, but she also does not want to talk to me much. She looks very lonely, has no friends, and lacks self-confidence.
建议你：
I suggest:
无私关爱陶冶 在学校，我会时刻关注，与她交谈，帮助她排班位，让同学们和她一起玩耍。
Unselfish love and cultivation. In school, I will always pay attention, talk with her, help her with seating arrangements, and encourage classmates to play with her.
主动谈心说服 课余时间，我会邀请她与我一起读书、讲故事、做手工。
Active heart-to-heart persuasion. In my spare time, I will invite her to read books, tell stories, and do crafts with me.
把握时机说服 慢慢地，我抓住了时机，与她沟通，让她明白，我们是不同性别，我们在一起是为了互相帮助，相互关心。
Seizing the opportunity for persuasion. Gradually, I seized the opportunity to communicate with her, making her understand that we are of different genders, and we are together to help and care for each other.

Fig. 3. Exemplary Case of PBChat Diagnosis

includes identifying the student as a female with health concerns, specifically a heart condition, and uncovering family dynamics where parents are often absent due to work, leading to a neglectful upbringing. The model also examines peer acceptance within the school environment, noting the student's social isolation. Subsequently, the system analyzes the student's behaviors to accurately diagnose signs of both behavioral and verbal withdrawal. By considering individual, family, and school factors, it identifies a lack of confidence and peer support as underlying causes, providing evidence for these diagnoses. Finally, the system recommends effective educational strategies to mitigate the student's issues, suggesting approaches such as unselfish love and cultivation, active heart-to-heart persuasion, seizing the opportunity for persuasion as strategies to address the behavioral concerns.

6 Conclusion

In this research, we endeavored to refine the diagnostic process for problem behaviors through the application of LLM technology. Building upon the Chat-GLM2 base model, we developed this PBChat, a specialized LLM tailored for diagnosing problem behaviors. Initially, we established a theoretical framework that encapsulates the various factors influencing students' problem behaviors. Following this, we curated a conversational diagnosis dataset derived from annotated case studies. Utilizing this dataset, we applied a fine-tuning methodology to craft an LLM specifically designed for problem behavior diagnosis. Our experimental evaluations of the model, employing both automated and manual methodologies, consistently demonstrated that our tailored model surpasses the capabilities of conventional LLMs in diagnosing problem behaviors.

Acknowledgements. This research is supported by the National Natural Science Foundation of China (No. 62177009), the Fundamental Research Funds for the Central Universities (No. 2022NTSS47).

References

1. Van den Akker, A.L., Prinzie, P., Deković, M., De Haan, A.D., Asscher, J.J., Widiger, T.: The development of personality extremity from childhood to adolescence: relations to internalizing and externalizing problems. J. Pers. Soc. Psychol. **105**(6), 1038 (2013)
2. Aydin, A.: Examining the mediating role of mindful parenting: a study on the relationship between parental emotion regulation difficulties and problem behaviors of children with asd. J. Autism Dev. Disord. **53**(5), 1873–1883 (2023)
3. Dan, Y., et al.: EduChat: a large-scale language model-based chatbot system for intelligent education. arXiv preprint arXiv:2308.02773 (2023)
4. Demol, K., Verschueren, K., Ten Bokkel, I.M., van Gils, F.E., Colpin, H.: Trajectory classes of relational and physical bullying victimization: links with peer and teacher-student relationships and social-emotional outcomes. J. Youth Adolesc. **51**(7), 1354–1373 (2022)
5. Dettmers, T., Pagnoni, A., Holtzman, A., Zettlemoyer, L.: QLoRA: efficient fine-tuning of quantized LLMs. arXiv preprint arXiv:2305.14314 (2023)
6. Du, Z., et al.: GLM: general language model pretraining with autoregressive blank infilling. In: Proceedings of the 60th Annual Meeting of the Association for Computational Linguistics (Volume 1: Long Papers), pp. 320–335 (2022)
7. Fitzer, S.A.: Psychosocial factors and E-cigarette use: an application of problem behavior theory. Ph.D. thesis, Old Dominion University (2022)
8. Fomby, P., Sennott, C.A.: Family structure instability and mobility: the consequences for adolescents' problem behavior. Soc. Sci. Res. **42**(1), 186–201 (2013)
9. Fu, J., et al.: KwaiYiiMath: Technical report. arXiv preprint arXiv:2310.07488 (2023)
10. Hosokawa, R., Katsura, T.: Effect of socioeconomic status on behavioral problems from preschool to early elementary school-a Japanese longitudinal study. PLoS ONE **13**(5), e0197961 (2018)
11. Hu, E.J., et al.: LoRa: low-rank adaptation of large language models. arXiv preprint arXiv:2106.09685 (2021)

12. Huang, L., et al.: A survey on hallucination in large language models: principles, taxonomy, challenges, and open questions. arXiv preprint arXiv:2311.05232 (2023)
13. Kazdin, A.E., et al.: Parent management training for conduct problems in children: enhancing treatment to improve therapeutic change. Int. J. Clin. Health Psychol. **18**(2), 91–101 (2018)
14. LaBelle, B.: Positive outcomes of a social-emotional learning program to promote student resiliency and address mental health. Contemp. Sch. Psychol. **27**(1), 1–7 (2023)
15. Leijten, P., Melendez-Torres, G., Gardner, F., Van Aar, J., Schulz, S., Overbeek, G.: Are relationship enhancement and behavior management "the golden couple" for disruptive child behavior? Two meta-analyses. Child Dev. **89**(6), 1970–1982 (2018)
16. Lin, C.Y.: Rouge: a package for automatic evaluation of summaries. In: Text Summarization Branches Out, pp. 74–81 (2004)
17. McGrath, P.J., Elgar, F.J.: Behavioral problems, effects of socio-economic status on (2015)
18. Penghe, C., Jiefei, L., Qi, X., Yu, L., Shengquan, Y.: An automatic annotation system and method for moral education cases based on meta-learning
19. Pinquart, M.: Associations of parenting dimensions and styles with externalizing problems of children and adolescents: an updated meta-analysis. Dev. Psychol. **53**(5), 873 (2017)
20. Spiller, A.N., Caro, K., Garay, J.A., Marcu, G.: Supporting behavior management with a classroom display providing immediate feedback to students. In: Proceedings of the 13th EAI International Conference on Pervasive Computing Technologies for Healthcare, pp. 159–168 (2019)
21. Sugai, G., Horner, R.H.: Defining and describing schoolwide positive behavior support. In: Sailor, W., Dunlap, G., Sugai, G., Horner, R. (eds.) Handbook of Positive Behavior Support, pp. 307–326. Springer, Boston (2009). https://doi.org/10.1007/978-0-387-09632-2_13
22. Sutherland, K., Conroy, M., McLeod, B., Granger, K., Broda, M., Kunemund, R.: Preliminary study of the effects of best in class-elementary on outcomes of elementary students with problem behavior. J. Posit. Behav. Interv. **22**(4), 220–233 (2020)
23. Zhao, W., et al.: Qiaoban: a parental emotion coaching dialogue assistant for better parent-child interaction (2023). https://github.com/HIT-SCIR-SC/QiaoBan
24. Yang, C., Chan, M.K., Ma, T.L.: School-wide social emotional learning (SEL) and bullying victimization: moderating role of school climate in elementary, middle, and high schools. J. Sch. Psychol. **82**, 49–69 (2020)
25. Yang, S., Zhu, X., Li, W., Zhao, H.: Associations between teacher-student relationship and externalizing problem behaviors among Chinese rural adolescent. Front. Psychol. **14** (2023)
26. Yoon, D.: Peer-relationship patterns and their association with types of child abuse and adolescent risk behaviors among youth at-risk of maltreatment. J. Adolesc. **80**, 125–135 (2020)
27. Yue, X., et al.: MAmmoTH: building math generalist models through hybrid instruction tuning. arXiv preprint arXiv:2309.05653 (2023)
28. Zarra-Nezhad, M., Moazami-Goodarzi, A., Aunola, K., Nurmi, J.E., Kiuru, N., Lerkkanen, M.K.: Supportive parenting buffers the effects of low peer acceptance on children's internalizing problem behaviors. Child Youth Care Forum **48**, 865–887 (2019)

Generating Situated Reflection Triggers About Alternative Solution Paths: A Case Study of Generative AI for Computer-Supported Collaborative Learning

Atharva Naik[✉], Jessica Ruhan Yin, Anusha Kamath, Qianou Ma, Sherry Tongshuang Wu, Charles Murray, Christopher Bogart, Majd Sakr, and Carolyn P. Rose

Carnegie Mellon University, Pittsburgh, PA 15213, USA
arnaik@andrew.cmu.edu
https://www.cmu.edu/

Abstract. An advantage of Large Language Models (LLMs) is their *contextualization* capability – providing different responses based on student inputs like solution strategy or prior discussion, to potentially better engage students than standard feedback. We present a design and evaluation of a proof-of-concept LLM application to offer students dynamic and contextualized feedback. Specifically, we augment an Online Programming Exercise bot for a college-level Cloud Computing course with ChatGPT, which offers students contextualized reflection triggers during a collaborative query optimization task in database design. We demonstrate that LLMs can be used to generate highly situated reflection triggers that incorporate details of the collaborative discussion happening in context. We discuss in depth the exploration of the design space of the triggers and their correspondence with the learning objectives as well as the impact on student learning in a pilot study with 34 students.

Keywords: Dynamic support for collaborative learning · Generative Artificial Intelligence · Code Generation

1 Introduction

For nearly two decades intelligent conversational agents have been employed to increase reflection and learning in Computer-Supported Collaborative Learning (CSCL) settings [11,18,25,27,31]. This line of work has yielded principles for the design of interactive collaborative scaffolding that increases learning impact over static forms of scaffolding [33], which is prevalent in the field of CSCL. In particular, the positive impact of situating reflection triggers in a specific conversational context has been established in earlier studies [2,6]. Recent studies [26] focusing specifically on Computer Science (CS) education suggest that collaboration support that shifts the focus of students more toward reflection and less towards the actual coding increases conceptual learning without harming the ability to

A. M. Olney et al. (Eds.): AIED 2024, LNAI 14829, pp. 46–59, 2024.
https://doi.org/10.1007/978-3-031-64302-6_4

write code in subsequent programming assignments. These past studies focused on manipulating the timing of the reflection triggers or the proportion of time dedicated to reflection versus programming. Recent advances in Generative AI (GenAI) and Large Language Models (LLMs) have enhanced AI capabilities for the evaluation of multimodal student input and real-time feedback, which has provoked intensive exploration of the space of application possibilities [13]. This technology opens up more options for adapting the specific content of reflection triggers from specific details of the students' work and discussion in context.

So far, less attention has been given to applying principles for designing effective dynamic support for CSCL using advances in GenAI. Most explorations are geared towards individual programmers, such as programming assistance for individual novice learners [12,14,15,24,32]. However, past success in developing intelligent forms of effective scaffolding for collaborative learning suggests that advances leading to the prevalence of these LLMs of code create a ripe area for exploration. The focus of this paper is the application of GenAI to dynamic support for reflection in learning through collaborative software development.

Our key contributions are outlined below:

- *Technical contribution:* Development of a novel prompt engineering approach to elicit contextually appropriate suggestions of alternative code contributions from code LLMs as a dynamic form of reflection trigger.
- *Learning resource development contribution:* Enhancement of a platform for online collaborative software development with dynamic support for reflection during software development.
- *Learning research contribution:* Testing the impact of LLM-constructed reflection triggers on student learning in an online collaborative SQL optimization activity, along with a thorough analysis of their impact on learning.

2   Related Work

2.1   Generative AI in Education

Since its December 2022 launch, ChatGPT has emerged as a leading GenAI technology [3,9]. Its accessibility has sparked innovations and debates on its role in education [7]. For students, its ability to process cross-domain knowledge is appealing [30], while educators see potential in content creation [35], personalized tutoring [29], as well as risks of enabling plagiarism [8] and dissemination of biased information [28]. Mixed sentiments surround GenAI's impact on academic integrity and the future of education [23]. Despite the potential risks, some education technologists are more optimistic and view GenAI as a student-centric technology for personalization and customized real-time feedback [7]. In our work, we provide personalized real-time intervention for collaborative learning, in the form of reflection triggers using ChatGPT.

2.2   Intelligent Support for Collaboration

Providing technological support for collaborative and discussion-based learning has long been the focus of CSCL [25]. Past studies highlight the benefits of interactive and context-sensitive support in group learning [17,18]. While static scaffolding like fixed prompts [33] and scripted roles [10] have been effective, contextualized interventions within specific conversational contexts [2,6] or perceived roles of students [11] have also shown positive outcomes. Studies like [17,18,25] have shown the effectiveness of discussion-based learning and conversational support using dialog agents. Finally [26,27] have shown the effectiveness of reflection-based learning for programming, showing that shifting the focus of students more towards reflection than actual coding can increase conceptual learning without harming the ability to write code [26]. However, these studies have not explored contextualized scaffolding based on multimodal input (like code entered by students), which is now possible with LLMs like ChatGPT and is the main focus of this work.

2.3   Intelligent Support for Programming in CS Education

In the realm of collaborative programming, most studies involve humans collaborating with intelligent agents in pair programming paradigms [20,21]. Some newer approaches take a more relaxed definition of pair programming as a collaborative setting where a human programmer receives AI assistance, often dubbed as pAIr programming [22]. Studies like [12,32] focus on program synthesizers, providing design recommendations to support novice programmers. Kazemitabaar et al. [14] analyzed the impact the Codex LLM assistance on 69 novice learners and found improvements in code authoring performance. Yilmaz and Karaoglan Yilmaz [34] found a significant and positive impact of ChatGPT programming assistance on the computational thinking, programming self-efficacy, and motivation of undergraduate students. However, in all of these studies, a sole human user pairs up with an AI agent, effectively making them individual learning scenarios. Our work tackles the more under-explored paradigm of facilitating collaborative learning in *mob programming*, where 3 to 5 students work on the same task while playing different roles [4], which is used more often in CSCL for advanced CS topics [26,27].

3   Learning Activity Design

Collaborative learning is most valuable for learning activities where there are multiple possible solution paths, and selecting from among them is less about finding the right answer than it is about evaluating complex sets of constraints and trade-offs [5,16]. Amid these activities, students benefit from exposure to each other's alternative points of view. Because CS is an engineering discipline, it is ripe with opportunities for evaluation of design trade-offs, especially in connection with advanced topics. We select a topic with important design trade-offs and a paradigm for orchestrating the collaboration that encourages sharing and challenging alternative perspectives.

3.1 Learning Task: SQL Optimization

Especially when large amounts of data are involved, database design offers a solution space with interesting trade-offs. We situate our investigation in the SQL optimization task, which involves organizing and modifying the database using techniques like datatype modification, index creation, and table joining (denormalization) for a given scenario or query load, to minimize query cost while satisfying a few constraints. The optimization centers around the following rubric dimensions:

Data Retrieval Efficiency: how quickly and efficiently the database can retrieve data and execute queries, and how optimization techniques/design can improve performance. It can benefit from techniques like denormalization and indexing in certain cases.

Write Performance: how effectively the database can handle insert, update, and delete operations, and how optimization techniques/design can affect performance. It can be hurt from indexing and denormalization.

Disk Storage: how efficiently the database uses disk storage, and how optimization techniques/design can reduce storage usage and improve performance.

Maintainability: how effectively the database design and optimization techniques enable the database to be maintained and updated over time, and how optimization techniques can simplify maintenance and prevent additional processing or complexity for developers.

3.2 Knowledge Resources: Primers and Learning Objectives

The specific target learning objectives are enumerated in Table 1. We provide students with three different primers to prepare them for the learning activity, which is the application of the primers in a concrete task.

Denormalization & Normalization: This primer covers concepts of normalization and denormalization, their benefits, and trade-offs. It also compares the read & write costs for both strategies along the rubric dimensions.

Data Types in MySQL: This primer covers MySQL data types and their trade-offs for comparisons like CHAR vs VARCHAR, INT vs CHAR and contexts where each type is helpful.

Indexing: This primer covers the trade-offs of indexing in databases as well as how they affect the performance of read and write operations. It also compares single-column and composite indexing strategies along the rubric dimensions.

3.3 Collaborative Learning Paradigm: Mob Programming

To intensify the opportunity for reflection, we bring additional perspectives into the discussion by adopting Mob Programming [4], where 3 to 5 students work synchronously in different roles: *navigator*, *driver* and *researcher*. The *driver* controls the keyboard and mouse, the *navigator* chooses what the driver works on, and the *researcher* finds related knowledge requested by the driver.

Table 1. The mapping of learning objectives (LOs) and primers.

Learning Objective (LO)	Primer
L1: Comparing benefits of a single column or a composite index	Indexing
L2: Discussing trade-offs of creating an index	Indexing
L3: Discussing trade-offs of using single-column and multi-column composite indexes in MySQL across various use cases.	Indexing
L4: Evaluating and comparing the complexity of updating data in normalized and denormalized tables	Denormalization
L5: Identifying use cases where denormalized or normalized tables would be preferred	Denormalization
L6: Comparing the performance of queries when using normalized vs. denormalized tables	Denormalization
L7: Evaluating different data types and determining the most appropriate choice for a given table field	Data type

In our deployment, the activity includes three tasks, and role assignment occurs at the onset of each such that the students cycle through all three roles. All the tasks involve aspects of all three optimization strategies but task1, task2, and task3 have a greater focus on datatype conversion, indexing (specifically composite indexing), and denormalization (and trade-offs related to rates of reads and writes of joined tables) respectively. The activity is orchestrated by an Online Programming Exercise bot or OPE_Bot, which does several housekeeping tasks like role assignment, hints, and asking reflection questions.

4   Method

We integrate our dynamic reflection triggers into a pre-existing cloud-based system shown in Fig. 3. The activity session for a group is triggered according to the schedule plan (schedule.json) by the Sail() platform [1]. The students run their MySQL commands on a Kubernetes[1] virtual machine via a modified Jupyter-Lab[2] (JLab) front-end, containing a chat window on the side for each group. We implement a mySQL LogScript to send the necessary context like the SQL commands entered by the students to the OpenAI Reflection Generator which determines when to intervene and provide personalized reflections (Fig. 1) by prompting ChatGPT with the appropriate context. We coordinate the activity for each group with an instance of the OPE_bot which is based on the Bazaar CSCL architecture [19]. When the students submit their solution, the assignment auto-grader computes a final score and sends it to the Sail() platform.

---

[1] https://kubernetes.io/.

[2] https://jupyterlab.readthedocs.io/en/stable/.

To realize the OpenAI Reflection Generator, we design components that determine when to intervene (reflection triggering), how to use the context like SQL commands to personalize the reflections (reflection personalization), validate the correctness of the reflections (reflection validation), and decide when to show them and how to space them apart (reflection scheduling). Each of these components is discussed in detail in Sects. 4.1, 4.2, 4.3, and 4.4 respectively.

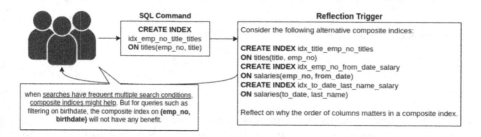

Fig. 1. An example of the COMPOSITE_IND_COL_ORDER reflection trigger along with a student response, demonstrating that our interventions make students think about the tradeoffs involved (underlined text) in the optimization.

4.1   Reflection Design and Triggering

To target the learning objectives from each of the primers we design five types of reflection triggers, which are triggered by specific command patterns based on regular expression-based matching as shown in Table 2.

For the data type primer, we designed the DATATYPE_COMPARISON reflection to be shown whenever students change the datatype of a column to facilitate the discussion of their choice. For the indexing primer, we designed two reflection triggers: the COMPOSITE_VS_MULTI_SINGLE reflection to encourage a comparison between composite and single column indices whenever the students create a single column index and the COMPOSITE_IND_COL_ORDER to facilitate discussion around the optimal ordering of columns whenever the students create a composite index. For the denormalization primer, the DENORMALIZATION_WHEN reflection encourages a discussion about the potential costs and benefits of denormalization for queries with inner joins whenever students use such queries and the TABLE_CHOICE_DENORMALIZATION reflection encourages a discussion about the choice of tables to be joined based on the expected reads and writes for each table, whenever the students create a denormalized table.

4.2   Reflection Personalization

To tailor the reflection triggers to the solution strategy of each group, we prompted ChatGPT (gpt-3.5-turbo-instruct) with the appropriate context,

Table 2. Five types of reflection triggers, the primers and learning objectives they address, and the student activities that trigger each type of reflection.

Reflection Type	Student Activity Trigger	Primer	LO
DATATYPE_COMPARISON	ALTER TABLE <tab>MODIFY <col><datatype>	Data type	L7
COMPOSITE_VS_MULTI_SINGLE	CREATE INDEX <ind>ON <tab>(<col>)	Indexing	L1, L3
COMPOSITE_IND_COL_ORDER	CREATE INDEX <ind>ON <tab>(<col_1>, ... <col_n>)	Indexing	L2
DENORMALIZATION_WHEN	SELECT <col_1>, ... <col_n>FROM <tab_1>INNER JOIN$_1$ <tab_2>ON <condition$_1$>...INNER JOIN$_n$ <tab_{n+1}>ON <condition$_n$>	Denormalization	L5, L6
TABLE_CHOICE_DENORMALIZATION	CREATE TABLE <tab>AS SELECT <col_1>, ... <col_n>FROM <tab_1>INNER JOIN$_1$ <tab_2>ON <condition$_1$>...INNER JOIN$_n$<tab_{n+1}>ON <condition$_n$>	Denormalization	L4

constructed from the SQL commands they entered to generate variations of the commands as alternative solutions. For instance, for the COMPOSITE_IND_COL_ORDER reflection trigger, we prompted ChatGPT to generate three alternative composite indices. So if the team comes up with the following indexing strategy CREATE INDEX dept_title_index ON em_dept_title (**dept_name, title**), then the following alternatives are suggested by ChatGPT to explore different column selections:
CREATE INDEX dept_title_index ON em_dept_title (**title, dept_name**),
CREATE INDEX dept_title_index ON em_dept_title (**emp_no, title**), and
CREATE INDEX dept_title_index ON em_dept_title (**title, emp_no**).

4.3 Reflection Validation

We validate the SQL generated by the model by using an SQL syntax checker[3] to ensure the generated alternatives are syntactically correct, along with simple regular expression-based correction for some commonly occurring errors (e.g. fixing incorrect joins). We implement static reflection triggers for each kind of dynamic reflection as a fallback strategy for reflections with unfixable errors.

4.4 Reflection Scheduling

To ensure that the students have ample opportunity to discuss the reflection triggers we try to space them apart by using a simple queue-based scheduling algorithm (Algorithm 1). Our method tries to enforce a time interval τ between consecutive reflections within a task.

To find the best value of τ to optimally space apart the reflections, we run simulations with some historical data from an older iteration of the Cloud Computing course. We compare different values of τ (Fig. 2) and find that while there are only minor differences: $\tau \geq 15s$ get almost the same performance. Hence, we

---

[3] https://pypi.org/project/sqlfluff/.

Algorithm 1. Queue-based Scheduling Algorithm.

1: **procedure** REFLECTIONSCHEDULE(r, r^t, Q, τ)   ▷ r is the triggered reflection, r^t is the reflection type and Q is the waiting queue
2:    **if** reflection of type $r^t \notin Q$ **then**   ▷ r^t type reflection hasn't been shown yet
3:        **if** HEAD(Q) triggered $> \tau$ seconds ago **then**   ▷ τ time elapsed from last reflection
4:            SHOW(p)
5:        **else**
6:            PUSH(r, Q)   ▷ add reflection r to waiting queue
7:    **else**
8:        **if** HEAD(Q) triggered $> \tau$ seconds ago **then**
9:            $r' \leftarrow$ POP(Q)   ▷ show last queued reflection
10:           SHOW(r')
11:   **return**

---

pick $\tau = 300s$ as a reasonable value. A possible reason for the similar performance is that for some sessions there is no opportunity to clear the queue till the last task (task3), and to avoid the possibility of the reflections getting skipped we decided to forego the scheduling algorithm for task3.

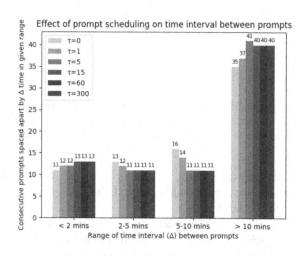

Fig. 2. The effect of reflection scheduling parameter τ on the range of time intervals Δ between consecutive reflections. Each point of the x-axis denotes a range of time intervals between consecutive reflections and the y-axis captures the number of reflections spaced apart by time interval Δ lying in that range.

5   Experimental Procedure and Study Design

We tested our intervention in a Cloud Computing course with 34 students who were assigned before the study to semester-long project groups of 3 students.

To test the impact of tailored reflection triggers during problem-solving, the groups were randomly assigned either to the experimental condition (in which they received the context-sensitive reflection triggers, or a control condition, in which they did not. In both conditions, to isolate the effect of the reflection triggers, the Online Programming Exercise bot (OPE-bot) played the role of a facilitator, providing the same task instructions, performing the same roles assignment strategy at the onset of each subtask, and providing the same task-relevant announcements at key times. Due to unexpected logistical issues, 22 students were assigned to the experimental condition while only 12 were assigned to the control condition.

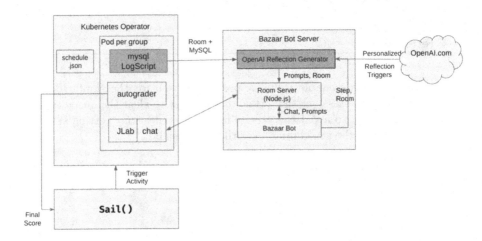

Fig. 3. The existing cloud infrastructure for the Online Programming Exercise (OPE) along with our newly added components (highlighted in red) for generating dynamic and personalized reflection triggers with ChatGPT (Color figure online)

To facilitate information sharing within groups, all students received the primers before the activity, with each student being assigned a specific primer but having access to the others.

At the beginning of the activity, each student individually took the pre-test, with 7 multi-part questions (27 points altogether) designed to test the individual learning objectives as well as combinations of them. The groups then engaged in the design activity for 80 min. Finally, the students individually took the post-test, which was identical to the pre-test.

6   Results

We evaluated the benefit for learning of the activity and compared success between conditions along two dimensions: (a) Task Completion, and (b) Learning Gains. Overall, we see that the insertion of tailored reflection triggers affected how students spent their time such that the completion rate per problem changed, but did not change the total amount learned from participating.

6.1 Task Completion

Responding to reflection triggers took time, which had a negative impact on the task completion rate at first, but then benefited task completion for the more difficult portion of the activity that came later. Analysis of task completion rates shows that out of the 34 students, 31 (91.17%) completed task 1 (19 in the experimental condition and 12 in the control condition), 29 (85.29%) completed task 2 (17 in the experimental condition and 12 in the control condition), and 20 (58.82%) completed task 3 (3 in the control condition and 17 in the experimental condition). Based on a chi-squared test, the differences across tasks were significant, but not over conditions. However, there was a significant interaction between task and condition such that students in the control condition had a higher task completion rate early, but the students in the experimental condition had a higher task completion rate on the harder tasks at the end.

6.2 Learning Gains

We further analyze the pre and post-test scores (Table 3) and do a factor analysis of the effect of the reflection triggers on learning, which demonstrates that the difference in task completion did not have a significant effect on student learning. The question here is whether and to what extent did students learn from the activity. Pre and post-test distributions (Fig. 4(a) and (b), respectively) show a gain in the score as evident from the mean and median quiz scores.

We computed an ANOVA with the normalized score as the dependent variable and with Phase (whether pre or post), Reflections, and Learning Objective (LO) as independent variables and additional pairwise and three-way interaction terms. **Students learned significantly between pre and post-phases**, the effect is small because of a large amount of variability between students: $F(1, 929) = 1.3, p < 0.05$, the effect size is 0.18σ. Furthermore, there was **no significant 2-way interaction** between LO and Phase, showing that students learned across all learning objectives. These results show that the OPE activity is useful for learning and are in agreement with the "Hypothesis 1" in [26], which had a similar OPE setup with reflection questions.

A question-level analysis reveals that Q7.2.3 (pre-quiz) and Q2, Q3, Q7.2.3, and Q7.3.3 (post-quiz) have less than 50% correct response rate. Q2 deals with indexing, while Q3 deals with denormalization. Q7.2.3 and Q7.3.3 are both yes/no questions dealing with composite indexing and data types respectively. A possible reason for the poor performance on Q2 and Q3 might be because the options are very long and hard to read, and the variation between them can be subtle.

The next question is whether and to what extent students benefited specifically from the tailored reflection triggers added in the experimental condition. Using the same ANOVA model, we tested the three-way interaction between LO, Phase, and Reflections. There was **no significant 3-way interaction** between LO, Phase, and Reflections, showing that the main effect of Phase did not depend

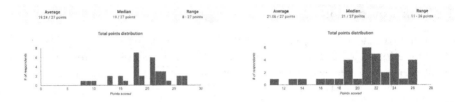

Fig. 4. Distribution of (a, left) pre-test and (b, right) post-test scores

on condition or learning objective. However, students had higher pretest scores in the test condition, which led to some spurious pairwise interactions.

To make a valid comparison across conditions, pretest scores were used as a covariate. Additional analysis of the residual mean across the learning objectives for each reflection trigger also did not demonstrate any significant effect. Thus, students learned during the activity, but the results of this study do not demonstrate a particular advantage yet for reflection triggers tailored using GenAI. We speculate that the lack of statistical significance could be due to the limitations outlined in Sect. 7, with engagement being the primary concern.

Table 3. Normalized mean pre and post-test scores (standard deviation in brackets) per topic/learning objective for the control and test groups. The results reveal an important confounding factor for our analysis - much higher pre-test scores for the test group which reduces possible gain.

	Control		Test	
	Pre	Post	Pre	Post
Integrated	65.1 (44.3)	73.7 (41.1)	65.9 (44.2)	71.3 (41.8)
Trade-offs of creating index	41.7 (51.5)	45.5 (52.2)	50.0 (51.3)	63.2 (49.6)
Complexity normalized vs. denormalized	16.7 (38.9)	63.6 (50.5)	30.0 (47.0)	42.1 (50.7)
Data types	77.1 (35.3)	77.2 (36.9)	80.3 (35.2)	83.6 (33.7)
Single vs composite	66.7 (49.2)	81.8 (40.5)	85.0 (36.6)	89.5 (31.5)
Performance normalized vs. denormalized	33.3 (49.2)	27.3 (46.7)	80.0 (41.0)	68.4 (47.7)

7   Future Work

In this paper, we present the first evaluation of a technique for personalizing reflection triggers using GenAI. From a technical perspective, the intervention worked as designed, however, due to the null effect when evaluated in comparison with the baseline collaboration support condition, we plan to explore several improvements. We identified three major issues requiring improvement.

Engagement: While collecting feedback from students and teaching assistants

(TAs), we learned that sometimes the reflection triggers were confusing and suboptimal. A comparison of the SQL command that set off the reflection triggers with the alternatives within them, revealed that if the students have the optimal solution, the alternatives by virtue of being different, are always suboptimal for the optimization context of the task. To remedy this we plan to generate alternative scenarios from the original task scenario for which the alternatives are more optimal to spark more meaningful discussion.

Prompting Context: Students and TAs gave the feedback that the reflection triggers sometimes felt unrelated to the discussion they were having in the chat window. To remedy this we plan to include elements from their chat messages to further personalize the reflection triggers and ensure coherence by matching them to the topic being discussed.

Readability: Another concern was the large size of the prompts for the chat window, which made the messages hard to engage with. In future experiments, we plan to break up the prompts into smaller readable chunks which are spaced apart in time to improve readability.

Acknowledgement. This work was funded in part by NSF grant DSES 2222762.

References

1. Sail(): The social and interactive learning platform. https://sailplatform.org/. Accessed 02 Feb 2024
2. Ai, H., Sionti, M., Wang, Y.C., Rosé, C.P.: Finding transactive contributions in whole group classroom discussions. In: Proceedings of the 9th International Conference of the Learning Sciences-Volume 1, pp. 976–983 (2010)
3. Carr, D.F.: ChatGPT topped 1 billion visits in February. https://www.similarweb.com/blog/insights/ai-news/chatgpt-1-billion/
4. Wikipedia contributors: Team programming. https://en.wikipedia.org/wiki/Team_programming#Mob_programming
5. Cress, U., Oshima, J., Rosé, C., Wise, A.F.: Foundations, processes, technologies, and methods: an overview of CSCL through its handbook. Int. Handbook Comput.-Supp. Coll. Learn., 3–22 (2021)
6. Cui, Y., Kumar, R., Chaudhuri, S., Gweon, G., Rosé, C.P.: Helping agents in VMT. In: Stahl, G. (eds.) Studying Virtual Math Teams. Computer-Supported Collaborative Learning Series, vol. 11, pp. 335–354. Springer, Boston (2009). https://doi.org/10.1007/978-1-4419-0228-3_19
7. Dai, Y., Liu, A., Lim, C.P.: Reconceptualizing chatgpt and generative AI as a student-driven innovation in higher education. Procedia CIRP **119**, 84–90 (2023). https://doi.org/10.1016/j.procir.2023.05.002, the 33rd CIRP Design Conference
8. Debby, R.E., Cotton, P.A.C., Shipway, J.R.: Chatting and cheating: ensuring academic integrity in the era of ChatGPT. Innov. Educ. Teach. Int., 1–12 (2023). https://doi.org/10.1080/14703297.2023.2190148
9. Duarte, F.: Number of ChatGPT users, January 2024. https://explodingtopics.com/blog/chatgpt-users

10. Fischer, F., Kollar, I., Stegmann, K., Wecker, C.: Toward a script theory of guidance in Computer-Supported collaborative learning. Educ. Psychol. **48**(1), 56–66 (2013)
11. Gweon, G., Rosé, C., Albright, E., Cui, Y.: Evaluating the effect of feedback from a CSCL problem solving environment on learning, interaction, and perceived interdependence, pp. 234–243, January 2007. https://doi.org/10.3115/1599600.1599645
12. Jayagopal, D., Lubin, J., Chasins, S.E.: Exploring the learnability of program synthesizers by novice programmers. In: Proceedings of the 35th Annual ACM Symposium on User Interface Software and Technology, UIST 2022. Association for Computing Machinery, New York (2022). https://doi.org/10.1145/3526113.3545659
13. Kasneci, E., et al.: ChatGPT for good? On opportunities and challenges of large language models for education, January 2023. https://doi.org/10.35542/osf.io/5er8f. www.osf.io/preprints/edarxiv/5er8f
14. Kazemitabaar, M., Chow, J., Ma, C.K.T., Ericson, B.J., Weintrop, D., Grossman, T.: Studying the effect of AI code generators on supporting novice learners in introductory programming. In: Proceedings of the 2023 CHI Conference on Human Factors in Computing Systems, CHI 2023, Association for Computing Machinery, New York (2023). https://doi.org/10.1145/3544548.3580919
15. Kazemitabaar, M., Hou, X., Henley, A., Ericson, B.J., Weintrop, D., Grossman, T.: How novices use LLM-based code generators to solve CS1 coding tasks in a self-paced learning environment. arXiv preprint arXiv:2309.14049 (2023)
16. Koschmann, T.: Computer Supported Collaborative Learning 2005: The Next 10 Years! Routledge, New York (2017)
17. Kumar, R., Rose, C.P.: Architecture for building conversational agents that support collaborative learning. IEEE Trans. Learn. Technol. **4**(1), 21–34 (2010)
18. Kumar, R., Rosé, C.P., Wang, Y.C., Joshi, M., Robinson, A.: Tutorial dialogue as adaptive collaborative learning support. In: Proceedings of the 2007 Conference on Artificial Intelligence in Education: Building Technology Rich Learning Contexts That Work, pp. 383–390 (2007)
19. Kumar, R., Rosé, C.P., Witbrock, M.J.: Building conversational agents with basilica. In: Johnston, M., Popowich, F. (eds.) Proceedings of Human Language Technologies: The 2009 Annual Conference of the North American Chapter of the Association for Computational Linguistics, Companion Volume: Demonstration Session, pp. 5–8. Association for Computational Linguistics, Boulder, June 2009. https://aclanthology.org/N09-5002
20. Kuttal, S.K., Myers, J., Gurka, S., Magar, D., Piorkowski, D., Bellamy, R.: Towards designing conversational agents for pair programming: accounting for creativity strategies and conversational styles. In: 2020 IEEE Symposium on Visual Languages and Human-Centric Computing (VL/HCC), pp. 1–11, August 2020. https://doi.org/10.1109/VL/HCC50065.2020.9127276
21. Kuttal, S.K., Ong, B., Kwasny, K., Robe, P.: Trade-offs for substituting a human with an agent in a pair programming context: the good, the bad, and the ugly. In: Proceedings of the 2021 CHI Conference on Human Factors in Computing Systems, CHI 2021. Association for Computing Machinery, New York (2021). https://doi.org/10.1145/3411764.3445659
22. Ma, Q., Wu, T., Koedinger, K.: Is AI the better programming partner? Human-Human pair programming vs. Human-AI pAIr programming. arXiv preprint arXiv:2306.05153 (2023). http://arxiv.org/abs/2306.05153
23. Petricini, T., Wu, C., Zipf, S.T.: Perceptions about generative AI and ChatGPT use by faculty and college students, August 2023. https://doi.org/10.35542/osf.io/jyma4. www.osf.io/preprints/edarxiv/jyma4

24. Prasad, S., Greenman, B., Nelson, T., Krishnamurthi, S.: Generating programs trivially: student use of large language models. In: Proceedings of the ACM Conference on Global Computing Education, CompEd 2023, vol. 1, pp. 126–132. Association for Computing Machinery, New York (2023). https://doi.org/10.1145/3576882.3617921

25. Rosé, C.P., Ferschke, O.: Technology support for discussion based learning: from computer supported collaborative learning to the future of massive open online courses. Int. J. Artif. Intell. Educ. **26**, 660–678 (2016)

26. Sankaranarayanan, S., et al.: Collaborative programming for work-relevant learning: comparing programming practice with example-based reflection for student learning and transfer task performance. IEEE Trans. Learn. Technol. **15**(5), 594–604 (2022). https://doi.org/10.1109/TLT.2022.3169121

27. Sankaranarayanan, S., et al.: Collaborative reflection "in the flow" of programming: designing effective collaborative learning activities in advanced computer science contexts. In: Proceedings of the 15th International Conference on Computer-Supported Collaborative Learning-CSCL 2022, pp. 67–74. International Society of the Learning Sciences (2022)

28. Sok, S., Heng, K.: ChatGPT for education and research: a review of benefits and risks. SSRN Electron. J., March 2023. SSRN: https://ssrn.com/abstract=4378735 or http://dx.doi.org/10.2139/ssrn.4378735

29. Sridhar, P., Doyle, A., Agarwal, A., Bogart, C., Savelka, J., Sakr, M.: Harnessing LLMs in curricular design: using GPT-4 to support authoring of learning objectives (2023)

30. Stokel-Walker, C., Van Noorden, R.: What ChatGPT and generative AI mean for science. https://doi.org/10.1038/d41586-023-00340-6

31. Tegos, S., Demetriadis, S., Karakostas, A.: Promoting academically productive talk with conversational agent interventions in collaborative learning settings. Comput. Educ. **87**, 309–325 (2015)

32. Vaithilingam, P., Zhang, T., Glassman, E.L.: Expectation vs. experience: evaluating the usability of code generation tools powered by large language models. In: Extended Abstracts of the 2022 CHI Conference on Human Factors in Computing Systems, CHI EA 2022. Association for Computing Machinery, New York (2022). https://doi.org/10.1145/3491101.3519665

33. Vogel, F., Kollar, I., Ufer, S., Strohmaier, A., Reiss, K., Fischer, F.: Scaffolding argumentation in mathematics with CSCL scripts: which is the optimal scripting level for university freshmen? Innov. Educ. Teach. Int. **58**(5), 512–521 (2021). https://doi.org/10.1080/14703297.2021.1961098

34. Yilmaz, R., Karaoglan Yilmaz, F.G.: The effect of generative artificial intelligence (AI)-based tool use on students' computational thinking skills, programming self-efficacy and motivation. Comput. Educ. Artif. Intell. **4**, 100147 (2023). https://doi.org/10.1016/j.caeai.2023.100147

35. Young, J.C., Shishido, M.: Evaluation of the potential usage of ChatGPT for providing easier reading materials for ESL students. In: Bastiaens, T. (ed.) Proceedings of EdMedia + Innovate Learning, pp. 155–162. Association for the Advancement of Computing in Education (AACE), Vienna (2023). https://www.learntechlib.org/primary/p/222496/

Automated Assessment of Encouragement and Warmth in Classrooms Leveraging Multimodal Emotional Features and ChatGPT

Ruikun Hou[1,3]([✉]) [iD], Tim Fütterer[1] [iD], Babette Bühler[1] [iD], Efe Bozkir[1,3] [iD],
Peter Gerjets[2] [iD], Ulrich Trautwein[1] [iD], and Enkelejda Kasneci[3] [iD]

[1] University of Tübingen, Geschwister-Scholl-Platz, 72074 Tübingen, Germany
{ruikun.hou,tim.fuetterer,babette.buehler,efe.bozkir,
ulrich.trautwein}@uni-tuebingen.de
[2] Leibniz-Institut für Wissensmedien, Schleichstrasse 6, 72076 Tübingen, Germany
p.gerjets@iwm-tuebingen.de
[3] Technical University of Munich, Arcisstrasse 21, 80333 Munich, Germany
{ruikun.hou,efe.bozkir,enkelejda.kasneci}@tum.de

Abstract. Classroom observation protocols standardize the assessment of teaching effectiveness and facilitate comprehension of classroom interactions. Whereas these protocols offer teachers specific feedback on their teaching practices, the manual coding by human raters is resource-intensive and often unreliable. This has sparked interest in developing AI-driven, cost-effective methods for automating such holistic coding. Our work explores a multimodal approach to automatically estimating *encouragement and warmth* in classrooms, a key component of the Global Teaching Insights (GTI) study's observation protocol. To this end, we employed facial and speech emotion recognition with sentiment analysis to extract interpretable features from video, audio, and transcript data. The prediction task involved both classification and regression methods. Additionally, in light of recent large language models' remarkable text annotation capabilities, we evaluated ChatGPT's zero-shot performance on this scoring task based on transcripts. We demonstrated our approach on the GTI dataset, comprising 367 16-min video segments from 92 authentic lesson recordings. The inferences of GPT-4 and the best-trained model yielded correlations of $r = .341$ and $r = .441$ with human ratings, respectively. Combining estimates from both models through averaging, an ensemble approach achieved a correlation of $r = .513$, comparable to human inter-rater reliability. Our model explanation analysis indicated that text sentiment features were the primary contributors to the trained model's decisions. Moreover, GPT-4 could deliver logical and concrete reasoning as potential teacher guidelines. Our findings provide insights into using multimodal techniques for automated classroom observation, aiming to foster teacher training through frequent and valuable feedback.

Keywords: Classroom observation · AI in Education · Teaching effectiveness · Multimodal machine learning · ChatGPT zero-shot annotation

1 Introduction

A comprehensive understanding of classroom interactions is crucial to deciphering the quality of teaching, providing hence the opportunity to foster an educational environment where learning thrives [25]. This understanding paves the way for interventions like real-time feedback, enriching the teaching and learning processes and empowering researchers to dissect teaching scenarios with heightened reliability and efficiency. These insights are particularly pivotal when assessing facets of teaching effectiveness, such as student support, where elements like classroom encouragement and warmth are not mere niceties but essential catalysts for effective teaching [20]. The ability to delve deeper into these components promises to enhance educational practices and refine teacher professional development programs, steering them toward fostering these nurturing classroom atmospheres. Traditionally, the task of capturing the nuances of teaching dynamics has involved human observers, employing structured classroom observation protocols like CLASS (Classroom Assessment Scoring System [21]). For this task, human observers watch lesson recordings and assign scores based on teaching effectiveness measures defined in the protocols. Whereas a human rating approach is valuable, it is fraught with multiple challenges [8]. Human-based observations are inherently subjective, often leading to higher-inference holistic-level assessments with low inter-rater agreement. Moreover, the manual nature of these assessments makes them resource-intensive in terms of time and cost.

Against this backdrop, there is growing interest in developing AI-driven approaches to automatically coding classroom observation protocols. Prior research addressed the task by either employing multimodal feature extraction together with supervised classifiers [13,22] or relying on advanced large language models (LLMs) [31,32]. In line with existing studies, our goal is to make an initial contribution to automatic evaluation approaches that reflect the eye of a highly trained human evaluator but overcome the limitations of human subjectivity and resource constraints. We focus on a specific aspect of teaching effectiveness, namely Encouragement and Warmth (EW), a significant component in the observation protocol of the Global Teaching Insights (GTI) study [18]. The component involves the provision of encouragement for students throughout their work, including positive comments and compliments, along with moments of shared warmth such as smiling and laughter [4]. This corresponds to the Positive Climate (PC) dimension in CLASS.

To assess classroom EW, we investigate the use of both multimodal models tailored to domain-specific data and LLMs' generative capabilities, aiming to harness each method's unique strengths. We first propose a supervised-learning approach based on multimodal representations of emotion and then apply the Shapley additive explanations (SHAP) [16] technique to examine the contributions of these explicit features. Additionally, we explore whether recent LLMs like ChatGPT, relying on their zero-shot annotation capabilities, can effectively score EW based on classroom discourse and reasonably interpret their decisions. Lastly, we evaluate the predictive performance of an ensemble approach that combines estimates from supervised models and ChatGPT.

2   Related Work

Recently, the success of machine learning has triggered a growing trend towards AI applications in classroom settings, such as analysis of student behavior [2,5] and engagement [11,27], classroom discourse [12,14], as well as teacher perception [28]. Specifically, a few recent studies have targeted the holistic analysis of automated teaching effectiveness coding within classroom observation protocols. They can be categorized into two strands: multimodal supervised methods [13,22] and LLM-based methods [31,32].

The first multimodal machine-learning system was proposed by James et al. [13], which employed visual, conversational, and acoustic features to identify whether the classroom climate is positive following the CLASS protocol. Their trained binary classifier yielded a $F1$-score of 0.77. Furthermore, Ramakrishnan et al. [22] presented a multimodal architecture to achieve a more fine-grained scoring of both PC and Negative Climate (NC) dimensions in CLASS. In line with the 7-point coding scale of CLASS, the prediction task was formulated as a 7-class classification problem. They utilized an ensemble model integrating visual and auditory pathways, achieving correlations between predictions and human ratings of $r = .55$ (PC) and $r = .63$ (NC).

In addition to multimodal approaches, recent research investigated using LLMs for transcript-based classroom observation scoring. Wang and Demszky [31] pioneered the employment of ChatGPT's zero-shot capabilities to score classroom transcripts across various dimensions of teaching effectiveness. They prompted GPT-3.5 with an entire transcript segment and a description of the respective dimension requiring rating. Based on 100 authentic transcript segments, ChatGPT estimates resulted in a weak correlation of $r = .04$ with human-coded scores regarding CLASS PC. Instead of using complete transcripts as prompts, Whitehill and LoCasale-Crouch [32] introduced an LLM-based approach focusing on utterance-level analysis. They leveraged zero-shot prompting with an LLM to distinguish individual teacher utterances and further trained a linear regressor on aggregated session-level features to assess the CLASS Instructional Support domain. Their best-performing model, employing features concatenated from the LLM and a Bag-of-Words method, reached a correlation of $r = .46$. Although this utterance-level method approached human inter-rater reliability, it lacked in grasping the semantic context in dialogues.

We propose leveraging both multimodal models and LLMs to predict classroom observation scores. Our work stands for a further contribution to this relatively unexplored domain, extending prior studies in the following aspects: (1) We focus on extracting interpretative features that explicitly constitute EW-related behavioral indicators, as opposed to utilizing low-level auditory features

[13, 22]. This enables us to (2) apply explainability frameworks to understand which behaviors contribute to model predictions, which is central in practical applications, such as teacher training. (3) Additionally, we explore regression methods that account for the ordering attribute of data labels compared to standard classification. Moreover, GPT-3.5 resulted in subpar performance for CLASS PC scoring [31]. As the recent GPT-4 model has demonstrated improved text understanding and generation capabilities as well as reduced hallucination [1], (4) we evaluate if GPT-4 surpasses its predecessor in achieving adequate zero-shot performance for this particular scoring task. (5) Furthermore, we investigate the potential of an ensemble method to boost predictive accuracy by leveraging the strengths of supervised models and ChatGPT's zero-shot approaches.

3 GTI Dataset

The dataset employed in this work stems from GTI [18], a large-scale classroom video study aiming to achieve a profound understanding of teaching and learning worldwide. Across eight participating countries, the study centers on a shared pedagogical topic in mathematics (quadratic equations) and emphasizes objective evidence on classroom practice by directly observing authentic lesson videos and instructional materials. A video observation protocol was developed to ensure consistent rating processes within the study. At a high level, the protocol consists of six *domains*: Classroom Management, Social-Emotional Support, Discourse, Quality of Subject Matter, Student Cognitive Engagement, and Assessment of and Responses to Student Understanding. Each domain comprises multiple *components* measuring the quality of distinct teaching constructs at higher inference levels. Human raters observed instructional videos divided into 16-min segments and assigned each segment a score on a 4-point scale for each component, guided by associated behavioral examples. The raters were required to attend dedicated training lessons and engage in several quality control checks, thus yielding heavy workloads to guarantee rating reliability. We focus on the EW component from the Social-Emotional Support domain, which captures behavioral characteristics comparable to the CLASS PC dimension. Particularly, Encouragement refers to using positive verbal and nonverbal cues to inspire students to begin or persist in tasks, such as reassurance for students' mistakes, positive comments, and compliments on their work, while Warmth is represented by, e.g., smiling, laughter, joking, and playfulness [4]. The score scale from one to four aligns with the occurrence frequency of these behaviors from no evidence to frequent instances throughout one segment.

We used the GTI data collected in Germany, involving 100 video-recorded math lessons, 50 recruited teachers, and over 1,140 students. Due to data protection regulations, we were only allowed to access 92 of the recordings. Each lesson lasted from 40 to 90 min. The lessons were videotaped simultaneously by two cameras at 25 FPS (frames per second): One tracked the teacher's movements (Fig. 1a), while the other was stationary and positioned to capture as many students as possible (Fig. 1b). We utilized the recordings from the latter camera, where the frontal faces of most participants were visible. In addition, GTI

(a) Teacher camera   (b) Student camera

Fig. 1. Classroom frame from two cameras, with people erased for privacy.

supplied lesson transcripts created by human transcribers, which we employed as the text modality in this work. In transcripts, timestamps and speakers were annotated following every conversation turn, where speakers were anonymized by their IDs, such as "L" for the teacher and "S01" for a student. The rating process involved 14 raters in Germany, with each lesson being annotated by two randomly assigned raters. Following the GTI protocol, we preprocessed the data by splitting lesson recordings and transcripts into 16-min segments. If the last segment of one recording spanned less than eight minutes, it was merged with the preceding segment. This resulted in 367 segments, serving as the dataset on which we built and evaluated our automated estimation methods.

4   Methodology

In this section, we elaborate on our proposed approach to automated estimation of EW scores (Fig. 2), including supervised learning methods based on multimodal features extraction (Sect. 4.1), ChatGPT zero-shot annotation (Sect. 4.2), and an ensemble model combining both paths (Sect. 4.3).

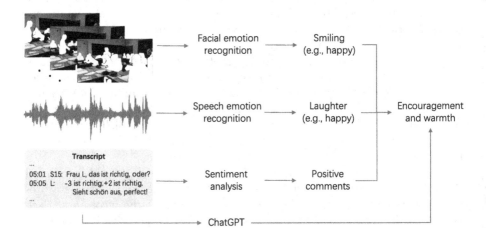

Fig. 2. Pipeline for multimodal estimation of EW scores.

4.1 Multimodal Supervised Models

We aimed to build a machine-learning approach to mimic a human's rating process to the greatest extent possible. Thus, we focused on the GTI rater training materials (e.g., coding rubrics and guidelines), which played an essential role in helping raters comprehend rating specifications. As indicated in the training materials, raters were required to pay attention solely to the dedicated behaviors listed in the EW definition (see Sect. 3). Hence, we extracted interpretable features by employing off-the-shelf techniques to represent the EW-associated behaviors. Notably, these behavioral cues are typically linked to the affective states of teachers and students. Therefore, we implemented facial emotion recognition to identify "smiling" in videos, applied speech emotion recognition (SER) to detect "laughter" in audio, and carried out text sentiment analysis to distinguish "positive comments" in classroom discourse.

Facial Emotion Recognition. We adopted a deep neural network architecture, EmoNet [30], which performs multi-task predictions of facial emotion in a single pass. Beyond the recognition of discrete emotion categories, EmoNet simultaneously estimates continuous valence (positive or negative) and arousal (excited or calm) values defined in the circumplex model of affect [23], enabling a more comprehensive depiction of human emotions. We utilized an EmoNet model pre-trained on AffectNet [17], a dataset containing a vast amount of facial images in the wild annotated with discrete and continuous emotion labels. To apply the model to our classroom dataset, we first down-sampled video segments from 25 to 2 FPS (i.e., 1920 frames for a 16-min segment), which considered the minimal variation between consecutive frames and reduced computational resources. We then employed RetinaFace [9], known for its competitive performance in crowded environments, to detect faces in each frame. Afterward, each face crop was input into EmoNet, which predicted valence and arousal values ranging from -1 to 1, along with a probability distribution over five emotions (neutral, happy, sad, surprise, fear). For each frame, we aggregated the predictions by averaging the valence and arousal values as well as the estimation scores over five discrete labels across all detected faces, yielding a 7D feature vector.

Speech Emotion Recognition. Recent work [19] tackled SER by employing transfer learning based on embeddings derived from pre-trained deep models, showing superior performance over conventional methods that relied on low-level acoustic features. Given that the spoken language in our dataset is German, we utilized XLSR proposed by Facebook AI [7] to extract cross-lingual deep embeddings from raw audio signals. XLSR is pre-trained in over 50 languages and built on top of wav2vec 2.0 [3], a transformer-based model trained on unlabeled data in a self-supervised manner for speech recognition. This approach enhances performance for low-resource languages. In particular, we used a publicly released XLSR model fine-tuned in German, mapping a speech signal to a 1024D latent embedding. We applied the model as a feature extractor to

EmoDB [6], a database consisting of 535 audio instances, each representing a German utterance categorized into seven discrete emotion labels (anger, boredom, disgust, fear, happiness, sadness, neutral). Based on the resulting features, we trained a two-layer Multi-Layer Perceptron (MLP) classifier on the randomly selected 80% of the EmoDB data and used the remainder for testing. The accuracy on the test set achieved 0.95. In our classroom setup, we divided the 16-min audio segments into 192 5-s windows without overlap due to the observation that over 95% of the EmoDB instances lasted under 5 s, with an average duration of 2.8 s. We then applied the XLSR model and the trained MLP classifier to infer emotions expressed in each audio window, yielding a 7D feature vector as a probability distribution over the seven emotion labels.

Sentiment Analysis. TextBlob[1] is a well-established toolkit supporting multiple natural language processing functionalities, such as tokenization and translation. It conducts lexicon-based sentiment analysis utilizing a predefined polarity dictionary. Prior research employed the toolkit for sentiment analysis of student feedback [24]. In this work, we applied TextBlob-de[2], a German language extension, to the transcript segments that were manually annotated regarding turn-taking between speakers. The tool assessed the sentiment of each teacher or student utterance by assigning a polarity score ranging from −1 to 1. In line with [24], we categorized utterances according to their polarity scores, labeling those above/equal to/below 0 as positive/neutral/negative, respectively. We generated a 4D feature vector for each transcript segment by counting positive, neutral, and negative utterances and computing a cumulative polarity score across all utterances. Since each lesson recording's final segment typically did not span 16 min, we normalized the feature vector by the respective segment duration.

Prediction. It is noteworthy that both facial and speech emotion features were temporal, whereas text sentiment features were generated in a segment-wise way. We aggregated the visual and auditory features for each segment by computing the mean along the temporal dimension. We then concatenated the segment-wise features from all three modalities into a single 18D vector. Afterward, we formulated the EW estimation problem as both classification and regression tasks. To compare the two approaches fairly, we trained an identical set of models: Random Forest (RF), Support Vector Machine (SVM), and MLP with two layers. As each segment was double-rated, we calculated the average of the two human ratings as the ground truth. For classification, we rounded the ground truth in the training set to the nearest integer, within the range of one to four. To guarantee generalization, all models were evaluated through stratified, lesson-independent 5-fold cross-validation, such that segments from the same lesson were always grouped into the same fold, and each fold maintained the original score distribution. The evaluation results were averaged across all test folds. Before model fitting, we

---

[1] https://textblob.readthedocs.io/.
[2] https://textblob-de.readthedocs.io/.

standardized the features using the mean and the standard deviation computed from the training data. Moreover, we carried out grid-search hyperparameter tuning to identify the best-performing configuration for each model.

4.2   ChatGPT Zero-Shot Annotation

Recent advances in LLMs introduce vast opportunities for their application in the education sector to enhance teaching and learning [15,26]. LLMs are typically trained on extensive text corpora and thus equipped with broad knowledge, enabling quick adaptation to new tasks without the need for retraining. Such zero-shot capability has shown notable effectiveness across various text annotation problems [10]. In the GTI study, transcripts play a significant role in rating processes (e.g., raters in many countries employed shorthand and highlighting tools directly on transcripts). Therefore, we explore the potential of ChatGPT to assess the EW component without requiring training, relying exclusively on classroom transcripts. In particular, we prompted ChatGPT with a transcript segment to rate, along with EW's definition, behavioral examples, and coding rubrics, as depicted in Fig. 3. Besides, we requested ChatGPT to reason its decision on the score assignment, as suggested by [31]. For this task, we employed two GPT models via the OpenAI API to compare their performance, namely *gpt-3.5-turbo-1106* and *gpt-4-1106-preview*. To consistently compare with those trained models, we evaluated ChatGPT's zero-shot performance on the same five test folds and averaged the results.

Consider the following math classroom transcript in German delimited by triple backticks. The teacher is labeled by a capital "L", while students are anonymized with IDs starting with a capital "S".

` ` ` { *Transcript* } ` ` `

Based on the transcript, your task is to rate Encouragement and Warmth, one of the components used to evaluate classroom interactions. Specifically, Encouragement is represented by the teacher's and students' using positive verbal cues that may inspire or motivate students to begin or keep trying to accomplish a task, e.g., reassuring students when errors are made, complimenting students' work, and making positive comments. Warmth indicates there are moments of shared warmth between the teacher and students and among students, e.g., smiling, laughter, joking, and playfulness.

Please rate this component on a scale of 1-4 (low-high), representing no/occasional/some/frequent occurrences of the afore-defined behaviors, respectively.

Format your answer as:
Rating: <the score on a 1-4 integer scale here>
Reasoning: <the reason for your decision here>

Fig. 3. Prompt for ChatGPT.

4.3   Ensemble Model

For text feature extraction, the sentiment analysis was performed at the level of individual utterances. ChatGPT analyzed an entire transcript, thus leveraging contextual information for a comprehensive understanding of classroom discourse [32]. Considering that each method might provide distinct and potentially complementary insights, we constructed an ensemble model that integrates the trained model with the ChatGPT zero-shot approach. The ensemble model computes the weighted average of estimates from both base models. Weights are allocated according to their evaluation performance on the training set in each fold. The unweighted averaging was also tested for comparison.

Table 1. Results of GTI Encouragement and Warmth estimation.

Approach	Model	Pearson r
Inter-Rater Reliability	Human Raters	**0.513 (0.028)**
Multimodal Classifier	RF	0.391 (0.041)
	SVM	0.375 (0.058)
	MLP	**0.392 (0.040)**
Multimodal Regressor	RF	0.429 (0.051)
	SVM	0.433 (0.041)
	MLP	**0.441 (0.039)**
ChatGPT Zero-Shot	GPT-3.5	0.027 (0.071)
	GPT-4	**0.341 (0.037)**
Ensemble	MLP Reg.+GPT-4 (Unweighted)	0.499 (0.033)
	MLP Reg.+GPT-4 (Weighted)	**0.513 (0.036)**

Notes. For humans, r corresponds to the average across raters. For automatic models, r is the average over 5 folds between estimates and mean human ratings. Standard error estimates are shown in parentheses, with the best model for each approach in bold.

5   Results

5.1   Model Performance

Similar to a recent study [32], we treat automatic estimation approaches as individual raters and explore the extent of their consistency with human raters in terms of Pearson correlation coefficient r. For this purpose, we first computed human inter-rater reliability (IRR) in a leave-one-rater-out fashion. In our context, there were 14 human raters conducting double ratings. We computed r for each rater by comparing their ratings with those assigned by other raters. The coefficients obtained were then averaged across all raters. Meanwhile, we report standard error estimates (standard deviation over all raters divided by $\sqrt{14}$). For model evaluation, we calculated the average of r between model predictions and human ratings over five test folds, as well as the corresponding standard error estimates. The results are summarized in Table 1. Except for GPT-3.5, the correlations for each model in each fold are statistically significant ($p < .05$).

For multimodal supervised approaches, the best-trained model is the MLP regressor ($r = .441$). Overall, it seems advantageous to formulate the problem as

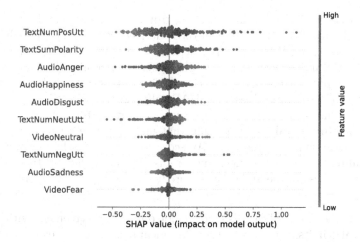

Fig. 4. SHAP summary plot for MLP regressor. Points depict SHAP values per feature per data sample. Features are ranked by their importance (sum of SHAP value magnitudes over all samples). The 10 most influential features are shown. (Color figure online)

a regression task. This is evident from the fact that classifiers achieve the highest Pearson r of .392, trailing behind all regressors in performance. Another interesting result is that the two-layer MLP outperforms the other two conventional models in both classification and regression tasks, underscoring the efficacy of contemporary neural networks. It is noteworthy that adding more layers did not enhance the performance, potentially attributed to the risk of overfitting inherent in deeper models when confronted with a limited dataset size. When turning to ChatGPT zero-shot results, a clear difference between the two model generations is identified. The estimates of GPT-3.5 show no significant correlation with human-rated scores. In contrast, GPT-4 has remarkably enhanced the capability to understand text, particularly achieving superior zero-shot performance over its predecessor to score EW in classroom discourse. It attains a Pearson r of .341 without requiring any training. Finally, the ensemble model, combining the best-performing MLP regressor with GPT-4, boosts the performance significantly. Using a simple averaging method raises the Pearson r to .499. When the averaging process considers the respective reliability of both base models, the correlation reaches a peak of .513, identical to that of the human IRR.

5.2   Model Explanation

Supervised Models. To explore which explicit features from which modality are influential in model decisions, we further applied Shapley additive explanation (SHAP) [16] to our MLP regressor. A SHAP value quantifies the impact of a feature on an individual prediction. In the cross-validation setting, we gathered SHAP values from each test fold and then created a summary plot involving the entire dataset. As illustrated in Fig. 4, the group of verbal and auditory features appears to be more informative than those derived from videos in enabling the regressor to predict EW scores. Specifically, the feature describing the number of positive utterances within a transcript segment contributes the most, followed by the overall polarity and three speech emotion features. By analyzing the relationship between feature values and SHAP values in the plot, it is apparent that high values (shown in red) in the two most important features are associated with positive SHAP values, suggesting that more positive sentiment cues contribute to increasing the predicted EW score. Additionally, the plot reveals the negative impact of detected anger and disgust as well as the positive impact of detected happiness in the audio on the prediction of higher EW scores.

ChatGPT. GPT-4's efficacy is evident not only in its agreement with human raters but also in its capability to provide logical reasoning. For example, GPT-4 assigned a score of 4 to a transcript consistently with the human ratings and explained its decision by identifying relevant evidence from the discourse:

> "[...] For instance, the teacher praises S15's work as 'sieht schön aus perfekt' *('looks beautiful perfect')* and encourages S04 by validating their thinking process. The teacher's tone is patient and nurturing, especially visible in exchanges like 'keine Panik' *('no panic')* [...] The teacher often uses humor, as seen in statements like 'bevor hier einer weint' *('before someone cries here')* [...]"

This aligns with the human rating procedure that measures the frequency of dedicated behaviors. Conversely, we observed that GPT-3.5 typically presented broad-level reasoning lacking concrete examples, which limited its explainability.

6   Discussion

Human raters achieve moderate agreement ($r = .513$) for EW coding in our dataset. Prior work [22] reported Pearson r of .38 and .42 for CLASS PC in two datasets they utilized. This indicates that coding teaching effectiveness requires holistic analysis and higher-level inference. Compared to human subjectivity, automatic tools benefit from providing more objective insights at scale [8,14]. To this end, we explore a novel approach to automated EW assessment. Our methods achieve correlations of .441 (the best-trained model) and .341 (GPT-4 zero-shot annotation) between estimates and human ratings. Combining both methods yields the highest predictive accuracy ($r = .513$) on par with human IRR. The findings from supervised models show that a set of multimodal, explicit, and low-dimensional features can effectively capture EW-relevant signals in a 16-min lesson segment. Unlike classifiers that treat categories independently, regressors account for the ordering attribute inherent in the data labels. For example, mistaking a true score of one as four incurs a higher penalty than confusing it with two. This may explain the observed superior performance of regressors. Another reason is that the ground-truth labels were rounded to integers used as classification categories, leading to a loss of information. Additionally, the SHAP analysis highlights the importance of text features, which can be interpreted by the EW coding rubrics where verbal cues constitute a large proportion of the associated behavioral indicators. Meanwhile, auditory features contribute more than visual features, which aligns with prior research findings [22]. Besides the competitive accuracy achieved by supervised models, GPT-4's ability to deliver persuasive reasoning showcases its potential as an easily accessible tool for teachers to obtain valuable feedback on classroom climate. Further, the efficacy of the ensemble approach suggests that the two base models may capture complementary information. Integrating GPT-4 zero-shot annotations with a specialized and shallow model does not require resource-intensive fine-tuning of LLMs yet enhances the final performance, providing insights into strategies for using LLMs both effectively and efficiently when addressing similar tasks.

Due to the use of distinct datasets and protocols, we note that the results are not comparable to prior studies [13, 22, 31, 32] which focused on CLASS dimensions. GTI EW exhibits similar behavioral indicators to CLASS PC, but their coding rubrics differ in aspects such as score scale (4- vs. 7-point). Our exploration contributes to a broader understanding of similar constructs across different protocols. Further, we explore methods for German speech and text processing, diverging from the common use of English data in existing research, which enriches the discussion on developing educational technologies in multilingual contexts. Moreover, transcripts not only provide more informative features for predictive models but also serve as a privacy-conscious modality compared to video and audio recordings. As classroom recordings are sensitive data involving minors, ethical considerations are of utmost importance. We emphasize that automated observation tools intend to streamline post hoc analysis and reduce the need for manual coding instead of being used for real-time classroom monitoring.

Our work is subject to several limitations. First, our analysis utilized manual transcripts. We could implement advanced speech recognition techniques to achieve a fully automatic system. In classroom environments, students' faces far away from the camera lead to limited spatial resolution, reducing the performance of emotion recognition models. Solutions could be applying super-resolution methods to improve face image quality. Due to the potential disagreement between distinct model explanation methods [29], it is valuable to validate various explainers on our EW scoring task. Another approach to feature importance analysis would be an ablation study to compare the performance of unimodality. Furthermore, one future direction is to adapt the proposed methods for the automated coding of additional GTI aspects, such as classroom discourse, to achieve a more comprehensive measurement of teaching effectiveness. As GPT-4 is a multimodal LLM capable of image understanding, another future study involves exploring whether the inclusion of classroom frames together with transcripts could enhance estimation accuracy. To further improve model generalizability, conducting cross-dataset evaluation, particularly employing data from diverse countries or cultures, would be part of future research.

7   Conclusions

We explore a machine-learning approach that harnesses multimodal supervised models and LLMs' zero-shot capabilities to automatically assess Encouragement and Warmth in classrooms. The results indicate that our approach achieves rating performance on par with human inter-rater reliability. We further show that verbal features contribute the most to model predictions, and GPT-4 provides specific evidence for its scoring decisions, outperforming GPT-3.5. Such AI-driven methods have the potential to replicate and augment human observational capabilities in the future, enabling frequent and valuable feedback for educational researchers regarding several facets of teaching effectiveness.

References

1. Achiam, J., et al.: GPT-4 technical report. arXiv preprint arXiv:2303.08774 (2023)
2. Ahuja, K., et al.: EduSense: practical classroom sensing at scale. Proc. ACM Interact. Mob. Wearable Ubiquitous Technol. 3(3), 1–26 (2019)
3. Baevski, A., Zhou, Y., Mohamed, A., Auli, M.: wav2vec 2.0: a framework for self-supervised learning of speech representations. Adv. Neural Inf. Process. Syst. 33, 12449–12460 (2020)
4. Bell, C., Qi, Y., Witherspoon, M., Barragan, M., Howell, H.: Annex A: Talis video training notes: holistic domain ratings and components. In: OECD (ed.) Global Teaching Insights: Technical report. OECD Publishing (2018)
5. Bühler, B., et al.: Automated hand-raising detection in classroom videos: a view-invariant and occlusion-robust machine learning approach. In: International Conference on Artificial Intelligence in Education, pp. 102–113 (2023)
6. Burkhardt, F., Paeschke, A., Rolfes, M., Sendlmeier, W.F., Weiss, B., et al.: A database of German emotional speech. In: Interspeech, vol. 5, pp. 1517–1520 (2005)
7. Conneau, A., Baevski, A., Collobert, R., Mohamed, A., Auli, M.: Unsupervised cross-lingual representation learning for speech recognition. arXiv preprint arXiv:2006.13979 (2020)
8. Demszky, D., Liu, J., Hill, H.C., Jurafsky, D., Piech, C.: Can automated feedback improve teachers' uptake of student ideas? Evidence from a randomized controlled trial in a large-scale online course. Educ. Eval. Policy Anal. (2023)
9. Deng, J., Guo, J., Ververas, E., Kotsia, I., Zafeiriou, S.: RetinaFace: single-shot multi-level face localisation in the wild. In: The IEEE Conference on Computer Vision and Pattern Recognition (CVPR), pp. 5203–5212 (2020)
10. Gilardi, F., Alizadeh, M., Kubli, M.: ChatGPT outperforms crowd-workers for text-annotation tasks. arXiv preprint arXiv:2303.15056 (2023)
11. Goldberg, P., et al.: Attentive or not? Toward a machine learning approach to assessing students' visible engagement in classroom instruction. Educ. Psychol. Rev. 33, 27–49 (2021)
12. Hunkins, N., Kelly, S., D'Mello, S.: "beautiful work, you're rock stars!": teacher analytics to uncover discourse that supports or undermines student motivation, identity, and belonging in classrooms. In: LAK22: 12th International Learning Analytics and Knowledge Conference, pp. 230–238 (2022)

13. James, A., et al.: Inferring the climate in classrooms from audio and video recordings: a machine learning approach. In: IEEE International Conference on Teaching, Assessment, and Learning for Engineering, pp. 983–988 (2018)
14. Jensen, E., Pugh, S., D'Mello, S.K.: A deep transfer learning approach to modeling teacher discourse in the classroom. In: LAK21: 11th International Learning Analytics and Knowledge Conference, pp. 302–312 (2021)
15. Kasneci, E., et al.: ChatGPT for good? On opportunities and challenges of large language models for education. Learn. Individ. Differ. **103**, 102274 (2023)
16. Lundberg, S.M., Lee, S.I.: A unified approach to interpreting model predictions. Adv. Neural Inf. Process. Syst. **30** (2017)
17. Mollahosseini, A., Hasani, B., Mahoor, M.H.: AffectNet: a database for facial expression, valence, and arousal computing in the wild. IEEE Trans. Affect. Comput. **10**(1), 18–31 (2017)
18. OECD: Global Teaching InSights: A Video Study of Teaching. OECD Publishing, Paris (2020)
19. Pepino, L., Riera, P., Ferrer, L.: Emotion recognition from speech using wav2vec 2.0 embeddings. In: Proceedings of the Interspeech, pp. 3400–3404 (2021)
20. Pianta, R.C., Hamre, B.K.: Conceptualization, measurement, and improvement of classroom processes: standardized observation can leverage capacity. Educ. Res. **38**(2), 109–119 (2009)
21. Pianta, R.C., La Paro, K.M., Hamre, B.K.: Classroom Assessment Scoring System^{TM}: Manual K-3. Paul H Brookes Publishing, Baltimore (2008)
22. Ramakrishnan, A., Zylich, B., Ottmar, E., LoCasale-Crouch, J., Whitehill, J.: Toward automated classroom observation: multimodal machine learning to estimate class positive climate and negative climate. IEEE Trans. Affect. Comput. **14**, 664–679 (2021)
23. Russell, J.A.: A circumplex model of affect. J. Pers. Soc. Psychol. **39**(6), 1161 (1980)
24. Sadriu, S., Nuci, K.P., Imran, A.S., Uddin, I., Sajjad, M.: An automated approach for analysing students feedback using sentiment analysis techniques. In: Mediterranean Conference on Pattern Recognition and Artificial Intelligence (2021)
25. Seidel, T., Shavelson, R.J.: Teaching effectiveness research in the past decade: the role of theory and research design in disentangling meta-analysis results. Rev. Educ. Res. **77**(4), 454–499 (2007)
26. Seßler, K., Xiang, T., Bogenrieder, L., Kasneci, E.: PEER: empowering writing with large language models. In: Viberg, O., Jivet, I., Muñoz-Merino, P., Perifanou, M., Papathoma, T. (eds.) EC-TEL 2023. LNCS, vol. 14200, pp. 755–761. Springer, Cham (2023). https://doi.org/10.1007/978-3-031-42682-7_73
27. Sümer, Ö., Goldberg, P., D'Mello, S., Gerjets, P., Trautwein, U., Kasneci, E.: Multimodal engagement analysis from facial videos in the classroom. IEEE Trans. Affect. Comput. **14**, 1012–1027 (2021)
28. Sümer, Ö., et al.: Teachers' perception in the classroom. In: Proceedings of the IEEE Conference on Computer Vision and Pattern Recognition Workshops (2018)
29. Swamy, V., Radmehr, B., Krco, N., Marras, M., Käser, T.: Evaluating the explainers: black-box explainable machine learning for student success prediction in MOOCs. arXiv preprint arXiv:2207.00551 (2022)
30. Toisoul, A., Kossaifi, J., Bulat, A., Tzimiropoulos, G., Pantic, M.: Estimation of continuous valence and arousal levels from faces in naturalistic conditions. Nat. Mach. Intell. **3**, 42–50 (2021)

31. Wang, R.E., Demszky, D.: Is chatGPT a good teacher coach? Measuring zero-shot performance for scoring and providing actionable insights on classroom instruction. arXiv preprint arXiv:2306.03090 (2023)
32. Whitehill, J., LoCasale-Crouch, J.: Automated evaluation of classroom instructional support with LLMs and BoWs: connecting global predictions to specific feedback. arXiv preprint arXiv:2310.01132 (2023)

Ruffle&Riley: Insights from Designing and Evaluating a Large Language Model-Based Conversational Tutoring System

Robin Schmucker[1]([✉]), Meng Xia[2], Amos Azaria[3], and Tom Mitchell[1]

[1] Carnegie Mellon University, Pittsburgh, PA 15213, USA
{rschmuck,mitchell}@cs.cmu.edu
[2] Texas A&M University, College Station, TX 77843, USA
[3] Ariel University, 4070000 Ariel, Israel

Abstract. Conversational tutoring systems (CTSs) offer learning experiences through interactions based on natural language. They are recognized for promoting cognitive engagement and improving learning outcomes, especially in reasoning tasks. Nonetheless, the cost associated with authoring CTS content is a major obstacle to widespread adoption and to research on effective instructional design. In this paper, we discuss and evaluate a novel type of CTS that leverages recent advances in large language models (LLMs) in two ways: First, the system enables AI-assisted content authoring by inducing an easily editable tutoring script automatically from a lesson text. Second, the system automates the script orchestration in a learning-by-teaching format via two LLM-based agents (Ruffle&Riley) acting as a student and a professor. The system allows for free-form conversations that follow the ITS-typical inner and outer loop structure. We evaluate Ruffle&Riley's ability to support biology lessons in two between-subject online user studies ($N = 200$) comparing the system to simpler QA chatbots and reading activity. Analyzing system usage patterns, pre/post-test scores and user experience surveys, we find that Ruffle&Riley users report high levels of engagement, understanding and perceive the offered support as helpful. Even though Ruffle&Riley users require more time to complete the activity, we did not find significant differences in short-term learning gains over the reading activity. Our system architecture and user study provide various insights for designers of future CTSs. We further open-source our system to support ongoing research on effective instructional design of LLM-based learning technologies.

Keywords: conversational tutoring systems · intelligent tutoring systems · authoring tools · conversation analysis · large language models

A. M. Olney et al. (Eds.): AIED 2024, LNAI 14829, pp. 75–90, 2024.
https://doi.org/10.1007/978-3-031-64302-6_6

1   Introduction

Intelligent tutoring systems (ITSs) are an transformative educational technology that provides millions of learners worldwide with access to learning materials and affordable adaptive instruction. ITSs can, in certain contexts, be as effective as human tutors [18] and can take on an important role in mitigating the educational achievement gap [12]. However, despite their potential, one major obstacle to the widespread adoption of ITS technologies, is the large costs associated with content development. Depending on the depth of instructional design and available authoring tools, preparing one hour of ITS content can take designers hundreds of hours [1]. This significant investment often necessitates that ITSs focus on core subject areas and cater to larger demographic groups, limiting the breadth of topics covered and the diversity of learners adequately served.

Conversational tutoring systems (CTSs) are a type of ITS that engages with learners in natural language. Various studies have confirmed the benefits of CTSs, across multiple domains, particularly on learning outcomes in reasoning tasks [27]. Still, many existing CTSs struggle to maintain coherent free-form conversations and understand the learners' responses due to limitations imposed by their underlying natural language processing (NLP) techniques [26]. In this paper, we introduce and evaluate a new type of CTS that draws inspiration from design principles of earlier CTSs [21,26] while leveraging recent advances in large language models (LLMs) to accelerate content authoring and to facilitate coherent free-form conversational tutoring. Our main contributions include:

- **LLM-based CTS Architecture**: We leverage LLMs to enable AI-assisted content authoring by generating an easily editable tutoring script from a lesson text, and to automate script orchestration in free-form conversation. The CTS features a learning-by-teaching format with two agents taking on the roles of a student (Ruffle) and a professor (Riley). The human learner engages with these agents, teaching Ruffle with support from Riley.
- **Evaluation of Learning Performance/Experience**: We report findings from two online user studies ($N = 200$) evaluating the effects of our LLM-driven CTS workflow on learning outcomes and user experience, comparing it to two simpler QA chatbots and reading activity.
- **Evaluation of Interaction/Conversation**: We study usage patterns and conversations and assess their relationships to learning outcomes. We further discuss directions for future system refinements and provide various insights related to the design and evaluation of LLM-based learning technologies.

2   Related Work

Conversational Tutoring Systems. Dialog-based learning activities are known to promote high levels of cognitive engagement and to benefit learning outcomes [5]. This motivated the integration of conversational activities into learning technologies. In their systematic review, Paladines and Ramirez [27] categorized the

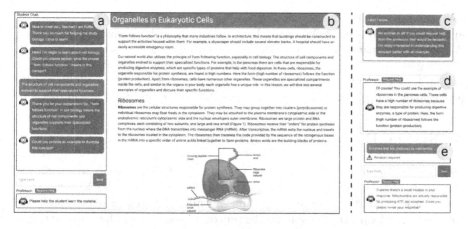

Fig. 1. UI of Ruffle&Riley. (a) Learners are asked to teach Ruffle (student agent) in a free-form conversation and request help as needed from Riley (professor agent). (b) The learner can navigate the lesson material during the conversation. (c) Ruffle encourages the learner to explain the content. (d) Riley responds to a help request. (e) Riley detected a misconception and prompts the learner to revise their response.

design principles underlying existing CTSs into three major categories: (i) expectation misconception tailoring (EMT) [26], (ii) model-tracing (MT) [31] and (iii) constraint-based modeling (CBM) [23]. While all three frameworks can benefit learning, they require designers to spend substantial effort configuring the systems for each individual lesson and domain [4]. Further, due to limitations of underlying NLP techniques, many CTSs struggle to maintain coherent free-form conversations, answer learners' questions, and understand learners' responses reliably [26]. In this context, this paper employs recent advances in NLP to facilitate free-form adaptive dialogues and AI-assisted content authoring.

Content Authoring Tools. One major obstacle to the widespread adoption of ITSs is the complexity of content authoring [8]. For early ITSs, the development ratio (i.e., the number of hours required to create one hour of instructional content) was estimated to vary between 200:1 and 300:1 [1]. This motivation the creation of content authoring tools (CATs) to facilitate ITS creation. ASSIST-ment Builder [30] was developed to support content authoring in a math ITS and enabled a development ratio of 40:1. For model tracing-based ITSs, example tracing [1] has proven itself as an effective technique that depending on the domain enables development ratios between 50:1 and 100:1. In the context of CTSs, multiple CATs have been developed for AutoTutor [4], but content authoring is still considered to be complex and labor intensive.

Recent advances in LLMs sparked a new wave of research that explores ways in which LLM-based technologies can benefit learners [14], for example via conversational agents [19]. Settings in which LLMs already have been found to be effective include question generation and quality assessment (e.g., [13,24,32]), feedback generation (e.g., [25, 28, 32]), question answering (e.g., [20]), automated grading (e.g., [3]), and helping teachers reflect on their teaching (e.g., [7, 22]). What sets this paper apart from the aforementioned works is that it does not focus on the generation of *individual* ITS components; instead, we propose a system that can automatically induce a *complete ITS workflow*, exhibiting the prototypical inner and outer loop structure [33], directly from a lesson text.

3   System Design and Architecture

Design Considerations. We approached the design of Ruffle&Riley with two specific goals in mind: (i) Facilitate an ITS workflow that provides learners with a sequence of questions (outer loop) and meaningful feedback during problem-solving (inner loop); (ii) Streamline the process of configuring the conversational agents for different lesson materials. We reviewed existing CTSs and identified expectation misconception tailoring (EMT) as a suitable design framework. EMT mimics teaching strategies employed by human tutors [11] by associating each question with a list of expectations and anticipated misconceptions. After presenting a question and receiving an initial user response, EMT-based CTSs provide inner loop support (goal (i)) by guiding the conversation via a range of dialogue moves to correct misconceptions and to help the learner articulate the expectations before moving on to the next question (outer loop). For a in-depth description of the EMT framework we refer to [10]. While EMT-based CTSs have been shown to be effective in various domains [26], they need to be configured in a labor-intensive process that requires instructional designers to define a *tutoring script* that specifies questions, expectations, misconceptions and other information for each lesson [4]. For us, tutoring scripts serve as a standardized format for CTS configuration that is easy to read and modify (goal (ii)).

User Interface. An overview of our user interface, together with descriptions of its key elements, is provided by Fig. 1. Inspired by the success of learning-by-teaching activities [9, 21], we decided to orchestrate the conversation in a learning-by-teaching format via two agents taking on the roles of a student (Ruffle) and a professor (Riley). While our design is similar to some CTSs in the AutoTutor family [26] that follow a trialogue format, one notable difference is that Riley solely serves as an assistant to the learner by providing support and correcting misconceptions. The two agents never talk directly with each other.

AI-Assisted Tutoring Script Authoring. Ruffle&Riley is capable of generating a tutoring script fully automatically from a lesson text by leveraging GPT-4 (Fig. 2). This involves a 4-step process: (i) A list of review questions is generated from the lesson text; (ii) For each question, a solution is generated based on question and lesson texts; (iii) For each question, a list of expectations is generated based on question and solution texts; (iv) The final tutoring script is compiled as a list of questions together with related expectations (Fig. 3). The first three steps are implemented via three separate prompts written in a

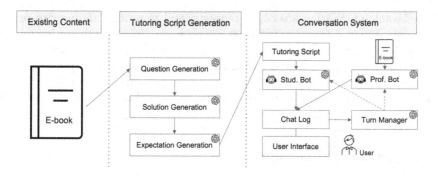

Fig. 2. System architecture. Ruffle&Riley generates a *tutoring script* automatically from a lesson text by executing three separate prompts that induce questions, solutions and expectations for the EMT-based dialog. During the learning process, the script is orchestrated via two LLM-based conversational agents in a free-form conversation that follows the ITS-typical inner and outer loop structure.

Topic 1: What does the principle "form follows function" mean in the context of cell biology? Provide an example to illustrate your answer.

Fact 1.1: "Form follows function" in cell biology means the structure of cell organelles supports their specialized functions.

Fact 1.2: An example is the high number of ribosomes in pancreas cells that produce digestive enzymes, supporting the cell's function of producing proteins.

Fig. 3. Tutoring script. To structure the conversational activity, Ruffle&Riley relies on a pre-generated script featuring a list of questions and related expectations for the EMT-based dialog. Tutoring scripts can be generated automatically from existing lessons text and offer instructional designers a convenient interface for system configuration.

way general enough to support a wide range of lesson materials. The resulting script can be easily modified and revised by instructional designers to meet their needs. Unlike traditional EMT-based CTSs, our tutoring scripts do not attempt to anticipate misconceptions learners might exhibit ahead of time (this is a difficult task even for human domain experts). Instead, we rely on GPT-4's ability to detect and respond to misconceptions in the learner's responses during the teaching process.

Tutoring Script Orchestration. The system automates the tutoring script orchestration by including descriptions of desirable properties of EMT-based conversations into the agents' prompts and captures the user's state solely via the chat log. The student agent receives the tutoring script as part of its prompt and is instructed to let the user explain the individual questions and to ask followups until all expectations are covered. Ruffle reflects on user responses to show understanding, provides encouragement to the user, and keeps the conversation on topic. In parallel, Riley's prompt contains the lesson text and instructions to offer relevant information after help requests, and to prompt the user to revise their response after detecting incorrect information. Both agents are instructed to keep the conversation positive and encouraging and to not refer to information outside the tutoring script and lesson text. The turn manager coordinates the system's queries to GPT-4. For details on design and function of Ruffle&Riley– including the exact prompts and logic of the conversational agents and tutoring script generation pipeline–we refer readers to our public GitHub repository.[1].

4   Experimental Design

We describe the experimental design shared by our two user studies. Studies and participant recruitment were approved by CMU's Institutional Review Board.

Learning Material. We adapted a Biology lesson on cell organelles from the OpenStax project [6]. We decided on this particular lesson because we expected participants to have low prior familiarity with the material to allow for a learning process. The lesson text is accessible to a general audience and covers 640 words.

Conditions. Similar to prior work [17], we construct conditions to compare the efficacy of our EMT-based CTS to reading alone and to a QA chatbot with limited dialog. To study potential differences, we equip the QA chatbot with content from different sources under two distinct conditions: one using content designed by a biology teacher and the other using LLM-generated content.

1. **Reading**: Participants study the material without additional support.
2. **Teacher QA (TQA)**: Participants study the material and can answer review questions presented by the chatbot. After submitting an answer, participants receive brief feedback about the correctness of their response and a sample solution. Questions and answers were designed by a human teacher.
3. **LLM QA (LQA)**: Same as TQA, but questions and answers were generated automatically by the LLM (Sect. 3).
4. **Ruffle&Riley (R&R)**: Participants study the material while being supported by the two conversational agents. The system is equipped with a LLM-generated tutoring script featuring the same questions as LQA.

Surveys/Questionnaires. We evaluate system efficacy from two perspectives: *learning performance* and *learning experience*. The *first study* gauges performance via a multiple-choice post-test after the learning session, consisting of five questions written by a separate biology teacher recruited via Upwork and two questions from OpenStax [6]. To evaluate the system more accurately and comprehensively, we conducted the *second study*. In particular, (1) we added a pre-test to assess students' prior knowledge and counterbalanced pre-test and post-test forms for different students; (2) we enriched the test questions from

---

[1] https://github.com/rschmucker/ruffle-and-riley.

only multiple-choice to two multiple-choice, three fill-in-the-blank and one free-form response question created to assess participants' deeper understanding of the taught concepts. This revision of the question format was informed by prior work which found the effects of CTSs to be less pronounced in recall-based test formats [10].

For both evaluations, learning experience is captured after post-test via a 7-point Likert scale survey that queries participants' perception of engagement, intrusiveness, and helpfulness of the agents, based on prior work [29]. To ensure data quality, we use two attention checks and one question asking whether participants searched for test answers online. Lastly, we included a demographics questionnaire to understand participants' age, gender, and educational background.

Recruitment. We recruited participants online via Prolific. Our criteria were: (i) located in the USA; (ii) fluent in English; (iii) possess at least a high-school (HS) degree. Participants were randomly assigned to conditions and free to drop out at any point. For each of the two studies, 100 participants completed the task.

5   Evaluation 1: Initial System Validation

We assess R&R's ability to facilitate a coherent and structured conversational learning activity. By comparing multiple conditions (Sect. 4), we explore hypotheses related to R&R's effects on learning performance and experience.

Hypotheses. We explore the following. H1: *Learning Outcomes*: R&R achieves higher test scores than baseline conditions (H1a); There is no significant difference between TQA and LQA (H1b). H2: *Learning experience*: R&R achieves higher ratings than baseline conditions in terms of learning experience metrics (H2a); There are no significant differences between TQA and LQA (H2b).

Participation. As shown in Table 1, 30 participants finished the reading condition, 17 finished TQA, 23 finished LQA, and 30 finished R&R. This imbalance was caused by the rotating condition assignment mechanism and participant drop-offs. After filtering participants who failed any of the attention check questions, or who did not rate "strongly disagree" when asked whether they looked up test answers, we were left with 58 (male: 33, female: 21, other: 4) out of the 100 participants (15 in reading, 7 in TQA, 15 in LQA, and 21 in R&R). The age distribution is: 18–25 (8), 26–35 (20), 36–45 (18), 46–55 (9), over 55 (3). The degree distribution is: HS or Equiv. (22), Bachelor's/Prof. Degree (25), Master's or Higher (11).

Learning Performance. The post-test consists of seven questions, each worth one point. The mean and standard error in post-test scores for each condition are shown by Table 1. A one-way ANOVA did not detect significant differences in post-test scores among the four conditions. Therefore, we find support for

H1b but not for H1a. Even though not significantly different, we observed that participants in R&R achieved somewhat higher scores (5.19 ± 0.25) than in TQA (4.14 ± 0.83). We find no significant differences in self-reported prior knowledge.

Learning Experience. Table 2 shows participants' learning experience and chatbot interaction ratings. We tested for significance ($p < 0.05$) using one-way ANOVA, followed by Bonferroni post-hoc analysis. We found no significant differences in self-reported engagement levels between the four conditions. However, among the three chatbot conditions, R&R was rated as significantly more helpful in aiding participants in understanding, remembering the lesson, and providing the support needed to learn. Further, R&R participants expressed more enjoyment than TQA and LQA participants. In addition, participants found R&R provided a significantly more coherent conversation than LQA. Interestingly, even though we expected R&R to be rated as more interrupting, we found no significant differences in perceived interruption among the chatbot conditions. Therefore, H2a is partially supported. In addition, we detected no significant differences in learning experience ratings between LQA and TQA. Thus, we cannot reject H2b.

Insights and Refinements. We found that R&R is positively received by its users (Table 2). Most importantly, the LLM-based system was able to facilitate coherent free-form conversations across an LLM-generated tutoring script featuring 5 questions and 17 expectations. Even though users were free to end the activity at any point, 17/21 users completed the entire script. Also, R&R yields significant learning experience improvements over more limited QA chatbots (TQA/LQA).

Table 1. Learning performance across different learning conditions.

Conditions	# of participants		Previous Knowledge	Learning Performance
	Before filtering 100	After filtering 58		Post-test Scores (i.e.,Multiple-Choice Questions)
Reading	30	15	2.53 ± 0.41	5.07 ± 0.33
Teacher Q/A	17	7	**3.0 ± 0.58**	4.14 ± 0.83
LLM Q/A	23	15	2.2 ± 0.3	4.67 ± 0.35
Ruffle & Riley	30	21	2.67 ± 0.43	**5.19 ± 0.25**

Table 2. Learning experience across different conditions. Symbol "*" indicates $p < 0.05$. Symbol "-" indicates that aspect was not asked in the respective condition.

Conditions	Learning Experience (1-strongly disagree, 7-strongly agree)						
	Engagement	Understanding	Remembering	Interruption	Coherence	Support	Enjoyment
Reading	4.33 ± 0.52	-	-	-	-	-	-
Teacher Q/A	5.0 ± 0.53	4.43 ± 0.65 *	4.43 ± 0.65 *	2.71 ± 0.64	5.43 ± 0.53	4.57 ± 0.57 *	3.71 ± 0.52 *
LLM Q/A	4.8 ± 0.47	4.4 ± 0.4 *	4.33 ± 0.42*	2.67 ± 0.45	4.8 ± 0.43*	4.0 ± 0.44*	4.0 ± 0.44*
Ruffle & Riley	**5.81 ± 0.3**	**5.81 ± 0.24**	**5.76 ± 0.22**	**2.19 ± 0.34**	**6.1 ± 0.21**	**5.9 ± 0.26**	**5.62 ± 0.31**

On the other side, R&R did not lead to significant improvements in learning performance over the reading activity. Further, the mean learning times varied largely between the conditions: reading (4 min), TQA (11 min), LQA (12 min), and R&R (18 min). Together, this motivated the second evaluation focused on R&R and reading with three revisions: (i) We addressed feedback about the student agent requesting similar information at different points in the conversation by trimming the tutoring script down to 4 questions and 12 expectations; (ii) we employed a test with deep understanding questions (see Sect. 4); (iii) we employed adaptive condition assignment probabilities to account for drop-offs.

6 Evaluation 2: Efficacy and Conversation Analysis

The *second study* focuses on R&R and reading–i.e., the conditions with highest test scores in the *first study*. We further conduct an in-depth analysis of conversations in R&R to explore usage patterns and their relations to learning performance.

Table 3. Learning performance for Ruffle&Riley and reading condition.

Conditions	# of participants		Pre-test Score	Post-test Score	Learning Gain	
	Before filtering 100	After filtering 72			Absolute	Normalized
Reading	50	38	1.37 ± 0.17	**3.53 ± 0.25**	**2.16 ± 0.25**	0.44 ± 0.07
Ruffle&Riley	50	34	**1.54 ± 0.23**	3.49 ± 0.28	1.94 ± 0.26	**0.47 ± 0.05**

Table 4. Learning experience for Ruffle&Riley and reading ("*" indicates $p < 0.05$).

Conditions	Learning Experience (1-strongly disagree, 7-strongly agree)				
	Engagement	Difficulty	Prior-Knowledge	Post-Knowledge	Remembering
Reading	5.05 ± 0.29	5.00 ± 0.29	1.74 ± 0.24 *	3.21 ± 0.32 *	4.26 ± 0.31
Ruffle&Riley	5.50 ± 0.24	4.74 ± 0.35	2.35 ± 0.26	4.00 ± 0.27	4.47 ± 0.27

Hypotheses. We explore H1: *Learning Outcomes*: R&R achieves higher test scores than the reading condition; H2: R&R achieves better ratings than reading in terms of engagement, understanding, remembering, and perceived difficulty.

Participation. As shown in Table 3, reading condition and R&R condition were each completed by 50 participants. After applying the same filter criteria as in the first evaluation, we were left with 72 (male: 29, female: 43) out of the 100 participants (38 in reading and 34 in R&R). The age distribution is: 18–25 (12), 26–35 (25), 36–45 (15), 46–55 (9), over 55 (11). The degree distribution is: HS or Equiv. (23), Bachelor's/Prof. Degree (33), Master's or Higher (16).

Learning Performance. The post-test consists of six questions, each worth one point. The mean and standard error in pre-test and post-test scores as well as derived absolute $(score_{post} - score_{pre})$ and normalized learning gain $(score_{post} - score_{pre})/(6 - score_{pre})$ measures are provided by Table 3. Comparing R&R and reading condition via one-sided t-tests, we do not detect significant differences $(p < 0.05)$ for any of the four measures. Thus, we do not find support for H1.

Learning Experience. Participants' learning experience ratings collected after the post-test are shown in Table 4. Different from the tested learning performance, the one-sided t-test unveiled that R&R users rated their pre- and post-activity knowledge (i.e., perceived knowledge) significantly higher than participants in the reading conditions. While R&R received advantageous scores in terms of overall engagement, remembering, and task difficulty, these differences could not be established as significant. Together, H2 is partially supported. Engagement scores might not be directly comparable due to large differences in learning times between the two conditions (R&R (20.8 min), reading (5.5 min)).

6.1 Analysing Interaction and Conversations

Interactions. We analyze interaction log data in R&R. First, 31/34 participants completed the full workflow. Users submitted on average 1.71 ± 0.45 help requests and received 1.77 ± 0.23 revision requests from Riley. Second, by evaluating temporal usage of conversation, scrolling, and help request features, we observe four distinct system usage patterns among the 31 completing participants (Fig. 4): (I) balanced feature usage; (II) conversation and reading only; (III) focus on conversation; (IV) focus on requesting help. Table 5 provides learning performance measures for each group. While sample sizes are too small to draw conclusions, we observe that the conversation-focused users (group (III)) achieve the highest learning gains. Further, the non-help seeking users (group (II) and (III)) achieve higher performance measures than the help-seeking users (group (I) and (IV)).

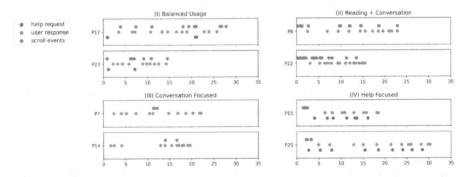

Fig. 4. Temporal Interaction Patterns. By visualizing the usage of text navigation, chat response, and help request features over time, we observe four distinct usage patterns.

Table 5. Learning performance of Ruffle&Riley users for each usage patterns.

Usage pattern	Num Users	Pre-Test	Post-Test	Absolute Gain	Relative Gain
Balanced	11	1.64 ± 0.48	3.27 ± 0.57	1.64 ± 0.32	0.45 ± 0.10
Read + Conv	13	1.62 ± 0.35	3.77 ± 0.47	2.15 ± 0.37	0.53 ± 0.09
Conv. Focused	4	0.62 ± 0.24	4.12 ± 0.52	3.50 ± 0.46	0.65 ± 0.09
Help Focused	3	0.67 ± 0.33	2.17 ± 0.44	1.50 ± 0.50	0.28 ± 0.09

Conversations. We analyze conversation log data in R&R. First, we evaluate the correct execution of the EMT-based dialogues. While all 31 participants went through all four questions in the tutoring script, we noticed that 7 conversations omitted 1 or 2 two expectations. Further, in 9 conversations, the student agent requested similar information at different points in the session, often when users wrote long responses covering multiple expectations at once. Another issue uncovered was that the system was often lenient towards user responses that only covered parts of one expectation (e.g., mention cellular respiration but do not explain its in- and outputs). An evaluation of *all* conversations verified the factual correctness of the GPT-4-based agents' responses. While we find that R&R facilitates coherent free-form conversational tutoring, future revisions are needed to enhance the system's ability to provide users with targeted feedback.

Second, we assess correlations between conversation features and participants' learning performance (Table 6). *Learning time* and the number of *words* in user explanations both show positive correlations to the performance measures. The number of submitted *help* and received *revision* requests exhibit negative correlations. In summary, the above analysis indicates that the way participants' engage with the system affects their learning outcomes.

Table 6. Pearson correlation analysis between conversation features and performance.

Feature	Pre-Test	Post-Test	Absolute Gain	Relative Gain
# User Messages	-0.06 ($p = 0.75$)	0.12 ($p = 0.54$)	0.22 ($p = 0.24$)	0.11 ($p = 0.54$)
# Help Requests	-0.14 ($p = 0.44$)	-0.35 ($p = 0.05$)	-0.32 ($p = 0.08$)	-0.34 ($p = 0.06$)
# Revisions	-0.35 ($p < 0.05$)	-0.20 ($p = 0.27$)	0.10 ($p = 0.58$)	-0.13 ($p = 0.50$)
# Words	0.24 ($p = 0.19$)	0.38 ($p = 0.04$)	0.26 ($p = 0.17$)	0.40 ($p = 0.02$)
Learning Time (min)	-0.17 ($p = 0.34$)	0.16 ($p = 0.35$)	0.39 ($p = 0.02$)	0.26 ($p = 0.13$)

7   Discussion and Future Work

Ruffle&Riley is a conversational tutoring system (CTS) that leverages recent advances in large language models (LLMs) to generate tutoring scripts automatically using existing lessons texts. These tutoring scripts define conversational learning activities which are orchestrated via two LLM-based agents (Ruffle&Riley) acting as a student and a professor. The human learner engages with the system by explaining a series of topics to the student (Ruffle) while being supported by the professor (Riley). Our user studies verified the system's ability to facilitate coherent free-form conversational tutoring. This highlights the potential of generative-AI-assisted content authoring for lowering resource requirements of CTS content development [4], promoting the design of learning activities meeting the needs of a wider diversity of learners and a broader range of subjects.

Our *first* study ($N = 100$), evaluates Ruffle&Riley's ability to support biology lessons comparing the system to QA chatbots offering limited feedback and a reading activity. In terms of learning experience, Ruffle&Riley users reported significantly higher ratings in terms of understanding, remembering, helpfulness of support and enjoyment. Still, corroborating prior research, the recall-focused multiple-choice post-test did not detect significant differences in learning outcomes between conversational tutoring and reading [10]. A *second* study ($N = 100$) compared Ruffle&Riley and reading condition using questions designed to assess deeper understanding. Again, we detected no significant differences in learning outcomes between the conditions even though Ruffle&Riley users required on average more time (20.8 min) than participants in reading condition (5.5 min).

We performed an in-depth evaluation of interaction and conversation log data to better understand how usage of Ruffle&Riley relates to learning outcomes. By studying temporal usage of conversation, scrolling and help request features we were able to identify four distinct system usage patterns. Interestingly, we found that users that focused on conversation and that did not request help achieved the highest learning gains, indicating a Doer effect [16]. The worst performing group, exhibited gaming behavior requesting help before each response. We further identified positive correlations between learning outcomes and number of *words* in users' explanations and overall learning *time*. Future work will explore revisions to nudge users towards active practice and to mitigate gaming behavior [2].

Reviewing the conversational logs, we found Ruffle&Riley to be receptive for partial explanations that miss important information (e.g., mention cellular respiration without explaining its in- and outputs) moving the conversation ahead too quickly. Future work will focus on enhancing the system's ability to provide feedback to help users elicit all information. While hallucination and biased outputs are well documented problems for LLM-based learning technologies [14], we highlight *affirmation of imprecise user responses* as additional challenge.

The present work is subject to several limitations: First, the system was evaluated via an online user study conducted via Prolific with adult participants exhibiting diverse demographics (e.g., age and education). Current findings focus on a broad population of online users and might not generalize to more specific populations (e.g., K12 students). Still, this environment enabled us to identify limitations of our system emphasizing the need for future research on instructional design principles [15] for LLM-based CTSs to improve effects on learning performance and efficiency [17] before running large-scale evaluations. Second, before evaluating Ruffle&Riley with younger learners, we need to certify safe and trustworthy system behavior adding to the system's existing mechanisms designed to ensure factual correctness of information that surfaces during conversations [14]. Relatedly, although initial investigations found the system able of facilitating coherent conversations on psychology and economics related lessons, future work is needed to assess its ability support different subjects. Third, participants only took part in a single learning session and might require time to adapt to the workflow. Fourth, while generative AI-based learning technologies like our GPT-4 based system show promise, they also incur regular costs due to API calls, raising important questions about equity and accessibility in educational contexts [14].

There are multiple exciting directions for extending Ruffle&Riley in future work: (i) Having the student agent take a test after the lesson can be an effective way to provide feedback on users' teaching [21]. (ii) Adaptive learning activity sequencing and personalization of agents and materials will become increasingly important when facilitating courses featuring multiple lessons. (iii) Multimodal generative-AI can serve as foundation for audio-visual interactions with the conversational agents and augment materials with illustrations and animations.

8   Conclusion

In this paper we introduced Ruffle&Riley, a novel type of LLM-based CTS, enabling AI-assisted content authoring and free-form conversational tutoring. We adopted Expectation Misconception Tailoring (EMT) [26] as design framework to facilitate structured conversational learning via pre-generated tutoring scripts. Importantly, scripts can be generated from existing lesson texts and can be revised by instructors based on their needs. We conducted two online users studies ($N = 200$) verifying Ruffle&Riley's ability to host conversational tutoring and to promote a positive learning experience. Our studies did not reveal significant differences in learning outcomes compared to the shorter reading activity, highlighting the importance of systematic evaluations of generative-AI-based learning technologies. An analysis of interaction log data motivates future system refinements to enhance Ruffle&Riley's ability to provide targeted feedback in response to imprecise user explanations. Our system architecture and experiments provide various insights for the design and evaluation of future CTSs.

Acknowledgement. We would like to thank Art Graesser for helpful suggestions and critiques during system development. We thank Microsoft for support through a grant from their Accelerate Foundation Model Academic Research Program. This research was supported by the AFOSR under award FA9550I710218.

Disclosure of Interests. The authors have no competing interests to declare.

References

1. Aleven, V., et al.: Example-tracing tutors: intelligent tutor development for non-programmers. Int. J. Artif. Intell. Educ. **26**, 224–269 (2016)
2. Baker, R.S.J., et al.: Adapting to when students game an intelligent tutoring system. In: Ikeda, M., Ashley, K.D., Chan, T.-W. (eds.) ITS 2006. LNCS, vol. 4053, pp. 392–401. Springer, Heidelberg (2006). https://doi.org/10.1007/11774303_39
3. Botelho, A., Baral, S., Erickson, J.A., Benachamardi, P., Heffernan, N.T.: Leveraging natural language processing to support automated assessment and feedback for student open responses in mathematics. J. Comput. Assist. Learn. **39**, 823–840 (2023)
4. Cai, Z., Hu, X., Graesser, A.C.: Authoring conversational intelligent tutoring systems. In: Sottilare, R.A., Schwarz, J. (eds.) HCII 2019. LNCS, vol. 11597, pp. 593–603. Springer, Cham (2019). https://doi.org/10.1007/978-3-030-22341-0_46
5. Chi, M.T., Wylie, R.: The ICAP framework: linking cognitive engagement to active learning outcomes. Educ. Psychol. **49**(4), 219–243 (2014)
6. Clark, M.A., Douglas, M., Choi, J.: Biology 2e. OpenStax (2018)
7. Demszky, D., Liu, J., Hill, H.C., Jurafsky, D., Piech, C.: Can automated feedback improve teachers' uptake of student ideas? Evidence from a randomized controlled trial in a large-scale online course. Educ. Eval. Policy Anal. (2023)
8. Dermeval, D., Paiva, R., Bittencourt, I.I., Vassileva, J., Borges, D.: Authoring tools for designing intelligent tutoring systems: a systematic review of the literature. Int. J. Artif. Intell. Educ. **28**, 336–384 (2018)
9. Duran, D.: Learning-by-teaching. Evidence and implications as a pedagogical mechanism. Innov. Educ. Teach. Int. **54**(5), 476–484 (2017)
10. Graesser, A.C., et al.: AutoTutor: a tutor with dialogue in natural language. Behav. Res. Meth. Instrum. Comput. **36**, 180–192 (2004)
11. Graesser, A.C., Person, N.K., Magliano, J.P.: Collaborative dialogue patterns in naturalistic one-to-one tutoring. Appl. Cogn. Psychol. **9**(6), 495–522 (1995)
12. Huang, X., Craig, S.D., Xie, J., Graesser, A., Hu, X.: Intelligent tutoring systems work as a math gap reducer in 6th grade after-school program. Learn. Individ. Differ. **47**, 258–265 (2016)
13. Jiao, Y., Shridhar, K., Cui, P., Zhou, W., Sachan, M.: Automatic educational question generation with difficulty level controls. In: Wang, N., Rebolledo-Mendez, G., Matsuda, N., Santos, O.C., Dimitrova, V. (eds.) AIED 2023. LNCS, vol. 13916, pp. 476–488. Springer, Cham (2023). https://doi.org/10.1007/978-3-031-36272-9_39
14. Kasneci, E., et al.: ChatGPT for good? On opportunities and challenges of large language models for education. Learn. Individ. Differ. **103**, 102274 (2023)
15. Koedinger, K.R., Booth, J.L., Klahr, D.: Instructional complexity and the science to constrain it. Science **342**(6161), 935–937 (2013)
16. Koedinger, K.R., Kim, J., Jia, J.Z., McLaughlin, E.A., Bier, N.L.: Learning is not a spectator sport: doing is better than watching for learning from a MOOC. In: Proceedings of the Second (2015) ACM Conference on Learning@Scale, L@S 2015, pp. 111–120. Association for Computing Machinery, New York (2015)
17. Kopp, K.J., Britt, M.A., Millis, K., Graesser, A.C.: Improving the efficiency of dialogue in tutoring. Learn. Instr. **22**(5), 320–330 (2012)
18. Kulik, J.A., Fletcher, J.D.: Effectiveness of intelligent tutoring systems: a meta-analytic review. Rev. Educ. Res. **86**(1), 42–78 (2016)
19. Labadze, L., Grigolia, M., Machaidze, L.: Role of AI chatbots in education: systematic literature review. Int. J. Educ. Technol. High. Ed. **20**(1), 56 (2023)

20. Lee, Y., Kim, T.S., Kim, S., Yun, Y., Kim, J.: DAPIE: interactive step-by-step explanatory dialogues to answer children's why and how questions. In: Proceedings of the 2023 CHI Conference on Human Factors in Computing Systems, pp. 1–22 (2023)

21. Leelawong, K., Biswas, G.: Designing learning by teaching agents: the Betty's brain system. Int. J. Artif. Intell. Educ. **18**(3), 181–208 (2008)

22. Markel, J.M., Opferman, S.G., Landay, J.A., Piech, C.: GPTeach: interactive ta training with GPT-based students. In: Proceedings of the 10th ACM Conference on Learning@Scale, pp. 226–236. ACM, New York (2023)

23. Mitrovic, A.: The effect of explaining on learning: a case study with a data normalization tutor. In: AIED, pp. 499–506 (2005)

24. Moore, S., Nguyen, H.A., Chen, T., Stamper, J.: Assessing the quality of multiple-choice questions using GPT-4 and rule-based methods. In: Viberg, O., Jivet, I., Muñoz-Merino, P., Perifanou, M., Papathoma, T. (eds.) EC-TEL 2023. LNCS, vol. 14200, pp. 229–245. Springer, Cham (2023). https://doi.org/10.1007/978-3-031-42682-7_16

25. Nguyen, H.A., Stec, H., Hou, X., Di, S., McLaren, B.M.: Evaluating ChatGPT's decimal skills and feedback generation in a digital learning game. In: Viberg, O., Jivet, I., Muñoz-Merino, P., Perifanou, M., Papathoma, T. (eds.) EC-TEL 2023. LNCS, vol. 14200, pp. 278–293. Springer, Cham (2023). https://doi.org/10.1007/978-3-031-42682-7_19

26. Nye, B.D., Graesser, A.C., Hu, X.: Autotutor and family: a review of 17 years of natural language tutoring. Int. J. AIED **24**, 427–469 (2014)

27. Paladines, J., Ramirez, J.: A systematic literature review of intelligent tutoring systems with dialogue in natural language. IEEE Access **8**, 164246–164267 (2020)

28. Pardos, Z.A., Bhandari, S.: Learning gain differences between ChatGPT and human tutor generated algebra hints. arXiv preprint arXiv:2302.06871 (2023)

29. Peng, Z., Liu, Y., Zhou, H., Xu, Z., Ma, X.: CReBot: exploring interactive question prompts for critical paper reading. Int. J. Hum.-Comput. Stud. **167**, 102898 (2022)

30. Razzaq, L., et al.: The assistment builder: supporting the life cycle of tutoring system content creation. IEEE Trans. Learn. Technol. **2**(2), 157–166 (2009)

31. Rosé, C.P.: Interactive conceptual tutoring in Atlas-Andes. In: Artificial Intelligence in Education: AI-Ed in the Wired and Wireless Future, pp. 256–266 (2001)

32. Ruan, S., et al.: BookBuddy: turning digital materials into interactive foreign language lessons through a voice chatbot. In: Proceedings of the Sixth ACM Conference on Learning@ Scale, pp. 1–4 (2019)

33. VanLehn, K.: The behavior of tutoring systems. Int. J. Artif. Intelli. Educ. **16**(3), 227–265 (2006)

Knowledge Tracing Unplugged: From Data Collection to Model Deployment

Luiz Rodrigues[1]([✉]) [ID], Anderson P. Avila-Santos[2] [ID], Thomaz E. Silva[1] [ID],
Rodolfo S. da Penha[3] [ID], Carlos Neto[3] [ID], Geiser Challco[4] [ID],
Ermesson L. dos Santos[1] [ID], Everton Souza[5] [ID], Guilherme Guerino[6] [ID],
Thales Vieira[1] [ID], Marcelo Marinho[5] [ID], Valmir Macario[5] [ID],
Ig Ibert Bittencourt[1,7] [ID], Diego Dermeval[1] [ID], and Seiji Isotani[7] [ID]

[1] Center for Excellence in Social Technologies, Federal University of Alagoas, Maceió,
Brazil
luiz.rodrigues@nees.ufal.br
[2] University of São Paulo, São Carlos, Brazil
[3] Federal University of Ceará, Fortaleza, Brazil
[4] Federal Rural University of the Semi-Arid Region, Pau dos Ferros, Brazil
[5] Federal Rural University of Pernambuco, Recife, Brazil
[6] State University of Paraná, Paraná, Brazil
[7] Harvard Graduate School of Education, Cambridge, USA

Abstract. Knowledge Tracing (KT) plays a pivotal role in Artificial
Intelligence in Education (AIED) by modeling and predicting learners'
mastery of skills over time. While AIED Unplugged aims to adapt AI
solutions for resource-constrained environments, integrating KT in such
scenarios is challenging due to limited digital interaction and the feasibility of exploring advanced algorithms. This paper introduces KT
Unplugged, addressing the gap in prior research by exploring and experimenting with creating, refining, and implementing state-of-the-art KT
models within resource-limited contexts. The contributions of this paper
are threefold. Firstly, we present and perform a procedure for simulating
data collection in unplugged contexts, resulting in a dataset and replicable methodology for future field studies. Secondly, an empirical study
focused on developing and validating KT models for resource-constrained
devices, employing sophisticated (deep learning) and classical (Bayesian)
algorithms. This contribution provides empirical evidence on the performance of KT unplugged, including a pre-trained model for numeracy
education. Finally, a technical study assesses the deployment of the pre-
trained model on disconnected, low-cost mobile devices, demonstrating
the technical feasibility of KT Unplugged with acceptable inference times
and maintained predictive power. By addressing the challenges of KT
integration in unplugged scenarios, this research opens new avenues for
personalized learning, adaptive instruction, and targeted interventions
in education settings with limited infrastructure.

Keywords: Deep Knowledge Tracing · Unplugged · Dataset · Mobile

A. M. Olney et al. (Eds.): AIED 2024, LNAI 14829, pp. 91–104, 2024.
https://doi.org/10.1007/978-3-031-64302-6_7

1   Introduction

Knowledge Tracing (KT) is a data-driven technique for modeling and predicting a learner's mastery of skills or concepts over time [3]. It employs statistical and machine learning algorithms to analyze students' responses to educational tasks, tracking their understanding and learning progress [7]. KT facilitates the creation of Intelligent Tutoring Systems (ITS), enabling personalized learning paths, adaptive instruction, and targeted interventions [8]. Hence, it is pivotal in Artificial Intelligence in Education (AIED).

Amid the growing significance of AIED, efforts to enhance its accessibility have sparked the concept of AIED Unplugged, aiming to adapt AI solutions for environments with limited infrastructure, such as low-cost hardware as well as restricted access to digital devices and the internet [5]. However, integrating KT in unplugged scenarios is challenging. Minimal digital interaction hampers detailed progress data collection, while limited internet access and hardware hinder the use of state-of-the-art algorithms (e.g., Deep KT, or DKT) [1,7, 10]. Thereby emerges the necessity for KT Unplugged, an approach tailored to resource-constrained contexts with limited data availability.

Nevertheless, as far as we know, prior research has not addressed this issue. On the one hand, studies on KT have evolved towards deep learning-based algorithms, aiming for higher predictive performance, which demands substantial data and processing power [1]. On the other hand, whereas studies on the recent field of AIED Unplugged have focused on designing ITS unplugged [14] or transcribing handwritten texts to provide feedback [4,11], they are yet to experiment with the design, development, and deployment of KT features. Therefore, a significant gap remains in exploring and experimenting with creating, refining, and implementing state-of-the-art KT models within these resource-limited contexts.

To address this gap, this paper presents three studies. First, we designed and executed a process that simulates the unplugged context within regular classrooms. In that context, teachers performed this procedure during real lessons, applying and collecting math activities in a setting close to AIED unplugged. Second, building upon the collected data, we developed and validated KT models optimized for the unplugged context. For this, we experimented with lightweight KT models inspired by state-of-the-art research, aiming for models optimized to low-cost, disconnected devices. Finally, we deployed and assessed the technical feasibility of running such a model on mobile devices without internet access. Thus, this paper presents three contributions to KT unplugged, particularly within the domain of numeracy (from K-2 to K-5 students):

1. A procedure simulating data collection in unplugged contexts conducive to KT unplugged applications (Sect. 3). This effort results in a dataset comprising interactions from multiple students, accompanied by a replicable methodology valuable for future field studies.
2. An empirical study on the development and validation of KT models targeting resource-constrained devices, including sophisticated and simpler algorithms, such as deep learning and Bayesian ones (Sect. 4). This contribution includes

empirical evidence on the performance of KT unplugged and a pre-trained KT model for numeracy education.

3. A technical study in which we deployed our model to assess its performance in disconnected, low-cost mobile devices (Sect. 5). This contribution demonstrates KT unplugged's technical feasibility based on real-time inference in disconnected, low-cost mobile phones while maintaining predictive power despite the optimization.

2   Background

2.1   Knowledge Tracing

KT stands as a fundamental technique in the realm of AIED. Aimed at modeling and predicting learners' mastery of specific concepts or skills over time, its primary goal is to understand and track individual student learning trajectories by analyzing their responses to educational tasks [3]. Data collection for KT involves tracking students' responses to educational tasks, such as in an online learning platform, gathering information on correct and incorrect answers, time taken, and the sequence of actions. Then, this data is used to create predictive models that find application in personalized learning, adaptive tutoring systems, and targeted interventions [1,8].

The principles behind KT involve utilizing varied algorithms [7]. Traditional KT algorithms, such as Bayesian Knowledge Tracing (or BKT), use probabilistic models to estimate the probability of a student mastering a skill based on their responses, providing insights into a student's grasp of a particular concept [15]. On the other hand, recent advancements in KT have shifted towards employing deep learning techniques, leading to the emergence of Deep Knowledge Tracing (DKT) [10]. DKT leverages neural network architectures to capture complex patterns in student learning behaviors, offering improved predictive accuracy and scalability [1].

However, DKT has challenges related to data availability and computational power. Implementing these sophisticated algorithms frequently requires extensive datasets containing intricate, detailed interactions from learners. Furthermore, the computational demands for performing real-time predictions using DKT models are substantial [1,7,13]. These resource-intensive requirements present practical limitations, especially in resource-constrained environments with limited access to abundant data and high computational power.

2.2   AIED Unplugged

As introduced in [5], AIED Unplugged expands AIED by tailoring it to tackle challenges prevalent in low and middle-income countries. Aimed at addressing fundamental issues, such as limited internet access and digital skills among users of AIED systems, this approach offers solutions acknowledging the digital divide prevalent in underserved contexts. This concept stems from the principles of

Jugaad innovation, characterized by resourcefulness and simplicity. It advocates for creative, accessible solutions utilizing available resources, aligning with AIED Unplugged's objective to make AIED systems practical and effective in low-resource environments.

Five key elements background the paradigm shift in AIED deployment, rendering it more accessible and relevant for diverse learners:

1. *Conformity*: Tailoring AI solutions to the existing infrastructure and pedagogical practices within a given setting, integrating seamlessly without disrupting the educational environment's functionality.
2. *Disconnect*: Designing AI solutions independent of constant internet connectivity yet capable of utilizing internet access once available for updates and data collection.
3. *Proxy*: Introducing intermediary interfaces bridging the gap between users with varying digital skills and the AI solution, ensuring accessibility for all.
4. *Multi-User*: Acknowledging shared access to hardware and software, necessitating AI solutions that accommodate multiple users without individual logins or extensive interaction recording.
5. *Unskillfulness*: Prioritizing user-friendly, uncomplicated interfaces that don't demand advanced digital skills, ensuring accessibility to a broad user base possessing basic technology like smartphones.

By leveraging AI to bridge educational gaps, AIED Unplugged seeks to revolutionize access for underserved communities without relying on extensive digital infrastructure. Emphasizing resourcefulness and simplicity, AIED Unplugged promises effective educational solutions to improve learning experiences in low-resource environments [5]. Accordingly, it raises the need for tailoring key AIED features, such as KT, to comply with the unplugged context.

2.3   Summary and Research Objectives

State-of-the-art KT encounters significant hurdles in resource-constrained settings. The demand for extensive, detailed interaction data and substantial computational power impedes the application of such sophisticated models within environments lacking abundant resources. We understand two major challenges in tailoring KT for the unplugged context.

First, KT unplugged must be able to operate with reduced data, compared to virtual learning environments, as collecting step-based, fine-grained data is likely to violate the *conformity* principle. The *multi-user* principle reiterates this context, as having a device available for such detailed data collection is unlikely. Moreover, if we consider that students need a *proxy* (e.g., teacher) for data collection and that teachers might lack the skills to interact with the digital devices that would enable data collection (i.e., the *unskillfulness* principle), this approach would require a significant amount of time and possibly violate the conformity principle. Second, the computational demand of advanced KT models, which are feasible for cloud and powerful hardware-based devices, likely violates the *disconnect* principle of unplugged contexts limited to low-cost devices such as mobile phones.

Addressing these challenges requires data collection strategies respecting the principles of *conformity* and *multi-user*. Solutions must consider simplified data gathering methodologies compatible with resource-constrained environments, potentially utilizing intermittent internet access for data updates. Moreover, developing lightweight KT models optimized for offline, low-cost devices, in line with the *disconnect* principle, becomes imperative, ensuring efficient computational processing without compromising prediction accuracy or violating the principles of AIED Unplugged.

Differently, recent KT algorithms demand extensive interaction data while collecting fine-grained, step-based data conflicts with the *conformity* and *multi-user* principles of AIED unplugged. Therefore, this paper sought a method and collected data relevant to KT unplugged within the context of numeracy education (Sect. 3). This objective focused on a method and a dataset that align with AIED Unplugged principles, ensuring a replicable methodology for efficient data gathering conducive to KT Unplugged applications and a reusable dataset.

Furthermore, we explore the generated dataset to implement KT unplugged by acknowledging the limited data availability. Based on our dataset, this paper sought to develop effective, lightweight KT models compatible with low-cost devices despite the limited data available (Sect. 4). This objective aimed to develop and validate models that balance computational efficiency and prediction accuracy. The implication involved providing a pre-trained model that maintains high performance while complying with AIED Unplugged principles, enabling effective numeracy education in resource-constrained settings.

Advanced KT models necessitate substantial computational resources for inference, conflicting with the *disconnect* principle in unplugged environments, limited to no or intermittent access to the internet and low-cost devices. Therefore, this paper sought to assess the performance of our lightweight pre-trained model in disconnected, low-cost mobile devices (Sect. 5). This objective demonstrated the technical feasibility of KT unplugged by complying with AIED Unplugged's *disconnected* principle, given its acceptable inference times in such devices while preserving predictive power despite optimizations.

3   Study 1: A Dataset for Knowledge Tracing Unplugged

This study aimed to design a method and generate a dataset for KT unplugged based on a simulation of using an AIED Unplugged system in the classroom. KT necessitates sequences of student interaction data with educational material, typically categorized as correct or incorrect responses. Then, to address the goal regarding the design and data collection process for KT Unplugged, we drew upon prior studies within the scope of AIED Unplugged across diverse domains.

Past research on AIED Unplugged for essay writing [4,5,11] has advocated a three-step procedure. Initially, students handwrite the essays, followed by teachers or managers scanning each individually. Subsequently, an AIED system grades the essays. A similar approach proposes an integrated system that scans multiple solutions from a single paper sheet, which is feasible due to the smaller space occupied by solutions to numeracy tasks compared to essays. It is later digitized for assessment by an ITS [9]. Another similar approach [14] suggests capturing solutions through photos instead of scanners, which enables capturing solutions immediately after students write them and then streamlining the feedback process.

Within this context, the iterative digitalization process yields a dataset encompassing multiple student interactions (e.g., handwritten essays or math task solutions). Once digitalized, these interactions generate sequences suitable for KT modeling. Importantly, all approaches consider that an AIED Unplugged system is already implemented and functioning. However, implementing such systems involves several technical challenges, including having data to develop and test the multiple AI-based systems required. Thus, we adapt those steps to design and execute a process that simulates gathering interaction data within the context of AIED unplugged. Particularly, we focus on numeracy tasks because they enable an easier generation of sequence data, simultaneously facilitating the capturing of multiple solutions. Nevertheless, the process might be replicated in other knowledge domains as well.

The proposed process is based on three steps. First, experts generate and print a list of numeracy exercises. This simulates teachers designing instructional material for students to work on during lessons. Second, teachers use the printed exercises during their regular lessons as part of their instructional practices. This simulates students interacting with instructional content recommended by the AIED in real contexts. Note that this process focuses on gathering data for KT before the AIED unplugged system is available, so the digitalization step suggested in related work is unnecessary. Instead, the third step is accomplished by human experts rather than the AIED Unplugged system. Then, teachers label students' solutions as correct or incorrect, as KT algorithms expect. By following this process, we capture the binary correctness of the student's responses and delve deeper into understanding the nature of their errors.

3.1 Execution

First, two experts in numeracy education accomplished step one. Both have over 20 years of experience working in the classroom as mathematics teachers and holding doctorates in mathematics education. Together, they designed 40 numeracy items that expected open-ended solutions and concerned eight knowledge components - KC (five for each). Opting for open-ended solutions in numeracy items allows a more comprehensive assessment of students' understanding. Open-ended questions encourage critical thinking, problem-solving, and creativity, providing insights into the correctness of the answer and the thought processes and reasoning behind it [12]. This approach aligns to capture a holistic view of learners' mastery rather than surface-level comprehension.

Furthermore, each item concerned a single KC. Focusing on a single KC facilitates analyzing and interpreting students' responses. It enables a more targeted evaluation of a specific skill or concept, facilitating a deeper understanding of individual proficiency in that particular area. This specificity enhances the precision of the knowledge tracing model, making it more effective in identifying and addressing specific learning needs.

The selected KCs align with the BNCC (Base Nacional Comum Curricular - Brazilian National Common Core) skills to ensure relevance and alignment with national educational standards. Addressing these components makes the knowledge tracing process directly applicable to the broader educational framework, promoting consistency and coherence in assessing and improving students' numeracy skills per established educational guidelines.

Second, our research team recruited numeracy teachers to accomplish step two. At first, we contacted school administrators, introduced the overall research project, and asked permission to present our research goal to teachers. Following their agreement, we contacted teachers and introduced the proposed pedagogical practice, and interested teachers agreed to include our suggestions in their daily pedagogical practices. Then, our research team printed the generated lists and handed them to teachers to use in their lessons as suitable in their daily routines. We asked teachers to use no more than one list in each class to facilitate collaboration. Next, we followed up with those teachers, and once they could incorporate the lists into their pedagogical practices, they returned the printed versions to our research team.

Third, the experts who accomplished the first step also realized the last. They revised each solution available within the lists received by the collaborating teachers, assigning either 1 or 0 when it was correct or incorrect, respectively. Additionally, they ensured the data had no identifying information and that it preserved the resolution sequence. An important issue is that, even if a student skipped an item, it was labeled as incorrect. Because such information indicates the student could not solve the item, we tagged it as incorrect so that the model could learn from this during training.

3.2   Results

Altogether, collaborating teachers sent us several solved lists, which summed up 1130 solutions from 224 students concerning the eight KCs selected. Table 1 describes our dataset for KT unplugged, while Table 2 demonstrates the distribution of answers correctness depending on each KC.

Table 1 demonstrates that the dataset has only five attributes. This is a key distinction compared to other datasets used for KT, which often feature dozens of features [1]. Although this might be replicated by removing additional features from other available datasets, ours was collected offline rather than through virtual learning environments. Hence, it adds a layer of ecological validity in simulating an unplugged context. Moreover, as each teacher was asked to use a single list concerning a single KC, there is an ecological imbalance in the number of answers per KC.

Table 1. Dataset description.

Feature	Description	Values
SID	Student identifier	1 to 224
OID	Resolution order identifier	1 to 5
PID	Problem identifier	Q1 to Q5 (for each KC)
COR	Whether the answer was correct	0 or 1
KC	Knowledge Component identifier according to BNCC	EF01MA06, EF01MA08, EF02MA05, EF02MA06, EF01MA02, EF02MA03, EF03MA03, EF03MA07

Table 2. Count of correct and incorrect answers for each knowledge component.

KC	0106	0108	0205	0206	0102	0203	0303	0307
Correct	66	206	108	42	107	117	70	74
Incorrect	19	29	92	38	103	23	25	11
Total	85	235	200	80	210	140	95	85
Correct %	0.78	0.88	0.54	0.53	0.51	0.84	0.74	0.87

4   Study 2: KT Unplugged Modeling

This study aimed to create a KT model for the unplugged context. For this, we first performed feature extraction. Despite the unplugged context imposing a limitation on data available for KT, the literature suggests that side information is likely to improve these models' performance [13]. Hence, we sought to extract new features based on the available data. Particularly, we calculated Item Response Theory (IRT) parameters, such as difficulty level, guessing attempts, and scale discrimination, as this approach is common for ITS [8]. We computed a Three-Parameter Logistic model using Marginal Maximum Likelihood [6]. With this, we gathered additional information concerning the dataset's 40 items, aiming to improve the KT modeling.

Second, we preprocessed our dataset. For this, we performed a windowing procedure, based on the order of resolution (i.e., OID), so that the preprocessed data would capture the temporal aspect of solving math tasks, which is prominent to KT. Consider the following hypothetical sequence: [0, 1, 2, 3, 4]. If the windowing procedure were to generate windows of size 2, they would be [0, 1], [1,2,2,3], and [3,4]. Here, considering our data features five solutions per student, we performed this procedure with window sizes of two, three, and four to investigate which option provided the best fit.

Third, we prepared our data for KT modeling. In this phase, we split our dataset into training (n = 904; 80%) and testing (n = 226; 20%) sets so we could assess the model's generalizability. Notably, when training models based on neural networks, such as DKT, a validation set is important to inform the optimization process [10]. Thereby, we saved 20% of the training set (n = 180) for validation when training our DKT models. Importantly, before the testing phase, the to-be-tested model was retrained with the complete training set.

Fourth, we implemented our baseline model. We used BKT, an approach well-established in the KT literature. This model was developed using Python and the *pyBKT* package [2]. As we wanted to develop a baseline model, we used the package's *standard BKT*, using its default values, as detailed in [2].

Fifth, we experimented with varied typologies of DLK, as this technique has shown potential to yield improved performance compared to BKT [1]. Here, data preparation included normalizing numeric features (i.e., those related to IRT parameters) using min-max normalization and preprocessing nominal ones (i.e., SID and KC) with either label encoding or embeddings. In seeking the best model, we conducted an extensive hyperparameter tuning using Keras Tuner[1], seeking the combination that provided the best predictive performance. The search space included layer types (Long Short-Term Memory - LSTM, Gated Recurrent Unit - GRU, Bidirectional LSTM, Simple Recurrent Neural Network - RNN, and Convolutional - CNN[2]) and number of neurons (8, 16, 32, and 64) aiming to explore a broad search space.

We did not explore more neurons or hidden layers to avoid seeking models with numerous parameters, as we expect the best model to run on a low-processing device (see Sect. 2). Notably, we used early stopping to prevent overfitting, setting it to stop training after five epochs with no improvements in the model's performance on the validation set. After the search was completed, we selected the best model based on the one with the highest Area Under the ROC Curve (AUC) on the validation set, as this metric is robust to class imbalance and useful for model comparisons [7].

Lastly, we assessed the modeling performance. For this, we compared the BKT model to the best DKT one. To achieve a holistic view of predictive power, we evaluated both models using Root Mean Squared Error (RMSE) and Mean Absolute Error (MAE), besides AUC, similar to related work [7]. Additionally, we assessed their average inference time based on 100 runs on the test set to evaluate their efficiency and potential for real-time application in educational settings. Furthermore, the sizes of the models were also assessed, providing insights into their computational and storage requirements.

4.1 Results

Table 3 presents the results of the hyperparameter search. It demonstrates the AUC of the best models for each analyzed window size and the setting that led to that performance. As the table shows, the model with the best performance is based on a window size of 3, embeddings, and an LSTM layer with 16 units, achieving an AUC of 0.98 on the validation set during the training process. Hence, it was selected as our best model and used in this paper's remaining analyses.

---

[1] https://keras.io/keras_tuner/.

[2] Our dataset features five resolutions per student, so we set kernel size to 4.

Table 3. Grid Search results based on predictions on the validation set. It shows the Area Under the Curve (AUC) of the best models for each window size (WS) analyzed, along with the training setting (i.e., layer type, number of units, and using embeddings or not) that achieved it.

	Embedding			One-hot encoding		
WS	AUC	Layer	Units	AUC	Layer	Units
2	0.95	LSTM	8	0.95	LSTM	16
3	0.98	LSTM	16	0.94	CNN	16
4	0.97	LSTM	8	0.96	LSTM	8

Building upon the previous result, Table 4 compares the predictive power of our baseline (i.e., BKT) and the best DKT model on the testing set. Moreover, the table compares these models regarding inference time and model size. It demonstrates the DKT model yielded a superior performance than BKT, our baseline, on AUC, RSME, and MAE. Conversely, BKT is shown to be faster and lighter than DKT by yielding smaller inference times and model size.

Table 4. Comparison of our baseline (i.e., BKT) and our best DKT model, which is based on embeddings and an LSTM layer with 16 units, in terms of predictive performance on the testing set as well as inference time (time) and model size (size).

Model	AUC	RMSE	MAE	Time	Size
BKT	0.77	0.42	0.36	0.038087	8 kb
DKT	0.95	0.33	0.18	0.072947	42 kb

AUC = Area Under the Curve; RMSE = Root Mean Squared Error; MAE = Mean Squared Error.

These results demonstrate the suitability of achieving reliable KT models within the unplugged context. Particularly, our findings suggest an advantage in predictive performance, whereas BKT was better regarding inference time and model size. Nevertheless, while the difference in predictive performance is substantial, the magnitude of differences in inference time and model size seems smaller.

5   Study 3: KT Unplugged Deployment

Following our previous findings, this study aimed to deploy and assess a KT unplugged model in disconnected, low-cost mobile devices.

Therefore, we first optimized our best model for mobile compatibility. We used Tensorflow Lite (TFLite)[3]. TFLite ensures mobile compatibility by quantizing models, reducing the precision of weights and activations, and optimizing

---

[3] https://www.tensorflow.org/lite.

for hardware accelerators. Furthermore, TFLite provides an interpreter for efficient execution on resource-constrained mobile devices, balancing reduced memory usage and acceptable accuracy. Hence, we used this approach to convert our best model to a mobile-compatible one.

Second, we prepared the dataset. Besides the preprocessing procedure of Study 2 (see Sect. 4), this phase has an additional step: each row of the preprocessed dataset was converted into a buffer to ensure they were compatible with the mobile application (see details below). This aimed to ensure all preprocessing was conducted before inputting data to the model, thereby increasing the reliability of measuring its inference time.

Third, we developed a testbed application. We used the Kotlin programming language[4] due to its compatibility with TFLite. The testbed is a simple mobile application that uses our optimized model to perform inferences on preprocessed data to assess its performance on a low-cost, disconnected device. Notably, the input of DKT models is three-dimensional (i.e., batch, sequences, and features), and converting standard data (e.g., in a CSV format) to this format so they can be inputted into a TFLite model is challenging. Another approach is performing this conversion beforehand by generating tensors, as described before. Therefore, we adopted this approach to focus on assessing the optimized model's inference time. Then, we repeatedly executed the testbed application for data collection based on a randomly selected sample of our data ($\approx 20\%$ of our test set).

5.1 Results

To increase our findings' validity, we executed the data collection process in two devices of less than 200 USD[5]: a Xiaomi Redmi Note 10 ($2 \times 2.2\,$GHz Kryo 460 Gold $+ 6 \times 1.7\,$GHz Kryo 460 Silver, 128 GB, 4 GB) and a Xiaomi Redmi Note 8 (Octa-core Max 2.01 GHz, 64 GB, 4 GB). Respectively, these devices took 139,6 (i.e., 0,29 per inference) and 155,3 (i.e., 0,33 per inference) milliseconds to perform 470 inferences each.

Furthermore, an interesting insight is that the quantization process did not affect the model's predictive performance, as it remained the same. Through additional investigations, we saw that differences in the original and quantized models' predictions would only appear after the fifth decimal case, explaining the equivalent predictive performance. Overall, these results demonstrate that our KT model fits the unplugged context, as it could perform several inferences per second and preserve its predictive performance while running in disconnected, low-cost devices.

6 Discussion

Amid the rising importance of AIED, efforts to make it more accessible led AIED Unplugged: adapting AI solutions for environments with limited infrastructure, including low-cost hardware and restricted access to digital devices and

---

[4] https://kotlinlang.org/.

[5] Based on Amazon's prices at the time of writing.

the internet [5]. Among AIED's several techniques, KT is a data-driven method used to model and predict a learner's skill or concept mastery over time [3]. KT is crucial for AIED as it powers ITS, allowing for personalized learning paths and targeted interventions [8].

Notably, integrating KT into unplugged scenarios poses challenges due to i) minimal digital interaction hindering the data collection process and i) limited internet access and hardware impeding the use of advanced algorithms like Deep KT (DKT). This leads to the need for KT Unplugged, a tailored approach for resource-constrained contexts with limited data availability. However, existing research has not addressed this issue, with studies on KT focusing on deep learning-based algorithms demanding substantial data and processing power [1, 4,11,14]. This context highlights a significant gap in exploring and experimenting with designing, developing, and deploying state-of-the-art KT models within these resource-limited contexts.

This paper addresses that need with a threefold contribution. Given the distinct nature of unplugged settings, our first contribution is a dataset (n = 1130) collected during real lessons that simulated the application of an AIED unplugged system. In contrast to most KT studies [1], this dataset is limited to minimal interaction data (i.e., response's correctness, resolution order, and student, problem, and KC identifiers). One might argue that disregarding additional features from existing datasets would replicate this situation. However, existing datasets are collected through interactions with virtual environments. Differently, our dataset is based on in-class, offline interactions. Hence, we contribute a dataset for KT unplugged research with increased ecological validity. Hence, these results demonstrate the feasibility of running KT models in disconnected, low-cost devices.

Our second contribution is a KT model for the unplugged scenario. KT research often explores detailed, step-by-step interaction data and additional information on some occasions [7]. On the other hand, we developed and validated a KT model based on the restricted information available in the unplugged scenario. To design a lightweight Deep Knowledge Tracing model, we explored additional information (i.e., IRT parameters), similar to recent KT research [13]. Our model's predictive performance on unseen data (AUC $\approx$0.9) beat our baseline (i.e., the traditional BKT) and was comparable to overall research on KT despite the limited information available. Thereby, we reveal the feasibility of achieving state-of-the-art performance on KT even within the restrictions of AIED unplugged.

Our third contribution concerns the technical feasibility of KT unplugged. Recently, state-of-the-art results on KT have been achieved by advanced deep learning models [1,7] and, given AIED unplugged's context, those are unlikely to fit low-cost devices with low-processing capabilities and no internet access to delegate the process. Therefore, we quantized our model, embedded it into a testbed application, and analyzed its inference times while running directly on a mobile device. Our results demonstrated the model could perform nearly 100 inferences in around 33 milliseconds. This number of inferences is sufficient for predicting a student's mastery in real-time and to run recommendations algorithms that would require predicting that of several students as well as [8]. Thus, we reveal the technical feasibility of deploying KT models in unplugged contexts.

Accordingly, these findings have three main implications. First, introducing a dataset tailored to unplugged settings contributes to developing more contextually relevant KT models, ensuring that findings align closely with real-world classroom dynamics. Second, the success of a KT model designed from minimal interaction data in unplugged scenarios expands the conventional reliance on detailed, step-by-step interaction data, suggesting that even with limited information, it is possible to achieve predictive performance comparable to traditional models, thereby opening avenues for more resource-efficient KT applications. Third, the technical feasibility exploration addresses the practicality of deploying KT models in resource-constrained environments. The successful deployment on a low-cost device with minimal processing capabilities suggests that advanced KT technologies can be applied in real-time, even in settings without internet access, a significant implication for expanding the reach of AIED to areas with limited infrastructure.

Importantly, one should interpret our findings in light of this study's limitations. This study focused on numeracy tasks, potentially limiting the generalizability of findings to other knowledge domains. Nevertheless, we chose this topic due to its significance to the changing world following the Covid-19 pandemic and the fact that focusing on a single topic is important for emerging research. Additionally, our dataset is prone to human subjectivity regarding human labeling. However, we mitigate this issue by relying on experienced math experts. Concerning the KT modeling, the limited available data limits its generalization, which we addressed with a systematic model development, validation, and testing process. Lastly, the deployment assessment was limited to a testbed application, which we mitigated by running multiple inference trials on two devices. Therefore, we call for future research to enhance the generalizability of our findings by studying multiple knowledge domains, employing larger and more varied datasets, and exploring real-world deployment scenarios beyond testbed applications.

Acknowledgments. We wish to express our gratitude to everyone involved in this national project, including researchers, policymakers, and teachers. This work was supported by the Brazilian Ministry of Education (MEC) - TED 11476. G. G. also received support from the Fundação Araucária and the Government of the State of Paraná/SETI.

References

1. Abdelrahman, G., Wang, Q., Nunes, B.: Knowledge tracing: a survey. ACM Comput. Surv. **55**(11), 1–37 (2023)
2. Badrinath, A., Wang, F., Pardos, Z.: pyBKT: an accessible python library of Bayesian knowledge tracing models. arXiv preprint arXiv:2105.00385 (2021)

3. Corbett, A.T., Anderson, J.R.: Knowledge tracing: modeling the acquisition of procedural knowledge. User Model. User-Adap. Inter. **4**, 253–278 (1994)
4. Freitas, E., et al.: Learning analytics desconectada: Um estudo de caso em análise de produçoes textuais. In: Anais do I Workshop de Aplicações Práticas de Learning Analytics em Instituições de Ensino no Brasil, pp. 40–49. SBC (2022)
5. Isotani, S., Bittencourt, I.I., Challco, G.C., Dermeval, D., Mello, R.F.: AIED unplugged: leapfrogging the digital divide to reach the underserved. In: Wang, N., Rebolledo-Mendez, G., Dimitrova, V., Matsuda, N., Santos, O.C. (eds.) AIED 2023. CCIS, vol. 1831, pp. 772–779. Springer, Cham (2023). https://doi.org/10.1007/978-3-031-36336-8_118
6. Kean, J., Reilly, J.: Item response theory. Handbook for clinical research: design, statistics and implementation, pp. 195–198 (2014)
7. Liu, T.: Knowledge tracing: a bibliometric analysis. Comput. Educ. Artif. Intell. **3**, 100090 (2022)
8. Nkambou, R., Mizoguchi, R., Bourdeau, J.: Advances in Intelligent Tutoring Systems, vol. 308. Springer, Heidelberg (2010). https://doi.org/10.1007/978-3-642-14363-2
9. Patel, N., et al.: Equitable access to intelligent tutoring systems through paper-digital integration. In: Crossley, S., Popescu, E. (eds.) ITS 2022. LNCS, vol. 13284, pp. 255–263. Springer, Cham (2022). https://doi.org/10.1007/978-3-031-09680-8_24
10. Piech, C., et al.: Deep knowledge tracing. Adv. Neural Inf. Process. Syst. **28** (2015)
11. Portela, C., et al.: A case study on AIED unplugged applied to public policy for learning recovery post-pandemic in Brazil. In: Wang, N., Rebolledo-Mendez, G., Dimitrova, V., Matsuda, N., Santos, O.C. (eds.) AIED 2023. CCIS, vol. 1831, pp. 788–796. Springer, Cham (2023). https://doi.org/10.1007/978-3-031-36336-8_120
12. Schukajlow, S., Krug, A., Rakoczy, K.: Effects of prompting multiple solutions for modelling problems on students' performance. Educ. Stud. Math. **89**(3), 393–417 (2015). http://www.jstor.org/stable/43590001
13. Song, X., Li, J., Cai, T., Yang, S., Yang, T., Liu, C.: A survey on deep learning based knowledge tracing. Knowl.-Based Syst. **258**, 110036 (2022)
14. Veloso, T.E., et al.: Its unplugged: leapfrogging the digital divide for teaching numeracy skills in underserved populations. In: Wang, N., Rebolledo-Mendez, G., Matsuda, N., Santos, O.C., Dimitrova, V. (eds.) Towards the Future of AI-augmented Human Tutoring in Math Learning 2023 - Proceedings of the Workshop on International Conference of Artificial Intelligence in Education co-located with The 24th International Conference on Artificial Intelligence in Education (AIED 2023). Springer, Cham (2023). https://doi.org/10.1007/978-3-031-36272-9
15. Yudelson, M.V., Koedinger, K.R., Gordon, G.J.: Individualized Bayesian knowledge tracing models. In: Lane, H.C., Yacef, K., Mostow, J., Pavlik, P. (eds.) AIED 2013. LNCS (LNAI), vol. 7926, pp. 171–180. Springer, Heidelberg (2013). https://doi.org/10.1007/978-3-642-39112-5_18

Grading Documentation with Machine Learning

Marcus Messer[1]([✉]), Miaojing Shi[2], Neil C. C. Brown[1],
and Michael Kölling[1]

[1] Department of Informatics, King's College London, London, UK
{marcus.messer,neil.c.c.brown,michael.kolling}@kcl.ac.uk
[2] College of Electronic and Information Engineering, Tongji University,
Shanghai, China
mshi@tongji.edu.cn

Abstract. Professional developers, and especially students learning to program, often write poor documentation. While automated assessment for programming is becoming more common in educational settings, often using unit tests for code functionality and static analysis for code quality, documentation assessment is typically limited to detecting the presence and the correct formatting of a docstring based on a specified style guide. We aim to investigate how machine learning can be utilised to aid in automating the assessment of documentation quality. We classify a large set of publicly available human-annotated relevance scores between a natural language string and a code string, using traditional approaches, such as Logistic Regression and Random Forest, fine-tuned large language models, such as BERT and GPT, and Low-Rank Adaptation of large language models. Our most accurate mode was a fine-tuned CodeBERT model, resulting in a test accuracy of 89%.

Keywords: Automated Grading · Assessment · Computer Science Education · Machine Learning · Large Language Models · Documentation · Programming Education

1 Introduction

Good documentation is important to any software project [2]. Technical documentation, which includes API reference manuals generated by programs like JavaDoc or PyDoc, offers details on the features and interfaces of a code base [3]. In this paper, we focus on technical documentation, such as JavaDoc or PyDoc, and refer to it simply as documentation. In conjunction with the code itself, the documentation quality affects how maintainable a code base is [1].

The challenge of writing good documentation and assessing documentation quality affects both professional and student programmers. Professional code is usually assessed in a code review. Correctness of the code can be assessed automatically through software testing, but documentation currently cannot be assessed automatically other than trivial checks for formatting. Students need to learn to write good documentation, but human graders must assess this manually. In addition to the presence and correct style of documentation, good documentation must also relate to the class or function it describes, be understandable, concise and consistent [25].

As student cohort sizes increase in educational settings, automated assessment is becoming more commonplace, with most automated assessment tools using unit tests to assess functionality and static analysis to assess the code [19]. Automatically assessing technical documentation is often limited to checking that documentation is present and matches a specific style guide, using tools such as CheckStyle[1], and does not assess documentation quality [19].

Some assignments manually assess the documentation quality alongside other aspects, including code readability and design. However, manual assessment of documentation quality is time-consuming, especially compared to the near-instantaneous feedback provided by automated assessment tools used to assess a submission's functionality. Automating the assessment of documentation quality could aid the teaching and learning of this important aspect of programming.

In order to automate the assessment of documentation quality, in this paper we will investigate how machine learning techniques can be used to classify how relevant a docstring is to a specific function. The relevance classification could be used as part of a grade for documentation quality alongside static analysis techniques for checking presence and style.

2   Related Work

A recent systematic literature review found only one automated assessment tool between 2017 and 2021 that discussed assessing documentation [19], called App-Grader. AppGrader was developed to provide an automatic assessment of Visual Basic applications, primarily focusing on the functionality of the student's submissions, including validating that specified form objects and common programming concepts are present. In terms of assessing documentation, AppGrader checks to see if functions, conditionals and loops have been comments preceding, on the same line or immediately following the statement. However, AppGrader only validates if the comment is present and not the quality of the content [12].

Validating the presence of documentation is common for many software linters aimed at professional development, including CheckStyle. However, to our knowledge, no professionally developed tools automatically check the documentation quality; thus it is typically part of manual code review.

To relieve developers from writing documentation, software engineering research has utilised neural networks to generate documentation from source code [18,24,28]. Similarly, other research has investigated how fine-tuned code-specific large language models can generate documentation and source code [7,8,10]. While our task focuses on evaluating the relevance of documentation to

---

[1] CheckStyle – A style guide enforcement utility: https://checkstyle.sourceforge.io/.

the corresponding source code, research into machine learning for code generation and summarisation shows that a model can learn the relationship between
natural language and code with sufficient data.

Within computer science education, machine learning approaches have been
used to assess the quality of a submission's functionality, including assessment
criteria that cannot typically be assessed with dynamic or static analysis [20].

One such tool utilised pre-trained image recognition models to assess the
image quality of a student's submission in an introductory computer graphics
course [21]. Another model was trained to assess the design quality of Python
submissions, using abstract syntax trees and a feed-forward neural network
to predict a score between 0 and 1 and provide personalised feedback based
on features within the individual program [26]. These machine learning-based
approaches to automatically assessing elements of programming assignments
that are typically manually graded show that machine learning can be adapted
for various traditionally manually graded criteria and, therefore, may also apply
to documentation relevance.

3   Methodology

3.1   Dataset Selection

We used the CodeSearchNet dataset from GitHub and Microsoft Research [15] to
train our machine learning models. The primary dataset consists of two million
source code and documentation pairs from many open-source repositories hosted
on GitHub, with languages including Python, Java and Go.

Listing 1.1. Example snippet from CodeSearchNet

```java
// NL Query: get the description of a http status code
// Relevance Score: 3
private static String toStatusString(int theStatusCode) {
    return Integer.toString(theStatusCode) + " " +
        defaultString(
            Constants.HTTP_STATUS_NAMES.get(theStatusCode)
        );
}
```

As part of the dataset, which focused on code and documentation[2] generation, the authors of the dataset collected an annotated subset of human-judged
relevance scores. The human experts judged a subset of realistic code/documen-
tation pairs. They rated the relevance between the documentation and the code
between zero and three inclusive, with zero being irrelevant and three being very
relevant. An exemplar snippet of the dataset can be found in Listing 1.1. We
filter the subset of code/documentation pairs to include only those written in
Java.

---

[2] In CodeSearchNet, documentation and natural language query are used interchangeably.

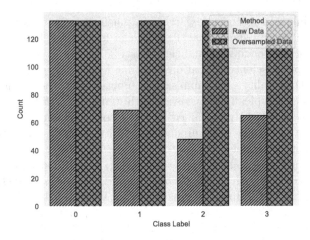

Fig. 1. The distribution of relevance scores before and after oversampling.

While educational programming datasets exist, such as Blackbox [5], FalconCode [11] or data from online judge tools such as HackerRank[3], they often focus on providing a grade for the submission's functionality, and do not contain any data on the documentation quality. CodeSearchNet does not focus on programming education and consists of code and documentation from professionally maintained open-source repositories. However, the large human-expert annotated subset provides an excellent source of ground truth to enable our research into how machine learning can be utilised to grade documentation quality, and to the best of our knowledge, no openly available educational dataset with documentation evaluation exists.

3.2   Pre-processing

Before training our machine learning models, we pre-processed the CodeSearchNet dataset. While the queries do not meet specific requirements for each language docstring requirements, the natural language queries provide a suitable equivalent for our purposes. We exclude any source code that contains non-ASCII characters to limit our model to classifying source code written in English and to limit issues with vectorisation using pre-trained large language models. Finally, we oversampled the minority classes by sampling with replacement, as the original distribution of classes was heavily skewed; see Fig. 1. Oversampling is a common way of balancing classes within a dataset to improve results when training a machine learning model. We chose not to undersample the data, another method for balancing classes within a dataset, as the size of the overall annotated dataset is relatively small.

---

[3] HackerRank: https://www.hackerrank.com/.

3.3 Vectorisation

Vectorisation transforms a string of tokens, typically a string of words forming a sentence into a vector, which is then passed to the machine learning model to train or predict a value. There are typically two common approaches to vectorisation: counting the distribution of tokens in a given set of documents or using pre-trained large language models to provide a vector for a specific token.

For counting the distribution of tokens in a given set of documents, we implemented both Bag Of Words and Term Frequency/Inverse Document Frequency (TFIDF), using the implementation provided by Scikit-Learn [22]. Bag of Words is the count of tokens within the training data, with the count of the words being used to generate embeddings to train our models. TFIDF is the proportion of the count of tokens in each string and the inverse of the document count [23].

In recent years, utilising pre-trained large language models has become one of the predominant methods for generating embedding vectors. These pre-trained models use a large dataset; for example, GPT-3 is trained with over 300 billion tokens [6]. The large dataset of pre-trained embeddings often increases accuracy, partially due to the lower likelihood of out-of-vocabulary errors: tokens that do not exist within the training data.

We used HuggingFace [27], which has a large collection of publicly available models, to implement pre-trained models to generate vector embeddings. We chose to use three commonly used pre-trained models, BERT [9], CodeBERT [10] GPT-3 [6]. These models are widely used in various applications from developing chatbots, such as ChatGPT, to code generation and summarization, in tools like GitHub CoPilot[4].

3.4 Model Development

After vectorisation, we trained a number of traditional machine learning models and fine-tuned large language models using 10-fold cross-validation and Optuna [4] for hyperparameter tuning. 10-fold cross-validation splits the training set into ten equal parts, or folds, of shuffled data. Nine of the ten folds are used to train the model, and one fold is withheld to validate the model's accuracy; this process is then repeated so each fold acts as the validation set. The idea of 10-fold cross-validation is to increase the training dataset size by not requiring withholding a large subset of validation data.

For our traditional machine-learning models, we used supervised learning models provided by SciKit-Learn [22], a comprehensive machine-learning library for Python. We trained a number of commonly used traditional machine learning models, including logistic regression, Bernoulli naive Bayes, k-nearest neighbours, random forest, and decision tree classifiers.

---

[4] GitHub CoPilot: https://github.com/features/copilot.

We chose to fine-tune two of the three pre-trained large language models we used to generate pre-trained embedding: BERT and CodeBERT, as these are some of the most common ones. We opted against fine-tuning a GPT 3.5 model as it was computationally too intensive. We use the embeddings from the corresponding pre-trained model in the vectorisation stage to fine-tune these models.

We explore how Low-Rank Adaption of Large Language Models (LoRA) can be used to fine-tune existing large language models. LoRA reduces the number of trainable parameters, which decreases the required GPU memory requirement and increases training throughput [14]. Using the same method as fine-tuning the full models, we fine-tune BERT and CodeBERT with LoRA. A summary of our experimentation parameters can be found in Table 1.

Table 1. A summary of our experiment parameters, all models were trained with the vectorisation methods stated, with Optuna optimising the hyperparameters.

Model Type	Vectorisation Method			Hyperparameters
	BERT	CodeBERT	GPT 3.5	
Logistic Regression	✓	✓	✓	
Bernoulli naive Bayes	✓	✓	✓	Smoothing: $\{0 - 1\}$
k-Nearest Neighbours	✓	✓	✓	n neighbours: $\{1 - 10\}$
Decision Tree	✓	✓	✓	Max Depth: $\{2 - 20\}$ Min Samples Leaf: $\{5 - 100\}$ Criterion: [Gini, Entropy]
Random Forest	✓	✓	✓	Max Depth: $\{2 - 32\}$x
BERT	✓			Learning Rate: $\{1 \times 10^{-6} - 1 \times 10^{-4}\}$ Batch Size: [16, 32] Epochs: [10, 50, 100]
CodeBERT		✓		Learning Rate: $\{1 \times 10^{-6} - 1 \times 10^{-4}\}$ Batch Size: [16, 32] Epochs: [10, 50, 100]
BERT with LoRA	✓			Learning Rate: $\{1 \times 10^{-6} - 1 \times 10^{-4}\}$ Batch Size: [16, 32] Epochs: [10, 50, 100] LoRA Rank: [8, 16, 32, 64] LoRA Modules: [query, value, key, dense]
CodeBERT with LoRA		✓		Learning Rate: $\{1 \times 10^{-6} - 1 \times 10^{-4}\}$ Batch Size: [16, 32] Epochs: [10, 50, 100] LoRA Rank: [8, 16, 32, 64] LoRA Modules: [query, value, key, dense]

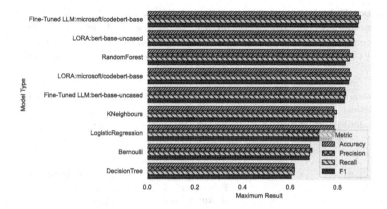

Fig. 2. The maximum results when testing the models against a withheld test set by model type. Precision, Recall and F1 are weighted by the number of true instances for each label.

4 Results

To train our traditional machine learning approaches, we used the traditional methods and the pre-trained large language models to generate the embeddings for the classification. For example, when fine-tuning CodeBERT, we used Code-BERT to tokenize and vectorise the input strings.

Figure 2 shows the maximum accuracy, weighted precision, recall and F1 scores for each of the types of models we trained, with the maximum being calculated after cross-validation and for each set of hyperparameters. Recall is a metric used to quantify the fraction of correctly classified positive instances; accuracy is the ratio of correct predictions over the total number of occurrences; precision is the ratio of correctly predicted positive instances over the total number of positive instances; and F1-Score is the harmonic mean between recall and precision values [13]. As we evaluate a multi-class classification problem, precision, recall and the F1 scores are average weighted by the number of true instances for each label, accounting for the label imbalance within our dataset.

Table 2 shows our top results, including vectorisation, model type and hyperparameters. Our top nine results used CodeBERT for vectorisation and classification with varying hyperparameters. As such, we chose to omit results four through nine, as the results were identical, with an accuracy of 89%. Other models that performed with an accuracy of more than 80% included Random Forest, with BERT embeddings, and a fine-tuned BERT model. Random Forest was the only traditional model to achieve an accuracy higher than 80%, with an accuracy of 87%. All our results can be found in our repository; see 'Data Availibility'.

Figure 3 shows the total runtime of the training and evaluation of our experiments as reported by Weights and Biases[5], grouped by model type. All models were trained using a high-performance cluster [16], with 32 CPU cores. LoRA and the fine-tuned approaches were additionally trained on a single Nvidia A100.

---

[5] Weights and Biases: https://wandb.ai/site.

Table 2. The 12 highest performing models, detailing the vectorisation method, classification method, hyperparameters and the performance of the models in terms of accuracy, weighted precision, recall, and F1-score. Results 4 to 9 have been omitted as they are all CodeBERT resulting in the same metric scores with varying hyperparameters.

Rank	Vectorisation Method	Classification Method	Hyperparameters	Acc.	Prec.	Recall	F1
1	CodeBERT	CodeBERT	Epochs: 50 Batch Size: 16 Learning Rate 5×10^{-5}	0.888	0.896	0.888	0.886
2	CodeBERT	CodeBERT	Epochs: 10 Batch Size: 16 Learning Rate 4×10^{-6}	0.888	0.896	0.888	0.886
3	CodeBERT	CodeBERT	Epochs: 50 Batch Size: 32 Learning Rate 7×10^{-5}	0.888	0.896	0.888	0.886
⋮	⋮	⋮	⋮	⋮	⋮	⋮	⋮
10	BERT	LoRA: BERT	Epochs: 10 Batch Size: 16 Learning Rate 1×10^{-5} LoRA Rank: 64 LoRA Modules: [query, value]	0.869	0.868	0.869	0.867
11	BERT	Random Forest	Max Depth = 12	0.850	0.865	0.850	0.834
12	BERT	LoRA: CodeBERT	Epochs: 100 Batch Size: 16 Learning Rate 8×10^{-5} LoRA Rank: 16 LoRA Modules: [value, dense]	0.850	0.856	0.850	0.848

5   Discussion

Overall, we showed that machine learning can be used to classify the relevance of natural language describing a corresponding function with an accuracy greater than 85%. While traditional machine learning approaches such as random forest can perform accurately, many traditional approaches fail to achieve an accuracy higher than 80%, with fine-tuned large language models often producing more accurate predictions.

Our most accurate model was a fine-tuned CodeBERT model with an accuracy of 89%. CodeBERT may have performed better than BERT, as it was trained to generate documentation/code pairs using the CodeSearchNet corpus. Meanwhile, BERT was trained for natural language tasks in English, not code-related.

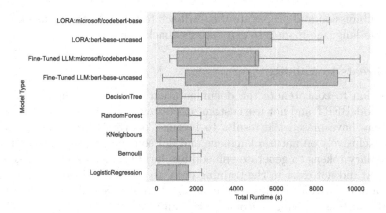

Fig. 3. The total runtime of each model's 10-fold cross-validation training and evaluation, as reported by Weights and Biases, grouped by model type. All approaches were trained on a high-performance cluster [16] with 32 CPU cores, and both the LoRA and Fine-Tuned approaches were additionally trained with a single Nvidia A100.

Our LoRA fine-tuned models performed better than most traditional models, with a LoRA BERT model having an accuracy of 87%. Though the LoRA models underperformed compared to the fully fine-tuned models, the required time to fine-tune and evaluate the model was significantly reduced on average, with the models completing processing approximately 30% faster for BERT and 20% faster for CodeBERT.

As large language models increase in parameter count, fully fine-tuning the parameters becomes less feasible [14], so techniques such as LoRA, which significantly reduce the computing power required to fine-tune an accurate model while sacrificing minimal performance in terms of accuracy, become necessary.

Our results show that a fine-tuned large language model can be used to accurately assign a grade for the relevance of a natural language description of a corresponding function. However, instructors should still validate the models' predictions using their judgement or allow for regrade requests. Regrade requests are common when automating grading using traditional approaches like unit tests or static analysis. The static analysis rules and test cases that instructors write can miss edge cases or not function as intended, providing an incorrect mark for students. While automated assessment tools do not always grade accurately, the benefits of faster feedback and decreased instructor workload, allowing instructors more time to support students, typically increase student satisfaction.

While our models can provide an accurate grade for the relevance of a documentation/code pair, they do not provide feedback on what was correct or what could be improved. Further research is required to determine how machine-learning approaches could be used to provide feedback to students – whether that is clustering similar submissions, allowing the instructor to provide feedback on a single submission and propagate the feedback to the rest of the cluster [17], or using large language models to generate feedback fully.

5.1 Threats to Validity

Our decision to concatenate the documentation and code strings in the same form as CodeBERT and not use custom processing steps for each large language model may have biased the results to favour CodeBERT. However, BERT is trained exclusively on natural language and not source code, so it has more out-of-vocabulary tokens to generate embeddings for. Out-of-vocabulary tokens are tokens that did not exist in the training data, so the pre-trained large language model cannot provide a known embedding. BERT greedily splits words into subwords, such as 'playing' becomes 'play' and '-ing', and uses these subwords to generate embeddings for out-of-vocabulary words [9].

The bias towards CodeBERT is more prominent when using GPT-3 as a tokenizer, as GPT-3 was trained on a dataset that contained both natural language and source code and was scraped from the internet. While there is a skew towards CodeBERT performing better than other large language models, it does not affect our overall results from this paper, showing that machine learning can be utilised to classify or grade the relevance of documentation of corresponding source code.

We trained our models on documentation and code provided by professionals and not students because, to our knowledge, no openly available education dataset contains annotations for the documentation quality. While the training data is written by professionals and not students in an educational setting, our results show that our methodology can be used to grade the relevance of documentation, and in the future, we aim to adapt our approach to a large educational dataset, similar to the dataset used to evaluate our current models against student submissions.

6   Conclusion

In this paper, we compared five traditional machine-learning approaches, two fine-tuned large language models and two adapter-based fine-tuned large language models to classify the relevance of natural language documentation to its corresponding source code. These documentation/code pairs were classified between zero and three, with zero being irrelevant and three being relevant, and the ground truth data is provided by human experts as part of the CodeSearch-Net corpus.

We train models with various vectorisation methods and perform hyperparameter tuning to train a model with an accuracy higher than 85%. The high accuracy of our model shows that a well-trained model could be used to grade the relevance of documentation to corresponding code. However, this is only a first step towards providing meaningful feedback to students, potentially by generating feedback or clustering similar submissions and propagating feedback from a single graded submission.

6.1 Future Work

In the future, we plan to explore how machine learning can be utilised to grade code quality by replicating our approach for this task and different tasks on an educational dataset we are developing, consisting of grades and feedback for correctness, maintainability, readability and documentation quality. We also plan to investigate how machine learning can be used to generate feedback and to develop a tool that provides grades and feedback for grading documentation quality using machine learning and static analysis.

Acknowledgements. We thank the King's College Teaching Fund for funding our study and CREATE [16] for providing the high-performance cluster we used to train and evaluate our models.

Data Availability Statement. All our raw data, data processing, model training and results can be found on GitHub (Data processing repository: https://github.com/ m-messer/Grading-Documentation-with-Machine-Learning).

References

1. Aggarwal, K., Singh, Y., Chhabra, J.: An integrated measure of software maintainability. In: Annual Reliability and Maintainability Symposium. 2002 Proceedings (Cat. No. 02CH37318), pp. 235–241 (2002). https://doi.org/10.1109/RAMS.2002. 981648
2. Aghajani, E., Nagy, C., Linares-Vásquez, M., et al.: Software documentation: the practitioners' perspective. In: Proceedings of the ACM/IEEE 42nd International Conference on Software Engineering, pp. 590–601. ICSE 2020. Association for Computing Machinery, New York, NY, USA (2020). https://doi.org/10.1145/3377811. 3380405
3. Aghajani, E., Nagy, C., Vega-Márquez, O.L., et al.: Software documentation issues unveiled. In: 2019 IEEE/ACM 41st International Conference on Software Engineering (ICSE), pp. 1199–1210 (2019). https://doi.org/10.1109/ICSE.2019.00122
4. Akiba, T., Sano, S., Yanase, T., et al.: Optuna: a next-generation hyperparameter optimization framework. In: Proceedings of the 25th ACM SIGKDD International Conference on Knowledge Discovery & Data Mining, KDD 2019, pp. 2623–2631. Association for Computing Machinery, New York, NY, USA (2019). https://doi. org/10.1145/3292500.3330701
5. Brown, N.C.C., Kölling, M., McCall, D., et al.: Blackbox: a large scale repository of novice programmers' activity. In: Proceedings of the 45th ACM Technical Symposium on Computer Science Education, SIGCSE 2014, pp. 223–228. Association for Computing Machinery, New York, NY, USA (2014). https://doi.org/10.1145/ 2538862.2538924
6. Brown, T., Mann, B., Ryder, N., et al.: Language models are few-shot learners. In: Larochelle, H., Ranzato, M., Hadsell, R., Balcan, M., Lin, H. (eds.) Advances in Neural Information Processing Systems, vol. 33, pp. 1877–1901. Curran Associates, Inc. (2020). https://proceedings.neurips.cc/paper%5Ffiles/paper/ 2020/file/1457c0d6bfcb4967418bfb8ac142f64a-Paper.pdf

7. Chen, M., Tworek, J., Jun, H., et al.: Evaluating large language models trained on code. arXiv preprint arXiv:2107.03374 (2021)
8. Clement, C.B., Drain, D., Timcheck, J., et al.: PyMT5: multi-mode translation of natural language and Python code with transformers. arXiv preprint arXiv:2010.03150 (2020)
9. Devlin, J., Chang, M.W., Lee, K., et al.: BERT: pre-training of deep bidirectional transformers for language understanding. arXiv preprint arXiv:1810.04805 (2019)
10. Feng, Z., Guo, D., Tang, D., et al.: CodeBERT: a pre-trained model for programming and natural languages. arXiv preprint arXiv:2002.08155 (2020)
11. de Freitas, A., Coffman, J., de Freitas, M., et al.: FalconCode: a multiyear dataset of Python code samples from an introductory computer science course. In: Proceedings of the 54th ACM Technical Symposium on Computer Science Education V. 1, SIGCSE 2023, pp. 938–944. Association for Computing Machinery, New York, NY, USA (2023). https://doi.org/10.1145/3545945.3569822
12. Gerdes, J.: Developing applications to automatically grade introductory visual basic courses. In: AMCIS 2017 Proceedings, August 2017. https://aisel.aisnet.org/amcis2017/ISEducation/Presentations/28
13. Hossin, M., Sulaiman, M.N.: A review on evaluation metrics for data classification evaluations. Int. J. Data Min. Knowl. Manage. Process (IJDKP) **5**, 1–11 (2015). https://doi.org/10.5121/ijdkp.2015.5201
14. Hu, E.J., et al.: LoRA: low-rank adaptation of large language models. arXiv preprint arXiv:2106.09685 (2021)
15. Husain, H., Wu, H.H., Gazit, T., et al.: CodeSearchNet challenge: evaluating the state of semantic code search. arXiv preprint arXiv:1909.09436 (2020)
16. King's College London: King's computational research, engineering and technology environment (CREATE) (2024). https://doi.org/10.18742/rnvf-m076
17. Koivisto, T., Hellas, A.: Evaluating CodeClusters for effectively providing feedback on code submissions. In: 2022 IEEE Frontiers in Education Conference (FIE), pp. 1–9 (2022). https://doi.org/10.1109/FIE56618.2022.9962751
18. LeClair, A., Haque, S., Wu, L., et al.: Improved code summarization via a graph neural network. In: Proceedings of the 28th International Conference on Program Comprehension, ICPC 2020, pp. 184–195. Association for Computing Machinery, New York, NY, USA (2020). https://doi.org/10.1145/3387904.3389268
19. Messer, M., Brown, N.C.C., Kölling, M., Shi, M.: Automated grading and feedback tools for programming education: a systematic review. ACM Trans. Comput. Educ. **24**(1), 1–43 (2024). https://doi.org/10.1145/3636515
20. Messer, M., Brown, N.C.C., Kölling, M., et al.: Machine learning-based automated grading and feedback tools for programming: a meta-analysis. In: Proceedings of the 2023 Conference on Innovation and Technology in Computer Science Education, vol. 1, pp. 491–497. ITiCSE 2023. Association for Computing Machinery, New York, NY, USA (2023). https://doi.org/10.1145/3587102.3588822
21. Muuli, E., et al.: Automatic assessment of programming assignments using image recognition. In: Lavoué, É., Drachsler, H., Verbert, K., Broisin, J., Pérez-Sanagustín, M. (eds.) EC-TEL 2017. LNCS, vol. 10474, pp. 153–163. Springer, Cham (2017). https://doi.org/10.1007/978-3-319-66610-5_12
22. Pedregosa, F., Varoquaux, G., Gramfort, A., et al.: Scikit-learn: machine learning in Python. J. Mach. Learn. Res. **12**, 2825–2830 (2011)
23. Sebastiani, F.: Machine learning in automated text categorization. ACM Comput. Surv. **34**(1), 1–47 (2002). https://doi.org/10.1145/505282.505283

24. Shi, E., Wang, Y., Du, L., et al.: On the evaluation of neural code summarization. In: Proceedings of the 44th International Conference on Software Engineering, ICSE 2022, pp. 1597–1608. Association for Computing Machinery, New York, NY, USA (2022). https://doi.org/10.1145/3510003.3510060

25. Treude, C., Middleton, J., Atapattu, T.: Beyond accuracy: assessing software documentation quality. In: Proceedings of the 28th ACM Joint Meeting on European Software Engineering Conference and Symposium on the Foundations of Software Engineering, ESEC/FSE 2020, pp. 1509–1512. Association for Computing Machinery, New York, NY, USA (2020). https://doi.org/10.1145/3368089.3417045

26. Walker, O., Russell, N.: Automatic assessment of the design quality of Python programs with personalized feedback. In: Proceedings of the 14th International Conference on Educational Data Mining, pp. 495–501 (2021)

27. Wolf, T., Debut, L., Sanh, V., et al.: HuggingFace's transformers: state-of-the-art natural language processing. arXiv preprint arXiv:1910.03771 (2020)

28. Zhang, J., Wang, X., Zhang, H., et al.: Retrieval-based neural source code summarization. In: Proceedings of the ACM/IEEE 42nd International Conference on Software Engineering, ICSE 2020, pp. 1385–1397. Association for Computing Machinery, New York, NY, USA (2020). https://doi.org/10.1145/3377811.3380383

Supporting Teaching-to-the-Curriculum by Linking Diagnostic Tests to Curriculum Goals: Using Textbook Content as Context for Retrieval-Augmented Generation with Large Language Models

Xiu Li[(✉)][iD], Aron Henriksson[(✉)], Martin Duneld, Jalal Nouri,
and Yongchao Wu

Stockholm University, NOD-huset, Borgarfjordsgatan 12, 16455 Stockholm, Sweden
{xiu.li,aronhen,xmartin,jalal,yongchao.wu}@dsv.su.se

Abstract. Using AI for automatically linking exercises to curriculum goals can support many educational use cases and facilitate teaching-to-the-curriculum by ensuring that exercises adequately reflect and encompass the curriculum goals, ultimately enabling curriculum-based assessment. Here, we introduce this novel task and create a manually labeled dataset where two types of diagnostic tests are linked to curriculum goals for Biology G7-9 in Sweden. We cast the problem both as an information retrieval task and a multi-class text classification task and explore unsupervised approaches to both, as labeled data for such tasks is typically scarce. For the information retrieval task, we employ state-of-the-art embedding model ADA-002 for semantic textual similarity (STS), while we prompt a large language model in the form of ChatGPT to classify diagnostic tests into curriculum goals. For both task formulations, we investigate different ways of using textbook content as a pivot to provide additional context for linking diagnostic questions to curriculum goals. We show that a combination of the two approaches in a retrieval-augmented generation model, whereby STS is used for retrieving textbook content as context to ChatGPT that then performs zero-shot classification, leads to the best classification accuracy (73.5%), outperforming both STS-based classification (67.5%) and LLM-based classification without context (71.5%). Finally, we showcase how the proposed method could be used in pedagogical practices.

Keywords: Teaching-to-the-Curriculum · Semantic Textual Similarity · Large Language Models · ChatGPT · Retrieval-Augmented Generation

1 Introduction

Curriculum is a structured plan that outlines the content, learning objectives, instructional methods, and assessments for a specific educational program [13],

A. M. Olney et al. (Eds.): AIED 2024, LNAI 14829, pp. 118–132, 2024.
https://doi.org/10.1007/978-3-031-64302-6_9

guiding instructors to achieve intended learning outcomes. Unlike teaching-to-the-test, which may prioritize rot memorization for exams and can result in superficial learning and teaching with a narrowed curriculum, teaching-to-the-curriculum encourages a broader exploration of topics, enhancing students' overall knowledge [1]. Therefore, in pedagogical practices, teaching-to-the-curriculum is essential, leading to more effective and lasting educational outcomes. However, the responsibility of aligning teaching materials with specific curriculum objectives currently rests solely on teachers, and fully implementing this approach requires challenging manual work. In light of these challenges, there is a need for automated support systems that can facilitate this process. Specifically, linking diagnostic test questions to curriculum goals can support many educational use cases such as curriculum-based assessment (CBA) and adaptive learning. For students, CBA directly addresses their understanding in relation to curriculum goals and allows for curriculum-oriented efforts to enhance their learning. For teachers, it allows them to gauge how well students have grasped the knowledge specified in the curriculum goals, while ensuring that instructional planning and diagnostic design are well aligned with curriculum goals [4]. Despite its vital role, there is currently little to no research exploring AI-based solutions for automatically linking learning materials to curriculum goals. Moreover, the techniques used for linking various types of learning materials are usually based on retrieval models, which filter items that are similar in terms of content features [11,16], and new techniques should be explored and adapted to enhance this process.

The rise of large language models (LLMs) like BERT [5] and GPT [14] has revolutionized natural language processing (NLP) and AI research, showcasing remarkable capabilities in tasks requiring nuanced language understanding and generation. The recent release of ChatGPT[1] has further fueled interest in LLMs. Despite their ability to acquire knowledge from vast datasets during pre-training and adapt to specific tasks through fine-tuning, LLMs still exhibit limited performance on knowledge-intensive or specialized tasks, primarily stemming from a lack of integration with external knowledge base [7]. To address the above issues, explicitly incorporating knowledge into pre-trained language models has been an emerging trend. Knowledge Enhanced Pre-trained Language Models (KE-PLMs) can employ knowledge graph (KG)-based and retrieval-based methods for this purpose [7]. KG-based KE-PLMs can inject structured knowledge from a KG into the transformer architecture [15] or incorporate entity embeddings using fusion strategies by either layerwise local/global fusion [18,19] or early/late fusion using traditional ensemble methods [10]. Retrieval-Augmented Generation (RAG) is a typical retrieval-based KE-PLM. It leverages external knowledge retrieved from specific domains or proprietary knowledge bases with LLM without retraining, and therefore optimizes the LLM output effectively [7,9]. Wu et al. [17] proposed a RAG-based educational question answering (EQA) system that enhances the EQA performance while providing retrieved educational context as an answer-grounded reference, thereby improving the trustworthiness of EQA systems. Lewis et al. [9] developed the RAG model by combining

---

[1] https://openai.com/blog/chatgpt. Last accessed 28 Jan. 2024.

the question with the top-K relevant Wikipedia article chunks retrieved from a retriever and feeding them into a BART generator. A Fusion-in-Decoder model [8] is similar to RAG, but feeds the encoder with the concatenation of the question and each top-K retrieved relevant passage, and then the decoder performs attention over the fusion of the resulting representations. These studies showcase the robust capabilities of the RAG model in enhancing the performance of NLP tasks.

In this paper, we introduce the task of automatically linking diagnostic test questions to curriculum goals and, to that end, create a manually labeled dataset for evaluation. We then explore and compare two unsupervised approaches since labeled data for this task is often not available and can be cumbersome and costly to create. We cast the task as an (i) information retrieval (IR) task and a (ii) multi-class text classification task. For both task formulations, we investigate different ways of using textbook content as a pivot for linking diagnostic test questions to curriculum goals, where textbook content that is relevant to a given question is retrieved automatically to provide additional context and help identify the most relevant curriculum goal. For the IR task, we use a pre-trained embedding model (ADA-002) for calculating semantic text similarity (STS), while for the classification task, we also explore the use of a generative LLM (ChatGPT) in a zero-shot setting in addition to the STS-based methods. The main contributions of this paper are listed below.

– We introduce the novel task of automatically linking diagnostic test questions to curriculum goals and create a manually labeled dataset for evaluation.
– We explore and compare two unsupervised approaches to this task, formulated either as an IR task or as a text classification task. For both tasks, we propose using textbook content as a pivot for providing additional context and improving performance on the task; to that end, we investigate automatically retrieving both narrow and broad contexts.
– We show that using an LLM (ChatGPT) in a zero-shot setting outperforms STS-based approaches; however, the best results are obtained when combining the two approaches in a RAG model, whereby STS is used for retrieving and providing the LLM with relevant textbook content.

2   Data and Methods

We first describe the process of creating a manually labeled dataset corresponding to the task of linking diagnostic test questions to curriculum goals, and report inter-annotator agreement (IAA) and Cohen's kappa as well as descriptive statistics for the resulting dataset. We then provide details of the various methods used for automatically identifying the most relevant curriculum goal for each diagnostic question, and finally describe different ways of using textbook content as a pivot for providing additional context to each question in an effort to improve linking accuracy.

2.1   Data and Annotation Process

The data used in this study comes from two sources. One source is the official curriculum data of Biology G7-9 from the Swedish National Agency for Education[2]. In the Swedish education system, the curriculum for a course is the official central learning objectives. The other source is from a Swedish digital textbook - Biology G7-9. We use two parts from the textbook: diagnostic tests and the instructive textbook content. Note that the curriculum goals, diagnostic tests and instructive textbook content are all in Swedish.

The curriculum data includes 11 curriculum goals. The textbook is divided into six chapters, each of which is divided into further sections and subsections, i.e. it follows a hierarchical structure of Chapter→Section→Subsection, which is what you would typically see in a Table of Content structure (TOCS) in paper textbooks. The diagnostic tests are designed to examine comprehensive mastery of each chapter. We use two types of diagnostic questions: (i) concept questions and (ii) mixed questions. Concept questions typically concern specific biological concepts, while mixed questions tend to be more comprehensive, and usually assess a broader scope of knowledge points than concept questions. The diagnostic questions come in various forms, e.g. fill-in-the-blank, multiple choice, and matching questions. In this study, we use diagnostic questions along with the correct answers. The statistics of the textbook data and curriculum data are described in Tables 1 and 2.

Table 1. Statistics of the textbook data.

Instructive Textbook Content	No.
Chapters	6
Sections	56
Subsections	585
Paragraphs	5834
Tokens per paragraph (avg.)	22
Diagnostic Questions	No.
Concept questions	100
Mixed questions	100

Table 2. Statistics of the curriculum data.

Curricula	No.
Goals	11
Study areas	3
Sentences	20
Tokens	324

To enable evaluation of the proposed methods, we first needed to create a manually labeled dataset where diagnostic questions are linked to the curriculum goals. Two annotators – one with a biotechnology background and one with a math background – were asked to select the most relevant curriculum goal for each diagnostic question. This process was carried out independently by the annotators in order not to influence each other. We then calculated

---

[2] https://www.skolverket.se/undervisning/grundskolan/laroplan-och-kursplaner-for-grundskolan/kursplaner-for-grundskolan. Last accessed 28 Jan. 2024.

IAA between the two annotators and obtained an agreement rate of 93% and a Cohen's kappa score of 0.91, indicating very high agreement [12]. To resolve disagreements between the annotators on 14 questions, we invited two experts to discuss and adjudicate, where a final agreement was reached through consensus.

2.2 Task Formulation

The task of automatically linking diagnostic test questions to curriculum goals is here formulated in two different ways, or in two different settings: (i) as an IR task, and (ii) as a text classification task.

In the IR setting, given a query corresponding to a diagnostic question, the task is to produce a ranked list of curriculum goals. The aim, then, is to rank the relevant curriculum goal as high as possible. For this task, we use pre-trained embeddings to represent both query and document vectors, and use STS to produce a similarity score which is used to rank the curriculum goals. The details of this approach are described in Sect. 2.3. For evaluation of the STS approaches in the IR task formulation, we report Recall@3 and mean reciprocal rank (MRR), where Recall@3 measures the proportion of queries where the correct curriculum goal is within the top 3 ranked outputs, while MRR gauges the model's ability to rank the relevant curriculum goal as high as possible [11].

In the text classification setting, given a diagnostic question, the task is to assign the class label corresponding to the relevant curriculum goal. For this task, we also evaluate the best-performing STS model in a classification setting by assigning the label with the highest similarity score to the embedded diagnostic question. We compare STS-based classification to prompting an LLM to carry out the classification in a zero-shot setting, and propose a RAG model to improve the performance. The details of the LLM-based classifiers are described in Sect. 2.4. For evaluation of the classification models, we report accuracy, which measures the proportion of correctly classified instances.

2.3 Semantic Textual Similarity

Semantic textual similarity measures the degree of similarity in meaning between two texts [2] and involves the following two steps: (i) transforming text into embedding vectors and (ii) calculating the distance between the embedding vectors as a proxy for semantic similarity. For (i), we use the pre-trained embedding model ADA-002[3] (`text-embedding-ada-002`) due to its demonstrated state-of-the-art performances for both English and Swedish STS-based tasks [6,10], and convenience for its embeddings-as-a-service API. The pre-trained embedding model is to retrieve embeddings for diagnostic questions, curriculum goals, as well as additional context in the form of textbook content. For (ii), we employ cosine similarity $Sim(v, u) = \frac{v \cdot u}{|v||u|}$ to calculate the similarity between texts, where v and u are the embedding vectors. We define $\vec{q_i} = \text{Embed}(q_i)$, where $\vec{q_i}$ is the embedding vector of diagnostic question $q_i \in Q$, Q is the diagnostic question

---

[3] https://platform.openai.com/docs/guides/embeddings. Last accessed 28 Jan. 2024.

set, and $\vec{c_j} = \mathrm{Embed}(c_j)$, where $\vec{c_j}$ is the embedding vector of context $c_j \in C$, C is the textbook content, and $\vec{y_k} = \mathrm{Embed}(y_k)$, where $\vec{y_k}$ is the embedding vector of a curriculum goal $y_k \in Y\{y_1, ..., y_k, ..., y_{11}\}$, Y is the set of curriculum goals.

In this study, we employ an STS-based approach for both formulations: (i) as a regular retrieval model in the IR task and (ii) as a classifier by assigning the label corresponding to the top-ranked curriculum goal. Furthermore, for both task formulations, we explore using textbook content as a pivot for linking diagnostic questions to curriculum goals and, to that end, STS is used for retrieving textbook content that is relevant for a given diagnostic question. See Sect. 2.5 for details regarding different ways of using textbook content to provide context – with varying scope – to diagnostic questions.

STS-based context retrieval is carried out by retrieving the textbook content with the highest similarity score to the diagnostic question, i.e. retrieves context $c_{max} \in C$ that approximates $\arg\max[Sim(\vec{c_j}, \vec{q_i}), c_j \in C]$ for q_i.

STS-based retrieval involves retrieving the curriculum goal $y_{max} \in Y$ according to the relevancy, in terms of which position the correctly labeled curriculum resides in the ranking list $Y^{\mathrm{ranked}}\{y_1^{\mathrm{ranked}}, ..., y_k^{\mathrm{ranked}}, ..., y_{11}^{\mathrm{ranked}}\}$. The query can be the diagnostic question, or a concatenation with textbook context.

STS-based classification. When *classifying without context*, the label corresponding to the curriculum goal with the highest similarity score to the diagnostic question is assigned. That is, q_i is classified to curriculum goal $y_{max} \in Y$ by approximating $\arg\max[Sim(\vec{q_i}, \vec{y_k}), y_k \in Y]$ for q_i. When *classifying with context*, the label corresponding to the curriculum goal with the highest similarity score to the context-enhanced diagnostic question representation is assigned. That is, q_i is classified to curriculum goal $y_{max} \in Y$ by approximating $\arg\max[Sim(\hat{c}_{max}, \vec{y_k}), y_k \in Y]$, where $\hat{c}_{max}$ is a context representative variation of c_{max}.

2.4 Large Language Model Classification

In addition to STS-based classification, we prompt an LLM to classify diagnostic questions into one of eleven curriculum goals in a zero-shot setting, i.e. without access to labeled examples for in-context learning. We hence regard the LLM-based classifier as a zero-shot one-number generator. The model only requires a prompt message with instructions, and is prompted to respond with a predicted class, i.e. the index number of the most relevant curriculum goal for a given diagnostic question. We further propose a RAG model that uses STS-based IR for retrieving relevant context to the diagnostic question in the form of textbook content, and compare it to the LLM without access to any additional context. The probability for the predicted curriculum number y_k follows the probability of the generator $p_\theta(y_k|q_i)$ without context, and $p_\eta(c_{max}|q_i) \cdot p_\theta(y_k|q_i, c_{max})$ with context for the RAG-based classifier [9]. In this study, we use ChatGPT as the LLM, specifically `gpt-3.5-turbo-0613` (GPT-3.5-TURBO)[4], which is the most

---

[4] https://platform.openai.com/docs/models. Last accessed 28 Jan. 2024.

capable and cost effective model in the GPT-3.5 family. We set the temperature=0 for more consistent and determined responses. Below, we describe the construction of the prompt and the proposed RAG model.

Prompt Message Construction. The prompt message contains detailed instructions for the task and specifies what the input and output are in order to prompt the LLM to understand the classification task and output the most relevant curriculum goal from $Y\{y_1, ...y_k, ..., y_{11}\}$ for diagnostic question q_i, either with or without textbook context C. We follow OpenAI's recommendations for prompt engineering regarding detailed and clear instructions, specified output format, and smart usage of delimiters[5]. As a result, we design our prompt message in two parts: system message and user message. The system message contains the high-level general instructions and task description, as well as the curriculum goals which are delimited by a specified delimiter. The user message is the query that contains specific examples with or without the retrieved contexts as input fed into the model and expected output structure.

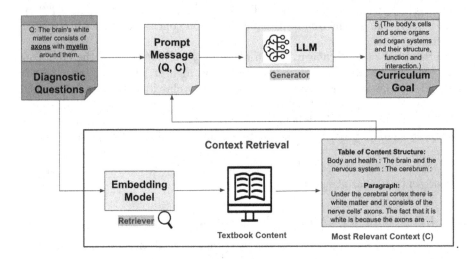

Fig. 1. Overall workflow of the proposed RAG-LLM for linking diagnostic questions to curriculum goals. The examples have been translated from Swedish.

RAG Model. This leverages *STS-based context retrieval* to retrieve relevant context or knowledge from external sources and then generates responses based on the retrieved information along with the rest of the prompt message. Therefore, a RAG model has two components: (i) a retriever and (ii) a generator. The workflow of our approach with the RAG-based LLM is illustrated in Fig. 1.

---

[5] https://platform.openai.com/docs/guides/prompt-engineering/. Last accessed 28 Jan. 2024.

The retriever is the component retrieving relevant and contextually rich information from a knowledge base to augment the generation of responses. A typical retriever aims to select the top-K relevant documents related to a query, and often concatenate them either before or after feeding the decoder (generator) [8,9]. In this study, we leverage STS-based context retrieval and the knowledge base is in the form of an instructive textbook. Due to its special content structure, we investigate the retrieval of context at two levels: paragraph-level and subsection-level (see Sect. 2.5). Specifically, we retrieve the most relevant textbook content, i.e., top-1, for a given diagnostic question. The retrieved textbook content acts as context, whether utilized independently or in conjunction with other pertinent texts such as the question and the corresponding TOCS, providing retrieval-augmented context for the generator.

The generator is responsible for producing responses, completions, or outputs based on the retrieved information. The generator takes into account the retrieval-augmented context from the retriever in order to generate contextually relevant responses. In our study, we use ChatGPT as the generator model, which is prompted to classify the diagnostic question into the most relevant curriculum goal, i.e. the generator is a multi-class classifier.

2.5   Using Textbook Context as a Pivot for Improved Linking

In addition to linking diagnostic questions directly to curriculum goals – whether in an IR or classification setting – we propose the use of textbook content as a pivot for providing additional context to diagnostic questions, aiming to improve performance on the task. To that end, we investigate STS-based context retrieval of textbook content with varying context scope, specifically *paragraph-level context* (narrow scope) and *subsection-level context* (broad scope). Furthermore, information such as TOCS and the inclusion of the diagnostic question itself can also be valuable when combined with paragraph or subsection context. To retrieve the embeddings for subsection-level context, we take the average of all the paragraph embeddings of a subsection or retrieve the embeddings of the whole subsection text. The respective maximum context window sizes for ADA and ChatGPT are sufficiently large to incorporate the whole subsection without risk of truncation (max token of ADA = 8191, GPT-3.5-TURBO = 4096; longest subsection = 1758, longest prompt = 528). The various context definitions are provided below.

Narrow context: paragraph-level textbook content
- The most relevant paragraph (denoted P)
- Concatenate "TOCS" and "the most relevant paragraph" (denoted T+P)
- Concatenate "diagnostic question" and "the most relevant paragraph" (denoted Q+P)
- Concatenate "diagnostic question", "TOCS" and "the most relevant paragraph" (denoted Q+T+P)

Broad context: subsection-level textbook content
- The most relevant subsection, i.e. entire subsection text (denoted S)
- Concatenate "TOCS" and "the most relevant subsection" (denoted T+S)
- Average paragraph embeddings of subsection (denoted AVG P of S)
- Average "TOCS + paragraph" embeddings of subsection (denoted AVG T+P of S)
- Concatenate "diagnostic question" and "the most relevant subsection" (denoted Q+S)
- Concatenate "diagnostic question", "TOCS" and "the most relevant subsection" (denoted Q+T+S)

3 Results

We present the results for the two task formulations, followed by an error analysis. Then, we showcase two practical use cases that build on the proposed results.

3.1 Information Retrieval Task

Table 3 shows the results of using the STS-based approach on the IR task.

Table 3. Results of the information retrieval task. (**Bold**: max. <u>Underline</u>: min.)

Context Level	Representation of diagnostic question	Concept Recall@3	Mixed Recall@3	Mean	Concept MRR	Mixed MRR	Mean
No Context	Q	0.860	0.780	<u>0.820</u>	0.713	0.734	<u>0.724</u>
Narrow Scope	P	0.870	0.840	0.855	0.758	0.769	0.763
	T+P	0.880	0.920	0.900	0.761	0.812	0.786
	Q+ P	0.910	0.870	0.890	0.779	0.760	0.769
	Q+T+P	0.920	0.900	0.910	0.754	0.775	0.765
Broad Scope	AVG P of S	0.880	0.900	0.890	0.762	0.792	0.777
	AVG T+P of S	0.860	**0.940**	0.900	0.776	**0.814**	**0.795**
	S	0.890	0.910	0.900	0.774	0.804	0.789
	T+S	0.910	0.920	**0.915**	0.773	0.807	0.790
	Q+S	**0.930**	0.900	**0.915**	**0.790**	0.800	**0.795**
	Q+T+S	0.910	0.890	0.900	0.771	0.792	0.782

Notation: Q = Diagnostic question. P = Paragraph of the textbook. S = Subsection of the textbook. T = Table of Content Structure of the textbook.

The Recall@3 and MRR results show that the STS-based approach consistently benefits from incorporating additional context from the textbook. The best results are all within the block of having textbook subsection as context, i.e. the broader contextual scope appears to enhance the performance of the STS-based IR model, helping it to rank the most relevant curriculum goal for the diagnostic question in a high position. Among the subsection-level context solutions, both AVG paragraph embeddings and the embedding of the entire subsection text yield top results, where using the entire subsection (Q+S) for embeddings yields the best performance w.r.t. both Recall@3 and MRR, while AVG paragraph embeddings (AVG T+P of S) achieve only the best MRR. The overall trend indicates that STS-based models benefit from incorporating context in the form of textbook content and that broader-scope context contributes to better performance compared to narrow-scope context.

3.2 Classification Task

Table 4 shows the results on the classification task, for both the zero-shot LLM-based classifiers and the STS-based classifiers, here evaluated using accuracy.

Table 4. Results of the classification task. (**Bold**: max. <u>Underline</u>: min.)

Classifier	Context Level	Context	Concept	Mixed	Mean
			Accuracy	Accuracy	
STS	No Context	–	55.0%	61.0%	<u>58.0%</u>
		P	64.0%	67.0%	65.5%
	Narrow	T+P	62.0%	69.0%	65.5%
	Scope	Q+ P	65.0%	63.0%	64.0%
		Q+T+P	61.0%	64.0%	62.5%
		AVG P of S	63.0%	67.0%	65.0%
		AVG T+P of S	65.0%	**70.0%**	**67.5%**
	Broad	S	64.0%	69.0%	66.5%
	Scope	T+S	64.0%	69.0%	66.5%
		Q+S	**66.0%**	69.0%	**67.5%**
		Q+T+S	63.0%	68.0%	65.5%
LLM	No context	–	72.0%	71.0%	71.5%
	Narrow	P	74.0%	**72.0%**	73.0%
	Scope	T+P	**76.0%**	71.0%	**73.5%**
	Broad	S	63.0%	62.0%	<u>62.5%</u>
	Scope	T+S	65.0%	70.0%	67.5%

Notation: P = Paragraph of the textbook. S = Subsection of the textbook. T = Table of Content Structure of the textbook.

As can be seen, the LLM-based approach yields overall superior performance to the STS-based approach. This demonstrates the power of the LLM-based model – here ChatGPT – compared to the embedding models in the classification formulation of the task. Moreover, the proposed RAG-based LLM outperforms the vanilla LLM, i.e. without access to any context. Specifically, the RAG model, where STS-based retrieval is used for retrieving the most relevant paragraph from the textbook and concatenated with the corresponding TOCS in order to augment the context for the LLM-based classifier leads to the best classification accuracy (73.5%).

Regarding the scope of the retrieved context, adding textbook context to the LLM-based classifier demonstrates that it benefits only from narrow, paragraph-level textbook context, and, in fact, deteriorates with the inclusion of broad, subsection-level context. This indicates that adding broader context does not help, but rather confuses, the LLM-based classifier.

When cross-comparing Tables 3 and 4, the results of STS-based retrieval and classification demonstrate a consistent pattern across metrics in terms of the best and worst performers. Additionally, the inclusion of TOCS in the context on average enhances LLM-based classification compared to the corresponding equivalent scope of context without TOCS. However, the outcome of this is unclear for STS-based IR and classification, as Recall@3, MRR, and accuracy show inconsistent results.

3.3 Error Analysis

Our evaluation of the curriculum linking system is contingent on the performance of two components: a context retriever and a classifier. The classifier takes in the output of the retriever to progress further. Therefore, the accuracy of the prediction for the entire system is based on the respective accuracy of the two components. However, the evaluation results reveal only the performance of the whole workflow and do not provide insights into the performance of each component or what the root cause of an error is. Specifically, we are unable to estimate the performance of the context retriever as we lack ground truth data and manually creating was deemed unfeasible. Instead, we focus the analysis on the factors that may reflect the impact of context retriever, and especially evaluating the robustness of the RAG system against noise. We use the results from the best model, i.e. the RAG-based ChatGPT classifier, enhanced with paragraph-level context, and compare its results to those of the ChatGPT-based classifier without context. Through manual inspection, we find that there are in total 29 diagnostic questions for which adding context to the ChatGPT-based classifier allows it to perform the classification correctly compared to no context. Among these 29 questions, there are 3 cases where the context retriever provides wrong contexts, yet these contexts still contribute to enhancing the classification results. Additionally, 26 diagnostic questions are misclassified after the inclusion of context. Within this set, 4 cases involve retrieved contexts that are deemed irrelevant. This indicates that irrelevant contexts can unexpectedly enhance accuracy, while relevant contexts can harm classification. This phenomenon is surprising, but we find other studies [3, 17] where the same phenomenon has been observed as well.

In order to examine how the classification results are distributed across curriculum goals, we demonstrate the results of the best model mentioned above in a confusion matrix in Fig. 2. The blue highlighted cells are the number and the percentage of diagnostic questions that are correctly classified. We see that diagnostic questions are not evenly distributed for each curriculum goal. Some

		Curriculum goal number										
		1	2	3	4	5	6	7	8	9	10	11	Total
Curriculum goal number	1	40 (83%)	1 (2%)	1 (2%)	1 (2%)	3 (6%)	1 (2%)	0 (0%)	0 (0%)	0 (0%)	1 (2%)	0 (0%)	48
	2	1 (50%)	1 (50%)	0 (0%)	0 (0%)	0 (0%)	0 (0%)	0 (0%)	0 (0%)	0 (0%)	0 (0%)	0 (0%)	2
	3	5 (19%)	0 (0%)	16 (62%)	3 (12%)	2 (8%)	0 (0%)	0 (0%)	0 (0%)	0 (0%)	0 (0%)	0 (0%)	26
	4	0 (0%)	0 (0%)	0 (0%)	5 (100%)	0 (0%)	0 (0%)	0 (0%)	0 (0%)	0 (0%)	0 (0%)	0 (0%)	5
	5	8 (12%)	1 (1%)	3 (4%)	1 (1%)	49 (73%)	0 (0%)	4 (6%)	1 (1%)	0 (0%)	0 (0%)	0 (0%)	67
	6	0 (0%)	2 (22%)	0 (0%)	0 (0%)	1 (11%)	6 (67%)	0 (0%)	0 (0%)	0 (0%)	0 (0%)	0 (0%)	9
	7	0 (0%)	0 (0%)	1 (14%)	1 (14%)	1 (14%)	0 (0%)	3 (43%)	1 (14%)	0 (0%)	0 (0%)	0 (0%)	7
	8	2 (10%)	0 (0%)	0 (0%)	1 (5%)	4 (20%)	0 (0%)	0 (0%)	13 (65%)	0 (0%)	0 (0%)	0 (0%)	20
	9	0 (0%)	0 (0%)	1 (8%)	0 (0%)	0 (0%)	0 (0%)	0 (0%)	0 (0%)	11 (92%)	0 (0%)	0 (0%)	12
	10	0 (0%)	0 (0%)	0 (0%)	0 (0%)	0 (0%)	0 (0%)	0 (0%)	0 (0%)	1 (33%)	2 (67%)	0 (0%)	3
	11	0 (0%)	0 (0%)	0 (0%)	0 (0%)	0 (0%)	0 (0%)	0 (0%)	0 (0%)	0 (0%)	0 (0%)	1 (100%)	1
	Total	56	5	22	12	60	7	7	15	12	3	1	200

Fig. 2. Confusion matrix for the classification results of the best model (RAG-LLM). (Color figure online)

curriculum goals contain only very few questions in the diagnostic test, e.g. curriculum goal 11 has only one diagnostic question. This implies that curriculum goals are empathized with different weights in the diagnostic test. We bold highlight the count/percentage of the curriculum goals that contain more than 20 questions and the misclassification rates are more than 10%. We see that the RAG-based classifier is confused when classifying to curriculum goal 3, where 19% are misclassified to curriculum 1, and 12% are misclassified to curriculum 4. Similarly, 20% questions that should be classified to curriculum 8 are misclassified to curriculum 5. We manually check the curriculum goals of these, and conclude that these curriculum goals are to some extent close in meaning, or even overlap. For example, both curriculum goal 3 and 4 cover knowledge around ecosystems. We also experienced the difficulty when annotating for these curriculum goals, which is part of the reason that caused the original 14 annotation disagreement. We thus conclude that as a take-away from this analysis, a good curriculum goal design plays an important role in this teaching-to-the-curriculum auto-linking task. We propose establishing a distinct delineation between curriculum goals to avoid any overlap or confusion in the knowledge scope.

3.4 Use Cases

There are many applications that can benefit from the results of the RAG-LLM that automatically links questions to curriculum goals. One use case involves including such curriculum-oriented learning analytics of students' performances in teacher dashboards. This component specifically focuses on curriculum-based behavior, and provides curriculum-oriented insights into students' progress, achievements, and overall class performance. Figure 3 is an example to show the performance of the whole class in terms of whether students answered correctly on the previous exercises/quizzes, and the performance is explicitly mapped to curriculum goals. With insights from the analytical graph, the teachers may adapt their teaching plan according to the mastery level of each curriculum goal.

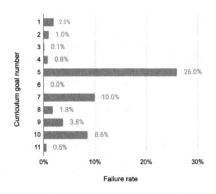 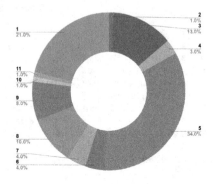

Fig. 3. Student performance: failure rate for different curriculum goals (mocked up)

Fig. 4. Distribution of diagnostic questions for 11 curriculum goals in a 100-question comprehensive diagnostic test for Biology G7-9.

In addition, when teachers design a diagnostic test, the distribution of the diagnostic questions with respect to curriculum goals can be visualized, as in Fig. 4, so that teachers get an overview of how well the test aligns with the curriculum goals, e.g. if and to what extent it encompasses the curriculum goals, or whether it focuses on certain curriculum goals. This can also be applied to individual students, providing insights into curriculum-oriented study behavior for adaptive learning purposes. Furthermore, linking questions to curriculum goals is also the foundation for curriculum-oriented content recommendation.

4 Conclusions

Our study introduced and addressed the novel task of automatically linking diagnostic test questions to curriculum goals, demonstrating its significance in supporting teaching-to-the-curriculum. We approached the task through two formulations of tasks: IR and classification. By utilization of textbook content as a pivot for additional context, we show that a RAG model, combining STS and LLM in a zero-shot setting, achieves the best classification accuracy, outperforming both the vanilla LLM without context and the STS-based approach. The results underscore the potential of AI in automating the linkage of learning materials to curriculum goals, offering valuable insights for educational use cases such as CBA and adaptive teaching and learning. Directions for future work include encompassing multiple disciplines and scaling the dataset to enable few-shot learning for improved accuracy, and investigating using LLM to auto-generate curriculum goal-oriented diagnostic tests. Furthermore, evaluations involving teachers and conducting interviews can provide further insights into the practical applicability of the proposed method.

References

1. Black, P., Wiliam, D.: Assessment and classroom learning. Assess. Educ. Principles Policy Pract. **5**(1), 7–74 (1998). https://doi.org/10.1080/0969595980050102
2. Cer, D., Diab, M., Agirre, E., Lopez-Gazpio, I., Specia, L.: SemEval-2017 Task 1: semantic textual similarity-multilingual and cross-lingual focused evaluation. arXiv preprint arXiv:1708.00055 (2017)
3. Cuconasu, F., et al.: The power of noise: redefining retrieval for RAG systems. arXiv preprint arXiv:2401.14887 (2024)
4. Darling-Hammond, L., Adamson, F., Abedi, J.: Beyond basic skills: the role of performance assessment in achieving 21st century standards of learning. In: International Conference on Applications of Natural Language to Information Systems, p. 52. Stanford Center for Opportunity Pollcy in Education (2010)
5. Devlin, J., Chang, M.W., Lee, K., Toutanova, K.: BERT: pre-training of deep bidirectional transformers for language understanding. arXiv:1810.04805 (2018)
6. Greene, R., Sanders, T., Weng, L., Neelakantan, A.: New and improved embedding model (2022)
7. Hu, L., Liu, Z., Zhao, Z., Hou, L., Nie, L., Li, J.: A survey of knowledge enhanced pre-trained language models. IEEE Trans. Knowl. Data Eng. **36**(4), 1413–1430 (2023)
8. Izacard, G., Grave, E.: Leveraging passage retrieval with generative models for open domain question answering. arXiv preprint arXiv:2007.01282 (2020)
9. Lewis, P., et al.: Retrieval-augmented generation for knowledge-intensive NLP tasks. Adv. Neural. Inf. Process. Syst. **33**, 9459–9474 (2020)
10. Li, X., Henriksson, A., Duneld, M., Nouri, J., Wu, Y.: Evaluating embeddings from pre-trained language models and knowledge graphs for educational content recommendation. Future Internet **16**(1), 12 (2024)
11. Li, X., Henriksson, A., Nouri, J., Duneld, M., Wu, Y.: Linking Swedish learning materials to exercises through an AI-enhanced recommender system. In: Milrad, M., et al. (eds.) Methodologies and Intelligent Systems for Technology Enhanced Learning, 13th International Conference, MIS4TEL 2023. LNNS, vol. 764, pp. 96–107. Springer, Cham (2023). https://doi.org/10.1007/978-3-031-41226-4_10
12. McHugh, M.L.: Interrater reliability: the kappa statistic. Biochemia medica **22**(3), 276–282 (2012)
13. Posner, G.J., Rudnitsky, A.N.: Course Design: A Guide to Curriculum Development for Teachers. Longman Publishers (1997)
14. Radford, A., Narasimhan, K., Salimans, T., Sutskever, I., et al.: Improving language understanding by generative pre-training (2018)
15. Ri, R., Yamada, I., Tsuruoka, Y.: mLUKE: the power of entity representations in multilingual pretrained language models. arXiv preprint arXiv:2110.08151 (2021)
16. Tarus, J.K., Niu, Z., Mustafa, G.: Knowledge-based recommendation: a review of ontology-based recommender systems for e-learning. Artif. Intell. Rev. **50**, 21–48 (2018)

17. Wu, Y., Henriksson, A., Duneld, M., Nouri, J.: Towards improving the reliability and transparency of ChatGPT for educational question answering. In: Viberg, O., Jivet, I., Muñoz-Merino, P., Perifanou, M., Papathoma, T. (eds.) Responsive and Sustainable Educational Futures. EC-TEL 2023. LNCS, vol. 14200, pp. 475–488. Springer, Cham (2023). https://doi.org/10.1007/978-3-031-42682-7_32
18. Zhao, Q., Lei, Y., Wang, Q., Kang, Z., Liu, J.: Enhancing text representations separately with entity descriptions. Neurocomputing **552**, 126511 (2023)
19. Zhong, Q., Ding, L., Liu, J., Du, B., Jin, H., Tao, D.: Knowledge graph augmented network towards multiview representation learning for aspect-based sentiment analysis. IEEE Trans. Knowl. Data Eng. **35**, 10098–10111 (2023)

VerAs: Verify Then Assess STEM Lab Reports

Berk Atil[✉], Mahsa Sheikhi Karizaki, and Rebecca J. Passonneau[✉]

The Pennsylvania State University, University Park 16801, USA
{bka5352,mfs6614,rjp49}@psu.edu

Abstract. With an increasing focus in STEM education on critical thinking skills, science writing plays an ever more important role. A recently published dataset of two sets of college level lab reports from an inquiry-based physics curriculum relies on analytic assessment rubrics that utilize multiple dimensions, specifying subject matter knowledge and general components of good explanations. Each analytic dimension is assessed on a 6-point scale, to provide detailed feedback to students that can help them improve their science writing skills. Manual assessment can be slow, and difficult to calibrate for consistency across all students in large enrollment courses with many sections. While much work exists on automated assessment of open-ended questions in STEM subjects, there has been far less work on long-form writing such as lab reports. We present an end-to-end neural architecture that has separate verifier and assessment modules, inspired by approaches to Open Domain Question Answering (OpenQA). VerAs first verifies whether a report contains any content relevant to a given rubric dimension, and if so, assesses the relevant sentences. On the lab reports, VerAs outperforms multiple baselines based on OpenQA systems or Automated Essay Scoring (AES). VerAs also performs well on an analytic rubric for middle school physics essays.

Keywords: Automated Assessment · Lab Reports · Analytic Rubrics

1 Introduction

Science writing plays an important role in science education, whether to prepare students for science careers, or to nurture a more informed citizenry. Informative, reliable and timely feedback on written work supports learning [13,28], which in turn is often facilitated through rubrics. A recent meta-review of rubric usage throughout the educational cycle across different subject areas found a positive effect on student learning and performance [29]. Yet rubrics are time-consuming for educators to develop and use. Further, when teaching assistants (TAs) apply rubrics, the results can be unreliable [32], reducing their benefit on learning. Automated support for assessment of writing has often addressed non-STEM automated essay scoring (AES; holistic scores) [3,8,12,45–47], or short answer

© The Author(s), under exclusive license to Springer Nature Switzerland AG 2024
A. M. Olney et al. (Eds.): AIED 2024, LNAI 14829, pp. 133–148, 2024.
https://doi.org/10.1007/978-3-031-64302-6_10

Pendulum D1: Is able to state the research question for reader clarity	
Points	Select One
1	Research question is included but incorrect. No mention of the three variables.
...	...
5	Research question is included and correct: *What affects the period of a pendulum?* Includes an explicit statement of the 3 variables: mass, angle of release, and string length.

Force & Motion D7: Is able to identify random errors and how they were or could be reduced.	
Points	Sum all that apply
1	Discusses one random error.
...	...
1	Includes one or more additional random or systematic errors.

Fig. 1. A rubric dimension from each of two lab reports, with different scoring strategies.

assessment in STEM [6,10,14,25,42,43] or non-STEM [27,44]. There has been far less work on automated support to apply analytic rubrics for long-form STEM writing. Our work addresses automated application of analytic rubrics.

Panadero et al. [29] define a rubric as setting expectations for student work through specification of evaluative criteria, and how to meet them. Their meta-review includes studies where rubrics lack a scoring strategy, as when the main goal is formative assessment, which occurs during a course while students are learning the material, to help them improve by the end of the course. A rubric is analytic if it specifies multiple criteria, or rubric dimensions. We designed an automated approach for rubric assessment of STEM writing, and evaluated it on college level physics lab reports that have a scoring strategy in the rubric, and on middle school essays where there is no scoring strategy.

Figure 1 illustrates a key challenge with the lab report rubrics: they can use different scoring strategies. The top of the figure shows part of the first rubric dimension for lab reports on the behavior of a pendulum. This is a criterion-based rubric where each point increment requires more explanation and correctness. The bottom of the figure shows part of the seventh dimension of a rubric for a report on Newton's second law. Here an inclusion-based criterion is used, and the scoring strategy is to sum all the points.

A second challenge is that in discursive science writing, it can be difficult to localize what part of a report is relevant for a given rubric dimension. As we discuss later in the paper, while human assessors can perform reliably on assigning a score for each dimension, they do not agree well on exactly which sentences address a given dimension.

Given a rubric with n dimensions and student lab reports, our assessment task is to generate a score for each dimension-report pair in the range $[0:5]$. Inspired by Open-Domain Question Answering, we propose VerAs[1], which has a verifier module to determine whether a report contains sentences relevant to a dimension, and a grader to score the relevant sentences selected by the verifier.

---

[1] The code for VerAs is available at https://github.com/psunlpgroup/VerAs.

We test its effectiveness on a published dataset of lab reports [30] against multiple baselines. Through ablations, we demonstrate the need for both modules, and the benefit of using an ordinal loss training objective for the grader. We provide detailed error analysis of performance differences across rubric dimensions. To demonstrate the generality of the architecture, we also report results on middle school physics essays where the grader module is not necessary. We present related work, the datasets, VerAs architecture, experiments and results.

2 Related Work

As noted in the introduction, AES and short answer assessment are active areas of research. In contrast, we find little work that attempts to automate rubric-based assessment for essays or lab reports. Ariely et al. [2] developed a method to detect biology concepts using convolutional neural networks in high school students' short explanation essays in Hebrew. Rahimi et al. [34] automated a rubric to assess students' use of evidence and organization of claims in source-based non-STEM writing. Ridley et al. [36] and Shibata & Uto [39] present neural models that assess specific traits to support holistic scores on a widely used dataset of non-STEM argumentative, narrative, and source-dependent essays [26]. Apart from [2], our work differs in its focus on rubric criteria for specific explanatory content, e.g., about energy, periodicity in a pendulum, or force and motion.

Our approach is inspired by Open Domain Question Answering (OpenQA), where the goal is to query multiple documents, some of which may contain no relevant information. Most OpenQA systems have two modules, a retriever to find relevant sentences, and a reader to extract the answer [17,18,23,24,37,40]. Izacard & Grave [19] combine Dense Passage Retrieval (DPR) [20] with a sequence-to-sequence transformer reader module. In later work, they propose FiD-KD to perform knowledge distillation as a way to compensate for training data that lacks labeled pairs of queries and documents with answers [18]. Similarly, Read+Verify [17] has a distinct module to assess whether a question-passage pair can provide an answer. VerAs processes sentences rather than passages, but also relies on a verifier module to first determine whether a lab report contains sentences relevant to a given rubric dimension. Similar to [17,18], we lack annotations on which sentences in a report, if any, are relevant to each rubric dimension.

Table 1. The top five rows for the college lab reports, and the bottom five for the essays, give the count in each data split, and mean length (sd) in sentences.

Split	Pendulum		Newton's 2nd Law		Both	
	N	Len_{sent}	N	Len_{sent}	N	Len_{sent}
Train	868	25.47 (12.95)	798	25.73 (12.88)	1,666	25,59 (12.92)
Val.	108	26.35 (13.57)	101	26.01 (14.19)	209	26,19 (13.88)
Test	102	26.51 (15.80)	106	25.48 (12.74)	208	25,98 (14.33)

Split	Essay 1		Essay 2		Both	
	N	Len_{sent}	N	Len_{sent}	N	Len_{sent}
Train	899	16.35 (8.82)	720	21.60 (11.59)	1619	18.69 (10.47)
Val	95	14.27 (7.40)	95	21.47 (12.48)	190	17.87 (10.87)
Test	99	19.15 (10.15)	56	31.00 (21.53)	155	23.43 (16.30)
Total	1,093		871		2,003	

3 Datasets

The college physics dataset consists of two sets of lab reports [30] from a curriculum designed to promote scientific reasoning skills. The first, about factors affecting the period of a pendulum, has a 7-dimension rubric. The second report, on Newton's Second Law, has an 8-dimension rubric. Each rubric dimension specifies precise criteria for each point increment on a six-point scoring scale, as illustrated in Fig. 1; the supplemental provides the complete rubrics. Each report has a ground truth score for each dimension from one of four trained raters. On random subsets of multiply labeled reports, raters had an average Pearson correlation of 0.72 on the 7 dimensions of the first report, and 0.69 on the 8 dimensions of the second report. The top half of Table 1 shows the size of the dataset splits (training, validation, test) and mean sentence lengths. As shown in Fig. 2, scores per dimension are highly skewed.

The middle school data consists of responses to two essay prompts from a unit on the physics of roller coasters [33], as shown in the bottom half of Table 1. The first essay rubric identifies six main ideas about energy and the law of conservation of energy. The second essay rubric adds two additional ideas about the relations of mass to speed, and height to speed. Only 159 of the essays have reliable manual labels indicating the presence of main ideas (Cohen's kappa = 0.77) (essay 1 test is entirely manual labels). The remaining labels are from an automated tool called PyrEval [15,41] whose accuracies on the two essays are 0.76 and 0.80, respectively, as reported below. For essay 2, there are reliable manual labels on 56 essays, corresponding to the essay 2 test set.

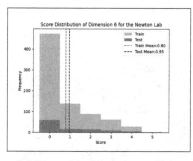 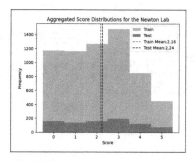

(a) Dimension 6 in the second lab.   (b) All dimensions in the second lab.

Fig. 2. For both lab reports, score distribution per dimension is highly skewed towards low or high scores, depending on the dimension difficulty, as in (a). The skew is less apparent when scores are aggregated across dimensions, as in (b).

4   VerAs Task and Architecture

VerAs treats each dimension of a rubric as a query, where the response to each query is a score in $[0:5]$. We make the simplifying assumption that at most a few sentences of a report are relevant for the assessment of a given dimension. To address the challenge that we lack labels on which sentences are relevant, we developed a pipeline with one module to select relevant sentences, and a subsequent module to apply the score. To address the challenge of the diversity and complexity of the dimensions (cf. Figure 1), each module has a dual encoder to learn better similarities of sentences to dimensions. As illustrated in Fig. 3, each sentence in a report is paired with each dimension and passed to the verifier, which in turn passes relevant sentences, the dimension, and the full report to the grader. The next two sections describe the verifier and grader in detail.

To develop a better understanding of the difficulty of the sentence selection process, the first author and a colleague independently selected relevant sentences for each dimension of 20 lab reports (10 from each of the two assignments), with no constraint on how many sentences to select. Both raters had access to the ground truth score on each dimension. We assess their agreement using Krippendorff's alpha combined with a distance metric developed for comparison of raters on set selection tasks [31]. Depending on the dimension and rater, the average number of selected sentences ranged from 1.3 to 8.4 ($\mu = 3.4, \sigma = 1.5$). Rater agreement on lab one was 0.54 and 0.43 on lab two. Thus the task is difficult for humans to achieve with consistency, and humans vary greatly in the number of relevant sentences they select. We attribute this in part to many sentences having multiple clauses where only part of a sentence might be relevant.

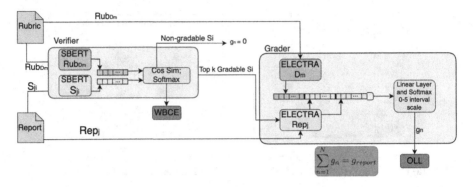

Fig. 3. VerAs: Using a dual encoder, the verifier assesses each report sentence (S_i) and rubric dimension (D_m) to forward the top k sentences to the grader, trained with weighted binary cross-entropy loss on whether the report receives a non-zero score. The grader also uses a dual encoder; it concatenates the top k sentences, D_m, and the full report Rep_j, trained with ordinal log loss as the training objective to assign a score.

4.1   Verifier

The verifier makes two decisions: deciding if a report should receive a non-zero score on a given dimension, and if so, determining which sentences are the most relevant. Because we only have labels on the first decision, and because the data is imbalanced (see Fig. 1), we use weighted binary cross-entropy as the loss function. To find relevant sentences, we learn representations for the sentences and rubric dimension to achieve meaningful similarity. Let the sentences in a lab report be denoted by $S = \{s_1, s_2, ..., s_n\}$. Given a rubric dimension q, we calculate the cosine similarity between the embeddings of the rubric dimension and each sentence as follows:

$$cos\_sim(q, s; \theta_q, \theta_s) = \frac{f(q; \theta_q)^T f(s; \theta_s)}{max(||f(q; \theta_q)||_2 ||f(s; \theta_s)||_2, \epsilon)} \tag{1}$$

where θ_q and θ_s are the parameters of our encoder functions for the rubric dimensions and report sentences respectively, and ϵ is a small value to avoid division by 0. A dual encoder [5] learns different embedding spaces for the rubric dimensions versus report sentences, using SBERT [35], which was designed to learn representations for semantic similarity comparisons more efficiently. After the calculation of pairwise cosine similarities, the top k similarities are averaged and converted to a probability using Eq. 2, as in [4]:

$$f_{softmax} = \frac{1}{1 + e^{-10(D - 0.5)}} \tag{2}$$

where D is the mean of the top k cosine similarity values.

4.2 Grader

Similar to the verifier, the grader relies on a dual encoder to learn more effective similarities of the encoded report (r) and top k relevant sentences (rel) with the rubric dimension. Inclusion of r provides a global context for rel, and potentially compensates for the possibility rel fails to include all the relevant sentences. This is likely given that k is fixed once VerAs is trained, whereas we found high variability in the number of sentences that human raters selected as relevant, across dimensions and reports.

The grader calculates the probability distribution P over scores (six classes) as:

$$P(q, r, rel; \beta_q, \beta_r, \phi) = f_{softmax}(f([g(q; \beta_q), g(r; \beta_r), g(rel; \beta_r)]; \phi)) \quad (3)$$

where f is a linear layer, g is the encoder, and β_q, β_r, and ϕ are the learned parameters of the encoders for the rubric dimension, report and prediction layers, respectively. We experiment with BERT [11] ELECTRA [9] and LongT5 [16] for the encoder function.

Cross entropy loss (CE) is not appropriate for our task, because the score classes are on an ordinal scale where the distance between pairs of values varies. Therefore, we use ordinal log loss (OLL) [7] as the grader's loss function:

$$L_{OLL\text{-}\alpha}(P, y) = -\sum_{i=1}^{N} \log(1 - p_i)\delta(y, i)^{\alpha} \quad (4)$$

given N classes, P as the model's estimated probability distribution, the true label y, a distance function δ, and a hyperparameter α. For δ, we use absolute distance.

5   Experiments

Experiments on the lab reports compare VerAs with multiple baseline models, plus the majority class baseline. We also perform five VerAs ablations to assess its components. We test only the VerAs verifier module on the essays, as explained further below.

All experiments use the Adam optimizer [21], and the same learning rates (0.001, 0.0001, 0.00001, 0.005, 0.0005, 0.00005) and batch sizes (4, 8, 16). We select the optimal hyperparameters given the validation loss, except for R^{2}BERT. Its loss automatically decreases each epoch because of its dynamic weight strategy, so we rely on the Spearman correlation instead. We tune α for OLL with 1, 1.5, 2, 2.5, and 3. Lastly, we try 1, 2, 3, 4, 20, and 25 as the top k parameter for both VerAs and FiD-KD. For FiD-KD, 25 is best, which is close to the average report length. For VerAs, 3 is best, which is close to the average of our two human raters (see Sect. 4).

5.1   Baselines

Two of the baselines are distinct models, and one is a variant of VerAs. The R²BERT [47] AES system predicts a total score given a report, without utilizing the rubric dimensions. It uses a BERT encoder followed by a linear layer to predict the score, scaling the scores to [0–1]. The loss is a dynamically weighted sum of a regression (MSE) and ranking loss (CE). We tune the learning rate, batch size, and truncation size of the report.

The second baseline is the OpenQA system that most directly inspires VerAs, FiD-KD [18] (see above). In the reader module, T5 encodes the question and passage. The concatenation of their vectors goes to the T5 decoder. The retriever uses a BERT-based bi-encoder to assess the similarity of a question-passage pair, similar to DPR [20]. Knowledge distillation is performed from the reader to the retriever, using reader attention scores as pseudo-labels to train the retriever. Here, we treat each rubric dimension as a question and each sentence as the passage. The class names ([0 : 5]) are spelled out, and a single time step decoding is carried out on this restricted vocabulary, as in [38].

The third baseline reimplements VerAs as a multi-task model: each rubric dimension becomes a separate problem, with a separate classification layer in the grader for each rubric dimension. The verifier module remains the same. VerAs$_{SEP}$ thus tests whether different classifiers are needed to handle the semantic diversity across rubric dimensions.

5.2   Ablations

The first ablation replaces the verifier with random selection of three sentences from each lab report (*Random Verifier*). The second and third ablations omit the verifier module altogether, with the grader receiving only the rubric dimension and report, using either a truncated report to meet the input length constraint (*W/o Verifier Trunc.*), or an average of embeddings of a moving window over the full report (*W/o Verifier Mov. Avg.*). In the fourth ablation, the input to the grader omits the report (*W/o Report*). The final ablation, VerAs$_{CE}$, uses cross entropy loss instead of OLL.

6   Results

We evaluate the performance of VerAs on the lab reports in two ways: on the total report score, which is the sum of the scores on each dimension, and also at the dimension level. In this section, we first present the evaluation metrics used here, then the two types of results, followed by error analysis. The final subsection presents results of the verifier module on the middle school physics essays.

Table 2. Total report score evaluations with 95% bootstrapped confidence intervals.

Model	MSE	$\alpha_{Interval}$	Weighted Acc.
	Comparison with baselines		
VerAs	**19.11 (19.09, 19.13)**	**0.77 (0.77, 0.77)**	**0.91 (0.91, 0.91)**
VerAs$_{SEP}$	23.27 (23.25, 23.29)	0.70 (0.70, 0.70)	0.90 (0.90, 0.90)
R^{2}BERT	27.05 (27.03, 27.07)	0.68 (0.68, 0.68)	0.89 (0.89, 0.89)
FiD-KD	27.46 (27.43, 27.48)	0.67 (0.67, 0.67)	0.89 (0.89, 0.89)
	Ablations		
Random Verifier	19.24 (19.23, 19.26)	0.69 (0.69, 0.69)	**0.91 (0.91, 0.91)**
W/o Verifier Trunc.	19.16 (19.14, 19.17)	0.67 (0.67, 0.67)	**0.91 (0.91, 0.91)**
W/o Verifier Mov. Avg.	20.65 (20.64, 20.67)	0.68 (0.68, 0.68)	0.90 (0.90, 0.90)
W/o Report	20.85(20.83, 20.87)	0.70 (0.70,0.70)	**0.91 (0.91, 0.91)**
VerAs$_{CE}$	24.29 (24.27, 24.32)	0.71 (0.71, 0.71)	0.89 (0.89, 0.89)

6.1 Evaluation Metrics

The total score on a report is the sum of the scores on each dimension. To evaluate the total score, we report Mean Squared Error (MSE), Krippendorff's alpha coefficient ($\alpha_{Interval}$), and weighted accuracy. MSE is the squared difference between the prediction and ground truth. Agreement coefficients like Krippendorff's alpha [22], which factor out agreements that could arise by chance, are most familiar with categorical decisions but use of an interval scale supports comparison of two numeric outcomes. Similarly, weighted accuracy takes the absolute distance between the ground truth and prediction into account; it is calculated as follows:

$$W_{acc} = \frac{\sum_{i=1}^{n} \frac{1-|g_n^i - y_n^i|}{max\ distance}}{n} \tag{5}$$

where g_n^i and y_n^i are the prediction and ground truth for the nth rubric dimension of report i, and max distance is the maximum absolute difference between the prediction and ground truth: 5 for the dimension level, 35 for the first lab, and 40 for the second.

The predicted total score could be correct without being correct on any one dimension, so we also evaluate how well the scores on each dimension agree. We report the Spearman correlation, which measures the distance between two rankings, by averaging the Spearman correlations of the per dimension predictions with the ground truth over all reports. We also report the average $\alpha_{Interval}$.

For the verifier, we evaluate its decision as to whether a report gets a non-zero grade, using accuracy, micro-averaged precision, recall and F1-score.

6.2 Results by Total Score and by Dimension

Table 2 shows that VerAs outperforms all of the baselines: by at least 17.9% on MSE, 8.0% on $\alpha_{Interval}$, and 0.8% on weighted accuracy. On total score,

Table 3. Average correlations across dimensions for each lab, along with the mean (std).

Model	Spearman Pend.	$\alpha_{Interval}$ Pend.	Spearman Newt.	$\alpha_{Interval}$ Newt.
Comparison with baselines				
VerAs	0.52 (0.36)	0.46 (0.35)	0.60 (0.30)	**0.54 (0.30)**
VerAs$_{SEP}$	**0.59 (0.33)**	**0.53 (0.36)**	**0.62 (0.27)**	**0.54 (0.28)**
FiD-KD	0.53 (0.36)	0.46 (0.36)	0.49 (0.37)	0.41 (0.32)
Ablations				
Random Verifier	0.45 (0.38)	0.27 (0.31)	0.48 (0.32)	0.35 (0.26)
W/o Verifier Trunc.	0.45 (0.39)	0.29 (0.32)	0.49 (0.32)	0.38 (0.26)
W/o Verifier Mov. Avg.	0.44 (0.37)	0.28 (0.30)	0.48 (0.34)	0.35 (0.27)
W/o Report	0.49 (0.34)	0.40(0.33)	0.58 (0.31)	0.52 (0.32)
VerAs$_{CE}$	0.42 (0.41)	0.33 (0.38)	0.44 (0.31)	0.37 (0.30)

VerAs$_{SEP}$ performs less well than VerAs, possibly because each classifier has only 1,666 examples instead of 12,460. Surprisingly, R^2BERT outperforms FiD-KD in two metrics although it uses a simpler architecture. VerAs also outperforms the ablations on MSE and $\alpha_{interval}$, especially when CE instead of OLL is used. The weighted accuracy results are uniformly high due to the extreme data skew, but show no sensitivity across models. Table 3, which gives the average per dimension correlations and agreement, shows VerAs$_{SEP}$ to have the highest performance, with VerAs outperforming FiD-KD.

6.3   Error Analysis of the Verifier's Binary Decision

With respect to overall performance, Table 4 shows that the verifier does a better job on lab 1, which is also easier for the students: Fig. 2b) shows the mean score on lab 2 to be 2.16 in the training data, compared to 2.95 on lab 1, but with relatively few zero scores on any dimension (see supplemental). We speculate that the verifier does better on lab 1 because the data is more balanced. On each dimension, verifier accuracy is often close to the majority class baseline. However, for dimension 6 on lab 1, and dimensions 2–5 on lab 2, it is lower than the majority class result; for dimensions 1, 6 and 8 on lab 2, the verifier accuracy is greater than the majority class baseline. In general, it provides good sentences, which is the more important responsibility of the verifier and through ablation studies, we show its effectiveness. There appears to be a relationship between the difficulty of the rubric dimension and the performance of the verifier for the second lab. We calculate the pearson and spearman correlations between the accuracy of the verifier and the average training ground truth scores for each rubric dimension and we get 0.96 and 0.78 respectively. However, there is no such correlation for the first lab.

Table 4. Verifier binary decision scores for the first (left) and second (right) lab.

	Pendulum						F & M				
Dim.	**Maj. Base.**	**Acc.**	**Prec.**	**Rec.**	**F1**	**Dim.**	**Maj. Base.**	**Acc.**	**Prec.**	**Rec.**	**F1**
1	0.90	0.90	0.96	0.93	0.95	1	0.92	0.95	0.98	0.97	0.97
2	0.95	0.96	0.97	0.99	0.98	2	0.88	0.82	1.00	0.80	0.89
3	0.98	0.99	0.99	1.00	1.00	3	0.90	0.84	0.98	0.84	0.90
4	0.96	0.97	0.98	0.99	0.98	4	0.85	0.75	0.97	0.72	0.83
5	0.93	0.93	1.00	0.93	0.96	5	0.84	0.82	0.96	0.82	0.88
6	0.98	0.94	1.00	0.94	0.97	6	0.56	0.61	1.00	0.13	0.23
7	0.76	0.75	0.96	0.71	0.81	7	0.92	0.92	0.97	0.94	0.95
						8	0.76	0.80	0.98	0.75	0.85
Overall	0.92	0.92	0.91	0.92	0.91	Overall	0.81	0.81	0.83	0.81	0.80

Table 5. Accuracies with confidence intervals on middle school essays.

	Essay 1		
Idea	**VerAs Verifier**	**FiD-KD**	**PyrEval**
1	0.68 (0.68, 0.69)	0.67 (0.67, 0.67)	0.65 (0.65, 0.65)
2	0.62 (0.61, 0.62)	0.70 (0.70, 0.70)	0.66 (0.65, 0.66)
3	0.67 (0.67, 0.67)	0.68 (0.68, 0.68)	0.69 (0.69, 0.69)
4	0.92 (0.91, 0.92)	0.95 (0.95, 0.95)	0.92 (0.91, 0.92)
5	0.85 (0.85, 0.86)	0.80 (0.80, 0.80)	0.85 (0.85, 0.86)
6	0.81 (0.81, 0.82)	0.78 (0.78, 0.79)	0.81 (0.81, 0.82)
Overall	0.76 (0.76, 0.76)	0.76 (0.76, 0.77)	0.76 (0.76, 0.76)
	Essay 2		
Idea	**VerAs Verifier**	**FiD-KD**	**PyrEval**
1	0.87 (0.87, 0.87)	0.84 (0.83, 0.84)	0.82 (0.82, 0.82)
2	0.93 (0.93, 0.93)	0.93 (0.93, 0.93)	0.93 (0.93, 0.93)
3	0.75 (0.74, 0.75)	0.73 (0.72, 0.73)	0.82 (0.82, 0.82)
4	0.93 (0.93, 0.93)	0.93 (0.93, 0.93)	0.93 (0.92, 0.93)
5	0.77 (0.76, 0.77)	0.82 (0.82, 0.82)	0.84 (0.83, 0.84)
6	0.80 (0.80, 0.81)	0.84 (0.84, 0.84)	0.77 (0.77, 0.77)
7	0.62 (0.62, 0.63)	0.57 (0.57, 0.57)	0.55 (0.55, 0.55)
8	0.73 (0.73, 0.73)	0.59 (0.59, 0.59)	0.78 (0.78, 0.79)
Overall	0.80 (0.80, 0.80)	0.78 (0.78, 0.78)	0.80 (0.80, 0.81)

6.4   Results on Middle School Essays

Like the lab report dataset, we have data for two middle school essay assignments, along with analytic rubrics for formative feedback, and where each rubric has a different number of dimensions (six for essay 1; eight for essay 2). Instead of dimensions that differ with respect to different aspects of an experiment, such as the research question, theoretical equation, or sources of error, each essay rubric dimension is an explanatory statement of one of the main ideas in the curriculum. These can be more general, such as how potential and kinetic energy in a roller coaster are related to one another, or more specific, such as an explanation of the law of conservation of energy. Instead of assessing each dimension on a scale, the essay feedback indicates only whether the student included a clear statement of one of the main ideas. As a result, the VerAs grader module plays no role. We include results of FiD-KD, and PyrEval.

PyrEval is a toolkit for assessing the content of short passages. From a small set of N reference passages it can automatically create a content model, called a pyramid, which is then used to detect similar content in unseen passages, all written to the same prompt. Content units (CUs) in the pyramid are sets of paraphrases extracted from the reference passages, where each CU has an importance weight equivalent to the number of reference passages that expressed that content. PyrEval can create content models from as few as 4 or 5 reference passages, and requires no training data.

Table 5 shows that all three models have the same overall accuracy on essay 1, while FiD-KD has slightly lower accuracy on essay 2. The per-dimension accuracies differ only slightly across models, and follow the same trend lines.

7   Conclusion

Our results show that formative assessment of longer forms of student writing, even those as complex as college-level lab reports with very detailed rubrics, can be handled by a neural network. VerAs performs very well on two sets of college level lab reports at applying a fine-grained analytic rubric, outperforming strong baselines. Ablations show that omitting the verifier module lowers MSE and $\alpha_{Interval}$ on the total report score. This indicates the verifier plays an important role despite the lack of labeled data for the verifier sentence selection task. Evaluation of how well each dimension is scored, however, shows that $VerAs_{SEP}$ outperforms VerAs. On a less complex essay dataset, VerAs, FiD-KD and a content assessment toolkit that requires no training perform equally well. Future work might focus on incorporating the score definitions in the rubrics and a better strategy to deal with the lack of labeled data for the sentence selection task. Additionally, large language models such as GPT-4 [1] can be prompted to have potentially noisy labels for the relevant sentences.

Limitations: VerAs needs to be retrained for new datasets, which reduces its generality. Future work might focus on this by using large language models. We test VerAs performance on two college physics lab reports and one middle school physics essay. Future work might test on other STEM domains such as biology.

Acknowledgements. We thank Sarkar Das, Vipul Gupta, Zhaohui Li, and Ruihao Pan for helpful discussions. The second author's work was supported by NSF DRK award 2010351.

References

1. Achiam, J.E.: GPT-4 technical report (2024). arXiv:2303.08774
2. Ariely, M., Nazaretsky, T., Alexandron, G.: Machine learning and Hebrew NLP for automated assessment of open-ended questions in biology. Int. J. Artif. Intell. Educ. 1–34 (2022). https://link.springer.com/article/10.1007/s40593-021-00283-x
3. Bai, H., Huang, Z., Hao, A., Hui, S.C.: Gated character-aware convolutional neural network for effective automated essay scoring. In: IEEE/WIC/ACM International Conference on Web Intelligence and Intelligent Agent Technology, pp. 351–359. ACM (2022). https://doi.org/10.1145/3486622.3493945
4. Bridle, J.: Training stochastic model recognition algorithms as networks can lead to maximum mutual information estimation of parameters. NIPS **2**, 1–7 (1989)
5. Bromley, J., Guyon, I., LeCun, Y., Säckinger, E., Shah, R.: Signature verification using a "Siamese" time delay neural network. In: Proceedings of the 6th International Conference on Neural Information Processing Systems, pp. 737-744. Morgan Kaufmann, San Francisco (1993)
6. Camus, L., Filighera, A.: Investigating transformers for automatic short answer grading. In: International Conference on Artificial Intelligence in Education (AIED), pp. 43–48 (2020). https://doi.org/10.1007/978-3-030-52240-7_8
7. Castagnos, F., Mihelich, M., Dognin, C.: A simple log-based loss function for ordinal text classification. In: Proceedings of the 29th International Conference on Computational Linguistics, pp. 4604–4609. International Committee on Computational Linguistics, Gyeongju (2022). https://aclanthology.org/2022.coling-1.407
8. Chen, Y., Li, X.: PMAES: prompt-mapping contrastive learning for cross-prompt automated essay scoring. In: Proceedings of the 61st Annual Meeting of the Association for Computational Linguistics, vol. 1: Long Papers, pp. 1489–1503. Association for Computational Linguistics, Toronto (2023). https://doi.org/10.18653/v1/2023.acl-long.83
9. Clark, K., Luong, M.T., Le, Q.V., Manning, C.D.: ELECTRA: pre-training text encoders as discriminators rather than generators. In: International Conference on Learning Representations (2020). https://openreview.net/forum?id=r1xMH1BtvB
10. Condor, A., Pardos, Z., Linn, M.: Representing scoring rubrics as graphs for automatic short answer grading. In: Rodrigo, M.M., Matsuda, N., Cristea, A.I., Dimitrova, V. (eds.) Artificial Intelligence in Education: 23rd International Conference, AIED 2022, Durham, UK, 27–31 July 2022, Proceedings, Part I, pp. 354–365. Springer, Heidelberg (2022)
11. Devlin, J., Chang, M.W., Lee, K., Toutanova, K.: BERT: pre-training of deep bidirectional transformers for language understanding. In: Burstein, J., Doran, C., Solorio, T. (eds.) Proceedings of the 2019 NAACL and HLT, pp. 4171–4186. Association for Computational Linguistics, Minneapolis (2019). https://doi.org/10.18653/v1/N19-1423
12. Do, H., Kim, Y., Lee, G.G.: Prompt- and trait relation-aware cross-prompt essay trait scoring. In: Findings of the Association for Computational Linguistics: ACL 2023, pp. 1538–1551. Association for Computational Linguistics, Toronto (2023). https://doi.org/10.18653/v1/2023.findings-acl.98
13. Evans, C.: Making sense of assessment feedback in higher education. Rev. Educ. Res. **83**(1), 70–120 (2013)
14. Filighera, A., Parihar, S., Steuer, T., Meuser, T., Ochs, S.: Your answer is incorrect... would you like to know why? introducing a bilingual short answer feedback dataset. In: Muresan, S., Nakov, P., Villavicencio, A. (eds.) Proceedings of the 60th

ACL, pp. 8577–8591. Association for Computational Linguistics, Dublin (2022). https://doi.org/10.18653/v1/2022.acl-long.587

15. Gao, Y., Sun, C., Passonneau, R.J.: Automated pyramid summarization evaluation. In: Proceedings of the 23rd Conference on Computational Natural Language Learning (CoNLL), pp. 404–418. Association for Computational Linguistics, Hong Kong (2019). https://doi.org/10.18653/v1/K19-1038. https://aclanthology.org/K19-1038

16. Guo, M., et al.: LongT5: efficient text-to-text transformer for long sequences. In: Findings of the Association for Computational Linguistics: NAACL 2022, pp. 724–736. Association for Computational Linguistics, Seattle (2022). https://doi.org/10.18653/v1/2022.findings-naacl.55. https://aclanthology.org/2022.findings-naacl.55

17. Hu, M., Wei, F., Peng, Y., Huang, Z., Yang, N., Li, D.: Read+ verify: machine reading comprehension with unanswerable questions. In: Proceedings of the AAAI Conference on Artificial Intelligence, vol. 33, pp. 6529–6537 (2019)

18. Izacard, G., Grave, E.: Distilling knowledge from reader to retriever for question answering. In: ICLR (2021). https://openreview.net/forum?id=NTEz-6wysdb

19. Izacard, G., Grave, E.: Leveraging passage retrieval with generative models for open domain question answering. In: Merlo, P., Tiedemann, J., Tsarfaty, R. (eds.) Proceedings of the 16th EACL, pp. 874–880. Association for Computational Linguistics, Online (2021). https://doi.org/10.18653/v1/2021.eacl-main.74

20. Karpukhin, V., et al.: Dense passage retrieval for open-domain question answering. In: Proceedings of the 2020 Conference on Empirical Methods in Natural Language Processing (EMNLP), pp. 6769–6781. Association for Computational Linguistics, Online (2020). https://doi.org/10.18653/v1/2020.emnlp-main.550. https://aclanthology.org/2020.emnlp-main.550

21. Kingma, D.P., Ba, J.: Adam: a method for stochastic optimization. arXiv preprint arXiv:1412.6980 (2014)

22. Krippendorff, K.: Computing Krippendorff's alpha-reliability. University of Pennsylvania Scholarly Commons (2011). https://repository.upenn.edu/asc_papers/43

23. Lee, J., Yun, S., Kim, H., Ko, M., Kang, J.: Ranking paragraphs for improving answer recall in open-domain question answering. In: Riloff, E., Chiang, D., Hockenmaier, J., Tsujii, J. (eds.) Proceedings of the 2018 EMNLP, pp. 565–569. Association for Computational Linguistics, Brussels (2018). https://doi.org/10.18653/v1/D18-1053

24. Lee, K., Chang, M.W., Toutanova, K.: Latent retrieval for weakly supervised open domain question answering. In: Proceedings of the 57th Annual Meeting of the Association for Computational Linguistics, pp. 6086–6096. Association for Computational Linguistics, Florence (2019). https://doi.org/10.18653/v1/P19-1612. https://aclanthology.org/P19-1612

25. Li, Z., Tomar, Y., Passonneau, R.J.: A semantic feature-wise transformation relation network for automatic short answer grading. In: Moens, M.F., Huang, X., Specia, L., Yih, S.W.t. (eds.) Proceedings of the 2021 EMNLP, pp. 6030–6040. Association for Computational Linguistics, Online and Punta Cana (2021). https://doi.org/10.18653/v1/2021.emnlp-main.487

26. Mathias, S., Bhattacharyya, P.: ASAP++: enriching the ASAP automated essay grading dataset with essay attribute scores. In: Proceedings of the Eleventh International Conference on Language Resources and Evaluation (LREC 2018). European Language Resources Association (ELRA), Miyazaki (2018). https://aclanthology.org/L18-1187

27. Mizumoto, T., et al.: Analytic score prediction and justification identification in automated short answer scoring. In: Proceedings of the Fourteenth Workshop on Innovative Use of NLP for Building Educational Applications, pp. 316–325. Association for Computational Linguistics, Florence (2019). https://doi.org/10.18653/v1/W19-4433, https://aclanthology.org/W19-4433

28. O'Donovan, B., Rust, C., Price, M.: A scholarly approach to solving the feedback dilemma in practice. Assess. Eval. High. Educ. **41**(6), 938–949 (2016)

29. Panadero, E., Jonsson, A., Pinedo, L., Fernández-Castilla, B.: Effects of rubrics on academic performance, self-regulated learning, and self-efficacy: a meta-analytic review. Educ. Psychol. Rev. **35**, article 113 (2023). https://doi.org/10.1007/s10648-023-09823-4

30. Passonneau, R.J., Li, Z., Atil, B., Koenig, K.M.: Reliable rubric-based assessment of physics lab reports: Data for machine learning (2022). https://doi.org/10.26208/BWE2-BR31

31. Passonneau, R.J.: Measuring agreement on set-valued items (MASI) for semantic and pragmatic annotation. In: Proceedings of the 5th International Conference on Language Resources and Evaluation (LREC 2006). ELRA, Genoa (2006)

32. Passonneau, R.J., Koenig, K., Li, Z., Soddano, J.: The ideal versus the real deal in assessment of physics lab report writing. Eur. J. Appl. Sci. **11**(2), 626–644 (2023). https://doi.org/10.14738/aivp.112.14406

33. Puntambekar, S., Dey, I., Gnesdilow, D., Passonneau, R.J., Kim, C.: Examining the effect of automated assessments and feedback on students' written science explanations. In: Blikstein, P., Van Aalst, J., Kizito, R., Brennan, K. (eds.) 17th International Conference of the Learning Sciences (ICLS 2023), pp. 1865–1866. International Society of the Learning Sciences (2023). https://repository.isls.org//handle/1/10060

34. Rahimi, Z., Litman, D.J., Correnti, R., Wang, E., Matsumura, L.C.: Assessing students' use of evidence and organization in response-to-text writing: Using natural language processing for rubric-based automated scoring. Int. J. Artif. Intell. Educ. **27**(4), 694–728 (2017)

35. Reimers, N., Gurevych, I.: Sentence-BERT: sentence embeddings using Siamese BERT-networks. In: Proceedings of the 2019 Conference on Empirical Methods in Natural Language Processing and the 9th International Joint Conference on Natural Language Processing (EMNLP-IJCNLP), pp. 3982–3992. Association for Computational Linguistics, Hong Kong (2019). https://doi.org/10.18653/v1/D19-1410. https://aclanthology.org/D19-1410

36. Ridley, R., He, L., Dai, X.Y., Huang, S., Chen, J.: Automated cross-prompt scoring of essay traits. In: Proceedings of the AAAI Conference on Artificial Intelligence, vol. 35, no. 15, pp. 13745–13753 (2021). https://doi.org/10.1609/aaai.v35i15.17620. https://ojs.aaai.org/index.php/AAAI/article/view/17620

37. Sachan, D., et al.: End-to-end training of neural retrievers for open-domain question answering. In: Zong, C., Xia, F., Li, W., Navigli, R. (eds.) Proceedings of the 59th ACL and the 11th IJCNL, pp. 6648–6662. ACL (2021). https://doi.org/10.18653/v1/2021.acl-long.519

38. Schick, T., Udupa, S., Schütze, H.: Self-diagnosis and self-debiasing: a proposal for reducing corpus-based bias in NLP. Trans. ACL **9**, 1408–1424 (2021)

39. Shibata, T., Uto, M.: Analytic automated essay scoring based on deep neural networks integrating multidimensional item response theory. In: Proceedings of the 29th ICCL, pp. 2917–2926. International Committee on Computational Linguistics, Gyeongju (2022). https://aclanthology.org/2022.coling-1.257

40. Singh, D., Reddy, S., Hamilton, W., Dyer, C., Yogatama, D.: End-to-end training of multi-document reader and retriever for open-domain question answering. Adv. Neural. Inf. Process. Syst. **34**, 25968–25981 (2021)
41. Singh, P., Passonneau, R.J., Wasih, M., Cang, X., Kim, C., Puntambekar, S.: Automated support to scaffold students' written explanations in science. In: Rodrigo, M.M., Matsuda, N., Cristea, A.I., Dimitrova, V. (eds.) Artificial Intelligence in Education, vol. 13355, pp. 660–665. Springer, Heidelberg (2022). https://doi.org/10.1007/978-3-031-11644-5_64
42. Sung, C., Dhamecha, T., Saha, S., Ma, T., Reddy, V., Arora, R.: Pre-training BERT on domain resources for short answer grading. In: Proceedings of the 2019 EMNLP and the 9th IJCNLP, pp. 6071–6075. Association for Computational Linguistics, Hong Kong (2019). https://doi.org/10.18653/v1/D19-1628
43. Takano, S., Ichikawa, O.: Automatic scoring of short answers using justification cues estimated by BERT. In: Kochmar, E., et al. (eds.) Proceedings of the 17th BEA Workshop, pp. 8–13. Association for Computational Linguistics, Seattle (2022). https://doi.org/10.18653/v1/2022.bea-1.2
44. Wang, T., Funayama, H., Ouchi, H., Inui, K.: Data augmentation by rubrics for short answer grading. J. Nat. Lang. Process. **28**(1), 183–205 (2021)
45. Wang, Y., Wang, C., Li, R., Lin, H.: On the use of BERT for automated essay scoring: joint learning of multi-scale essay representation. In: Proceedings of the 2022 Conference of the North American Chapter of the ACL (NAACL), pp. 3416–3425. Association for Computational Linguistics (2022). https://doi.org/10.18653/v1/2022.naacl-main.249
46. Xie, J., Cai, K., Kong, L., Zhou, J., Qu, W.: Automated essay scoring via pairwise contrastive regression. In: Proceedings of the 29th International Conference on Computational Linguistics, pp. 2724–2733. International Committee on Computational Linguistics, Gyeongju (2022). https://aclanthology.org/2022.coling-1.240
47. Yang, R., Cao, J., Wen, Z., Wu, Y., He, X.: Enhancing automated essay scoring performance via fine-tuning pre-trained language models with combination of regression and ranking. In: Findings of EMNLP 2020, pp. 1560–1569. ACL, Online (2020). https://doi.org/10.18653/v1/2020.findings-emnlp.141

Evaluating the Effectiveness of Comparison Activities in a CTAT Tutor for Algorithmic Thinking

Amanda Keech and Kasia Muldner[✉]

Department of Cognitive Science, Carleton University, Ottawa, Canada
{amandakeech,kasiamuldner}@cunet.carleton.ca

Abstract. Prior work has shown that novice programmers find algorithmic thinking challenging. However, to date there is little research on how to help students learn this skill. We investigated if comparison of programming examples helps novice programmers improve skills related to generation and modification of algorithms. As a testbed for our research, we implemented two versions of a CTAT tutoring system: one that presented pairs of examples and asked students to compare them (comparison tutor), and a second that presented examples sequentially (sequential tutor). Findings from an experimental study ($N = 57$) showed that ignoring condition, students learned from the tutors. The comparison tutor significantly improved procedural knowledge but with the caveat that Bayesian statistics did not provide strong evidence for the alternative hypothesis. In contrast, there was no significant effect of tutor version on procedural flexibility or transfer questions and Bayesian statics provided substantial evidence for the null hypothesis.

Keywords: Example-Comparison Tutor · Algorithmic Thinking · Procedural Flexibility

1 Introduction

On the first day of our introductory programming class, we ask our students if they know how to sort a stack of books in alphabetical order. They all say yes. However, when we ask them to describe the sorting procedure, their answers lack sufficient detail. Identifying the detailed sequence of steps needed to perform a task is referred to as algorithmic thinking [12, 13]. This is a core skill students need for programming activities [26]. A related skill is flexibility, which refers to knowing multiple ways to solve a given problem [23]. This is particularly important in the context of programming activities as there is almost always more than one way to compose an algorithm. Since programming concepts and classes are now advocated at earlier stages of schooling, and for all students (not only computer science majors), research is needed on how to help students learn algorithmic thinking.

We designed and evaluated a CTAT [1] tutoring system that aims to help students learn algorithmic thinking through comparison activities involving pairs of programming examples. To evaluate the tutor, we compared it to a version that provided examples in isolation and did not prompt for their comparison. We describe the methods and results after presenting the related work.

The majority of research on example comparison involved paper-based math activities with middle or high school students [21–23]. To illustrate with an example, Rittle-Johnson et al. [23] conducted a study with novice middle-school students, who studied examples showing step-by-step solutions to algebra problems; the solution steps were labelled with the rules that generated them. There were several conditions in the study. The *comparison* group's booklet showed a pair of side-by-side examples on each page; the two examples in each pair showed two different ways to solve the same problem. The *sequential* group studied the same examples but their booklets showed only one example per page. (There was a third condition where the examples in a pair showed two different problems.) All groups also solved problems. The comparison group had significantly higher procedural flexibility posttest scores, while the sequential group had the lowest. Conceptual performance was similar in all conditions. Recently, Durkin et al. [5] conducted a meta-review of math comparison activities, and concluded that comparison of multiple solutions to the same problem was more effective at improving procedural flexibility than other controls (that most commonly involved viewing examples in isolation); comparison was also effective at supporting procedural and conceptual knowledge. Durkin et al. proposed that comparison is effective because it makes the structural example features salient. Recommendation based on this review included that (1) examples be presented side-by-side and their steps labelled, and (2) activities include open-ended prompts encouraging comparison.

Studies in other domains have also demonstrated the benefit of comparison [7,8]. There is less work on comparison activities in the programming domain, but with notable exceptions. We begin with work that did not involve tutoring systems. Price et al. [19] used a 2×2 design varying the presence and absence of comparison and self-explanation prompts for programming problems. For each problem, participants first generated their own solution as part of homework. Once they submitted their solution, they were shown the instructor's solution and one of four possible items, namely either (1) a prompt to compare the solutions, (2) a prompt to explain the instructor's solution, (3) prompts to both compare and explain the solutions, or (4) no prompts. Immediately after, participants completed an isomorphic problem. A benefit of comparison was not found: students who explained the instructor's solution without any comparison had the highest performance on the isomorphic problem. Ma et al. [15] also tested the utility of comparison programming activities, here with fourth grade students in the context of their classroom. A block-based programming language was used (Scratch). The comparison group was given a sheet of paper with two examples, showing two different Scratch programs. Students were asked to run the programs and compare them; their teacher then went over each example and answered questions. The sequential group studied a single example, with teacher guidance. After the example study, both groups wrote a Scratch program. This process was repeated 3 times (study example(s), solve problem). The comparison group performed significantly better on the posttest. However, it's not possible to determine if this benefit was due to comparison or due to the fact that the comparison group studied twice as many examples (with teacher feedback).

In the context of AIED, there are various tutoring systems that include examples [3,16,17], including in the programming domain [6,9,11,14]. The majority of this research is focused on example design (e.g., to test if example prompts improve learning [6,14]) and/or content (e.g., to test if examples with errors improve learning over standard examples [16]). While there is some research testing when examples should be presented (e.g., before or after a problem) [11,18,24], to the best of our knowledge no work has evaluated example comparison activities involving algorithms in the context of a tutoring system. The closest exception is Weinman et al. [25]'s study involving Parsons problems. Parsons problems consist of scrambled lines of code that students are tasked with arranging into the correct order. Thus, all the syntax is already provided and correct, allowing students to focus on a program's logic (i.e., the algorithm). In the study, students first either worked on Parsons problems or code-writing exercises (*algorithm exposure* phase). All students subsequently wrote programs and their solutions were analyzed to identify the algorithms used. The Parsons group's programs used more algorithms from the exposure phase, suggesting this group learned more algorithms. This study did not evaluate the effect of comparison activities.

In sum, examples are commonly integrated into tutoring systems and other activities. Research in math education suggests that showing examples side-by-side is more beneficial than individually. However, this approach is yet to be tested in the context of a tutoring system targeting algorithmic thinking.

2   Comparison Vs. Sequential CTAT Tutor

We built two versions of a tutoring system for algorithm-related activities using the Cognitive Tutor Authoring Tools (CTAT) [1]. The two versions were called the *comparison tutor* and the *sequential tutor.*

2.1   Comparison Tutor

The comparison tutor presented pairs of examples side-by-side; the examples in each pair corresponded to two Python programs that produced identical output but included one or more differences, such as the starting value of a counter variable (e.g., Fig. 1a), the order of statements in the loop (e.g., Fig. 1b), and the condition related to the break statement (e.g., Fig. 1a,b,c). By varying the starting value of counter, the goal was to illustrate that there are many ways to initialize a counter variable, and it is the relation between the starting value and the condition related to the break that determines the number of loop iterations. The goal of varying the order of statements in the loop was to highlight that it was the relative order of the statements that mattered, not just their position in the loop. Overall, these variations aimed to emphasize that loop logic (e.g., number of iterations) depends on the starting value of a counter variable, the condition associated with the break statement, and the order of statements in the loop.

The comparison tutor presented each pair of examples on a separate screen (referred to as the 'comparison screen' below). To illustrate, the top half of Fig. 2 shows a comparison screen with one of the example pairs. Both programs in the pair use a loop to calculate the sum of the numbers from 1 to 5 and print that sum after the loop ends.

	Program A	Program B

(a)

```
counter = 0
num = 4
while True:
    if counter >2:
        break
    counter = counter + 1
    num = num * 2
    print (num)
```

```
counter = 1
num = 4
while True:
    if counter >3:
        break
    counter = counter + 1
    num = num * 2
    print (num)
```

(b)

```
counter = 5
while True:
    print (counter)
    counter = counter -1
    if counter <0 :
        break
```

```
counter = 5
while True:
    print (counter)
    if counter <1:
        break
    counter = counter -1
```

(c)

```
counter = 1
total = 0
while True:
    total = total + counter
    counter = counter + 1
    if counter > 5:
        break
print (total)
```

```
counter = 0
total = 0
while True:
    if counter > 4:
        break
    counter = counter + 1
    total = total + counter
print (total)
```

Fig. 1. Sample example pairs used in the comparison tutor (see Fig. 2 for details on how the pairs were presented).

However, these programs are not identical (counter is initialized to a different value; the condition uses a different value; the order of the steps is different). A comparison screen included a shuffle activity for each program, shown directly below the program, that encoded the algorithmic steps implemented in the corresponding program. The steps were shown as series of blocks, each labelled with a plain-English translation of a corresponding program component (typically a line). For instance, in Fig. 2, the first green block below the program on the left (Program A) has the label 'initialize variable counter to 1'. This block represents the first step implemented in the program. Taken together, the series of blocks comprised the algorithm for the program above. The blocks related to Program A in Fig. 2 are shown in correct order for the purpose of illustration, but the tutor initially showed the blocks in random order (see blocks below Program B, Fig. 2). Students were tasked with re-organizing the blocks into the correct order (i.e., one that matched the logic of the program).

The shuffle activity aimed to promote algorithmic thinking by encouraging reflection about the *steps* in a program at a level above the program syntax. This activity is similar to Parsons problems [4]. Weinman et al. [25] reported that students were able learn programming patterns (i.e., algorithms) by solving Parsons problems more

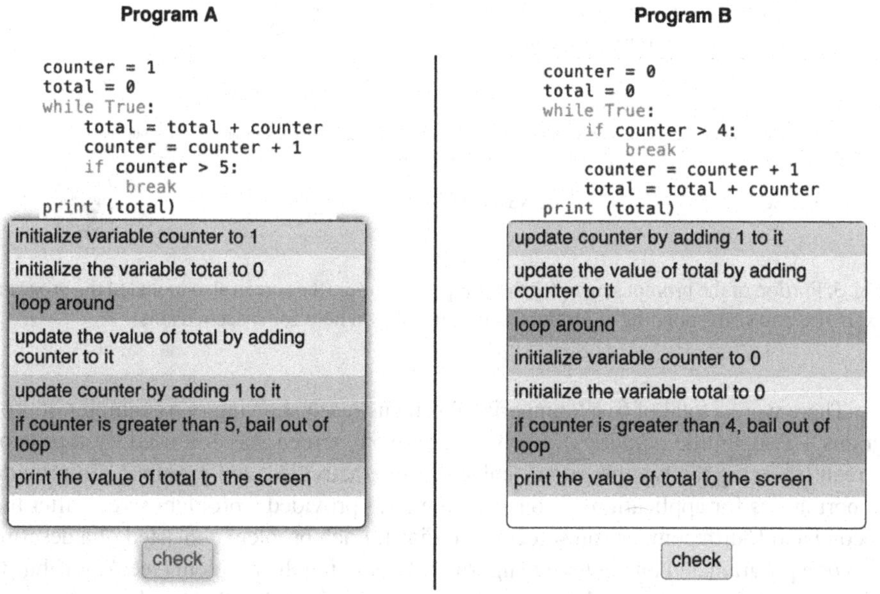

Program A

```
counter = 1
total = 0
while True:
    total = total + counter
    counter = counter + 1
    if counter > 5:
        break
print (total)
```

initialize variable counter to 1
initialize the variable total to 0
loop around
update the value of total by adding counter to it
update counter by adding 1 to it
if counter is greater than 5, bail out of loop
print the value of total to the screen

check

Program B

```
counter = 0
total = 0
while True:
    if counter > 4:
        break
    counter = counter + 1
    total = total + counter
print (total)
```

update counter by adding 1 to it
update the value of total by adding counter to it
loop around
initialize variable counter to 0
initialize the variable total to 0
if counter is greater than 4, bail out of loop
print the value of total to the screen

check

Fig. 2. A comparison screen with (1) two Python program examples, top; (2) a shuffle activity that showed the algorithmic steps for each program using blocks - the blocks were randomly shuffled and participants had to re-order them to match the program logic. For illustration purposes, the blocks for Program A have been ordered and the check button was pressed (green highlight around blocks indicates correct ordering); the blocks for Program B are shuffled and awaiting to be ordered. (Color figure online)

effectively than by writing code. The other function served by the blocks in the shuffle activity was to facilitate comparison of the two programs, following the advice from prior work that examples include labels so that comparison can take place at a level above syntax [5] (here, a label is a step in the algorithm). The tutoring system provided hints for the shuffle activities for the first example pair in case participants needed help while they were becoming familiar with the activities.

When students were done ordering the blocks for a given program, they pressed the "check" button and the tutor provided immediate feedback for correctness. If all of the blocks were in the correct order, then a green outline appeared around the box containing the blocks and the blocks locked in place. Otherwise, a red outline appeared, and the blocks remained unlocked so the student could try again. The tutor did not provide feedback on which blocks were incorrect. Once both shuffle activities were correctly solved, students moved to the next screen by pressing a done button.

After a comparison screen, the next screen showed the same example pair and corresponding algorithm blocks below each program (now in correct order) and provided a series prompts about the two examples, see Fig. 3 (only the prompts are shown here - the program and solved shuffle activity not shown due to space limitations). The prompts were designed to encourage students to reflect on the algorithms.

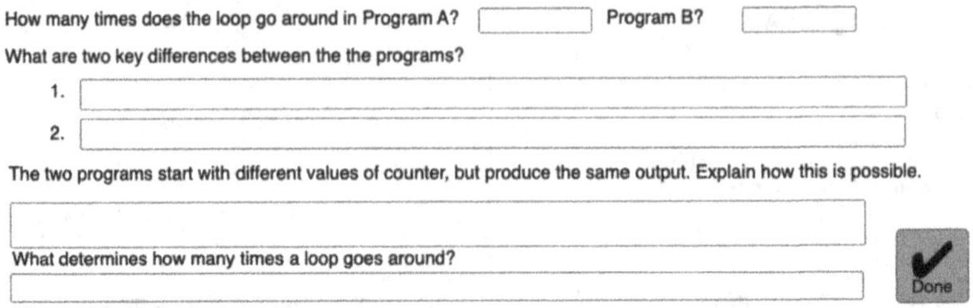

Fig. 3. Portion of the prompt screen for the comparison tutor (the screen also included the program examples and corresponding algorithm blocks, not shown here for space reasons).

There were a total of five 'comparison' screens (each showing two examples of programs + two shuffle activities). Each 'comparison' screen was followed by a prompt screen (showing the two programs, solved shuffle activities, and prompts). To provide opportunities for application of concepts, the tutor provided a problem screen after the second and fourth comparison screens, see Fig. 4. Each problem consisted of a description of a program and one potential algorithm for solving the problem (see Algorithm 1, Fig. 4). Students were tasked with creating a second algorithm that produced the same output. This was done through a shuffle activity (initially an incorrectly ordered set of blocks, see Algorithm 2, Fig. 4) and two text boxes used to fill-in-the blanks in the algorithm. If an incorrect algorithm was submitted, the tutor generated a message that the solution was incorrect and to try again. If an algorithm duplicating the provided one was submitted, the tutor generated a message that the original algorithm had to be modified. The problem could be attempted as many times as needed for a correct solution to be generated. If the submitted algorithm was both correct and different from the provided algorithm, then the tutor generated a message that the algorithm was correct, and a done button became visible, allowing students to move to the next screen. An answer that varied either the order of statements and/or the initial value of the counter was considered correct for both problems. Error checking was implemented based on expression matching in CTAT.

In sum, there was a total of 13 screens in the comparison tutor. The first screen provided instructions on how to use the tutor, followed by the first comparison screen, the corresponding prompt screen, the second comparison and prompt screens, and so on (recall there were five comparison screens and five corresponding prompt screens, along with two problem screens).

2.2 Sequential Tutor

We created a second version of the tutor called the sequential tutor. This tutor was populated with the same examples, shuffle activities, prompt screens, and two practice problems as the comparison tutor described above (activities were presented in the same order). However, in the sequential tutor, each example was shown on its own, separate screen that also included the corresponding shuffle activity for that program; the prompt screens had questions about only one example and did not ask participants

Suppose we want to write a program that prints "python is fun" to the screen 3 times and "programming is fun" to the screen once. On the left is one possible solution (Algorithm 1). Produce a second algorithm for this program by re-arranging the blocks in Algorithm 2 and filling in the missing values in the blocks using the boxes on the far right. When you are done, click on the "check all" button. Note: the second algorithm must be different.

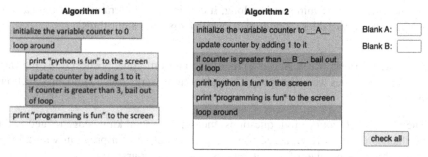

Fig. 4. Sample problem requiring generation of a second algorithm, different from Algorithm 1.

to compare anything (as was the case for the comparison tutor, these were shown after each example screen). Thus, the sequential tutor had 22 screens (10 for the examples with corresponding shuffle activities, 10 prompt screens, and 2 practice problems). The total number of prompts was the same as in the comparison tutor and the prompts that didn't require comparison were the same in both conditions (e.g., *'How many times does the loop repeat?'* and *'What determines how many times a loop goes around?'*). In place of the comparison prompts, in the sequential condition there was a question about a feature of the example (e.g., *'What is the condition that stops the loop?'*).

3   The Current Study: Method

We evaluated the comparison tutor by comparing it to the sequential tutor in an experimental study we now describe. The study was reviewed and approved by the university Ethics board. We had the following research question:

RQ: *Do students learn more from the comparison tutor that showed examples side-by-side, as compared to the sequential version that showed examples in isolation?*

3.1   Participants

The participants were 57 university students (38 females, 18 males, and one unreported). To be eligible, participants must not have taken any university-level programming classes, with one exception. Participants were still eligible if they took an introductory programming class designed for individuals with no programming experience (it is offered in our university to bachelor of arts majors). While the class does cover concepts like loops, students require additional practice after completing it to fully understand the concepts. Participants were recruited from SONA (an online recruitment system available to students in a first-year university class) (compensation: 2% course credit), and from social media posts and word of mouth (compensation: $25).

3.2 Materials

Instructional Video. An instructional video lesson was created to provide key foundations. The video was 18 min and 56 s long. It consisted of a narrator describing a slide show and covered variables, conditionals, while loops, and algorithms.

Pretest and Posttest. We designed a pretest and an isomorphic posttest (only the variable names and values were changed). The test sections were grouped into three sections as follows:

- The *procedural* section (four questions) measured general knowledge related to the mechanics of loops (e.g., asked users to predict output of a program with a loop; asked about the number of loop iterations in a given program).
- The *procedural flexibility* section (five questions) focused on algorithm modification. Three questions provided one completed algorithm (in the form of ordered algorithm blocks), along with a description of its output, and asked for the generation of a second, different algorithm that produced the same output. To do so, participants were given a new set of algorithmic blocks, initially labelled with the text for the completed algorithm. Algorithm generation was realized by dragging these algorithmic blocks into a solution area and changing their content as needed to produce the new algorithm. We included one algorithm and asked participants to generate a second one (rather than asking participants to generate two algorithms) because generation is cognitively demanding and so needs scaffolding, particularly for novice populations like ours; here, the scaffolding was provided by the presence of the first, completed algorithm. The other two questions in this section provided a description of a program, one completed algorithm in the form of blocks, and a proposed change to the algorithm (e.g., changing the initial value of counter from 1 to 0); the questions asked what other changes were needed to make the revised algorithm produce the same output as the original.
- The *transfer* section corresponded to a question asking for the generation of an algorithm for a program that prints certain numbers in a sequence (we refer to it as a transfer question because unlike activities in the tutor, scaffolding in the form of an existing algorithm was not provided).

The tests were implemented using Google Documents; questions that required shuffling of blocks were created using Google Drawings.

3.3 Design and Procedure

A between-subject design was used with two conditions, namely comparison tutor vs. sequential tutor. Participants were assigned to conditions using a round-robin fashion. All sessions were conducted one-on-one over Zoom by the first author and lasted up to two hours per participant. The researcher had their camera on while speaking, but otherwise off to avoid participants feeling watched.

The same procedure was used for both conditions (with the exception of the version of the tutor that they interacted with, comparison tutor vs.sequential tutor). At the beginning of each session, participants were welcomed to the study and sent a link to a consent form and given time to read it over before signing. After signing the form to indicate consent, a brief overview of the study was given. The study started with the instructional video providing background on key programming concepts. This was shown to participants by the researcher sharing their screen and sound, after advising participants to pay close attention. After the lesson video, the researcher shared their screen again, and demonstrated how to open and edit the Google Drawings (needed for the pretest). Participants were then sent a link to the pretest (17 min). After the pretest, participants filled out a demographic questionnaire. They then completed several brief tasks (a questionnaire and a reading, which were part of another project). Participants were then given the option of taking a short break. Next, the tutoring system was introduced. The researcher explained that the first screen of the tutor would demonstrate the basic functions of the tutor, and asked participants if they required any clarification; they also informed participants that subsequent screens of the tutor would have instructions on what to do. Participants were asked to share their screen while working with the tutor and informed that they would have a maximum of 40 min to complete what they could in that time. Finally, participants were sent a link to the posttest. As with the pretest, every participant received the same version of the test and was given as maximum of 17 min to complete it. Finally, participants were thanked for their participation and given compensation (course credit or money).

3.4 Statistical Analyses

The data was analyzed using both descriptive and inferential statistics (all analysis was done using SPSS software). For the inferential statistics, we used Null Hypothesis Significance Testing (NHST) and Bayesian statistics using Bayes Factor (BF). In the latter, the *"likelihood of the data is considered under both the null and alternative hypotheses, and these probabilities are compared via the Bayes factor. The Bayes factor is a ratio that contrasts the likelihood of the data fitting under the null hypothesis with the likelihood of fitting under the alternative hypothesis"* [10]. The ratio can be computed in either direction (i.e., to show the results from the perspective of the alternative hypothesis or the null hypothesis). Since this method can provide evidence for either model (as opposed to NHST that can only provide evidence for the alternative model), there have been calls to present results from both frameworks (NHST, Bayesian), so that complimentary evidence can be compared [10, 20].

We follow the convention to report the Bayes factor for the more likely model: BF_{01} when the null model is more likely (no conditional difference) and BF_{10} when the alternative model is more likely (conditional difference exists), stating which direction we are reporting. Of note, the other BF factor can be calculated simply by inverting the reported one. When either Bayes factor is close to 1, this indicates lack of evidence for either model being superior. As the Bayes factor increases, it provides mounting evidence for the target model (null or alternative, depending on the way the ratio is set up). We follow the guidelines in [10] to interpret Bayes factors, as follows: $BF = 1$–3 provides anecdotal evidence for the corresponding model; $BF = 3$ - 10 provides substantial evidence for the corresponding model; and $BF > 100$ provides decisive evidence. To obtain the Bayes factor, we used SPSS with the default settings (i.e., Rouder's method).

Table 1. Mean and standard deviation for pretest%, posttest% and normalized gain for each question type and condition.

Question Type		Comparison Tutor $n = 29$ M (SD)	Sequential Tutor $n = 28$ M (SD)
Procedural	Pretest (%)	46.9 (28.8)	53.3 (30.7)
	Posttest (%)	60.5 (27.7)	64.3 (24.1)
	Normalized Gain	.39 (.39)	.18 (.32)
Procedural Flexibility	Pretest (%)	39.8 (24.0)	34.2 (26.5)
	Posttest (%)	54.0 (22.6)	51.8 (22.4)
	Normalized Gain	.20 (.45)	.22 (.27)
Transfer	Pretest (%)	32.4 (35.4)	15.7 (24.9)
	Posttest (%)	50.0 (37.7)	51.43 (34.9)
	Normalized Gain	.28 (.45)	.39 (.40)

4   Results

As noted above, our primary research question was if the comparison tutor produced more learning. Learning was operationalized by normalized gain based on change from pretest to posttest (details below). Each test was graded out of 24.5 points, using a pre-defined rubric. The majority of the points came from questions measuring various aspects of algorithmic thinking (corresponding to the procedural flexibility and transfer questions, see Sect. 3.2).

We conducted separate analyses for each question category, as results can be obscured if data is aggregated. The descriptive statistics are shown in Table 1 for each type of question (procedural, procedural flexibility, transfer). As expected given the novice population recruited, the pretest scores were low. The scores were similar for the comparison and sequential groups for the procedural and flexibility questions (procedural: $t(55) = 0.8$, $p = .42$, $d = 0.2$; $BF_{01} = 3.7$; flexibility: $t(55) = 0.8$, $p = .41$, $d = 0.2$; $BF_{01} = 3.6$). By chance the comparison group had higher pretest scores for the transfer question, $t(55) = 2.1$, $p = .04$, $d = 0.5$, but the Bayes factor evidence did not support this pattern because the alternative model was only very slightly more likely, $BF_{10} = 1.25$ and the evidence was weak for either model.

Collapsing across condition, participants did learn from using the tutoring systems, as indicated by the significant improvement from pretest to posttest ($M_{gain} = 17.1\%$, $SD = 18.9$), $t(56) = 6.9$, $p < .001$, with a large effect, $d = 0.9$; $BF_{10} > 100$.

To measure the effect of condition on learning, we used normalized gain [2], calculated as follows:

$$\frac{posttest(\%) - pretest(\%)}{100\% - pretest(\%)}$$

Normalized gain characterizes how much a student learned relative to how much they could have learned. This is accomplished by an adjustment of the gain score with pretest scores, enabling a more fair comparison between groups, particularly in situations where there are differences in pretest scores. The data were considered normal if skewness and kurtosis were in the range of [-1,1]. This was the case except for the procedural flexibility category, because one outlier was skewing the distribution. After its removal, the distribution was normal and so we proceeded with the analysis.

The normalized gain scores are in Table 1. For the procedural questions, the comparison tutor group had significantly higher normalized gain than the sequential tutor group, $t(48) = 2.1$, $p = .048$, $d = 0.3$. However, the Bayes factor only reported the alternative hypothesis to be slightly more likely than the null, $BF_{10} = 1.3$. There was no significant difference in the procedural flexibility normalized gain, $t(54) = 0.3$, $p = .80$, $d = 0.08$; the Bayes factor showed the null model was almost five times as likely, $BF_{01} = 4.8$, which is substantial evidence for the null hypothesis [10]. There was also no significant difference for the transfer normalized gain, $t(52) = 0.9$, $p = .35$, $d = 0.26$, with substantial evidence for the null model, $BF_{01} = 3.3$.

Additional analyses. To gain insight into the types of modifications participants made to the algorithm in the tests, we performed a high-level analysis of the first three procedural flexibility questions on the pretest and posttest (we selected these questions as they afforded the most freedom to implement changes). We coded the answers for three broad categories of changes: (1) counter initialization, (2) order, which included changes to the order of steps in the algorithm as well as related changes (e.g., condition that breaks the loop); (3) other (e.g., changing the value that counter is increased by). The coding of the data was categorical, i.e., we identified the presence or absence of each type of change in a given answer (we did not count the *number* of changes); correctness was not considered in the coding. The categories were not mutually exclusive and so a given solution could at most have three 'present' labels. We subsequently calculated the percentage of participants who used a given modification strategy, taking into account all three questions (but only including in the analysis answers that had modifications corresponding to the target categories, to exclude answers that were blank, duplicated the existing algorithm, etc.). Table 2 shows the results. Changes to the order of statements were the most common and a similar percentage of participants used them in the pretest compared to the posttest for the comparison group (the sequential group decreased slightly from pre to posttest). A much smaller percentage of participants changed the initial value of the counter variable, despite the fact that this type of change is arguably simpler to implement. (Recall that both initialization and order strategies could be implemented as the three strategies are not mutually exclusive).

5   Discussion and Future Work

Prior research involving mathematics activities demonstrated that comparison of examples improves learning over presenting examples in isolation. We tested this approach in the context of a CTAT tutoring system that presented examples aimed to encourage algorithmic thinking. In our study, each comparison activity involved two different

Table 2. Percentage of participants who performed a target modification (init, order, other) in the first 3 flexibility questions on the pretest and posttest. *Note:* Init = counter initialization.

	Pretest			Posttest		
	Init	Order	Other	Init	Order	Other
Comparison tutor	11.1%	79.6%	13.0%	19.5%	77.9%	6.5%
Sequential tutor	17.4%	80.4%	15.2%	23.5%	72.1%	8.8%

algorithms for solving a problem involving a while loop. While both algorithms provided the solution to the same problem (a loop program), and both included the same set of steps, the algorithms were not identical. Differences included initial variable values, logic of conditionals to break the loop, and order of the algorithmic steps. Thus, the differences went beyond the surface level, because they could not be resolved with superficial pattern matching of the type described in [17]. We chose to focus on algorithms for a single construct, namely while loops, based on our teaching experiences. In particular, when our students first learn about while loops, some believe that aspects like order of steps is fixed, i.e., they are not flexible in their algorithmic thinking. When we demonstrate several algorithms for the same while loop problem, students' shift in understanding is evident from their responses.

The comparison tutor group had significantly higher normalized gain scores on procedural questions about while loops (with the caveat that the Bayesian analysis did not confirm this result). However, there was no significant difference between the two tutor-version groups (comparison, sequential) for the procedural flexibility gain scores that measured various aspects of algorithmic thinking, with a very small effect. Bayesian statistics provided substantial evidence for the null model, confirming the frequentist statistics pattern. This latter result was surprising, given prior research showing that example comparison improves procedural flexibility over sequential example presentation [5]. One potential explanation relates to the length of the intervention. Much of the prior work took place in students' classrooms over several days. Due to the logistics of laboratory studies, the present study had to be constrained to one two-hour session, limiting the number of examples and testing opportunities (e.g., we did not include a delayed posttest as doing so is logistically very challenging in a lab study).

Another possibility for a lack of differences between the two tutor groups could be due to engagement with the prompts that followed each example pair or example (depending on condition). As a coarse measure of engagement with the prompts, we analyzed length of responses. Note that in both conditions, one third of the prompts could have been answered with just one character (i.e., prompts asking about the number of loop iterations). The average length of a response per prompt was longer for the comparison group than the sequential group (comparison group: $M = 37.7$ characters, $SD = 18.1$; sequential group: $M = 19.2$ characters, $SD = 14.1$). This suggests that engagement with the prompts does not offer an explanation for the lack of a comparison effect. As a next step, we plan to analyze the content of the prompt answers to obtain more fine-grained insight into how students answered the prompts.

Prior research has shown comparison activities are only beneficial if the examples are designed for a novice population [23]. We did aim to design the examples for novices, but perhaps they were too challenging for pure novices, as indicated by the fact that the posttest scores for all three question categories were low. To see if experience interacted with example presentation, we divided the participants into two groups based on their programming experience, as follows: zero prior experience ($n = 41$) or some prior experience ($n = 16$) (note that *some experience* in all but one case corresponded to either a high-school class or tinkering). We did not find evidence that participants with no prior experience were overloaded by comparison activities. For the procedural question category, this group descriptively had almost double the normalized gain when given the comparison tutor as compared to the sequential tutor. For the procedural flexibility section, the gains were very similar between the two tutor versions for the no-prior experience group. The one test section the comparison tutor group had slightly lower gain scores was for the transfer question. Thus, in general, we do not have evidence that comparison was too cognitively overloading.

In sum, we found that comparison activities improve procedural knowledge (here, about the mechanics of while loops) but more research is needed on how this approach can be used to improve flexible algorithmic thinking. In our study, the posttest scores were low, and anecdotally, one of the culprits was that students did not provide evidence of code tracing their algorithms to determine if they functioned correctly. As the next step, we will extend the instructional activities in the tutor to cover this skill, and evaluate the tutor with a talk aloud study to gain insight into how students reason when interacting with the tutor.

Acknowledgements. This work was supported by an NSERC Discovery grant.

References

1. Aleven, V., et al.: Example-tracing tutors: intelligent tutor development for non-programmers. Int. J. Artif. Intell. Educ. **26**(1), 224–269 (2016)
2. Coletta, V., Steinert, J.: Why normalized gain should continue to be used in analyzing preinstruction and postinstruction scores on concept inventories. Phys. Rev. Phys. Education Res. **16** (02 2020)
3. Conati, C., Vanlehn, K.: Toward computer-based support of meta-cognitive skills: a computational framework to coach self-explanation. Int. J. Artif. Intell. Educ. **11**, 398–415 (2000)
4. Du, Y., Luxton-Reilly, A., Denny, P.: A review of research on parsons problems. In: Proceedings of the Twenty-Second Australasian Computing Education Conference, pp. 195-202. ACE'20, ACM (2020)
5. Durkin, K., Star, J.R., Rittle-Johnson, B.: Using comparison of multiple strategies in the mathematics classroom: lessons learned and next steps. ZDM **49**, 585–597 (2017)
6. Fabic, G.V.F., Mitrovic, A., Neshatian, K.: Evaluation of parsons problems with menu-based self-explanation prompts in a mobile python tutor. Int. J. Artif. Intell. Educ. **29**(4), 507–535 (2019)
7. Gadgil, S., Nokes-Malach, T.J., Chi, M.T.: Effectiveness of holistic mental model confrontation in driving conceptual change. Learn. Instr. **22**(1), 47–61 (2012)
8. Gentner, D.: Structure-mapping: a theoretical framework for analogy. Cogn. Sci. **7**(2), 155–170 (1983)

9. Hosseini, R., et al.: Improving engagement in program construction examples for learning python programming. Int. J. Artif. Intell. Educ. **30**(2), 299–336 (2020)

10. Jarosz, A.F., Wiley, J.: What are the odds? A practical guide to computing and reporting Bayes factors. J. Probl. Solving **7**, 2–9 (2014)

11. Jennings, J., Muldner, K.: When does scaffolding provide too much assistance? A code-tracing tutor investigation. Int. J. Artif. Intell. Educ. **31**(4), 784–819 (2021)

12. Kátai, Z.: The challenge of promoting algorithmic thinking of both sciences- and humanities-oriented learners. J. Comput. Assist. Learn. **31**, 287–299 (2015)

13. Knuth, D.E.: Algorithmic thinking and mathematical thinking. Am. Math. Mon. **92**(3), 170–181 (1985)

14. Kumar, Amruth N..: An Evaluation of Self-explanation in a Programming Tutor. In: Trausan-Matu, S., Boyer, K.E., Crosby, M., Panourgia, K. (eds.) ITS 2014. LNCS, vol. 8474, pp. 248–253. Springer, Cham (2014). https://doi.org/10.1007/978-3-319-07221-0_30

15. Ma, N., Qian, J., Gong, K., Lu, Y.: Promoting programming education of novice programmers in elementary schools: a contrasting cases approach for learning programming. Educ. Inf. Technol. **28**(7), 9211–9234 (2023)

16. McLaren, B.M., van Gog, T., Ganoe, C., Karabinos, M., Yaron, D.: The efficiency of worked examples compared to erroneous examples, tutored problem solving, and problem solving in computer-based learning environments. Comput. Hum. Behav. **55**, 87–99 (2016)

17. Muldner, K., Conati, C.: Scaffolding meta-cognitive skills for effective analogical problem solving via tailored example selection. Int. J. Artif. Intell. Educ. **20**(2), 99–136 (2010)

18. Najar, A.S., Mitrovic, A., McLaren, B.M.: Learning with intelligent tutors and worked examples: selecting learning activities adaptively leads to better learning outcomes than a fixed curriculum. User Model. User-Adap. Inter. **26**(5), 459–491 (2016)

19. Price, T.W., Williams, J.J., Solyst, J., Marwan, S.: Engaging students with instructor solutions in online programming homework. In: Proceedings of the CHI Conference on Human Factors in Computing Systems, pp. 1–7. ACM (2020)

20. Quintana, D.S., Williams, D.R.: Bayesian alternatives for common null-hypothesis significance tests in psychiatry: a non-technical guide using JASP. BMC Psychiatry **18**(1), 178 (2018)

21. Rittle-Johnson, B., Star, J.R.: Compared with what? The effects of different comparisons on conceptual knowledge and procedural flexibility for equation solving. J. Educ. Psychol. **101**(3), 529–544 (2009)

22. Rittle-Johnson, B., Star, J.R.: Does comparing solution methods facilitate conceptual and procedural knowledge? An experimental study on learning to solve equations. J. Educ. Psychol. **99**, 561–574 (2012)

23. Rittle-Johnson, B., Star, J.R., Durkin, K.: Developing procedural flexibility: are novices prepared to learn from comparing procedures? Br. J. Educ. Psychol. **82**(Pt 3), 436–55 (2012)

24. van Gog, T., Kester, L., Paas, F.: Effects of worked examples, example-problem, and problem-example pairs on novices' learning. Contemp. Educ. Psychol. **36**(3), 212–218 (2011)

25. Weinman, N., Fox, A., Hearst, M.A.: Improving instruction of programming patterns with faded Parsons problems. In: Proceedings of the CHI Conference on Human Factors in Computing Systems. CHI '21, ACM (2021)

26. Xie, B., et al.: A theory of instruction for introductory programming skills. Comput. Sci. Educ. **29**(2–3), 205–253 (2019)

Automated Long Answer Grading
with RiceChem Dataset

Shashank Sonkar[(✉)], Kangqi Ni, Lesa Tran Lu, Kristi Kincaid,
John S. Hutchinson, and Richard G. Baraniuk

Rice University, Houston, USA
shashank.sonkar@rice.edu

Abstract. This research paper introduces a new area of study in the
field of educational Natural Language Processing (NLP): Automated
Long Answer Grading (ALAG). Distinguishing itself from traditional
Automated Short Answer Grading (ASAG) and open-ended Automated
Essay Grading (AEG), ALAG presents unique challenges due to the complexity and multifaceted nature of fact-based long answers. To facilitate the study of ALAG, we introduce RiceChem, a specialized dataset
derived from a college-level chemistry course, featuring real student
responses to long-answer questions with an average word count notably
higher than typical ASAG datasets. We propose a novel approach to
ALAG by formulating it as a rubric entailment problem, employing natural language inference models to verify whether each criterion, represented by a rubric item, is addressed in the student's response. This formulation enables the effective use of large-scale datasets like MNLI for
transfer learning, significantly improving the performance of models on
the RiceChem dataset. We demonstrate the importance of rubric-based
formulation in ALAG, showcasing its superiority over traditional score-
based approaches in capturing the nuances and multiple facets of student responses. Furthermore, we investigate the performance of models
in cold start scenarios, providing valuable insights into the data efficiency
and practical deployment considerations in educational settings. Lastly,
we benchmark state-of-the-art open-sourced Large Language Models
(LLMs) on RiceChem and compare their results to GPT models, highlighting the increased complexity of ALAG compared to ASAG. Despite
leveraging the benefits of a rubric-based approach and transfer learning
from MNLI, the lower performance of LLMs on RiceChem underscores
the significant difficulty posed by the ALAG task. With this work, we
offer a fresh perspective on grading long, fact-based answers and introduce a new dataset to stimulate further research in this important area.
The code and dataset can be found at https://github.com/luffycodes/
Automated-Long-Answer-Grading.

Keywords: Automated Long Answer Grading · Rubric-based
Grading · Natural Language Inference · Large Language Models

S. Sonkar and K. Ni—Equal contribution.

A. M. Olney et al. (Eds.): AIED 2024, LNAI 14829, pp. 163–176, 2024.
https://doi.org/10.1007/978-3-031-64302-6_12

1   Introduction

The field of educational Natural Language Processing (NLP) has traditionally focused on short answer grading and open-ended essay grading. This paper explores an innovative, comparatively unexplored domain: Automated Long Answer Grading (ALAG). Unlike open-ended essays, which are assessed on traits such as coherence and originality [13], long answers are fact-based and require a different, more nuanced grading approach. Traditional ASAG methods [3,5,25], which utilize a 5-way classification system categorizing answers as 'correct, partially correct, contradictory, irrelevant, or not in the domain', are not entirely suitable for grading long answers. This is due to the fact that long responses can simultaneously exhibit characteristics of multiple categories, rendering the 5-way ASAG classification formulation ineffective.

To enable a comprehensive study of ALAG, we curated a unique dataset, RiceChem, which consists of 1264 long answer responses from a college-level chemistry course. *RiceChem includes 1264 long answers, each graded against a subset of 27 rubric items, resulting in 8392 data points.* The dataset is characterized by an average word count of 120, significantly higher than existing datasets such as SciEntsBank [8], Beetle [8], and Texas 2011 [21], which have average word counts of 13, 10, and 18 respectively. This stark difference in word count makes RiceChem an apt dataset for the exploration of ALAG.

In view of the shortcomings of traditional ASAG methods for ALAG, we redefine the problem as a rubric entailment task. In this novel formulation, each rubric item serves as a specific criterion that the student's answer should fulfill. We utilize natural language inference models to determine whether each rubric is entailed in the response, enabling a thorough and nuanced grading.

We set baselines for the ALAG task on the RiceChem dataset by fine-tuning encoder models such as BERT [7], RoBERTa [18], and BART [16]. Our findings highlight the increased complexity and challenge of ALAG, even when the grading is facilitated by the rubric-based approach. We demonstrate the importance of rubric-based formulation in ALAG, showcasing its superiority over traditional score-based approaches in capturing the nuances and multiple facets of student responses. Furthermore, we investigate the performance of models in cold start scenarios, providing valuable insights into the data efficiency and practical deployment considerations in educational settings.

Lastly, we benchmark state-of-the-art open-sourced Large Language Models (LLMs) [1,12,17,26–29,33] on RiceChem and compare their results to GPT models [4], highlighting the increased complexity of ALAG compared to ASAG. Despite leveraging the benefits of a rubric-based approach, the lower performance of LLMs on RiceChem, compared to ASAG's SciEntsBank, underscores the significant difficulty posed by the ALAG task.

Contributions. This work marks one of the first attempts, to our knowledge, at tackling Automated Long Answer Grading (ALAG) in the field of educational NLP. Our contributions are threefold. First, we provide a unique dataset,

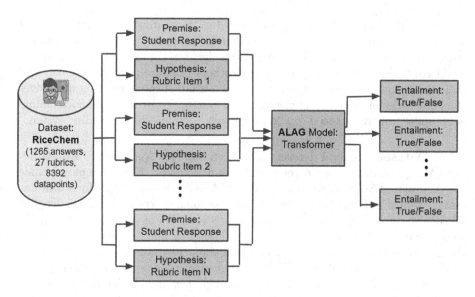

Fig. 1. Schematic illustration of the Automated Long Answer Grading (ALAG) using the RiceChem dataset. The figure highlights our novel approach of formulating ALAG as a rubric entailment problem, where each student response (premise) is paired with a corresponding rubric item (hypothesis). These pairs are then processed by a fine-tuned ALAG-transformer model, which predicts whether the response entails the rubric item. The use of rubrics in RiceChem allows for a detailed, point-by-point evaluation, making the grading process interpretable by design.

RiceChem, designed specifically for ALAG, to encourage further research in this crucial area of educational NLP. Second, we propose a new grading formulation, tailored to address the unique complexities of long answers. Finally, we present a comprehensive evaluation of state-of-the-art models, including LLMs, on the ALAG task, highlighting the challenges and opportunities for future research in this domain.

2   Related Work: ALAG Vs ASAG/AEG

ASAG Datasets. Several ASAG datasets have been developed for different scales and domains, including SciEntsBank, Beetle, and Texas2011. These have been invaluable in advancing automated grading, however, their utility in the context of ALAG is limited. One of the reasons for this is the significant disparity in the average word count of responses in ASAG datasets (ranging from 10 to 18 words) compared to those in our RiceChem dataset (with an average word count of 120).

Furthermore, the grading systems in these ASAG datasets are based on a classification approach, where answers are categorized into 'correct, partially correct, contradictory, irrelevant, or not in the domain' in the case of the 5-way categorization. *While this system may be suitable for short answers, it fails to capture the complexity of long answers.* In a long answer, multiple facets of an answer can simultaneously fall into different categories, leading to a situation where the existing classification system is insufficient. For example, a long answer could contain elements that are correct, others that are partially correct, and still others that are irrelevant, leading to an overlap of categories that challenges the boundaries of the existing classification system. To rectify this shortcoming, we propose a more nuanced grading approach - a rubric-based grading system.

AEG Datasets. Several datasets have been built for Automated Essay Grading (AEG), encompassing diverse scales and domains. Key examples include the Automated Student Assessment Prize (ASAP) dataset [11], which contains essays from 8 different prompts; the International Corpus of Learner English [9] consisting of essays from higher education students of varying English proficiency levels, evaluated on coherence, lexical richness, and grammatical accuracy; the Cambridge Learner Corpus [23], which comprises examination scripts from candidates taking Cambridge ESOL examinations, assessed on content, communicative quality, organization, and language use; and the TOEFL11 dataset [2], which includes essays evaluated on development, organization, and language use, written by English learners for the Test of English as a Foreign Language (TOEFL) exam.

All these datasets, despite their diversity, have grading criteria notably different from those required for ALAG. *While essays emphasize attributes such as originality, coherence, and lexical richness, long answers necessitate a more fact-based grading approach, focusing on the accuracy and completeness of the information.*

3   Dataset and Method

In this section, we first introduce our unique RiceChem dataset, followed by the problem formulation of the ALAG task (Fig. 2). We also provide an overview of our proposed ALAG system, with all its components illustrated in Fig. 1.

3.1   RiceChem Dataset

To enable the exploration of the ALAG task, we have developed the RiceChem dataset. This dataset not only serves as a valuable resource for researchers working on ALAG, but also paves the way for the creation of more reliable and interpretable grading systems that can provide meaningful feedback to students through the use of rubrics.

RiceChem features *4 exam questions, 27 rubric items, and 1264 graded student responses* from a college-level chemistry course. Multiple teaching assistants graded the student responses against individual rubric items by assigning a TRUE or FALSE label. *There are a total of 4880 TRUE labels and 3512 FALSE labels.* Each rubric item holds a designated point value, and the final score for a response is determined by aggregating the scores of rubric items that are correctly answered.

Question: When studying the emission sources within the Milky Way, a satellite detected interplanetary clouds containing silicon atoms that have lost five electrons. b) The ionization energies corresponding to the removal of the third, fourth, and fifth electrons in silicon are 3231, 4356, and 16091 kJ/mol, respectively. Using core charge calculations and your understanding of Coulomb's Law, briefly explain 1) why the removal of each additional electron requires more energy than the removal of the previous one, and 2) the relative magnitude of the values observed. This question can be answered reasonably in around 150 words or fewer.

Rubric Items:

1. Correctly cites decreased electron electron repulsion
2. Relates decreased e/e repulsion to decreased potential energy
3. 3rd and 4th electrons ionized feels same core charge
4. 3rd and 4th electrons ionized from n=3 shell and has same radius
5. 5th electron- ionized from n=2 shell and feels higher core charge
6. 5th e ionized from n =2 shell and has smaller radius
7. Correctly explains relationship of potential energy to ionization energy
8. Partially explains relationship between potential energy and ionization energy

Student Response: The removal of each additional electron requires more energy than the removal of the previous electron due to the decrease in electron-electron repulsion. The more electrons present in the valence shell, the further each individual electron is pushed away from nucleus, as the negative charges of the electrons are pushing against each other. As there are less electrons in the outer shell, there will be a stronger pull on the electron from the positively charged nucleus. The ionization energy of the fifth electron in silicon is significantly larger while the ionization energies of the third and fourth electrons of silicon are much more similar due to the fifth electron being located in a lower electron shell that is closer to the nucleus. As silicon has 4 electrons in its valence shell, after the first four electrons are removed, the next electron (the fifth) will have to be removed from a lower energy level (closer to nucleus), which means smaller radius, greater attraction from nucleus, and a larger ionization energy. The third and fourth electrons removed have similar ionization energy values as they are both from the 3s^2 sub-shell of Silicon.

Fig. 2. An example from our RiceChem dataset showing a question, rubric items, and a student response. Underlined rubric items have been correctly answered by the student.

3.2   Automated Long Answer Grading (ALAG)

Let an inference model $M : (P, H) \to L$ be give, which takes a premise P and a hypothesis H as inputs and predicts a label $L \in \{True, False\}$ indicating whether P *entails* H. To formulate grading as an inference problem, a student response R and a rubric item I can be treated as the premise and hypothesis respectively such that $(R, I) \xrightarrow{M} L$.

Our ALAG approach implements the aforementioned formulation by training a language model to predict the entailment of a rubric item from a student response. The predictions can effectively pinpoint the correctly addressed rubric items in the student response, thus providing automated feedback.

Table 1. Baseline performance of base and large transformer language models on the RiceChem dataset for the ALAG task. The figures in parentheses next to the model names represent the number of model parameters (in millions). We report the mean and standard deviations across 5 runs. The large models exhibit superior performance compared to the base models (except the least performing BERT model), indicating that the task complexity requires more sophisticated models for RiceChem.

Model	Accuracy	Precision	Recall	F1
RoBERTa-base (125M)	83.0 (0.7)	0.830 (0.020)	0.883 (0.020)	0.856 (0.004)
RoBERTa-large (355M)	**84.1** (0.9)	**0.840** (0.009)	0.891 (0.015)	**0.864** (0.008)
BART-base (140M)	83.6 (1.2)	0.832 (0.021)	0.892 (0.014)	0.861 (0.009)
BART-large (406M)	83.9 (0.9)	0.833 (0.009)	**0.897** (0.007)	**0.864** (0.007)
BERT-base (110M)	82.8 (1.1)	0.828 (0.019)	0.882 (0.018)	0.854 (0.008)
BERT-large (340M)	82.5 (0.5)	0.825 (0.009)	0.879 (0.015)	0.851 (0.005)

4 Experiments

In this section, we provide a comprehensive overview of our experimental setup and results. We begin by detailing the training procedure for transformer language models on the RiceChem dataset and introducing the evaluation metrics used throughout our experiments. Next, we present the benchmarking results of various transformer models, including BERT, RoBERTa, and BART, on the ALAG task. We then highlight the importance of entailment-based and rubric-based formulation in ALAG and demonstrate its superiority over traditional score-based approaches. Furthermore, we investigate the performance of these models in cold start scenarios, where limited labeled data are available, and discuss the implications for practical deployment in educational settings. Finally, we evaluate the performance of state-of-the-art open-sourced Large Language Models (LLMs) on RiceChem and compare their results to GPT models, showcasing the increased complexity of ALAG compared to ASAG.

4.1 Experimental Setup

To fine-tune the transformer models on RiceChem, we pre-process the data by employing an 80-10-10 train-validation-test split. Specifically, for each question, we randomly select 80% of student responses for training, 10% for validation, and 10% for testing to ensure three splits of responses are disjoint.

We conduct the experiments using the Hugging Face transformers library [32]. The training process uses an NVIDIA A100-PCIE-40GB GPU. During training, we use the AdamW optimizer [19], setting the initial learning rate to $2e^{-5}$. Each update is performed with a mini-batch size of 16, and the model is trained for a maximum of 10 epochs. The hyper-parameters α and β are set to

Table 2. Performance comparison of large models and their MNLI fine-tuned versions on the RiceChem dataset. RoBERTa and BART show accuracy increases by 3.2% and 1.8%, and F1 score increases by 2.8% and 1.4%. This highlights the benefit of formulating ALAG as an entailment problem, enabling the use of the large MNLI dataset for performance enhancement.

Model	Accuracy	F1
RoBERTa-large	84.1	0.864
RoBERTa-large-mnli	**86.8**	**0.888**
BART-large	83.9	0.864
BART-large-mnli	85.4	0.876

0.9 and 0.999 respectively. After the training, we select the model with the highest validation F1 score as the best model for evaluations. For the experimental baselines, we employ a comprehensive set of evaluation metrics, including accuracy, precision, recall, and F1 score. To ensure robustness, we report the averages and standard deviations of metrics across 5 runs with 5 different seeds.

4.2 Benchmarking on Discriminative Models

We assess the performance of state-of-the-art discriminative language models, such as BERT, RoBERTa, and BART, as benchmarks on the RiceChem dataset. In Table 1, we compare the results achieved by both base and large models. Notably, the large models outperform their base counterparts, demonstrating the advantages of employing more advanced models for this task. However, for the BERT model, it is not the case which can be attributed to the instability of fine-tuning it [22].

4.3 The Value of Entailment Formulation in ALAG

In Table 2, we compare the performance of language models and their MNLI fine-tuned counterparts on the RiceChem dataset. The results demonstrate a significant improvement in both accuracy and F1 score when the models are fine-tuned on the MNLI (Multi-Genre Natural Language Inference Corpus) dataset [31], highlighting the value of formulating ALAG as an entailment problem.

By framing ALAG as an entailment task, we enable the use of the MNLI dataset, which contains a diverse set of premise-hypothesis pairs covering a wide range of topics and linguistic variations. The MNLI dataset, with its 4 million examples, provides a rich source of linguistic knowledge and reasoning capabilities that can be effectively transferred to the ALAG task.

The entailment formulation allows us to leverage the models pre-trained on the MNLI dataset, which have already acquired a strong understanding of the entailment relationship between premises and hypotheses. By fine-tuning these models on the RiceChem dataset, we can efficiently transfer the learned knowledge and adapt it to the specific domain of long answer grading.

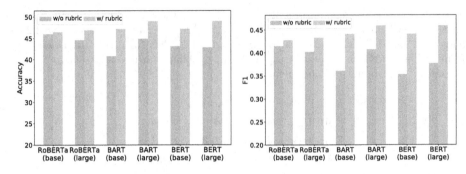

Fig. 3. Comparisons between the traditional score based grading approach and rubric-based ALAG approach on the RiceChem dataset. Rubric-based ALAG offers an average increase of 9.2% in accuracy and an average increase of 15.4% in F1 score, proving that breaking down grading into smaller rubric items helps models focus on smaller parts of the task instead of doing the entire task altogether. The improvement is evident across all models regardless of their number of parameters.

The performance gains observed in Table 2 underscore the effectiveness of this transfer learning approach. The RoBERTa model, when fine-tuned on MNLI, exhibits a 3.2% increase in accuracy and a 2.8% increase in F1 score compared to its vanilla counterpart. Similarly, the BART model shows a 1.8% increase in accuracy and a 1.4% increase in F1 score. These improvements demonstrate the successful transfer of knowledge from the MNLI dataset to the ALAG task, enabled by the entailment formulation.

4.4   The Importance of Rubric-Based Formulation in ALAG

The use of rubrics in automated grading has been shown to lead to performance improvements in short answer grading [20] and essay grading [15]. Our experiments confirm that this holds true for ALAG as well. However, the importance of rubric-based formulation in ALAG is even more pronounced due to the complexity and multifaceted nature of long answers. To illustrate this, we compare our rubric-based ALAG approach to a traditional score-based approach. In the score-based approach, we pre-process the RiceChem dataset by structuring the data into sentences (student responses) and labels (scores), with the language model aiming to predict an integer score from 0 to 8. On the other hand, our rubric-based ALAG formulation breaks down the grading process into smaller, more manageable components, allowing the model to focus on specific aspects of the answer as defined by the rubric items.

Figure 3 demonstrates the superiority of the rubric-based approach, with an average improvement of 9.2% in accuracy and 15.4% in F1 score compared to the traditional score-based approach. This significant performance gain highlights the importance of leveraging rubrics in ALAG. By decomposing the complex task of grading long answers into smaller, well-defined rubric items, the model can more effectively capture the nuances and multiple facets of student responses.

Table 3. Performance of RoBERTa-Large-MNLI on unseen questions. The model is trained on three questions and evaluated on the remaining question, demonstrating its ability to generalize to new questions without prior training data.

Unseen	Accuracy	Precision	Recall	F1
Q1	65.9	0.703	0.717	0.705
Q2	68.7	0.704	0.584	0.629
Q3	66.7	0.649	0.644	0.633
Q4	60.6	0.892	0.611	0.717

It is worth noting that creating high-quality rubrics is a challenging task that requires careful consideration and effort. However, this effort needs to be invested only once, and the benefits of a well-designed rubric can be reaped repeatedly in the automated grading process. The rubric serves as a comprehensive framework that guides the model in assessing the key aspects of the answer, ensuring a more accurate and reliable grading outcome.

Moreover, the use of rubrics in ALAG enhances the interpretability and transparency of the grading process. By aligning the model's predictions with specific rubric items, educators and students can gain a clearer understanding of the strengths and weaknesses of the answers, facilitating targeted feedback and improvement.

4.5   Benchmarking on Cold Start Scenarios

In educational settings, it is common to encounter situations where limited labeled data is available for training automated grading models, especially when dealing with new courses, subjects, or question types. Therefore, it is crucial to assess the performance of models in cold start settings and understand how their performance evolves as more training data becomes available. The analysis in this section provides valuable insights into the data efficiency of the models and helps determine the minimum amount of labeled data required to achieve satisfactory grading results.

We begin by evaluating the performance of the RoBERTa-Large-MNLI model on unseen questions, simulating a scenario where the model is fine-tuned on some questions but is applied to grade responses for a completely new question without any prior training data. For this type of investigation, we trained the model on three questions in the dataset, and used the remaining unseen question for testing. As shown in Table 3, the model demonstrates a reasonable level of generalization, with accuracy ranging from 60.6% to 68.7% and F1 scores ranging from 0.629 to 0.717 across the four questions. This suggests that the model, fine-tuned on similar types of questions, has acquired some level of transferable knowledge that enables it to handle unseen questions to a certain extent, which can be valuable in educational settings where labeled data for new questions may be scarce.

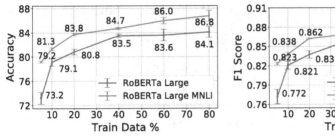
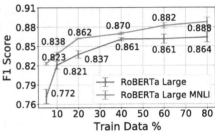

(a) Accuracy of RoBERTa models   (b) F1 Score of RoBERTa models

Fig. 4. Performance of RoBERTa-Large and RoBERTa-Large-MNLI models with varying amounts of training data, ranging from 5% to 80%. The models show consistent improvement in accuracy and F1 score as more labeled data becomes available, with diminishing returns after 40% for RoBERTa-Large and 20% for RoBERTa-Large-MNLI.

Next, we investigate the performance of the RoBERTa-Large model and its MNLI fine-tuned version as the amount of training data increases from 5% to 80%. Figure 4 illustrates the trends in accuracy and F1 score for both models. As expected, there is a consistent improvement in performance as more training data becomes available. For RoBERTa Large, the accuracy increases from 73.2% to 84.1%, and the F1 score rises from 0.772 to 0.864. Similarly, for the MNLI fine-tuned version, the accuracy improves from 79.2% to 86.8%, and the F1 score increases from 0.823 to 0.888.

Interestingly, the performance gains exhibit diminishing returns, particularly after 40% of the training data for RoBERTa-Large and 20% for RoBERTa-Large-MNLI. This observation suggests that the models can achieve competitive grading results even with a relatively small amount of labeled data, and the benefits of additional data become less pronounced beyond a certain point. Moreover, the standard deviations for both accuracy and F1 score remain within 1.12% across different seeds, indicating the reliability and consistency of the models' performance.

These findings have important implications for practical deployment scenarios in educational institutions. The analysis provides guidance on the trade-off between labeling effort and performance gains, allowing educators and administrators to make informed decisions based on their specific requirements and resource constraints when implementing automated grading systems.

4.6 Benchmarking on Large Language Models

The rapid advancements in LLMs have led to significant improvements in various natural language processing tasks. To assess the potential of these models in the context of ALAG, we evaluate the zero-shot performance of several LLMs on the RiceChem dataset, as presented in Table 4. For GPT models [4], we used the following hyperparameters: a temperature of 1.0, a frequency penalty of 0,

Table 4. Zero-shot performance comparison of Large Language Models (LLMs) on the RiceChem dataset for the Automated Long Answer Grading (ALAG) task. The results highlight the complexity of ALAG, with even the best-performing model, GPT-4, achieving an accuracy of 70.9% and an F1 score of 0.689. The performance gap between LLMs on ALAG and their performance on Automated Short Answer Grading (ASAG) tasks emphasizes the unique challenges posed by grading long, fact-based answers.

LLM	Accuracy	Precision	Recall	F1
Phi 2 [17]	58.65	64.15	9.16	16.04
Gemma 7B 1.1 IT [29]	60.05	58.71	24.53	34.6
Mistral 7B Instruct-v0.2 [12]	61.9	60.59	33.15	42.86
OLMo 7B Instruct [10]	60.39	57.5	31.0	40.28
Zephyr 7B Beta [30]	57.03	50.6	11.32	18.5
Vicuna 13B v1.5 [6]	59.58	**81.08**	8.09	14.71
Qwen1.5 14B Chat [1]	58.65	63.64	9.43	16.43
Qwen1.5 32B Chat [1]	62.37	60.44	36.66	45.64
Yi 34B Chat [33]	58.19	53.22	24.53	33.5
GPT3.5 [24]	67.1	59.1	45.7	51.6
GPT4 [14]	**70.9**	63.9	**74.8**	**68.9**

and a presence penalty of 0. For open-sourced LLMs [1,6,10,12,17,29,30,33], we considered the log probabilities of the 'True' and 'False' tokens to determine if the student answered the rubric item correctly or not. By comparing these log probabilities, we obtained the model's predicted label.

Despite the impressive success of LLMs in many domains, the RiceChem dataset proves to be a formidable challenge. The best-performing model, GPT-4, achieves an accuracy of 70.9% and an F1 score of 0.689, highlighting the complexity of the ALAG task. This performance is particularly striking when compared to the results of GPT models on ASAG tasks.

Previous research has shown that GPT models can achieve an F1 score of 0.74 on the SciEntsBank dataset for ASAG without the use of rubrics [14]. In contrast, GPT-4 obtains a lower F1 score of 0.69 on RiceChem, despite the significant beneficial impact of rubrics typically observed in ASAG [20] and AEG [15] tasks. This discrepancy underscores the increased difficulty of ALAG compared to ASAG.

It is worth noting that the actual difference in complexity between ASAG and ALAG may be even more substantial than the five-point difference in F1 scores suggests. The use of rubrics in RiceChem provides a structured framework for grading, which is expected to enhance model performance. However, even with this advantage, GPT-4 still struggles to match its performance on ASAG tasks without rubrics.

The results in Table 4 also reveal the varying performance of different LLMs on the RiceChem dataset. While GPT-4 and GPT-3.5 stand out as the top performers, other models such as Qwen1.5 32B Chat [1] and Mistral [12] show promising results, with F1 scores of 0.456 and 0.429, respectively. These findings indicate that the choice of LLM architecture and training methodology can have a significant impact on the model's ability to handle the complexities of ALAG.

In summary, the benchmarking of LLMs on the RiceChem dataset highlights the unique challenges posed by the ALAG task. The performance gap between GPT models on ASAG and ALAG tasks, even with the benefit of rubrics, emphasizes the need for further research and development of specialized models and techniques to effectively tackle the complexities of grading long, fact-based answers. As LLMs continue to evolve, it will be crucial to explore their potential in the context of ALAG and develop strategies to harness their capabilities for improving automated grading systems in educational settings.

5   Conclusion

In this paper, we introduce the novel task of Automated Long Answer Grading (ALAG) and present the RiceChem dataset, specifically designed to facilitate research in this domain. Our rubric-based formulation of ALAG provides a nuanced and pedagogically sound approach to evaluating long answers, offering a more comprehensive assessment compared to traditional ASAG methods. Through extensive experiments, we demonstrate the importance of rubric-based formulation, the value of entailment formulation, and the challenges posed by cold start scenarios. Furthermore, our benchmarking of state-of-the-art models, including LLMs, confirms that ALAG poses a significantly greater challenge compared to ASAG. We believe this work will stimulate and inspire further research in this crucial area of educational NLP, contributing to the development of advanced models capable of handling the complexities and intricacies of the ALAG task.

Acknowledgments. This work was supported by NSF grant 1842378, ONR grant N0014-20-1-2534, AFOSR grant FA9550-22-1-0060, a Vannevar Bush Faculty Fellowship, and ONR grant N00014-18-1-2047.

References

1. Bai, J., et al.: Qwen technical report. arXiv preprint arXiv:2309.16609 (2023)
2. Blanchard, D., Tetreault, J., Higgins, D., Cahill, A., Chodorow, M.: Toefl11: A corpus of non-native english. ETS Research Report Series **2013**(2), i–15 (2013)
3. Bonthu, Sridevi, Rama Sree, S.., Krishna Prasad, M.. H.. M..: Automated short answer grading using deep learning: a survey. In: Holzinger, Andreas, Kieseberg, Peter, Tjoa, A Min, Weippl, Edgar (eds.) CD-MAKE 2021. LNCS, vol. 12844, pp. 61–78. Springer, Cham (2021). https://doi.org/10.1007/978-3-030-84060-0_5
4. Bubeck, S., et al.: Sparks of artificial general intelligence: early experiments with GPT-4. arXiv preprint arXiv:2303.12712 (2023)

5. Burrows, S., Gurevych, I., Stein, B.: The eras and trends of automatic short answer grading. Int. J. Artif. Intell. Educ. **25**, 60–117 (2015)
6. Chiang, W.L., et al.: Vicuna: an open-source chatbot impressing gpt-4 with 90%* chatgpt quality (2023). https://lmsys.org/blog/2023-03-30-vicuna/
7. Devlin, J., Chang, M.W., Lee, K., Toutanova, K.: Bert: pre-training of deep bidirectional transformers for language understanding. arXiv preprint arXiv:1810.04805 (2018)
8. Dzikovska, M., et al.: Semeval-2013 task 7: the joint student response analysis and 8th recognizing textual entailment challenge. In: Proceedings of the Second Joint Conference on Lexical and Computational Semantics, vol. 2, pp. 263–274 (2013)
9. Granger, S., Dagneaux, E., Meunier, F., Paquot, M., et al.: International corpus of learner English, vol. 2. Presses universitaires de Louvain Louvain-la-Neuve (2009)
10. Groeneveld, D., et al.: Olmo: accelerating the science of language models. arXiv preprint arXiv:2402.00838 (2024)
11. Hamner, B., Morgan, J., Lynnvandev, M.S., Ark, T.V.: The hewlett foundation: automated essay scoring (2012). https://kaggle.com/competitions/asap-aes
12. Jiang, A.Q., et al.: Mistral 7b. arXiv preprint arXiv:2310.06825 (2023)
13. Klebanov, B.B., Madnani, N.: Automated Essay Scoring. Springer, Heidelberg (2022)
14. Kortemeyer, G.: Performance of the pre-trained large language model gpt-4 on automated short answer grading. arXiv preprint arXiv:2309.09338 (2023)
15. Kumar, R., Mathias, S., Saha, S., Bhattacharyya, P.: Many hands make light work: using essay traits to automatically score essays. arXiv preprint arXiv:2102.00781 (2021)
16. Lewis, M., et al.: Bart: denoising sequence-to-sequence pre-training for natural language generation, translation, and comprehension (2019)
17. Li, Y., Bubeck, S., Eldan, R., Del Giorno, A., Gunasekar, S., Lee, Y.T.: Textbooks are all you need ii: phi-1.5 technical report. arXiv preprint arXiv:2309.05463 (2023)
18. Liu, Y., et al.: Roberta: a robustly optimized BERT pretraining approach. CoRR arxiv:1907.11692 (2019)
19. Loshchilov, I., Hutter, F.: Decoupled weight decay regularization (2019)
20. Marvaniya, S., Saha, S., Dhamecha, T.I., Foltz, P., Sindhgatta, R., Sengupta, B.: Creating scoring rubric from representative student answers for improved short answer grading. In: Proceedings of the 27th ACM International Conference on Information and Knowledge Management, pp. 993–1002 (2018)
21. Mohler, M., Bunescu, R., Mihalcea, R.: Learning to grade short answer questions using semantic similarity measures and dependency graph alignments. In: Proceedings of the 49th Annual Meeting of the Association for Computational Linguistics: Human Language Technologies, pp. 752–762 (2011)
22. Mosbach, M., Andriushchenko, M., Klakow, D.: On the stability of fine-tuning bert: misconceptions, explanations, and strong baselines (2021)
23. Nicholls, D.: The cambridge learner corpus: error coding and analysis for lexicography and elt. In: Proceedings of the Corpus Linguistics 2003 Conference, vol. 16, pp. 572–581. Cambridge University Press Cambridge (2003)
24. Ouyang, L., et al.: Training language models to follow instructions with human feedback. Adv. Neural. Inf. Process. Syst. **35**, 27730–27744 (2022)
25. Ramesh, D., Sanampudi, S.K.: An automated essay scoring systems: a systematic literature review. Artif. Intell. Rev. **55**(3), 2495–2527 (2022)
26. Sonkar, S., Chen, X., Le, M., Liu, N., Basu Mallick, D., Baraniuk, R.: Code soliloquies for accurate calculations in large language models. In: Proceedings of the 14th Learning Analytics and Knowledge Conference, pp. 828–835 (2024)

27. Sonkar, S., Liu, N., Mallick, D., Baraniuk, R.: Class: a design framework for building intelligent tutoring systems based on learning science principles. In: Findings of the Association for Computational Linguistics: EMNLP 2023, pp. 1941–1961 (2023)
28. Sonkar, S., Ni, K., Chaudhary, S., Baraniuk, R.G.: Pedagogical alignment of large language models. arXiv preprint arXiv:2402.05000 (2024)
29. Team, G., et al.: Gemma: open models based on gemini research and technology. arXiv preprint arXiv:2403.08295 (2024)
30. Tunstall, L., et al.: Zephyr: direct distillation of LM alignment. arXiv preprint arXiv:2310.16944 (2023)
31. Williams, A., Nangia, N., Bowman, S.R.: A broad-coverage challenge corpus for sentence understanding through inference. CoRR arxiv:1704.05426 (2017)
32. Wolf, T., et al.: Huggingface's transformers: state-of-the-art natural language processing (2020)
33. Young, A., et al.: Yi: open foundation models by 01. ai. arXiv preprint arXiv:2403.04652 (2024)

Knowledge Tracing as Language Processing: A Large-Scale Autoregressive Paradigm

Bojun Zhan[1], Teng Guo[1], Xueyi Li[1], Mingliang Hou[2], Qianru Liang[1], Boyu Gao[1], Weiqi Luo[1], and Zitao Liu[1(✉)]

[1] Guangdong Institute of Smart Education, Jinan University, Guangzhou, China
zbj0613@stu2022.jnu.edu.cn,
{tengguo,liangqr,bygao,lwq,liuzitao}@jnu.edu.cn
[2] TAL Education Group, Beijing, China

Abstract. Knowledge tracing (KT) is the process of modelling students' cognitive states to forecast their future academic performance, using their historical learning interactions as a reference. Recent scholarly investigations have introduced a range of deep learning-based knowledge tracing (DLKT) methodologies, which have demonstrated considerable potential in their outcomes. Considering the excellent performance of large models in various domains, we have explored the possibility of migrating their architecture to the KT domain. We posit that the efficacy of the large language model (LLM) can be largely attributed to the utilization of an auto-regressive Transformer decoder, which facilitates the learning of comprehensive representations and the processing of extensive data. Hence, we propose a DLKT model, *LLM-KT*, which is based on the LLM architecture. This model addresses the long-term dependency between students' historical interactions and their subsequent performance through a stack of Transformer decoders. To fully utilize the potential of large models, we evaluated our model capabilities on EdNet, which is currently the world's largest real KT dataset. Through a series of quantitative and qualitative experimental analyses, we answer two key questions: (1) is it feasible to apply the LLM-like architectures in the KT domain? (2) can the continuous extension of models improve prediction performance in KT? To encourage reproducible research, we make our data and code publicly available at https://github.com/ai4ed/AIED2024-LLM-KT.

1 Introduction

Knowledge tracing (KT) is a sequence prediction task that uses data from a student's previous learning interactions to model the current student's knowledge state and then predict future performance on the next question (as illustrated in Fig. 1). The potential of these predictive capabilities, when combined with high-quality educational resources and instruction, could significantly enhance and expedite the learning process for students.

© The Author(s), under exclusive license to Springer Nature Switzerland AG 2024
A. M. Olney et al. (Eds.): AIED 2024, LNAI 14829, pp. 177–191, 2024.
https://doi.org/10.1007/978-3-031-64302-6_13

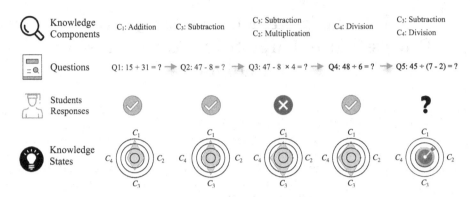

Fig. 1. An illustration of the KT problem.

Research related to KT has been underway since the 1990 s. To the best of our understanding, Corbett and Anderson were the pioneers in this field, being the first to estimate a student's current knowledge in relation to each individual knowledge component (KC) [7]. A KC is a description of a mental structure or process that a learner uses, alone or in combination with other KCs, to accomplish steps in a task or a problem[1]. Since then, many attempts have been made to solve the KT problem, such as probabilistic graphical models and factor analysis based models [13,20]. Recently, with the rapid development of deep learning techniques, impressive progress has been made in proposing deep learning-based knowledge tracing (DLKT) models [2,5,8,12,15,18–20,29]. Such as autoregressive-based deep sequential KT models [9,20], memory-based KT models [28,29], attention-based KT models [5,8,19] and graph-based KT models [18].

Recently, large language models (LLMs), such as GPT4 [1] and Gemini [24] have gained substantial attention due to their robust predictive capabilities across various domains. The success behind LLMs significantly benefits from rich representations learning via autoregressive Transformer decoders. The stack of Transformer decoders in models like GPT4 and Gemini performs exceptionally well, one crucial reason being its deep-layered architecture and self-attention mechanisms. These layers enable complex, contextual language processing, while self-attention provides flexibility in understanding word relationships [21]. Enhanced by stable training techniques and a scalable architecture, LLMs demonstrate proficient handling of diverse linguistic patterns, which is indicative of their robustness in natural language processing [1,24].

When training DLKT models, each student interaction sequence is represented by student responses to discrete questions or to KCs sequences, which are very similar to word sequences in LLM. Given this context, a compelling question arises drawing inspiration from the success of LLMs: can the mechanisms driving LLMs be effectively applied in the domain of KT? The essence

---

[1] A KC is a generalization of everyday terms like concept, principle, fact, or skill.

of this question revolves around whether a large-scale autoregressive paradigm, similar to that in LLMs, can capture the underlying patterns in the evolution of student knowledge in KT tasks. To the best of our knowledge, we are the first to propose this question in KT field. However, experimentation with large-scale models presents an enormous challenge, necessitating not only efficient model architectures but also imposing stringent demands on both data quality and computational resources. These prerequisites add layers of difficulty to this task, making it a highly demanding endeavor.

In light of this context, our study is dedicated to exploring this pivotal question and addressing the associated challenges. Specifically, we choose to use the decoder-only Transformer as our building block and design a large model for KT tasks, i.e., *LLM-KT*. We first proposed an interaction encoder to learn the representation of learning interactions from the student sequences. And then to capture the relevance among students' historical interactions and their performance on the next questions, we use a stack of Transformer decoders to handle long-term dependencies. We choose to use EdNet [6] as our training and testing dataset which is the largest publicly available KT dataset in this world. Earlier studies on KT have only selected partial samples from the EdNet dataset. In contrast to previous studies, we train and test different versions of sizes of *LLM-KT* with the complete set of EdNet dataset to fully utilize the learning capabilities of large models. 80% of student sequences were used for training and validation, while a random sample of 40,000 student sequences was selected from the remaining 20% for model evaluation. The method is evaluated on a benchmark dataset and compared with 18 previous methods under a strict KT evaluation protocol [13]. The experimental results show that *LLM-KT* has an average performance enhancement of 9.7% over the existing 18 DLKT models. However, the performance improvement levels off as we keep increasing the model size.

2   Related Work

2.1   Deep Learning Based Knowledge Tracing

Recently, deep neural networks have been increasingly used in the KT domain due to their ability to handle large amounts of data and effectively learn representations. Piech et al. presented DKT, which pioneered the application of deep learning in KT tasks by employing an LSTM layer to encapsulate students' knowledge states and predict students' response performances [20]. Since then, many methods have tended to use deep learning techniques to solve KT problems [8,12,18,29]. For example, Zhang et al. incorporated a static matrix to store the relationships among KCs and a dynamic matrix to track students' knowledge state [29]. Pandey et al. utilized a self-attention mechanism to capture relations between exercises and the students' responses [19]. Ghosh et al. presented AKT, which utilizes two self-attention modules to extract the underlying relationships of questions and interactions respectively, and explicitly model students' forgetting behaviors via a monotonic attention mechanism [8]. Choi et al. employed the Transformer-based encoder-decoder architecture to respectively capture the sequences of students' exercises and responses [5]. Nakagawa et al. conceptualized the knowledge structure as a graph, redefining the KT task as a time-series, node-level classification problem within the context of Graph Neural Networks [18]. Liu et al. proposed a simple but tough-to-beat KT model by additionally capturing the individual differences among questions with the same KCs using the ordinary dot-product attention function [12].

2.2 Pretrained Large Models

The advent and subsequent surge in popularity of GPT have led to a notable increase in the utilization of pretrained LLMs across diverse research fields [16]. GPT's remarkable proficiency in generating text akin to human language underscores the potential and efficacy of LLMs in tasks related to natural language processing. These pretrained models, with their inherent ability to extract language semantics and structure from extensive amounts of unlabelled text data, have prompted a paradigm shift in how researchers approach language-related tasks. Pretrained Large Models are not only developed on language tasks, they have been shown to succeed in many domains. For example, Xiao et al. proposed a framework called SpeedyFeed, which efficiently trains news recommenders of superior quality [26]. Chen et al. developed a new pretraining model, IPT, for low-level visual tasks, which outperforms the current state-of-the-art methods on various low-level benchmarks [3].

3   Data

As the aim is to train a KT model based on the LLM-like architecture, this study requires a dataset with sufficient size to match the model's extensive parameters. EdNet[2] is an extensive dataset of educational information gathered from Korean students learning English via Santa, an AI-powered tutoring system. To ensure a systematic and organized presentation of diverse activities, EdNet offers four datasets with varying degrees of abstraction. For this research, we used the most basic dataset, which is a log documenting students' attempts to complete exercises. This log contains 131,441,538 interactions from 784,309 students collected over 2 years.

We scrutinized a range of real-world datasets such as Assistment2009, Algebra2005, Bridge2006, NeurIPS, XES3G5M and EdNet. Among these, EdNet notably surpassed the others in terms of data size, as illustrated in Table 1. Therefore, it was ultimately chosen for our empirical investigation. The abundance of this data can provide a more diverse and comprehensive array of information, which is advantageous for training robust and generalizable models. Furthermore, the superior average number of KCs present in EdNet suggests a broader coverage of knowledge areas. This could potentially enhance the complexity and richness of the model. Despite having a lower count of questions

---

[2] https://github.com/riiid/ednet.

compared to other datasets, the sheer volume of interactions can still offer a plethora of contexts for each question, thereby augmenting the model's capacity to comprehend and generate various responses. In response to these benefits, we decided to conduct further experiments around EdNet.

Table 1. Dataset statistics of 6 datasets.

	Assistment2009	Algebra2005	Bridge2006	NeurIPS	XES3G5M	EdNet
# of Students	4,217	574	1,146	4,918	18,066	784,309
# of Interactions	346,860	809,694	3,679,199	1,382,727	5,549,635	131,441,538
# of questions	26,688	210,710	207,856	948	7,652	13,169
# acg KCs	1.1969	1.3634	1.0136	1.0148	1.164	2.2611

4   The LLM-KT Model

The objective of KT models can be articulated as the prediction of a student's likelihood to correctly answer a question, denoted as q_*. This is based on the chronologically ordered sequence of a student's past interactions, represented as $S = \{s_j\}_{j=1}^{T}$, where T signifies the total number of interactions. Each interaction is represented as a 4 tuple s, i.e., $s = < q_j, \{c | c \in \mathcal{N}_q\}, r_j, t_j >$, where q_j, $\{c\}$, r_j, t_j correspond to the distinct question, the associated KC set, the student's response[3], and the timestamp of the student's response, respectively. The term $\mathcal{N}_q$ denotes the set of KCs that are correlated with the question q_j. We would like

Table 2. Notations and descriptions.

Notations	Description
s_j	a student's past interaction
q_j	a question ID
c_j	a KC ID
d_t	a dataset ID
$\mathcal{N}_q$	a set of KCs that are associated with a question
r_j	the ground-truth label of student response of q_j
t_j	the timestamp of the student's response
$\hat{r}_j$	the output probability that a student can answer q_j correctly
$\mathbf{q}_t$	the latent representation of question
$\bar{\mathbf{c}}_t$	the latent representation of KCs associated with question
$\mathbf{d}_t$	the latent presentation of dataset
$\bar{\mathbf{a}}_t$	the latent representation of response
$\mathbf{e}_t$	the latent representation of interaction
$\mathbf{x}_t$	the latent representation of combination of question, KCs and dataset
$\mathbf{h}^{(l)}$	the student's knowledge state matrix
S	a set of a student past T interaction s_j

---

[3] The response r_j takes a value in the set $\{0,1\}$, where 1 signifies that the student answered correctly, and 0 indicates otherwise.

to estimate the probability $\hat{r}_*$ of the student's future performance on arbitrary question q_*. For easier follow, the mathematical notations used in this paper are summarized in Table 2.

4.1 Interaction Encoder

In real-world educational scenarios, the question bank is usually much bigger than the set of KCs, for example, the number of questions is more than 60 times larger than the number of KCs in the well-cited EdNet dataset (described in Table 1). Therefore, to effectively learn and fairly evaluate the DLKT models from such highly sparse question-response data, most research studies such as DKT [20] and SAKT [19] mitigate the data sparsity problem by employing KCs to index questions, i.e., setting $q_j = c_j$. All questions containing the same KC are defined as the same question. However, this conversion forces the DLKT models to follow the homogeneous assumption, thus ignoring the different learning effects due to the individualized nature of different questions with the same KCs. On the other hand, question-centric models such as qDKT [23] completely ignore the relationships between questions and KCs and use only the question sequence to track students' knowledge states. Some research studies such as AKT [8] have used problem-specific difficulty representations to capture individual differences between questions with the same KCs. In this study, we choose to use question sequences to track students' knowledge states. Unlike qDKT, we use additional dataset side information, meanwhile capturing the intrinsic relations between questions and KCs at a more fine-grained level. The interaction encoder approach in different studies is shown in Table 3.

Table 3. Illustrations of different interaction encoding approaches. M, N represent the total number of KCs and questions in the dataset, respectively. $\mathbf{m}^q$ denotes the representation of difficulty of question q_t which contains the KC c_j. $\oplus$ is the element-wise addition operator. $\odot$ is the element-wise product operator.

Model	Interaction Encoding Method
DKT/SAKT	$\mathbf{e}_t \in \{0,1\}^{2M}$
qDKT	$\mathbf{e}_t \in \{0,1\}^{2N}$
AKT	$\mathbf{e}_t = \mathbf{c}_j \oplus \mathbf{m}^q \odot \mathbf{q}_t \oplus \mathbf{r}_t$
LLM-KT	$\mathbf{e}_t = \mathbf{q}_t \oplus \bar{\mathbf{c}}_t \oplus \mathbf{d}_t \oplus \mathbf{r}_t$

Specifically, each question sequences interaction e_t is represented as an aggregation of question, response, the corresponding set of KCs, and the dataset ID, i.e.,

$$\mathbf{x}_t = \mathbf{q}_t \oplus \bar{\mathbf{c}}_t \oplus \mathbf{d}_t; \quad \mathbf{e}_t = \mathbf{x}_t \oplus \mathbf{r}_t$$
$$\mathbf{q}_t = \mathbf{W}^q \cdot \mathbf{e}_t^q; \quad \mathbf{r}_t = \mathbf{W}^a \cdot \mathbf{a}_t^q; \quad \mathbf{d}_t = \mathbf{W}^d \cdot \mathbf{d}_t^q$$

where $\mathbf{e}_t^q \in \mathbb{R}^{N \times 1}$ is the one-hot vector indicating the corresponding question q_t. $\mathbf{a}_t^q \in \mathbb{R}^{2 \times 1}$ is the one-hot vector indicating whether the question q_t is answered correctly. $\mathbf{d}_t^q \in \mathbb{R}^{20 \times 1}$ is the one-hot vector indicating the corresponding dataset ID which is used to distinguish future downstream finetune datasets. $\mathbf{W}^q \in \mathbb{R}^{d \times N}$, $\mathbf{W}^a \in \mathbb{R}^{d \times 2}$, and $\mathbf{W}^d \in \mathbb{R}^{d \times 20}$ are learnable linear transformation operations. N is the total number of questions in EdNet dataset. $\oplus$ is the element-wise addition operator. $\cdot$ is the standard matrix/vector multiplication. $\bar{\mathbf{c}}_t$ is the average embedding of the associated KCs to the question q_t, i.e.,

$$\bar{\mathbf{c}}_t = \frac{1}{|C_{q_t}|} \sum_{j=1}^{M} \mathbf{c}_j * \mathbb{I}(c_j \in \mathcal{N}_{q_t}); \quad \mathbf{c}_j = \mathbf{W}^c \cdot \mathbf{e}_t^c$$

where $\mathbf{c}_j \in \mathbb{R}^{d \times 1}$ is one of the latent representations of the related KC to question q_t. $\mathbf{e}_t^c \in \mathbb{R}^{M \times 1}$ is the one-hot vector indicating the related KC from EdNet dataset. $\mathbf{W}^c \in \mathbb{R}^{d \times M}$ is learnable linear transformation operations. M is the total number of KCs in the question bank. C_{q_t} is the size of $\mathcal{N}_{q_t}$. $\mathbb{I}(\cdot)$ is the indicator function. The framework of the interaction encoder and the forward procedure is shown in Fig. 2.

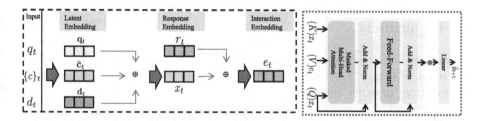

Fig. 2. The structure of interaction encoder (left) and forward procedure (right).

4.2 Auto-Regressive Training Module

We use a stack of Transformer decoders to capture student knowledge states dynamically, i.e.,

$$\mathbf{h}^{(0)} = \mathbf{E} \oplus \mathbf{P}; \quad \mathbf{h}^{(l)} = \textbf{Tranformer\_block}(\mathbf{h}^{(l-1)}) \forall l \in [1, L]$$

where $\mathbf{E} = (\mathbf{e}_1, ..., \mathbf{e}_T)$ denote the representations of T past interactions of a student. The position embedding is represented by $\mathbf{P}$. The symbol L signifies the number of layers. The knowledge state embedding matrix of a student by T past interactions is represented by $\mathbf{h}^{(l)} \in \mathbb{R}^{T \times d}$. It is important to note that in order to estimate the knowledge states of students through their historical interactions, we utilize QC embeddings $\mathbf{x}_t = \mathbf{q}_t \oplus \bar{\mathbf{c}}_t \oplus \mathbf{d}_t$ for mapping both queries and keys of the self-attention mechanism, and interaction embedding $\mathbf{e}_t$ for mapping values.

4.3   Prediction Module

We employ a two-layer fully connected network as a mechanism to refine the state of knowledge and utilize ReLU activation function and sigmoid activation function, denoted as $\sigma(\cdot)$ to forecast a student's performance on the subsequent question q_{t+1}:

$$\mathbf{y}_{t+1} = \text{ReLU}(\mathbf{W}_1 \cdot [\mathbf{h}_{t+1}^{(l)}; \mathbf{x}_{t+1}] + \mathbf{b}_1)$$
$$\hat{r}_{t+1} = \sigma(\mathbf{w}^\top \cdot \text{ReLU}(\mathbf{W}_2 \cdot \mathbf{y}_{t+1} + \mathbf{b}_2) + b)$$

where $\mathbf{W}_1 \in \mathbb{R}^{d \times 2d}$, $\mathbf{W}_2 \in \mathbb{R}^{d \times d}$, $\mathbf{w} \in \mathbb{R}^{d \times 1}$, $\mathbf{b}_1 \in \mathbb{R}^{d \times 1}$, $\mathbf{b}_2 \in \mathbb{R}^{d \times 1}$ and b are trainable parameters.

All learnable parameters in *LLM-KT* are trained in an end-to-end way by minimizing the binary cross entropy loss between predicted probability $\hat{r}_t$ and the ground-truth label r_t:

$$\mathcal{L} = -\sum_{t=1}^{T} \left(r_t \log \hat{r}_t + (1 - r_t) \log(1 - \hat{r}_t) \right)$$

5   Experiment

In this section, the experimental setup and the corresponding results are described in detail. Then we conduct experiments with the aim of answering the following research questions:

- **RQ1**: Is it feasible to apply the LLM-like architectures in the field of KT?
- **RQ2**: Are larger models uniformly better in KT domain?

5.1   Baselines

To comprehensively and systematically evaluate the performance of *LLM-KT*, we carefully classify the existing DLKT models into 4 categories and compared *LLM-KT* against selected 18 baselines as follows:

Deep Sequential KT Models that utilize an autoregressive framework to dynamically monitor the knowledge states of students, selected baselines are as follows:

- DKT [20]: it utilizes an LSTM layer to simulate the learning processes of students.
- DKT+ [28]: it enhances the original DKT model by tackling the issues of reconstruction and inconsistency.
- DKT-F [17]: it enhances original DKT by considering students' forgetting behaviors.

- qDKT [23]: it's a variant of DKT that models a student's response to individual questions over a period of time.
- LPKT [22]: it constructs a learning cell to emulate the pedagogical processes of students for the purpose of assessing their knowledge states.
- IEKT [14]: it concurrently evaluates the knowledge states of students through the modules of cognitive understanding and knowledge acquisition.
- QIKT [4]: it enhances the interpretability of students' knowledge modeling by collectively learning its representation through a question-centric knowledge acquisition module and a problem-solving module.
- AT-DKT [11]: it introduces two auxiliary learning tasks, namely the question tagging prediction task and the individualized prior knowledge prediction task, to enhance the prediction performance of the DKT model.

Memory Augmented KT Models that capture latent relations between KCs and student knowledge states via memory networks, selected baselines are as follows:

- DKVMN [29]: it integrates a static matrix to preserve the relationships among KCs, and a dynamic matrix to monitor the evolving state of a student's knowledge.
- DeepIRT [27]: it incorporates DKVMN and item response theory to enhance the interpretability of the prediction output of DKVMN.

Attention-Based KT Models that capture dependencies between interactions via the attention mechanism, selected baselines are as follows:

- SAKT [19]: it employs self-attention to discern the relevance between historical interactions and KCs.
- SAINT [5]: it is a Transformer-based model for KT that encodes questions and responses in the encoder and decoder respectively.
- AKT [8]: it utilizes three self-attention modules to estimate the relevance between questions and historical interactions, and explicitly models a student's forgetting behavior through a monotonic attention mechanism.
- simpleKT [12]: it investigates the ordinary dot-product attention-based KT models by capturing the individual differences among questions that cover the same set of KCs.
- sparseKT [10]: it incorporates a k-selection module to only pick items with the highest attention scores to improve the robustness and generalization of the attention-based DLKT approaches.

Other KT Models that do not belong to the above categories, selected baselines are as follows:

- ATKT [9]: it introduces adversarial perturbations into the student interaction sequences to improve the generalization capability.

- GKT [18]: it casts the knowledge structure as a graph and reformulates the KT task as a time series node-level classification problem in GNN.
- HawkesKT [25]: it utilizes the Hawkes process to model temporal cross-effects in student historical interactions.

5.2 Experiment Setting

As mentioned above, EdNet is used as the dataset for both the training and testing phases. Interactions that do not conform to the 4 tuple interaction representation or any type of information is missing are systematically excluded. Furthermore, student sequences with less than three trials are also removed. To optimize computational efficiency, student interaction sequences with a length exceeding 200 are truncated according to the methodology proposed by Liu et al. [13].

To investigate whether the larger the model, the better it performs in the KT domain, we train and benchmark five $LLM\text{-}KT$ models ranging from 334 million parameters to 7 billion parameters. The corresponding model sizes and architectures are summarized in Table 4. n_{params} is the total number of trainable parameters. n_{layer} is the number of Transformer decoder blocks. d_{model} is the size of the embedding dimension. d_{head} is the number of dimension of each attention head and d_{ff} is the dimension of feedforward layer.

Table 4. The model sizes and architectures of trained $LLM\text{-}KT$.

Model Name	n_{params}	n_{layers}	d_{model}	d_{head}	d_{ff}
$LLM\text{-}KT$-334M	334M	24	1024	16	1024
$LLM\text{-}KT$-794M	794M	32	1536	24	2560
$LLM\text{-}KT$-1.8B	1.8B	40	2560	40	2560
$LLM\text{-}KT$-4B	4B	64	3600	40	3600
$LLM\text{-}KT$-7B	7B	72	4160	40	4160

5.3 Implementation Details

We used 80% of the student sequences for training and validation, and a random sample of 40,000 student sequences was selected from the remaining 20% for model evaluation. We selected ADAM as the optimizer to train our model. The number of training epochs is set to 50. All the models are implemented in PyTorch and are trained on a cluster of Linux servers with the NVIDIA RTX A800 GPU devices. We used 2, 4, and 8 cards to train models with sizes 334 M, 794 M, and 1.8 B respectively. Conversely, for the larger models of sizes 4 B and 7 B, we employed 16 cards during the training process. Owing to constraints related to computational resources and cost, it is impractical to conduct an exhaustive grid search for hyperparameters in models comprising billions of parameters. Hence, the batch size is configured at 512, accompanied by the implementation of gradient accumulation. The hyperparameters learning rate and dropout rate are searched from [0.0001, 0.00001], [0.1, 0.2] for fair comparison in various model sizes.

5.4   Results

To answer RQ1, we compare *LLM-KT* performance with SOTA baselines. The overall performance of baselines and *LLM-KT* are shown in Table 5.

Overall Performance. From the above results, we find the following results: (1) In comparison to the aforementioned 18 benchmark models, *LLM-KT*-334M consistently achieves the highest rank in terms of AUC scores. The results unequivocally dispel the preliminary skepticism and confirm that the LLM-like modelling architectures can be successfully adapted to the KT domain, substantially enhancing prediction accuracy, contingent upon the availability of ample data. (2) *LLM-KT* always significantly outperforms attention-based DLKT models. Specifically, it exhibits notable enhancements in AUC scores, with an average increase of 9.7%. This observation substantiates the efficacy of the interaction encoder and the autoregressive training methodology employed within the Transformer decoder framework proposed by our study. (3) The observed AUC performance differential between *LLM-KT* and other models such as QIKT, IEKT, and LPKT is relatively modest, averaging an increment of merely 1.8%. This outcome may be ascribed to the architectural simplicity inherent in *LLM-KT*. In stark contrast to the more elaborate models, which incorporate additional modules upon an initial RNN architecture, *LLM-KT* is characterized by its utilization of a straightforward autoregressive decoder-only architecture, suggesting that a less complex structure does not substantially compromise and may even be conducive to, effective performance in KT applications.

Impact of Sizes of *LLM-KT* In order to answer RQ2, we varied the parameter sizes of *LLM-KT*, specifically 334 M, 784 M, 1.8 B, 4 B, and 7 B. The AUC scores of the model are displayed in Fig. 3.

From the above results, we can see that: (1) The model's predictive performance demonstrates a diminishing return and reaches a plateau with the escalation in model parameters. This provides a response to RQ2, affirming that in the domain of KT, an increase in model size does not invariably translate to enhanced performance. (2) When the model parameters reached 4B, there was a significant drop in model prediction performance. During training, we observed that the loss and the AUC were decreasing simultaneously (shown in Fig. 4), which we interpreted as overfitting. We suggest that model size and data scale are complementary to each other. Although EdNet is the largest KT dataset, it is still small compared to a mix of datasets containing trillions of tokens to train an LLM in the NLP domain. Therefore, there is no consistent performance boost when we increase the size of model parameters.

Table 5. The overall evaluation performance in terms of AUC and Accuracy of all the baseline models and *LLM-KT*. The Usage of Questions/KCs means whether the information of KCs and questions is used in the interaction encoder.

Method	Model Type	Usage of Questions	Usage of KCs	AUC	Accuracy
DKT [20]	Sequential	No	Yes	0.6649	0.6775
DKT+ [28]	Sequential	No	Yes	0.6579	0.6756
DKT-F [17]	Sequential	No	Yes	0.6768	0.6834
DKVMN [29]	Memory	No	Yes	0.6518	0.6689
ATKT [9]	Other	No	Yes	0.6675	0.6786
GKT [18]	Other	No	Yes	0.6469	0.6676
SAKT [19]	Attention	No	Yes	0.6539	0.6722
SAINT [5]	Attention	Yes	Yes	0.7047	0.6709
AKT [8]	Attention	Yes	Yes	0.7065	0.6788
HawkesKT [25]	Others	Yes	Yes	0.7376	0.7113
DeepIRT [27]	Memory	No	Yes	0.6485	0.6701
LPKT [22]	Sequential	Yes	Yes	0.7758	0.7321
IEKT [14]	Sequential	Yes	Yes	0.7787	0.7335
qDKT [23]	Sequential	Yes	No	0.7689	0.7273
AT-DKT [11]	Sequential	Yes	Yes	0.6902	0.6892
simpleKT [12]	Attention	Yes	Yes	0.6990	0.6699
QIKT [4]	Sequential	Yes	Yes	0.7715	0.7292
sparseKT [10]	Attention	Yes	Yes	0.6930	0.6698
LLM-KT-334M	Attention	Yes	Yes	**0.7933**	**0.7420**
LLM-KT-794M	Attention	Yes	Yes	0.7928	0.7415
LLM-KT-1.8B	Attention	Yes	Yes	0.7900	0.7401
LLM-KT-4B	Attention	Yes	Yes	0.7761	0.7312
LLM-KT-7B	Attention	Yes	Yes	0.7762	0.7312

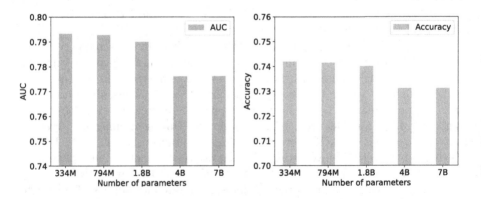

Fig. 3. Impact of sizes of *LLM-KT*.

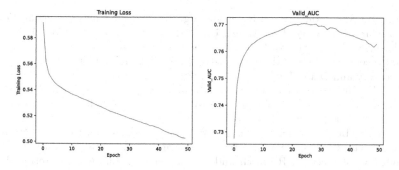

Fig. 4. Training Loss and Valid_AUC during *LLM-KT*-4B training.

6 Conclusion

In this paper, we propose a large-scale autoregressive paradigm for the KT domain by using a stack of Transformer decoders. To the best of our knowledge, we are the first to apply an LLM-like architecture in the KT domain. The experimental results on EdNet, both quantitative and qualitative, demonstrate that the LLM-like architecture is suitable for application in the KT domain. However, it is important to note that the prediction performance tends to level off as the model grows, indicating that the architecture may not be as effective for larger models.

Acknowledgements. This work was supported in part by National Key R&D Program of China, under Grant No. 2022YFC3303600 and in part by Key Laboratory of Smart Education of Guangdong Higher Education Institutes, Jinan University (2022LSYS003).

References

1. Achiam, J., et al.: Gpt-4 technical report. arXiv preprint arXiv:2303.08774 (2023)
2. Carlon, J.: A shap-inspired method for computing interaction contribution in deep knowledge tracing. In: Artificial Intelligence in Education: 24th International Conference, pp. 460–465 (2023)
3. Chen, H., et al.: Pre-trained image processing transformer. In: Proceedings of the IEEE/CVF Conference on Computer Vision and Pattern Recognition, pp. 12299–12310 (2021)
4. Chen, J., Liu, Z., Huang, S., Liu, Q., Luo, W.: Improving interpretability of deep sequential knowledge tracing models with question-centric cognitive representations. In: Proceedings of the AAAI Conference on Artificial Intelligence. vol. 37, pp. 14196–14204 (2023)
5. Choi, Y., et al.: Towards an appropriate query, key, and value computation for knowledge tracing. In: Proceedings of the Seventh ACM Conference on Learning@Scale, pp. 341–344 (2020)
6. Choi, Y., et al.: Ednet: A large-scale hierarchical dataset in education. In: Artificial Intelligence in Education: 21st International Conference, pp. 69–73 (2020)

7. Corbett, A.T., Anderson, J.R.: Knowledge tracing: modeling the acquisition of procedural knowledge. User Model. User-Adap. Inter. **4**(4), 253–278 (1994)

8. Ghosh, A., Heffernan, N., Lan, A.S.: Context-aware attentive knowledge tracing. In: Proceedings of the 26th ACM SIGKDD International Conference on Knowledge Discovery and Data Mining, pp. 2330–2339 (2020)

9. Guo, X., Huang, Z., Gao, J., Shang, M., Shu, M., Sun, J.: Enhancing knowledge tracing via adversarial training. In: Proceedings of the 29th ACM International Conference on Multimedia, pp. 367–375 (2021)

10. Huang, S., Liu, Z., Zhao, X., Luo, W., Weng, J.: Towards robust knowledge tracing models via k-sparse attention. In: Proceedings of the 46th International ACM SIGIR Conference on Research and Development in Information Retrieval, pp. 2441–2445 (2023)

11. Liu, Z., et al.: Enhancing deep knowledge tracing with auxiliary tasks. In: Proceedings of the ACM Web Conference 2023, pp. 4178–4187 (2023)

12. Liu, Z., Liu, Q., Chen, J., Huang, S., Luo, W.: simpleKT: a simple but tough-to-beat baseline for knowledge tracing. In: International Conference on Learning Representations (2023)

13. Liu, Z., Liu, Q., Chen, J., Huang, S., Tang, J., Luo, W.: PYKT: A python library to benchmark deep learning based knowledge tracing models. In: Thirty-sixth Conference on Neural Information Processing Systems (2022)

14. Long, T., Liu, Y., Shen, J., Zhang, W., Yu, Y.: Tracing knowledge state with individual cognition and acquisition estimation. In: Proceedings of the 44th International ACM SIGIR Conference on Research and Development in Information Retrieval, pp. 173–182 (2021)

15. Lu, Y., Wang, D., Meng, Q., Chen, P.: Towards interpretable deep learning models for knowledge tracing. In: International Conference on Artificial Intelligence in Education, pp. 185–190 (2020)

16. Min, B., et al.: Recent advances in natural language processing via large pre-trained language models: a survey. ACM Comput. Surv. **56**(2), 1–40 (2023)

17. Nagatani, K., Zhang, Q., Sato, M., Chen, Y.Y., Chen, F., Ohkuma, T.: Augmenting knowledge tracing by considering forgetting behavior. In: The World Wide Web Conference, pp. 3101–3107 (2019)

18. Nakagawa, H., Iwasawa, Y., Matsuo, Y.: Graph-based knowledge tracing: modeling student proficiency using graph neural network. In: 2019 IEEE/WIC/ACM International Conference on Web Intelligence, pp. 156–163. IEEE (2019)

19. Pandey, S., Karypis, G.: A self-attentive model for knowledge tracing. In: 12th International Conference on Educational Data Mining, pp. 384–389. International Educational Data Mining Society (2019)

20. Piech, C., et al.: Deep knowledge tracing. In: Twenty-eighth Conference on Neural Information Processing Systems (2015)

21. Pu, S., Yudelson, M., Ou, L., Huang, Y.: Deep knowledge tracing with transformers. In: Bittencourt, I.I., Cukurova, M., Muldner, K., Luckin, R., Millán, E. (eds.) Artificial Intelligence in Education: 21st International Conference, AIED 2020, Ifrane, Morocco, July 6–10, 2020, Proceedings, Part II, pp. 252–256. Springer International Publishing, Cham (2020). https://doi.org/10.1007/978-3-030-52240-7_46

22. Shen, S., et al.: Learning process-consistent knowledge tracing. In: Proceedings of the 27th ACM SIGKDD Conference on Knowledge Discovery and Data Mining, pp. 1452–1460 (2021)

23. Sonkar, S., Waters, A.E., Lan, A.S., Grimaldi, P.J., Baraniuk, R.G.: qDKT: question-centric deep knowledge tracing. In: Proceedings of The 13th International Conference on Educational Data Mining, pp. 677–681 (2020)

24. Team, G., et al.: Gemini: a family of highly capable multimodal models. arXiv preprint arXiv:2312.11805 (2023)
25. Wang, C., et al.: Temporal cross-effects in knowledge tracing. In: Proceedings of the 14th ACM International Conference on Web Search and Data Mining, pp. 517–525 (2021)
26. Xiao, S., et al.: Training large-scale news recommenders with pretrained language models in the loop. In: Proceedings of the 28th ACM SIGKDD Conference on Knowledge Discovery and Data Mining, pp. 4215–4225 (2022)
27. Yeung, C.K.: Deep-IRT: Make deep learning based knowledge tracing explainable using item response theory. arXiv preprint arXiv:1904.11738 (2019)
28. Yeung, C.K., Yeung, D.Y.: Addressing two problems in deep knowledge tracing via prediction-consistent regularization. In: Proceedings of the Fifth Annual ACM Conference on Learning at Scale, pp. 1–10 (2018)
29. Zhang, J., Shi, X., King, I., Yeung, D.Y.: Dynamic key-value memory networks for knowledge tracing. In: Proceedings of the 26th International Conference on World Wide Web, pp. 765–774 (2017)

Can GPT4 Answer Educational Tests? Empirical Analysis of Answer Quality Based on Question Complexity and Difficulty

Luiz Rodrigues[1]([✉]) [ID], Filipe Dwan Pereira[2,4] [ID], Luciano Cabral[3,4] [ID], Geber Ramalho[5] [ID], Dragan Gasevic[6] [ID], and Rafael Ferreira Mello[4] [ID]

[1] Federal University of Alagoas, Maceió, Brazil
luiz.rodrigues@nees.ufal.br
[2] Federal University of Roraima, Boa Vista, Brazil
[3] Federal Institute of Pernambuco, Jaboatão dos Guararapes, Brazil
[4] CESAR School, Centro de Estudos e Sistemas Avançados do Recife, Recife, Brazil

[5] Federal University of Pernambuco, Recife, Brazil
[6] Monash University, Clayton, Australia

Abstract. While recent advancements in Large Language Models (LLMs) suggest their potential to tackle these challenges, limited research exists on how well LLMs respond to open-ended questions with varying difficulty and complexity. This paper addresses this gap by comparing GPT4's performance with human counterparts, considering question difficulty (assessed through Item Response Theory – IRT) and complexity (categorized based on Bloom's taxonomy levels) using a dataset of 7,380 open-ended questions related to high school topics. Overall, the results indicate that GPT4 surpasses non-native speakers and demonstrates comparable performance to native speakers. Moreover, despite facing challenges in tasks involving basic recall or creative thinking, GPT4's performance notably improves with increasing question difficulty. Therefore, this paper contributes empirical evidence on GPT4's effectiveness in addressing open-ended questions, enhancing our understanding of its potential and limitations in educational settings. The findings offer valuable insights for practitioners and researchers seeking to incorporate LLMs into educational practices, such as assessment, virtual assistant and feedback.

Keywords: Large Language Models · GPT4 · Question-answering · Assessment

1 Introduction

Recent advancements in Large Language Models (LLMs) have attracted researchers' attention to explore their potential in the educational context [22].

A. M. Olney et al. (Eds.): AIED 2024, LNAI 14829, pp. 192–205, 2024.
https://doi.org/10.1007/978-3-031-64302-6_14

LLMs hold the promising potential to empower intelligent tutoring systems (ITSs) as LLMs might equip ITSs with question-answering based features that go beyond developers' expectations [19,22]. For instance, they might enable automating question-answering systems (e.g., student coaching and detailed, open-ended feedback provision) to guide learners throughout the learning process [6,19]. While prior research has explored similar tasks, recent literature highlights the need for updating educational innovations with state-of-the-art LLMS [22].

However, the literature on LLMs for question-answering remains limited. Previous studies, which are mainly focused on GPT-based LLMs, have either focused on assessing their effectiveness in answering multiple-choice questions [17] or depend on contextual reading material to enable LLMs to respond to open-ended questions [20]. Furthermore, these studies lack a comprehensive assessment of LLMs' performance in answering open-ended questions of varied difficulty levels and complexity, as well as how their language capabilities compare to native and non-native speakers [13,17,20]. This comparison is prominent to ensure the responsible and ethical deployment of LLMs within educational technology, where it is crucial to mitigate potential biases inherent in AI-based systems [21].

Therefore, this paper presents an empirical study assessing the capacity of LLMs to answer open-ended questions in an educational context, comparing their performance with human counterparts and examining variations based on question difficulty and complexity. For this, we rely on a dataset featuring 7,380 open-ended questions answered either by native and non-native speakers or GPT4, in which question complexity is categorized according to levels of Bloom's taxonomy [1]. Moreover, we employ Item Response Theory (IRT) to gauge question difficulty and acquire a deeper, particular understanding regarding each question [5]. Therefore, this paper contributes empirical evidence on the efficacy of a state-of-the-art LLM in addressing open-ended questions of varying complexities, offering insights for practitioners and researchers into its potential and limitations in educational settings.

2   Literature Review

LLMs have rapidly advanced, becoming powerful tools for diverse Natural Language Processing (NLP) tasks [23]. Built upon the transformer architecture [18], with more than 175 billion parameters [4], LLMs capture contextual dependencies through self-attention mechanisms. Recent studies indicate their potential to adapt to various tasks without new training [16], facilitating knowledge transfer [4]. The literature reports an advancement in question answering emerged with the introduction of GPT-4 [15]. GPT-4 demonstrates heightened capabilities in comprehending and responding to intricate queries, significantly enhancing contextual understanding and language generation [4].

The effectiveness of LLMs in Question Answering (QA) has been demonstrated by recent studies, particularly those built on the GPT architecture, for automating the question-answering process [6,19,22]. In Divya [6], seven pre-trained embedding models were compared to assess their similarity with student responses. Using regression models, the researchers predicted scores for short-answer questions within the Mohler dataset, employing RMSE and Pearson correlation coefficients for evaluation. Another investigation [19] explored the potential of generative AI as an automated teacher coach, focusing on scoring transcript segments, identifying instructional highlights, and providing actionable suggestions for student reasoning. Expert evaluations indicated that ChatGPT's insights were relevant but lacked novelty, aligning with existing teacher actions in 82% of cases. In a comprehensive scoping review [22], 118 peer-reviewed papers were analyzed to understand the current state of using LLMs for automating educational tasks, revealing practical and ethical challenges including updating existing innovations with advanced LLMs, such as GPT-4.

In Rosol et al. [17], the efficacy of two LLMs, ChatGPT (GPT-3.5) and GPT-4, was extensively evaluated during the Polish Medical Final Examination (MFE). The assessment, encompassing three MFE editions (Spring, Autumn 2022, and Spring 2023) in English and Polish, focused on model accuracy and the correlation between answer correctness and various metrics. GPT-4 consistently outperformed GPT-3.5 across all MFE iterations, irrespective of the examination language. A K-12 education study [13] introduced a hybrid automatic question-answering system, combining Knowledge-Based Question Answering (KB-QA) and Information Retrieval-based Question Answering (IR-QA) methodologies. The system's empirical evaluation, involving over 9,000 questions, highlighted its remarkable average accuracy rate exceeding 70%, emphasizing the hybrid approach's effectiveness in processing educational questions and its potential for enhancing learning experiences in the K-12 sector. In de Winter [20], GPT-3.5's performance in Dutch national high school exams was examined. ChatGPT achieved an average score of 7.3, closely aligning with the national student average of 6.99, demonstrating its proficiency. GPT-4 surpassed this with an average score of 8.3.

These studies provided valuable contributions towards understanding how LLMs contribute to QA. However, available research is limited by relying on small datasets, focusing on a specific domain, lacking investigations on how question complexity and difficulty affect LLMs' performance, or shedding light into how their performance compare to those of humans (e.g., native and non-native speakers) [13,17,20]. This is important because comprehensive insights into the generalizability and limitations of LLMs across diverse contexts are crucial for advancing their utility. Addressing these gaps will enhance our understanding of the relationship between LLMs, question characteristics, and human performance, fostering more effective applications and informed development. Therefore, there is a lack of understanding of how state-of-the-art LLMs, such as GPT-4, perform in QA given questions of varied topics, complexities, and difficulty levels.

3 Method

The objective of this study was to assess the capacity of GPT-4 to answer open-ended questions of varied topics to understand its performance in questions of varied difficulty and complexity levels. In addressing this gap, we explored Bloom's taxonomy and Item Response Theory (IRT) as measures of complexity and difficulty, respectively. Analysis of the model's responses with Bloom's taxonomy levels – spanning from basic knowledge recall to higher-order cognitive skills like synthesis and evaluation – provided a detailed evaluation of its understanding and proficiency in varied complexity levels. All this contributed to a more holistic and informative assessment of the model capabilities [1]. Furthermore, incorporating IRT [14] enhanced the precision of the assessment process and provided a principled means to understand the questions' difficulty, ensuring that the assessment accurately gauges one's cognitive abilities. Hence, this approach enables the comprehensive assessment of GPT-4's capabilities.

3.1 Dataset

We compiled an extensive dataset of 7,380 responses to 738 open-ended questions. The dataset was generated through the Clickworker[1] crowdsourcing platform, employing a methodology consistent with prior research [2,9].

First, crowd workers were tasked with formulating questions ("Generated Question") at various Bloom's taxonomy levels and focusing on specific subjects ("topic"). They were given detailed instructions explaining the concepts associated with each taxonomy level, supplemented by examples. Each generated question comprised 5 to 20 words, resulting in 738 questions distributed evenly across topics and Bloom's taxonomy levels. In total, the dataset covers three distinct topics related to high school topics: Biology, Earth Sciences, and Physics.

Second, we generated responses for each question through a dual approach. Initially, we engaged crowd workers to create eight responses per question, considering language proficiency (native and non-native speakers) and gender (male and female) as stratification criteria. Additionally, we employed the GPT-4 model to generate two more answers for each question automatically. Thus, our dataset included 10 answers for each question, comprising two responses from each category: native-female speakers, non-native-female speakers, native-male speakers, non-native-male speakers, and GPT-4. Moreover, the dataset is evenly distributed across the six levels of Bloom's taxonomy (Remembering, Understanding, Applying, Analyzing, Evaluating, and Creating), with 1,230 (16,7%) instances for each level.

Lastly, crowd workers were assigned the task of assessing the answers using a grading scale that ranged from 0 to 5. The assessment criteria were clearly defined: a score of 5 indicated an *excellent* answer, 4 denoted a *very good* answer,

---

[1] https://www.clickworker.com/.

3 was assigned to a *good* answer, 2 represented an *acceptable but somewhat simplistic and lacking in detail* answer, 1 was given to a *slightly unclear* response, and 0 was reserved for answers deemed *incorrect or not matching the question*. Throughout this evaluation, the assessors were directed to consider three fundamental aspects: the completeness of the content, the stylistic presentation, and the quality of argumentation. This methodical evaluation process was crafted to ensure a thorough assessment of each response, aiming at enhancing the reliability and usefulness of our dataset.

3.2 GPT Model and Prompt

A fundamental aspect when leveraging the capabilities of GPT-4 is the creation of well-structured prompts [8]. The formulation of prompts plays a critical role in optimizing GPT-4's performance, ensuring that the generated responses align precisely with the intended goals and requirements of the task at hand [8].

In our scenario, the prompt was crafted based on the guidelines employed in the dataset creation's second phase, encompassing criteria for answer evaluation and output format. Additionally, we incorporated the directive 'think step by step' to enhance the performance of GPT-4 [11]. The conclusive prompt for GPT-4 QA was:

- **Instruction**: Think step by step to answer the question.
- **Criteria**: A good output should be coherent, include the main concepts related to the question, and present a clear argumentation in the response.
- **Output format**: The answer should have up to 100 words in length, and should be in English only.
- **Data input**: Question: [question]. Answer:

3.3 Data Analysis

The data analysis process was based on Hierarchical Linear Modeling (HLM) because our dataset had a hierarchical/nested structure [10]. This technique is suitable for this type of data as it acknowledges and controls for the relationship of data within a particular group, which would violate the assumptions of classical regression analysis [7]. For this, HLM estimates fixed and random effects (or coefficients). *Fixed* effects concern the overall structure of the regression model and are independent of their groupings. *Random* effects capture the variance between groups, estimating how fixed effects vary depending on their groupings. For instance, while *difficulty* is a fixed effect, the levels of Bloom's taxonomy might concern the random effects and indicate how difficulty changes depending on each level.

The first step in HLM is data preparation. Here, we first calculated the IRT parameters. For this, we used the Georgia Tech Item Response Theory package[2]. As our dependent variable (i.e., answer *evaluation*) was in the Likert scale, the

---

[2] https://eribean.github.io/girth/.

straightforward approach was to use a polynomial approach that estimates the difficulty of each *point* of that scale (e.g., the difficulty of an evaluation 1, 2, and so on for a given question). However, when a given question has no answer with a particular evaluation, its difficulty cannot be estimated [5]. Therefore, we also extracted IRT-based difficulty using a binary approach. For this, our coding procedure was correct for evaluations equal to three or more and incorrect otherwise, considering that the highest score is five.

Respectively, we estimated difficulty parameters using the partial credit and Rasch models for polynomial and binary approaches [5]. In both cases, we used marginal maximum likelihood. Additionally, we acknowledge the different topics in our dataset by estimating the IRT parameters separately. When a parameter could not be estimated in the polynomial approach, we fulfilled it with the mean value of similar ones, following standard, experimental practices in regression modeling [10]. Lastly, we standardized the difficulty values using the z-score to ensure they had a mean of 0. This is important for regression models as they help with model convergence and coefficient interpretation [7].

The subsequent step in HLM is model development. For this, we mainly used two R^3 packages: *lme4* and *lmerTest* [3,12]. Following literature suggestions, we adopted a top-down approach to optimize the model development process [10]. Importantly, we estimated IRT parameters based on two approaches (i.e., polynomial and binary) to comply with our dataset's characteristics. In practice, however, those estimates might overlap, yielding similar features that might complicate the HLM rather than improve it [7]. Thereby, we decided to develop two models, one for each IRT approach, and then compare them to understand which one provides the best fit.

In our top-down model development, we first fitted a *full model* for the binary-IRT approach. Full models are often defined based on the data available and the researchers' assumptions [10]. Accordingly, our full model might be represented as follows:

$$e_{ij} = \beta_0 + \beta_1 \cdot s_{ij} + \beta_2 \cdot d_{ij} + \beta_3 \cdot s_{ij} \cdot d_{ij} + \gamma_{0j} + u_{0j} \cdot s_{ij} + v_{0j} \cdot d_{ij} + \epsilon_{ij}$$

where:

- e_{ij} is the evaluation score for the ith observation within the jth group.
- $\beta_0, \beta_1, \beta_2$, and β_3 are fixed effect coefficients for the intercept, s (speaker), d (difficulty), and their interaction, respectively.
- γ_{0j} represents the random intercept for the jth grouping variable (question_id, bloom, topic).
- u_{0j} and v_{0j} are the random slopes for s (speaker) and d (difficulty), respectively, within the jth grouping variable.
- s_{ij} and d_{ij} are the values of s (speaker) and d (difficulty) for the ith observation within the jth group.
- ϵ_{ij} represents the residual error term.

---

[3] https://www.r-project.org/.

This full model features fixed effects concerning the one who answered the question (GPT4, native speaker, or non-native speaker) and the question's difficulty (after standardization), as estimated based on the IRT. As difficulty's effect on answer quality might change depending on who answered it, the full model features and interaction between the fixed effects. Concerning the random effects, the full model features those for questions, levels of Bloom's taxonomy, and question topics so that their intercepts can vary from one group to another. Additionally, to further acknowledge the grouping structure, the full model allowed the slopes of Bloom levels and topics to vary. Hence, this model was expected to capture how IRT-based difficulty affects the quality of GPT4's answers, compared to humans, while controlling for variations due to the question's topic and level of Bloom's taxonomy.

Next, we used lmerTest's *step* function to perform backward elimination of non-significant coefficients. This procedure firstly assesses all random effects, then repeats the process for the fixed ones [12]. At each step, it evaluates whether removing a coefficient significantly impacts model fit based on Likelihood Ratio Tests (LRT), the recommended procedure in the literature [10]. Here, we followed the default parameters for term removal of 90% and 95% confidence levels for random and fixed coefficients, respectively. Notably, this procedure ensures that coefficient removal does not violate the hierarchy of terms in the model. Thereby, it optimizes the model development process while following literature standards.

Finally, we achieved the best model for the binary approach. This model only featured parameters that significantly affected their fit. Then, we repeated this procedure for the polynomial alternative. The single difference was that it featured five difficulty coefficients, one for each point of the Likert scale, instead of a single one. Finally, we compared the best models of both approaches, using the same comparison approach, to define our final model.

4   Results

This section presents the results of the backward feature elimination process for both IRT approaches, which leads to our final model. Then, the section analyzes the best model to understand which/how factors such as speaker nature, question difficulty and topic, and Bloom's levels affect answer evaluation.

4.1   Model Development

Tables 1 and 2 show the results of the top-down modeling process for the binary IRT approach. Table 1 demonstrates that the only random effect to affect model fit significantly was that of Bloom levels according to the speaker that answered the question ($p < 0.001$). In terms of fixed effects, Table 2 shows that both speaker and difficulty, but not their interaction, significantly affected the model fit (both p-values < 0.01). Therefore, the best model for the binary approach featured speaker and difficulty as fixed coefficients and random effects for levels of Bloom's taxonomy and speaker nature.

Table 1. Backward elimination of non-significant coefficients for random effects of the binary IRT approach.

Model	Chisq	Df	P-value
(1 | bloom)	0.00	1	1.00
(1 | topic)	0.00	1	1.00
speaker in (speaker | topic)	2.24	5	0.82
(1 | topic)1	0.00	1	1.00
(1 | question$_i$$d$)	2.07	1	0.15
speaker in (speaker | bloom)	68.23	5	< 0.01

Table 2. Backward elimination of non-significant coefficients for fixed effects of the binary IRT approach.

Factor	Sum Sq	Mean Sq	NumDF	DenDF	F value	P-value
speaker:difficulty	5.20	2.60	2	7369.30	1.46	0.23
speaker	163.26	81.63	2	9.90	45.79	< 0.01
difficulty	17.12	17.12	1	7372.80	9.61	0.00

Tables 3 and 4 demonstrate the results of the top-down modeling process for the polynomial IRT approach. Table 3 demonstrates that similar to the other approach, the only random effect to affect model fit significantly was that of Bloom levels according to the speaker that answered the question ($p < 0.001$). Concerning fixed effects, Table 4 shows that only the interaction between the speaker and the difficulty of achieving an evaluation equal to 5 was statistically significant ($p = 0.004$). Therefore, the best model for the polynomial approach featured the speaker, difficulty 5, and their interaction as fixed coefficients, as well as random effects for levels of Bloom's taxonomy according to the speaker's nature.

Table 3. Backward elimination of non-significant coefficients for random effects of the polynomial IRT approach.

Model	Chisq	Df	P-value
(1 | topic)	0.00	1	1.00
(1 | bloom)	0.00	1	1.00
speaker in (speaker | topic)	1.05	5	0.96
(1 | topic)1	0.00	1	1.00
(1 | question$_i$$d$)	2.67	1	0.10
speaker in (speaker | bloom)	49.22	5	< 0.01

Table 4. Backward elimination of non-significant coefficients for fixed effects of the polynomial IRT approach.

Factor	Sum Sq	Mean Sq	NumDF	DenDF	F value	P-value
speaker:dif2	1.47	0.73	2	7369.8	0.41	0.66
dif2	0.21	0.21	1	7369.6	0.12	0.73
speaker:dif1	1.73	0.87	2	7368.9	0.49	0.61
dif1	0.01	0.01	1	7368.6	0.01	0.94
speaker:dif3	6.19	3.10	2	7367.1	1.74	0.18
dif3	0.46	0.46	1	7372.1	0.26	0.61
speaker:dif4	5.29	2.65	2	7348.7	1.49	0.23
dif4	4.36	4.36	1	7373.9	2.45	0.12
speaker:dif5	19.31	9.66	2	4539.0	5.42	< 0.01

Lastly, we compared the best models of both approaches. The LRT yielded a non-significant difference ($\chi^2 = 0.4534$; Degrees of freedom = 2; p-values = 0.7972). Hence, there was no evidence that the model based on the polynomial approach, which was more has more coefficients, provided a better fit than the one based on the binary approach. Therefore, we proceeded with our analysis using the binary IRT-based model.

4.2   Model Assessment

Our best model is summarized in Tables 5 and 6, demonstrating its fixed and random coefficients, respectively. Note that the intercept concerns GPT4, as it was the model's reference level. Hence, the model's coefficients (e.g., native speakers) unveil how their evaluations compared to those of GPT4. Additionally, recall that difficulty was standardized to mean 0. Therefore, to facilitate interpreting our final model, we discuss its insights with Figs. 1 and 2.

Table 5. Summary of the best model, which is based on the binary IRT approach, after performing backward elimination of non-significant coefficients for both fixed and random effects.

Variable	Estimate	Std. Error	df	t value	P-value
(Intercept)	3.69	0.16	6.00	22.49	< 0.01
speakernative	−0.35	0.16	6.00	−2.23	0.07
speakernon-native	−0.65	0.15	6.00	−4.27	0.01
difficulty	−0.05	0.02	7372.76	−3.10	0.00

Figure 1 illustrates how the model's random effects compare. It demonstrates that the role of coefficients is in Creating and Remembering questions. Overall,

Table 6. Random effects of the best model, demonstrating how they change for native and non-native speakers, compared to GPT4 (i.e., the intercept/reference level), for varying levels of Bloom's taxonomy.

Group	GPT4/Intercept	Native	Non-native
Remembering	−0.10	0.52	0.50
Understanding	0.27	−0.15	−0.15
Applying	0.37	−0.27	−0.26
Analyzing	0.18	−0.32	−0.31
Evaluating	0.08	−0.25	−0.24
Creating	−0.79	0.47	0.47

GPT4 (i.e., the intercept, given it is the model's reference level) had a reduced performance on those questions compared to native and non-native speakers. In contrast, the random coefficient of GPT4 seems to have outperformed the questions of the remaining levels. Hence, these findings suggest an advantage for GPT4 in most levels of Bloom's taxonomy, except Remembering and Creating ones.

Figure 2 helps understand how these differences behaved for questions of varied difficulty levels measured by IRT, demonstrating how answer evaluation changed depending on the speaker who answered it for the different levels of Bloom's taxonomy. Interestingly, the figure shows GPT4's evaluations improved as question difficulty increased, while native speakers followed a similar trend with a reduced slope. In contrast, non-native speakers' evaluation decreases as question difficulty increases.

Furthermore, the figure reiterates differences in the intercept depending on Bloom's taxonomy. While GPT4 seems to have outperformed or achieved comparable evaluations on all levels, its performance was worse than those of both speaker and non-native speakers for questions with small difficulty levels in the Remembering and Creating levels. Conversely, non-native speakers were outperformed by both GPT4 and native ones at all levels for the most difficult questions. Therefore, while GPT4 seems to have performed the best on questions with high difficulty levels, it was unable to outperform humans in Remembering and Creating questions with below-the-average levels of difficulty.

5 Discussion

Our study delved into the capacity of Large Language Models (LLMs), particularly GPT4, in responding to open-ended questions within the educational context. The results revealed insights into the comparative performance of GPT4, native speakers, and non-native speakers across questions of various complexity (based on the levels of Bloom's taxonomy) and difficulty (based on Item Response Theory) levels, as discussed next.

Overall, our study reveals that, when controlling for question complexity and difficulty, GPT4 surpasses non-native speakers and exhibits comparable performance to native speakers (see Table 5). While GPT4's capabilities aligned closely with those of native speakers, it fell short of surpassing their proficiency. This finding aligns with the notion that LLMs, leveraging their extensive training data, can effectively handle complex and nuanced tasks, demonstrating their overall potential for integration in educational technology.

Nevertheless, question complexity played an important role in moderating that overall advantage. Our findings demonstrate that, despite GPT4 generally performing well across different Bloom's levels, it faced difficulties under the categories of Remembering and Creating (see Table 6 and Fig. 1). For those particular levels, both native and non-native speakers seem to have had an advantage over GPT4. These results suggest that GPT4 struggled to match human proficiency in tasks requiring simple recall or creative generation, emphasizing the importance of understanding the specific complexity of particular questions when exploring LLMs.

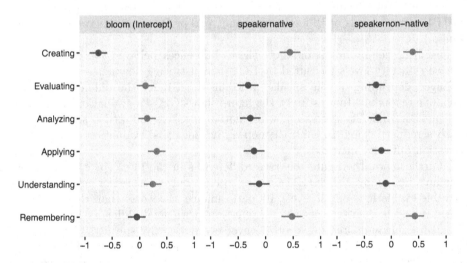

Fig. 1. Random effects of the best model, demonstrating how they change for native and non-native speakers, compared to GPT4 (i.e., the intercept/reference level), for varying levels of Bloom's taxonomy.

Furthermore, we found question difficulty was another important factor. Based on our findings (see Fig. 2), GPT4's evaluations exhibited an interesting trend, improving with increasing question difficulty. This contrasted with the trend observed for non-native speakers, whose evaluations tended to decrease with rising question difficulty. Native speakers demonstrated more stability in their evaluations as question difficulty heightened. This finding revealed GPT4's ability to excel, particularly on highly challenging questions, despite encountering challenges at specific complexity levels, highlighting the importance of discerning between the complexity and difficulty of tasks posed to LLMs.

In summary, this paper contributes empirical evidence comparing the performance of GPT4 and human speakers (both native and non-native) in answering open-ended questions, considering question complexity and difficulty. In contrast to previous research, which has focused on multiple-choice questions [17] or contextual supplementation for open-ended questions [13,20], our study uniquely assessed LLMs' performance across questions of diverse difficulty levels and complexity based on IRT and levels of Bloom's taxonomy, respectively. By addressing this gap in the literature, we expanded the understanding of how GPT4 performed on open-ended question-answering and offered insights for future research and technological integration.

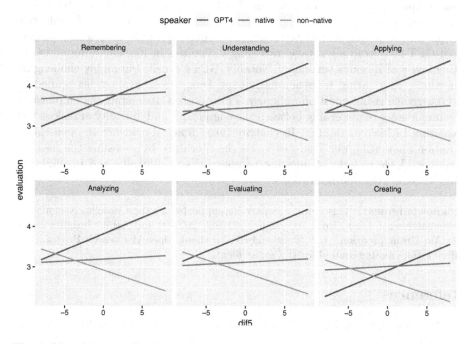

Fig. 2. Visualization of how an answer evaluation changes, depending on the speaker, Bloom's level and question difficulty, based on our best model.

Particularly, our findings have the following implications. First, they highlight the need to evaluate LLMs like GPT4 within the context of open-ended questions, considering the relationship between question complexity and difficulty. Second, our study brings attention to GPT4's strengths, notably its proficiency on highly challenging questions, as well as its challenges in tasks involving Remembering and Creating levels of Bloom's taxonomy. This detailed understanding is essential for those aiming to utilize LLMs in educational settings effectively. Furthermore, our research adds to the existing literature by providing a comprehensive assessment of LLMs' performance across diverse difficulty levels and complexity in open-ended question scenarios.

It is crucial to acknowledge the limitations of our study in interpreting its findings. Our dataset assessed English language questions of three particular topics (Biology, Earth Sciences, and Physics), which mostly concern high school subjects, limiting the generalizability of findings to beyond this setting. Further investigations could explore cross-linguistic variations in LLM performance. Furthermore, the study was limited to a single LLM (GPT4) and a single prompt. Future research endeavors could delve into investigating other LLMs as well as exploring prompt engineering to understand how they affect LLMs' proficiency, especially in tasks involving basic recall and creative thinking.

In conclusion, our study demonstrated that GPT4, when controlling for question complexity and difficulty, surpassed non-native speakers and exhibited performance comparable to native speakers, aligning with the potential for effective integration into educational technology. However, question complexity moderated GPT4's overall advantage, revealing difficulties in Remembering and Creating levels of Bloom's taxonomy. Notably, GPT4 excelled in highly challenging questions, contrasting with non-native speakers, whose evaluations tended to decrease with rising question difficulty. This detailed understanding emphasizes the importance of discerning between the complexity and difficulty of tasks when exploring LLMs' capabilities. In contributing empirical evidence to open-ended question answering, our study expands the literature by revealing the need to evaluate LLMs in light of question complexity and difficulty and highlighting GPT4's strengths and limitations.

Acknowledgment. This study was partialy supported by the Conselho Nacional de Desenvolvimento Científico e Tecnológico (310888/2021-2) and Acuity Insights under the Alo Grant program. Also, we would like to thank OpenAI research Program for offering the credits required for this experiment.

References

1. Anderson, L.W., Sosniak, L.A.: Bloom's Taxonomy. Univ. Chicago Press, Chicago (1994)
2. Basu, S., Jacobs, C., Vanderwende, L.: Powergrading: a clustering approach to amplify human effort for short answer grading. Trans. Assoc. Comput. Linguist. **1**, 391–402 (2013)
3. Bates, D., Mächler, M., Bolker, B., Walker, S.: Fitting linear mixed-effects models using lme4. arXiv preprint arXiv:1406.5823 (2014)
4. Brown, T.B., et al.: Language models are few-shot learners (2020)
5. Cai, L., Choi, K., Hansen, M., Harrell, L.: Item response theory. Annu. Rev. Stat. Appl. **3**, 297–321 (2016)
6. Divya, A., Haridas, V., Narayanan, J.: Automation of short answer grading techniques: comparative study using deep learning techniques. In: 2023 Fifth International Conference on Electrical, Computer and Communication Technologies (ICECCT), pp. 1–7. IEEE (2023)

7. Gelman, A., Hill, J.: Data Analysis Using Regression and Multilevel/Hierarchical Models. Cambridge University Press, Cambridge (2006)
8. Hackl, V., Müller, A.E., Granitzer, M., Sailer, M.: Is GPT-4 a reliable rater? Evaluating consistency in GPT-4 text ratings. arXiv preprint arXiv:2308.02575 (2023)
9. Horbach, A., Stennmanns, S., Zesch, T.: Cross-lingual content scoring. In: Proceedings of the Thirteenth Workshop on Innovative Use of NLP for Building Educational Applications, pp. 410–419 (2018)
10. Hox, J.J., Moerbeek, M., Van de Schoot, R.: Multilevel Analysis: Techniques and Applications. Routledge, London (2010)
11. Kojima, T., Gu, S.S., Reid, M., Matsuo, Y., Iwasawa, Y.: Large language models are zero-shot reasoners. Adv. Neural. Inf. Process. Syst. **35**, 22199–22213 (2022)
12. Kuznetsova, A., Brockhoff, P.B., Christensen, R.H.B.: lmerTest package: tests in linear mixed effects models. J. Stat. Softw. **82**(13), 1–26 (2017). https://doi.org/10.18637/jss.v082.i13
13. Liu, Y., Xu, B., Yang, Y., Chung, T., Zhang, P.: Constructing a hybrid automatic Q&A system integrating knowledge graph and information retrieval technologies. In: Foundations and Trends in Smart Learning. LNET, pp. 67–76. Springer, Singapore (2019). https://doi.org/10.1007/978-981-13-6908-7_9
14. Lord, F., Novick, M.: Statistical Theories of Mental Test Scores. Addison-Wesley Series in Behavioral Sciences: Quantitative Methods, Information Age Publishing, Incorporated (2008)
15. OpenAI: GPT-4 technical report (2023)
16. Raffel, C., et al.: Exploring the limits of transfer learning with a unified text-to-text transformer (2023)
17. Rosoł, M., Gasior, J.S., Łaba, J., Korzeniewski, K., Młyńczak, M.: Evaluation of the performance of GPT-3.5 and GPT-4 on the Polish medical final examination. Sci. Rep. **13**(1), 20512 (2023)
18. Vaswani, A., et al.: Attention is all you need. In: Neural Information Processing Systems (2017). https://api.semanticscholar.org/CorpusID:13756489
19. Wang, R., Demszky, D.: Is ChatGPT a good teacher coach? Measuring zero-shot performance for scoring and providing actionable insights on classroom instruction. In: Kochmar, E., et al. (eds.) Proceedings of the 18th Workshop on Innovative Use of NLP for Building Educational Applications (BEA 2023), pp. 626–667. Association for Computational Linguistics, Toronto, Canada (2023). https://doi.org/10.18653/v1/2023.bea-1.53
20. de Winter, J.C.F.: Can ChatGPT pass high school exams on English language comprehension? Int. J. Artif. Intell. Educ. (2023)
21. Xia, Q., Chiu, T.K., Zhou, X., Chai, C.S., Cheng, M.: Systematic literature review on opportunities, challenges, and future research recommendations of artificial intelligence in education. Comput. Educ. Artif. Intell. 100118 (2022)
22. Yan, L., et al.: Practical and ethical challenges of large language models in education: a systematic scoping review. Br. J. Educ. Technol. **n/a**(n/a). https://doi.org/10.1111/bjet.13370
23. Yenduri, G., et al.: GPT (generative pre-trained transformer) - a comprehensive review on enabling technologies, potential applications, emerging challenges, and future directions (2023)

Understanding Gender Effects in Game-Based Learning: The Role of Self-Explanation

J. Elizabeth Richey[1]([✉]), Huy A. Nguyen[2], Mahboobeh Mehrvarz[2],
Nicole Else-Quest[3], Ivon Arroyo[4], Ryan S. Baker[5], Hayden Stec[2], Jessica Hammer[2],
and Bruce M. McLaren[2]

[1] University of Pittsburgh, Pittsburgh, PA 15260, USA
jer177@pitt.edu
[2] Carnegie Mellon University, Pittsburgh, PA 15213, USA
[3] University of North Carolina at Chapel Hill, Chapel Hill, NC 27599, USA
[4] University of Massachusetts Amherst, Amherst, MA 01003, USA
[5] University of Pennsylvania, Philadelphia, PA 19104, USA

Abstract. We conducted a 2 × 2 study comparing the digital learning game *Decimal Point* to a comparable non-game tutor with or without self-explanation prompting. We expected to replicate previous studies showing the game improved learning compared to the non-game tutor, and that self-explanation prompting would enhance learning across platforms. Additionally, prior research with *Decimal Point* suggested that self-explanation was driving gender differences in which girls learned more than boys. To better understand these effects, we manipulated the presence of self-explanation prompts and incorporated a multidimensional gender measure. We hypothesized that girls and students with stronger feminine-typed characteristics would learn more than boys and students with stronger masculine-typed characteristics in the game with self-explanation condition, but not in the game without self-explanation or in the non-game conditions. Results showed no advantage for the game over the non-game or for including self-explanation, but an analysis of hint usage indicated that students in the game conditions used (and abused) hints more than in the non-game conditions, which in turn was associated with worse learning outcomes. When we controlled for hint use, students in the game conditions learned more than students in the non-game tutor. We replicated a gender effect favoring boys and students with masculine-typed characteristics on the pretest, but there were no gender differences on the posttests. Finally, results indicated that the multidimensional framework explained variance in pretest performance better than a binary gender measure, adding further evidence that this framework may be a more effective, inclusive approach to understanding gender effects in game-based learning.

Keywords: Digital Learning Game · Gender Studies · Self-Explanation · Hints

1 Introduction

Digital learning games can promote learning through playful, engaging, and highly interactive interfaces, but depending on the features of games (e.g., narrative, design style, pacing), they may not be equally effective for all learners [1, 2]. *Decimal Point,*

© The Author(s), under exclusive license to Springer Nature Switzerland AG 2024
A. M. Olney et al. (Eds.): AIED 2024, LNAI 14829, pp. 206–219, 2024.
https://doi.org/10.1007/978-3-031-64302-6_15

a digital math game designed to teach students about decimal numbers and operations, has produced consistent gender effects favoring girls [3]. In prior research seeking to understand the source of these consistent gender effects, we have found evidence that differences in girls' and boys' response patterns to self-explanation prompts seem to be at least partially responsible for the gender differences in learning outcomes [3]. Specifically, girls have demonstrated fewer errors and less gaming the system [4] on self-explanation questions in the game, and those behaviors in turn have mediated the gender effect on learning outcomes [5].

In this paper, we seek to advance our understanding of the gender effect in *Decimal Point* in two ways: first, we conducted an experiment to directly test whether self-explanation was responsible for the gender effect by varying whether students were prompted to self-explain as they played the game or solved equivalent problems in a non-game platform. Second, we sought to expand our understanding of gender differences in gameplay and learning by incorporating a multidimensional gender framework. This framework holds that gender includes multiple separate but interrelated dimensions, including aspects of identity as well as activities and interests [6]. It has the potential to reveal more nuanced gender-related characteristics that might more directly explain differences in how students play and learn from digital games. Additionally, a multidimensional gender representation promises to be a more inclusive approach to understanding gender, as it will allow us to include students of all genders in our analyses–instead of limiting analyses to students who fit within binary gender identity categories–while capturing more complex aspects of gender along a continuous spectrum. In this paper, we focus on the gender dimension of students' self-reports of gender-typed occupational interests, activities, and traits [7].

1.1 Enhancing Learning Through Digital Learning Games

There is ample evidence that digital learning games can engage students and support learning [8–11]. Educational technology researchers have embraced game-based learning by building games for a variety of domains [12–14]. Games are thought to support learning through engagement [15] and flow [16], in which learners focus their full attention on game play, potentially taking focus away from negative thoughts or emotions about the instructional content. For struggling learners, game-based learning can be a particularly effective way to engage with the material [17].

Decimal Point is a digital math learning game that has produced better learning outcomes than comparable non-game instruction. In a study with 153 middle school students who either played *Decimal Point* or learned with a content-equivalent online tutor [12], *Decimal Point* students learned more than the tutored students, with relatively high effect sizes (immediate posttest: $d = 0.65$; delayed posttest: $d = 0.59$). The *Decimal Point* students also reported enjoying their experience significantly more than the tutored students, according to a post-game questionnaire. An analysis of learning outcomes by binary gender identity revealed that girls learned significantly more from the game than boys [18]. We subsequently performed experiments with different versions of the game and consistently uncovered the same finding: girls learned significantly more from the game than boys regardless of alterations [3, 19].

Some prior work has explored the role of self-explanation in game-based learning [20], and we were also interested in the impact self-explanation could have on learning with the game. As a result, the game incorporates a series of multiple-choice self-explanation questions designed to address misconceptions and promote conceptual understanding of the decimal number concepts in the game [21]. When investigating sources of the gender difference in learning outcomes, we discovered that boys' and girls' game-play behaviors and error rates especially differed on the self-explanation steps [3]. Specifically, girls tended to have lower error rates and exhibited fewer behaviors suggesting they were trying to take advantage of the affordances of the learning system to get the right answer without thinking, a behavior referred to as "gaming the system" [4]. Further analyses indicated that rates of gaming the system partially mediated the effect of gender on learning, suggesting that students' interactions with the self-explanation prompts were at least in part driving the gender effect in *Decimal Point* [5].

1.2 Gender, Math, and Digital Learning Games

Although girls and boys tend to perform equally well in math, gender differences often emerge in motivation, emotions, and perceptions around math. For example, girls report greater anxiety towards math and less self-confidence in their math abilities [22, 23].

Digital math games could provide a valuable tool for promoting more equitable engagement in math. Games may be especially effective for girls to the degree that they promote enjoyment and reduce salient cues likely to trigger stereotype threat, which occurs when being reminded of social group stereotypes impairs the performance of members of that group [24]. In the context of math, even implicit cues like labeling the nature of a task as mathematical can trigger stereotype threat for women and reduce their performance [25]. Embedding math practice within the context of a digital learning game might therefore reduce stereotype threat by de-emphasizing the mathematical nature of the task.

Digital learning games in math appear to be effective for all genders [26], despite some evidence of broader gender-based differences in game preferences [27]. In fact, other math digital learning games have uncovered gender differences in learning outcomes based on different game features [1], lending evidence that digital learning games may affect some aspects of learning for girls and boys differently.

One key to better understanding how and why digital learning games might produce gender differences involves taking a more comprehensive view of gender. All prior research investigating gender differences in digital game learning has adopted a binary view of gender, sorting learners into binary categories of boys and girls and typically excluding anyone outside the gender binary due to small numbers. However, it is likely that any effect of gender on learning reflects gender-related differences in behaviors, interests, and experiences rather than binary gender identity itself. As a result, a measure of gender that captures these nuanced, multifaceted factors may be a more powerful predictor of learning differences. It could also better illuminate which aspects of gender (e.g., interests, activities) are most predictive of differences in learning behaviors and outcomes.

We address this gap in prior research and aim to better understand gender differences in *Decimal Point* by incorporating a multidimensional gender framework and measuring gender-typed occupational interests, activities, and traits, which we will refer to as "gender-typed characteristics" [7]. We also experimentally test the evidence that self-explanation prompts are producing the gender effect observed in *Decimal Point*. In this research, we explored the following questions:

RQ1: Will the learning platform (game vs. non-game control) and the presence or absence of self-explanation questions affect learning outcomes?

RQ2: Will the learning platform (game vs. non-game control) and the presence or absence of self-explanation questions interact with gender to explain *gender-based* differences in learning outcomes?

RQ3: Will the multidimensional gender framework predict variance in learning outcomes better than a measure of binary gender identity?

For the first research question, we hypothesized that we would replicate previous studies showing a learning advantage for students in the game condition compared to the non-game tutor [12], regardless of self-explanation condition. We also hypothesized that self-explanation questions would lead to greater learning across both learning platforms, which we have not previously tested. We did not predict an interaction between self-explanation and learning platform, as we expected each to contribute an additive effect to learning.

For the second research question, we hypothesized that removing the self-explanation prompts would eliminate the gender effect in the game based on prior analyses suggesting that the self-explanation prompts were driving gender differences. Specifically, we hypothesized that girls and students with stronger feminine-typed traits would learn more than boys and students with stronger masculine-typed traits *only in the game + self-explanation condition*. We did not expect to see gender differences in the game condition without self-explanation, as this condition would be missing the learning component hypothesized to be creating gender differences. In other words, if the self-explanation step in the game was creating the gender effect, then we would no longer expect to see a gender difference in the game when self-explanation was removed. We did not expect to see a gender difference in the non-game platform, regardless of the presence or absence of self-explanation prompts, as prior research revealed no gender differences in the non-game [18].

For the third research question, if students differed in test performance by gender, we expected that the continuous measures of gender (i.e., gender-typed occupational interests, activities, and traits) would explain more variance in test performance than a binary measure of gender identity (i.e., boy or girl). This is because gender-typed characteristics reflect more nuanced aspects of gender, which are likely more closely related than binary gender identity to any motivation or emotion that might in turn predict students' learning in a math game. This prediction is also consistent with preliminary work showing that multidimensional measures of gender explain differences in game preferences better than binary gender [27].

2 Method

2.1 Participants

We conducted our research in eight elementary and middle schools in a mid-sized U.S. city and the surrounding suburban and rural areas. A total of 576 students from fifth- and sixth-grade classes participated in the study, but a technical issue resulted in data not being recorded for 90 students from one school. An additional 100 students were excluded from analyses because they did not complete all study and test materials. The remaining 386 students were assigned to one of four conditions: a game with self-explanation, a game without self-explanation, a non-game with self-explanation, or a non-game without self-explanation. Given the potential distraction of having some students play a game while others worked with a non-game tutor within a classroom, we randomly assigned students to game conditions at the classroom level and to the self-explanation conditions at the individual level. Students ranged in age from 10 to 13 years old ($M = 10.85$, $SD = 0.65$); 201 identified as male, 182 identified as female, one identified as trans or non-binary, and two preferred not to disclose their gender identity.

2.2 Materials and Procedure

Materials consisted of a pretest, posttest, and delayed posttest, as well as different versions of *Decimal Point* or the non-game tutor and a series of pre- and post-intervention questionnaires. Instructional materials were created using an open use authoring suite [28]. We presented all materials in an established online learning management system using the HTML/JavaScript framework [29].

Learning Materials. Learning materials consisted of either *Decimal Point* or a non-game tutor designed to be equivalent to *Decimal Point* in the instructional content but without the game elements or playful design. Both *Decimal Point* and the non-game tutor varied in whether they included self-explanation prompts or not.

Decimal Point is a web-based single-player game that uses an amusement park metaphor to teach middle school students about decimal numbers, as shown in Fig. 1 [12]. *Decimal Point* is made up of 24 mini-games, with each mini-game focusing on a specific type of problem-solving task. In total, there are 48 decimal problems, with two problems in each of the 24 mini-games. These problems cover various tasks such as ordering decimals, placing them on a number line, completing sequences, sorting them into "buckets" based on magnitude, and adding decimals. Players must provide the correct answer for each problem in order to progress in the game. The aim is to play through all the mini-games in sequence. The game provides immediate accuracy feedback and allows students to retry problems until they find the correct solution. The game also incorporates hints, which are designed to support students' learning progress and help them when they become stuck on a problem [30]. Students were provided with three levels of on-demand hints while solving the problems: Level 1 hints offered general reminders about relevant decimal concepts and operations; Level 2 hints provided more detailed suggestions about solution steps; and Level 3 hints provided the answer. The

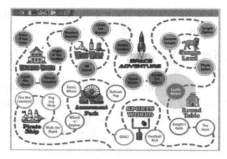

Fig. 1. The main map in *Decimal Point,* depicting the amusement park game narrative.

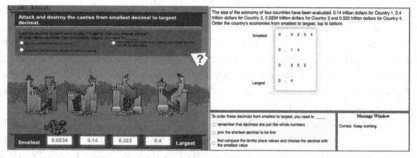

Fig. 2. Self-explanation prompts at the end of the *Castle Attack* mini-game in *Decimal Point* (left) and at the end of an equivalent sorting problem in the non-game tutor (right).

hints could be accessed by selecting the "Hint" button and could be navigated using the "Previous" and "Next" buttons.

After solving each pair of mini-game problems, students in the self-explanation condition were prompted to self-explain using multiple-choice questions (Fig. 2), which encouraged them to think more deeply about the concepts and misconceptions targeted by the problems. Prior work has found multiple-choice self-explanation prompts to be more effective than open-ended self-explanation prompts in digital learning games, possibly because this format is less disruptive to game flow or because it introduces less cognitive load to respond [20]. In the game condition without self-explanation, the game moved on to the map without any prompting for self-explanation.

The non-game version of *Decimal Point* uses the same web-based learning management system as the game version. In the non-game version, students solve decimal problems using a conventional user interface without the game features or narrative, as shown in Fig. 2 [12]. The non-game condition presents the same five types of decimal problems in the same order, with a total of 48 problems. As in the game condition, students received immediate feedback on accuracy, could access the same three levels of on-demand hints, and had to answer correctly to progress to the next problem. In the condition with self-explanation, students were prompted to self-explain decimal concepts and misconceptions with the same multiple-choice questions used in the game condition. In the non-game condition without self-explanation, the tutor moved on to the next problem set without prompting for self-explanation.

Learning Assessment Tests. We assessed students' knowledge of relevant decimal concepts using three isomorphic tests administered immediately before the learning materials, immediately after the learning materials, and approximately one week after students completed the learning materials. Tests were counterbalanced to account for any differences in difficulty. Each test contained 43 items targeting students' procedural and conceptual knowledge about decimal number operations. Some items contained multiple parts, and students could earn a total of 52 points.

Questionnaires. Questionnaires were administered immediately before students began the learning intervention (after the pretest) and immediately after they concluded the learning intervention (before the posttest). After the pretest, students responded to a series of demographic questions concerning their age, grade level, gender identity, and race. They also completed an adapted version of the Children's Occupational Interests, Activities, and Traits - Personal Measure (COAT-PM) to measure students' gender-typed characteristics [7]. This survey assesses children's interests, activities, and traits in relation to masculine- and feminine-stereotyped norms on a four-point Likert-type scale, with 18 items each in the *occupation, activity,* and *traits* subscales. Items in the *occupation* subscale targeted stereotypically gendered professions like "hairstylist" (feminine) and "construction worker" (masculine), and students reported their interests in the targeted professions on a scale from 1 (not at all) to 4 (very much). Items in the *activity* subscale targeted stereotypically gendered activities such as "making jewelry" (feminine) and "going fishing" (masculine), and students reported the frequency with which they engaged in these activities on a scale of 1 (never) to 4 (very often). Items in the *traits* subscale targeted self-perceptions of stereotypically gendered personal characteristics such as "gentle" (feminine) and "adventurous" (masculine). All subscales were averaged together to produce scales of feminine-typed characteristics ($\alpha = 0.85$) and masculine-typed characteristics ($\alpha = 0.81$).

After completing the learning intervention, students responded to a series of surveys targeting their engagement, enjoyment, and emotions, including affective and behavioral/cognitive engagement [31]; dimensions of meaning, mastery, and challenge from the Player Experience Inventory [32]; situational interest [33]; the enjoyment dimension of the Achievement Emotions Questionnaire [34]; evaluation apprehension and test anxiety [24]; and state anxiety [35]. Due to space constraints, we do not report results related to these questionnaires.

3 Results

To assess whether students learned from the intervention materials, we conducted repeated-measures analyses of variance (ANOVAs) examining changes from pretest to posttest and pretest to delayed posttest. Results indicated a large effect from pretest to posttest, $F = 97.88$, $p < .001$, $\eta^2_p = .20$, and between pretest and delayed posttest, $F = 128.33$, $p < .001$, $\eta^2_p = .25$, , indicating that students generally learned from the intervention materials (see Table 1 for test performance means by condition and test).

Table 1. Average test score by gender and condition.

	N	Pretest M (SD)	Posttest M (SD)	Delayed Posttest M (SD)
Game conditions	203	22.32 (12.10)	25.40 (12.19)	26.12 (12.91)
Game +SE	105	21.36 (11.56)	24.61 (12.14)	25.07 (13.29)
Game −SE	98	23.35 (12.63)	26.26 (12.25)	27.26 (12.45)
Non-game conditions	183	20.83 (10.20)	23.32 (10.66)	24.34 (11.10)
Non-game +SE	96	20.93 (10.67)	23.43 (10.36)	24.67 (10.73)
Non-game −SE	87	20.72 (9.71)	23.21 (11.03)	23.99 (11.53)
+SE conditions	201	21.15 (11.12)	24.04 (11.31)	24.88 (12.11)
−SE conditions	185	22.11 (11.40)	24.82 (11.76)	25.72 (12.11)
Girls	182	20.00 (9.70)	23.71 (10.34)	24.52 (11.42)
Boys	201	23.27 (12.27)	25.28 (12.40)	26.20 (12.60)

3.1 Condition Effects on Learning (RQ1)

These results contradicted prior research showing a significant learning advantage for students playing *Decimal Point* compared to the non-game tutor [12]. To understand why the results might be different, we conducted a post hoc analysis of students' use of hints, as the hint feature was added in recent years after the initial research showing an advantage for the game compared to the non-game tutor.

To test whether hint requests mediated the effect of learning platform on test performance, we built mediation models with the learning platform (game or non-game tutor) as an independent variable, the number of hint requests during intervention as a mediator, and the posttest and delayed posttest scores as the dependent variables. The confidence interval of the indirect effect was estimated at the 0.05 significance level via bias-corrected non-parametric bootstrapping with 2000 iterations [36]. Based on the mediation results (Fig. 3), we found that the effect of the learning platform on posttest performance was mediated by the number of hint requests. The regression coefficient between the learning platform (with the game coded as 1) and number of hint requests was positive and significant, while the coefficient between the number of hint requests and posttest score was negative and significant. In other words, students in the game tended to request more hints, which in turn was associated with worse learning outcomes. The bootstrap procedures indicated a significant indirect effect ($ab = -0.87$, 95% CI $[-1.45, -0.29]$, $p < .001$).

Similar findings were observed in the mediation model predicting delayed posttest scores, with a significant indirect effect of the number of hint requests ($ab = -0.97$, 95% CI $[-1.65, -0.41]$, $p < .001$; see Fig. 3). On the other hand, the direct effects of the game on posttest and delayed posttest performance, without considering the mediator, were positive and significant.

We built regression models predicting posttest and delayed posttest scores with pretest scores, number of hint requests, and learning platform (game coded as 1) to

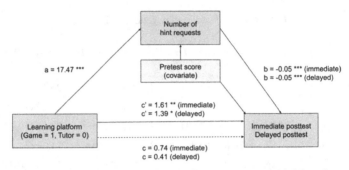

Fig. 3. The mediation pathway from learning platform to posttest and delayed posttest performance through hint usage behavior. (*) $p < .05$, (**) $p < .01$, (***) $p < .001$.

assess the unique contributions of each factor. When predicting posttest performance, number of hints requested was negatively associated with test performance ($\beta = -0.05$, $p < .001$); however, when controlling for hint requests and pretest ($\beta = 0.63, p < .001$), the game platform was associated with higher posttest scores ($\beta = 1.61, p = .001$). A regression model predicting delayed posttest performance with pretest, hints requests, and learning platform showed similar results: the number of hint requests was negatively associated with delayed posttest scores ($\beta = -0.06, p < .001$), but when controlling for pretest ($\beta = 0.62, p < .001$) and hint requests, the game platform was associated with higher delayed posttest scores ($\beta = 1.39, p = .018$).

In other words, while the game did lead to better learning when controlling for hint requests, students playing the game also requested more hints than those using the tutor, which in turn reduced their learning. Likely due to these conflicting trends, the total effect of the learning platform on test performance was not significant.

3.2 Binary Gender Differences in Learning (RQ2)

We sought to understand how different dimensions of gender related to learning outcomes across instructional conditions. First, we examined differences in pretest, posttest, and delayed posttest scores based on binary gender identity; given the small number of students identifying as non-binary or trans or declining to indicate gender ($N = 3$), we excluded these students from analyses using binary gender identity. A one-way ANOVA revealed a significant effect of gender identity on pretest performance ($F = 8.27, p = .004$, $\eta_p^2 = .021$), with boys receiving higher pretest scores than girls. Table 1 reports mean scores on all tests based on binary gender identity.

We also tested the interaction of binary gender with instructional condition on test scores. A series of three-way ANCOVAs examined the effects of learning platform, self-explanation condition, and binary gender on posttest and delayed posttest scores while controlling for pretest. On the posttest, results indicated no main effect of gender identity ($F = 2.28, p = .13, \eta_p^2 = .006$) and no interactions between gender and learning platform ($F = 0.63, p = .43, \eta_p^2 = .002$) or gender and self-explanation ($F = 1.25, p = .26, \eta_p^2 = .003$). The analysis of delayed posttest also revealed no main effect of gender identity ($F = 3.03, p = .08, \eta_p^2 = .008$) and no interactions between gender identity and learning platform ($F = 0.05, p = .82, \eta_p^2 = .001$) or self-explanation prompts ($F = 2.86, p = .09, \eta_p^2 = .008$). Our predictions that girls would learn more, but only in the game condition with self-explanation, were not supported.

3.3 Predicting Learning Differences with Gender-Typed Characteristics (RQ3)

Next, we examined the correlations between binary gender identity and gender-typed characteristics. Results showed that gender identity, where "female" was coded as 1 and "male" coded as 0, was strongly, positively correlated with feminine-typed characteristics ($r = 0.58, p < .001$) and moderately, negatively correlated with masculine-typed characteristics ($r = -0.34, p < .001$). Feminine-typed characteristics were weakly, positively correlated with masculine-typed characteristics ($r = 0.20, p < .001$). Given the correlation coefficients, while the three gender dimensions were moderately correlated, they were not redundant.

We also analyzed test performance using the measure of gender-typed interests, activities, and traits. A regression model predicting pretest scores based on masculine-typed and feminine-typed characteristics revealed that masculine-typed characteristics were a significant, positive predictor ($\beta = 3.64, p = .003$) while feminine-typed characteristics were a significant, negative predictor ($\beta = -4.42, p < .001$).

To compare the predictive value of binary gender identity and gender-typed characteristics, we built a regression model predicting pretest scores based on binary gender identity (with "female" coded as 1 and "male" coded as 0), masculine-typed characteristics, and feminine-typed characteristics. Results showed that masculine-typed characteristics ($\beta = 4.01, p = .008$) and feminine-typed characteristics ($\beta = -4.96, p = .004$) were significant predictors, while binary gender identity was not ($\beta = 0.75, p = .66$). These results suggest the measures of gender-typed characteristics may better explain variance in pretest scores than binary gender identity.

To understand interactions between the multiple dimensions of gender and learning conditions, we built regression models predicting posttest and delayed posttest scores that included pretest score as a covariate and the following predictor variables: learning platform (game or non-game), self-explanation prompt (with or without self-explanation), masculine-typed characteristics, feminine-typed characteristics, and their interactions with the learning platform and self-explanation conditions. Results showed that none of the variables significantly predicted posttest scores or delayed posttest scores. These results are consistent with the lack of effects revealed when examining the interaction of binary gender with the learning conditions.

We also examined self-explanation errors by gender, as well self-explanation performance as a mediator between gender and learning outcomes. We considered only students who were prompted to self-explain in the game (n = 105) and non-game (n = 96) conditions. A two-way ANCOVA assessing the effects of the learning platform and gender identity on the number of self-explanation errors, with pretest score as covariate, showed a significant main effect of gender ($F = 7.53, p = .007, \eta_p^2 = .037$), with girls ($M = 30.35, SD = 14.35$) making fewer self-explanation errors than boys (M = 35.74, SD = 13.12). The effects of learning platform ($F = 0.38, p = .54, \eta_p^2 = .002$) and its interaction with gender identity ($F = 0.23, p = .63, \eta_p^2 = .001$) were not significant.

We also built a regression model predicting the number of self-explanation errors based on gender-typed characteristics and their interactions with the learning platform, using pretest performance as covariate. Our results showed that masculine-typed characteristics were a marginally significant, positive predictor of self-explanation errors ($\beta = 4.42$, $p = .06$), while feminine-typed characteristics were a significant, negative predictor of self-explanation errors ($\beta = -6.05, p = .008$). The learning platform and its interaction with gender-typed characteristics were not significant predictors. In other words, both binary gender identity and gender-typed characteristics predicted self-explanation errors in similar patterns, with girls and students with stronger feminine-typed characteristics making fewer errors.

4 Discussion and Conclusion

This research uncovered several important findings, with clear implications for future research with *Decimal Point* as well as recommendations for research on gender and digital learning games more generally. First, we failed to replicate prior research showing a learning advantage for the game compared to a non-game tutor [12]. However, our post hoc analyses uncovered a promising explanation for this result. In the time since the original study, we added hints to *Decimal Point* to help students avoid getting stuck and increase opportunities to learn from errors [30]. Based on hint use patterns in the game and non-game platforms, however, it appears that using too many hints is generally harmful to learning, which is consistent with other research on hint use [3, 37]. Students in the game tended to overuse hints *more* than students in the non-game, and this overuse of hints seemed to negate the benefits of learning with the game. As additional evidence, when we analyzed learning outcomes from the game vs. non-game and controlled for hint use, students in the game performed better than students in the non-game condition on the posttest and delayed posttest.

Second, we failed to replicate the gender effect on learning outcomes, although we replicated results showing that boys outperformed girls on the pretest. Girls have learned more than boys across many previous studies with *Decimal Point*, although this was not the case in the non-game tutor [3, 18]. We are unsure why this result failed to replicate in the game condition with self-explanation. It is possible that the 2x2 design was underpowered to detect interaction effects with gender, especially after a technical error caused data loss for some students. This unexpected result merits additional investigation, particularly given how seemingly unrelated changes to the game (e.g., the addition of hints) affected other outcomes. We replicated results that girls made fewer errors on self-explanation steps compared to boys, which contributes to evidence that the self-explanation step may be responsible for gender differences in learning with *Decimal Point*.

Third, we have found evidence that incorporating multiple dimensions of gender is a fruitful avenue for better understanding gender differences in learning. While binary gender was correlated with the measures of gender-typed characteristics (strongly with feminine-typed characteristics and moderately with the masculine-typed characteristics), results indicated that between about 40 and 65 percent of the variance in masculine- and feminine-typed occupational interests, activities, and traits was *not* explained by binary gender. These more nuanced aspects of gender, which are captured on a spectrum and target a wide range of interests and activities, are likely to more directly shape individual learning experiences that, in turn, predict students' math motivation and achievement in different learning environments. Critically, the measures of gender-typed occupational interests, activities, and traits explained differences on the pretest better than binary gender; when the masculine- and feminine-typed characteristics were included in a regression model predicting pretest performance, binary gender was no longer a significant predictor.

Results strongly support the use of multiple dimensions of gender in future research investigating gender differences in game-based learning. This is a novel contribution, as we know of no other research that has adopted such an approach to studying gender in the context of digital learning games. This approach is likely to provide a better explanation for gender-based differences in learning behaviors and outcomes compared to binary measures of gender identity, which in turn may illuminate clearer recommendations for making digital learning games more equitable and beneficial for all students. Our results also reveal the importance of examining each design choice in terms of its impact on learning behaviors and outcomes. Specifically, our results suggest that adding hints was particularly detrimental to students in the game conditions, who used (and likely abused) hints more than students in the non-game conditions. Future research should explore why students might be more likely to overuse hints in a digital learning game compared to a non-game tutor.

Acknowledgments. This work was supported by the National Science Foundation Award #DRL-2201796. The opinions expressed are those of the authors and do not represent the views of NSF. Thanks to Jimit Bhalani, John Choi, Kevin Dhou, Darlan Santana Farias, Rosta Farzan, Jodi Forlizzi, Craig Ganoe, Rick Henkel, Scott Herbst, Grace Kihumba, Kim Lister, Patrick Bruce Gonçalves McLaren, and Jon Star for important contributions to the development and early experimentation with *Decimal Point*.

Disclosure of Interests. The authors have no competing interests to declare that are relevant to the content of this article.

References

1. Arroyo, I., Burleson, W., Tai, M., Muldner, K., Woolf, B.P.: Gender differences in the use and benefit of advanced learning technologies for mathematics. J. Educ. Psychol. **105**(4), 957 (2013). https://doi.org/10.1037/a0032748
2. Dele-Ajayi, O., Strachan, R., Pickard, A., Sanderson, J.: Designing for all: exploring gender diversity and engagement with digital educational games by young people. In: 2018 IEEE Frontiers in Education Conference (FIE), pp. 1–9 (2018)
3. Nguyen, H., Hou, X., Richey, J.E., McLaren, B.M.: The impact of gender in learning with games: a consistent effect in a math learning game. Int. J. Game-Based Learn. **12**(1), 1–29 (2022). https://doi.org/10.4018/IJGBL.309128
4. Baker, R.S., Corbett, A.T., Koedinger, K.R., Wagner, A.Z.: Off-task behavior in the cognitive tutor classroom: when students "game the system. In: Proceedings of the SIGCHI Conference on Human Factors in Computing Systems, pp. 383–390 (2004). https://doi.org/10.1145/985692.985741

5. Baker, R.S., et al.: Gaming the system mediates the relationship between gender and learning outcomes in a digital learning game (under review)
6. Hyde, J.S., Bigler, R.S., Joel, D., Tate, C.C., van Anders, S.M.: The future of sex and gender in psychology: five challenges to the gender binary. Am. Psychol. **74**(2), 171 (2019). https://doi.org/10.1037/amp0000307
7. Liben, L.S., Bigler, R.S., Ruble, D.N., Martin, C.L., Powlishta, K.K.: The developmental course of gender differentiation: conceptualizing, measuring, and evaluating constructs and pathways. Monogr. Soc. Res. Child Dev. **67**(2), 1–183 (2002)
8. Clark, D.B., Tanner-Smith, E., Killingsworth, S.: Digital games, design, and learning: a systematic review and meta-analysis. Rev. Educ. Res. **86**(1), 79–122 (2016). https://doi.org/10.3102/0034654315582065
9. Mayer, R.E.: Computer games in education. Annu. Rev. Psychol. **70**, 531–549 (2019). https://doi.org/10.1146/annurev-psych-010418-102744
10. Hussein, M.H., Ow, S.H., Elaish, M.M., Jensen, E.O.: Digital game-based learning in K-12 mathematics education: a systematic literature review. Educ. Inf. Technol. **27**, 2859–2891 (2022). https://doi.org/10.1007/s10639-021-10721-x
11. Schöbel, S., Saqr, M., Janson, A.: Two decades of game concepts in digital learning environments–a bibliometric study and research agenda. Comput. Educ. **173**, 104296 (2021). https://doi.org/10.1016/j.compedu.2021.104296
12. McLaren, B.M., Adams, D.M., Mayer, R.E., Forlizzi, J.: A computer-based game that promotes mathematics learning more than a conventional approach. Int. J. Game-Based Learn. **7**(1), 36–56 (2017). https://doi.org/10.4018/IJGBL.2017010103
13. Hooshyar, D., Malva, L., Yang, Y., Pedaste, M., Wang, M., Lim, H.: An adaptive educational computer game: effects on students' knowledge and learning attitude in computational thinking. Comput. Human Behav. **114**, 106575 (2021). https://doi.org/10.1016/j.chb.2020.106575
14. Lester, J.C., Ha, E.Y., Lee, S.Y., Mott, B.W., Rowe, J.P., Sabourin, J.L.: Serious games get smart: intelligent game-based learning environments. AI Mag. **34**(4), 31–45 (2013). https://doi.org/10.1609/aimag.v34i4.2488
15. Gee, J.P.: What Video Games have to Teach Us About Learning and Literacy. Palgrave Macmillan, New York (2007)
16. Czikszentmihalyi, M.: Flow: The Psychology of Optimal Experience. Harper & Row, New York (1990)
17. Ronimus, M., Eklund, K., Pesu, L., Lyytinen, H.: Supporting struggling readers with digital game-based learning. Educ. Technol. Res. Dev. **67**, 639–663 (2019). https://doi.org/10.1007/s11423-019-09658-3
18. McLaren, B., Farzan, R., Adams, D., Mayer, R., Forlizzi, J.: Uncovering gender and problem difficulty effects in learning with an educational game. In: André, E., Baker, R., Hu, X., Rodrigo, M.M.T., du Boulay, B. (eds.) AIED 2017. LNCS (LNAI), vol. 10331, pp. 540–543. Springer, Cham (2017). https://doi.org/10.1007/978-3-319-61425-0_59
19. McLaren, B.M., Richey, J.E., Nguyen, H.A., Mogessie, M.: A digital learning game for mathematics that leads to better learning outcomes for female students: further evidence. In: Proceedings of the 16th European Conference on Game Based Learning (ECGBL 2022), pp. 339–348 (2022). https://doi.org/10.34190/ecgbl.16.1.794
20. Johnson, C.I., Mayer, R.E.: Applying the self-explanation principle to multimedia learning in a computer-based game-like environment. Comput. Human Behav. **26**(6), 1246–1252 (2010). https://doi.org/10.1016/j.chb.2010.03.025
21. Nokes, T.J., Hausmann, R.G., VanLehn, K., Gershman, S.: Testing the instructional fit hypothesis: the case of self-explanation prompts. Instr. Sci. **39**, 645–666 (2011). https://doi.org/10.1007/s11251-010-9151-4

22. Else-Quest, N.M., Hyde, J.S., Linn, M.C.: Cross-national patterns of gender differences in mathematics: a meta-analysis. Psychol. Bull. **136**(1), 103 (2010). https://doi.org/10.1037/a00 18053

23. Else-Quest, N.M., Mineo, C.C., Higgins, A.: Math and science attitudes and achievement at the intersection of gender and ethnicity. Psychol. Women Q. **37**(3), 293–309 (2013). https://doi.org/10.1177/0361684313480694

24. Spencer, S.J., Steele, C.M., Quinn, D.M.: Stereotype threat and women's math performance. J. Exp. Soc. Psychol. **35**(1), 4–28 (1999). https://doi.org/10.1006/jesp.1998.1373

25. Doyle, R.A., Voyer, D.: Stereotype manipulation effects on math and spatial test performance: a meta-analysis. Learn. Individ. Differ. **47**, 103–116 (2016). https://doi.org/10.1016/j.lindif. 2015.12.018

26. McLaren, B.M., Nguyen, H.A.: Digital learning games in Artificial Intelligence in Education (AIED): a review. In: du Boulay, B., Mitrovic, A., Yacef, K. (eds.) Handbook of Artificial Intelligence in Education, pp. 440–484 (2023). https://doi.org/10.4337/9781800375413. 00032

27. Nguyen, H., Else-Quest, N., Richey, J.E., Hammer, J., Di, S., McLaren, B.M.: Gender differences in learning game preferences: results using a multi-dimensional gender framework. In: Proceedings of 24th International Conference on Artificial Intelligence in Education (AIED 2023), pp. 553–564 (2023). https://doi.org/10.1007/978-3-031-36272-9_45

28. Aleven, V., et al.: Example-tracing tutors: Intelligent tutor development for non-programmers. Int. J. Artif. Intell. Educ. **26**, 224–269 (2016). https://doi.org/10.1007/s40593-015-00882

29. Aleven, V., Mclaren, B.M., Sewall, J., Koedinger, K.R.: A new paradigm for intelligent tutoring systems: example-tracing tutors. Int. J. Artif. Intell. Educ. **19**(2), 105–154 (2009)

30. Lester, J.C., Spain, R.D., Rowe, J.P., Mott, B.W.: Instructional support, feedback, and coaching in game-based learning. In: Handbook of Game-Based Learning, pp. 209–237 (2020)

31. Ben-Eliyahu, A., Moore, D., Dorph, R., Schunn, C.D.: Investigating the multidimensionality of engagement: affective, behavioral, and cognitive engagement across science activities and contexts. Contemp. Educ. Psychol. **53**, 87–105 (2018). https://doi.org/10.1016/j.cedpsych. 2018.01.002

32. Abeele, V.V., Spiel, K., Nacke, L., Johnson, D., Gerling, K.: Development and validation of the player experience inventory: a scale to measure player experiences at the level of functional and psychosocial consequences. Int. J. Hum. Comput. Stud. **135**, 102370 (2020). https://doi. org/10.1016/j.ijhcs.2019.102370

33. Linnenbrink-Garcia, L., et al.: Measuring situational interest in academic domains. Educ. Psychol. Meas. **70**(4), 647–671 (2010). https://doi.org/10.1177/0013164409355699

34. Pekrun, R., Goetz, T., Frenzel, A.C., Barchfeld, P., Perry, R.P.: Measuring emotions in students' learning and performance: the Achievement Emotions Questionnaire (AEQ). Contemp. Educ. Psychol. **36**(1), 36–48 (2011). https://doi.org/10.1016/j.cedpsych.2010.10.002

35. Chung, B.G., Ehrhart, M.G., Holcombe Ehrhart, K., Hattrup, K., Solamon, J.: Stereotype threat, state anxiety, and specific self-efficacy as predictors of promotion exam performance. Group Organ. Manag. **35**(1), 77–107 (2010). https://doi.org/10.1177/1059601109354839

36. Hayes, A.F., Rockwood, N.J.: Regression-based statistical mediation and moderation analysis in clinical research: observations, recommendations, and implementation. Behav. Res. Ther. **98**, 39–57 (2017). https://doi.org/10.1016/j.brat.2016.11.001

37. Muir, M., Conati, C.: An analysis of attention to student – adaptive hints in an educational game. In: Cerri, S.A., Clancey, W.J., Papadourakis, G., Panourgia, K. (eds.) ITS 2012. LNCS, vol. 7315, pp. 112–122. Springer, Heidelberg (2012). https://doi.org/10.1007/978-3-642-30950-2_15

Calcium Regulation Assignment: Alternative Styles in Successfully Learning About Biological Mechanisms

Marco Kragten[1]([✉]) [ID] and Bert Bredeweg[1,2] [ID]

[1] Faculty of Education, Amsterdam University of Applied Sciences, Amsterdam, Netherlands
{m.kragten,b.bredeweg}@hva.nl

[2] Faculty of Science, Informatics Institute, University of Amsterdam, Amsterdam, Netherlands

Abstract. We have developed a pedagogical approach wherein learners acquire systems thinking skills and content knowledge by constructing qualitative representations. In this paper, we focus on how learners learn about the biological mechanisms of calcium regulation by constructing such a representation, how they interact with the software, and the effect on learning outcomes. The software contains various functionalities to support learners, and a workbook guides them through the process. Cluster analysis of learners' use of the software categorizes them into three styles, which we have labelled: exploratory, comprehensive, and efficient. Learning outcomes are evaluated through pre- and post-tests and show overall improvement on systems thinking skills and content knowledge. No significant differences in outcome are observed between the interaction styles of learners. This implies that constructing qualitative representations effectively increases learners' systems thinking skills and understanding of calcium regulation, regardless of their interaction style.

Keywords: Qualitative representations · Interaction style · System thinking

1 Introduction

Understanding system behavior is crucial in the field of biology education [1]. While modeling offers a promising approach for learning about dynamic systems [2], it is acknowledged as a challenging method [3, 4]. Learners often tend to concentrate on observable outcomes like graphs or numerical data, potentially overlooking the essential underlying mechanisms of the system [4]. To address this, an effective pedagogical approach must guide learners to focus on the fundamental biological mechanisms. Moreover, there is a risk that learners may not complete the model, introduce errors, and run non meaningful simulations, impeding deeper understanding [5, 6]. Therefore, learners require appropriate support during the modeling process to enable them to construct the complete model and run simulations that contribute to their understanding of dynamic systems. In this context, qualitative representations emerge as a well-suited solution to address these challenges and foster a more comprehensive understanding of dynamic biological systems.

A. M. Olney et al. (Eds.): AIED 2024, LNAI 14829, pp. 220–234, 2024.
https://doi.org/10.1007/978-3-031-64302-6_16

Qualitative representations and their accompanying algorithms have a long history [7]. For the work presented in this contribution we use the Garp3 approach [8]. Qualitative representations are non-numerical descriptions of (physical) systems and their behavior [9]. These methods allow computer programs to reason about the behavior of systems, without precise quantitative information, but using qualitative descriptions of causal relations (e.g., proportional positive), values (e.g., zero, positive) and changes (e.g., decreasing, constant, increasing) of quantities. A key feature of qualitative representations is the ability to capture the conceptual notions needed to understand the working of systems explicitly. Despite not using numerical information, the details captured in such representations can be simulated showing the system dynamics.

The explicit nature of the representation facilitates the implementation of miscellaneous scaffolds to aid learners during their constructive learning effort [10, 11]. This support facilitates that in principle all learners should be able to successfully complete the model. However, finding a balance is essential; the lesson should neither be too open-ended nor become a rigid recipe, but rather actively engage learners as much as needed. To foster active engagement and effective learning, it's essential to create learning environments that accommodate individual preferences for modelling [12].

This paper is organized as follows. Section 2 describes the calcium regulation assignment. Section 3 presents the design of the study that focuses on how learners interact with the software. As previously noted, an effective approach should afford learners some flexibility in their interaction with the software, allowing them to engage in their preferred manner. As such, we are interested whether we can detect alternative styles and their effect on learning outcomes. Section 4 presents the results obtained during the study. Section 5 concludes and discusses the paper.

2   Calcium Regulation Assignment

A workbook guides the learners in constructing the qualitative representation. The workbook contains explanatory texts and diagrams for explaining calcium regulation and the vocabulary of the representation. The representation is created step-by-step and ingredient types and support features are introduced when they are relevant. Each step generally consists of (i) processing new information about a part of the calcium regulation and how this can be expressed in the vocabulary of the representation, (ii) extending the representation, (iii) simulating, and (iv) reflecting. The learners are provided guidance but are required to think about how to represent the information correctly. For instance, they need to determine which quantities are causally related, specify the direction of the causal relationship, and identify the type of the causal relationship. The workbook includes exercises which prompt learners to reflect on the behavior of the system they have created thus far. For instance, they may encounter embedded answer questions like 'If calcium levels in the blood are *smaller than/equal to/larger than* the norm, what will happen to calcitonin levels: *decrease/stay constant/increase*?'.

In the following subsections we first discuss the vocabulary of qualitative representations and the support available. Next, we discuss the specific content knowledge learning goals regarding calcium regulation. Finally, we give a detailed description of how learners construct the qualitative representation of calcium regulation.

2.1 Representation

Learners used the Dynalearn software [13] to construct the qualitative representation of calcium regulation. The vocabulary deployed in the Dynalearn software is summarized in Table 1. It distinguishes *entities* and *quantities*. Entities can be structurally related to each other by a *configuration*. Quantities have a derivative (δ) and can have causal relationships with other quantities. Possible *values* of a quantity are described using the notion of a *quantity space*, which consists of a range of possible point and interval values that represent distinct states of the system.

There are two types of causal relationships: *influence* (I) which represents a primary cause of change, due to a process being active, and *proportionality* (P) which propagates change. *In/equalities* can be used to specify ordinal relationships between quantities. *Calculi* allow qualitative calculations of quantity values, leading to the derivation of a new value.

Table 1. Qualitative vocabulary for representing systems (and simulating their behavior).

ID	Type	Description
I	Entity	Physical objects (or abstract concepts) that make up the system
II	Quantity	Measurable features of entities
III	Configuration	Structural relationships between entities
IV	Quantity value	Points and intervals that represent quantity measures
V	Derivative (δ)	Direction of change of a quantity (dec., std., or inc.)
VI	Quantity space	Possible values that a quantity can take on
VII	Influence (I)	Quantity relationship that represents primary cause of change
VIII	Proportionality (P)	Quantity causal relationship that propagates changes
IX	Exogenous influence	Continuous external effect on a quantity (dec., std., or inc.)
X	In/equality	Ordinal relationships between quantities ($<, \leq, =, \geq, >$)
XI	Calculus	Qualitative addition (A + B = C) and subtraction (A-B = C)
XII	Scenario	Initial settings to start a simulation
XIII	State	Possible behavior of the system (simulation result)
XIV	State graph	Set of states and their transitions (simulation result)

A qualitative representation can be simulated to determine the behavior of the represented system (direction of change of the quantities & possible states) under the specified initial settings (*scenario*). A *state graph* presents the possible *states* of the system. Learners can click the states to inspect the behavior of the system.

Support. There are four functionalities in the software that support the learners with constructing the representation and simulating its behavior:

- The **norm-based support** detects differences between the learners' representation and a predefined norm representation [10, 11]. An incorrect ingredient will be highlighted in red, and a red question mark will be shown on the right-hand side of the canvas. Clicking on the question mark provides a short hint about the incorrect ingredient, e.g., 'Causal dependency: between wrong quantities?'. The support system does not provide the correct answer, encouraging learners to reflect upon their mistakes, which in turn promotes the learning process.
- The **progress bar** is located at the bottom of the canvas. It provides information about how many ingredients must be created, how many ingredients already have been created, and how many of these ingredients are (in)correct.
- The **scenario advice** assesses whether all necessary initial settings have been fulfilled for running a simulation. In the case they have not, a blue exclamation mark will be displayed on the right-hand side of the canvas. Learners can click on it to receive a hint about the required initial settings, e.g., 'Assign initial change quantity?'.
- The built-in **video support** informs learners on how to add ingredients to the representation. There is a short video for each ingredient type that learners can click for viewing.

2.2 Content Knowledge Learning Goals

The assignment focuses on calcium regulation, a topic commonly addressed in biology curricula at the upper secondary education level. The regulation of calcium levels in the human body is determined by two hormones: calcitonin and parathyroid hormone (PTH). The parathyroid glands, located on the thyroid gland, produce PTH in response to low blood calcium levels. PTH acts on the bones, kidneys, and intestines to regulate calcium levels. In the bones, PTH stimulates the release of calcium from the bone matrix, increasing blood calcium levels. Within the kidneys, PTH promotes the reabsorption of calcium from the urine back into the bloodstream, preventing excessive loss of calcium in the urine. Simultaneously, PTH facilitates the absorption of dietary calcium in the intestines. Contrastingly, calcitonin, produced by the thyroid gland, acts as a counterregulatory hormone to PTH. When blood calcium levels rise, calcitonin inhibits bone resorption, reabsorption by the kidneys and absorption in the intestines.

2.3 Constructing the Qualitative Representation of Calcium Regulation

This section describes how learners construct the representation step-by-step. Figure 1 shows the target representation for calcium regulation. The entities are *Blood, Thyroid, Parathyroid, Bones, Kidneys, Intestines,* and *Skin. Thyroid* is structurally related to *Parathyroid* via the configuration *has*. The quantities are *Ca2+* (assigned to *Blood* and *Bones*), *Calcitonin, Norm,* and *Difference* (assigned to *Thyroid*), *PTH* (assigned to *Parathyroid*), *Resorption Ca2+* (assigned to *Kidneys*), *Uptake Ca2+* (assigned to *Intestines*), *Sunlight,* and *Vitamin D* (assigned to *Skin*). The representation comprises 12 causal relationships of type proportionality and two of type influence. There is one equality and one calculus. Three quantities have a quantity space, notably {+} assigned to *Ca2+* from *Blood*, {norm} assigned to *Norm* from *Blood*, and {−, 0, +} assigned to *Difference* from *Thyroid*. The total number of ingredients amounts to 40.

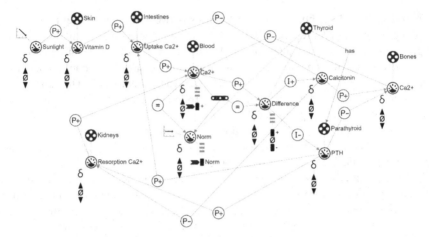

Fig. 1. Target qualitative representation of calcium regulation.

Effect of Calcitonin. The first part of the assignment informs learners about calcitonin and the relationship with calcium levels in the blood and in the bones. Learners are informed how to represent this part of the system by using ingredient types: entity, quantity, derivative, and proportional relationship (Table 1: I, II, V, and VIII). They are asked to create the entities *Thyroid, Bones* and *Blood,* to add the corresponding quantities *Calcitonin* and *Ca2+* (2 times) and to add the correct causal relationships (P+ and P−) between these quantities. The first simulation can now be run. Learners are informed about using an exogenous influence as an initial setting and how to interpret simulation results (Table 1: IX, XII, XIII, and XIV). They are guided to simulate the representation with successively using a decreasing, steady and increasing exogenous influence. Figure 2 shows the representation so far and the simulation results with an increasing exogenous influence acting on *Calcitonin.* When *Calcitonin* increases, the *Ca2+* of *Bones* increases and the *Ca2+* of *Blood* decreases, due to the proportional relationships between these quantities. Exercises in the workbook support learners in understanding the behavior of this part of the system.

Effect of PTH. Next, learners are told about the effect of PTH, a hormone produced by the parathyroid glands which are part of the thyroid, on the levels of calcium in the bones and blood. They are informed that configurations can be added to describe how entities are structurally related (Table 1: III). Learners are asked to add the *Parathyroid gland* to the representation, to structurally relate it with *Thyroid* by adding the configuration *has,* and to add the quantity *PTH.* They are asked to add the correct causal relationship between *PTH* and the quantities already in the representation. They can investigate the behavior of this part of the system by running simulations under different initial settings and are asked to complete related exercises.

Effect of Calcium Deviation. The next part focusses on what happens with the release of calcitonin and PTH if the level of calcium in the blood deviates from normal. Learners are first informed about what happens if the level of calcium in the blood is too high. If calcium is too high, then the thyroid is stimulated to produce more calcitonin.

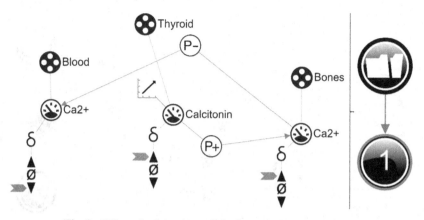

Fig. 2. Effect of calcitonin on Ca2+ levels in bones and blood.

Deviations thereby trigger a process, introducing the notion of causal relationships of type influence, quantity spaces and values (Table 1: IV, VI, and VII). To learn about these notions, learners add a quantity *Difference* to represent deviations and a quantity space with only the values zero (0) and positive (+). They then are asked to add the correct causal relationship (I+) between *Difference* and *Calcitonin*. Note that in this case, the casual relationship actually describes the ratio between production and degradation of calcitonin. If the value is positive (+) than more calcitonin is being produced than degraded. Learners then are required to add the value negative (−) to the quantity space of *Difference* and are asked to simulate the representation with several initial settings and complete exercises.

Negative Feedback. The negative feedback mechanism responsible for sustaining the system in a state of homeostatic equilibrium can now be added to the representation. The production of the hormones calcitonin and PTH that results from deviations from the norm provides negative feedback on calcium levels in the blood and thereby the difference with the norm decreases. For example, if there is a negative difference with the norm (calcium levels in blood are too low), more calcium will be released from the bones, resulting in an increase of calcium in the blood. The difference of calcium levels in the blood compared to the norm will then decrease. To include the notion of feedback, learners are asked to add the correct causal relationship (P+) between *Ca2+* of *Blood* and *Difference*. A simulation can now be run with the representation describing the feedback mechanism. Figure 3 shows state 1 of the simulation results with initial settings: *Difference* is positive (+) and the exogenous influence acting on *Difference* is removed. The latter is required because the change of *Difference* is now determined by the change of *Ca2+* of *Blood*. The simulation results in two consecutive states. In the first state *Calcitonin* is increasing and *PTH* is decreasing because the value of *Difference* is positive (+). As a result, *Ca2+* of *Bones* is increasing, *Ca2+* of *Blood* is decreasing and *Difference* is decreasing towards zero (0). In the second state (not shown), *Difference* is zero (0) and there is no more change in the system, indication that homeostatic equilibrium is reached. Exercises guide learners to interpret the two consecutive states of the state graph (Table 1: XIV).

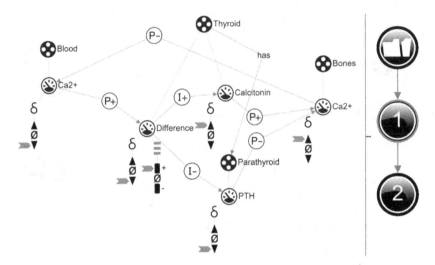

Fig. 3. Negative feedback maintains homeostatic equilibrium.

Explicating the Norm. Next, learners are guided to refine the representation by explicitly including the norm and that the difference from the norm can be expressed as an inequality between levels of calcium in the blood and the norm (Table 1: X). Learners are asked to add the quantity *Norm* and a quantity space with value *Norm*. They are also asked to add a quantity space with value positive (+) to *Ca2+* of *Blood*. Note that, while not essential for describing the behavior of the system, these quantity spaces enhance clarity for learners. They help noticing that the inequalities and calculus in the representation apply to the values of these quantities (their amount) rather than to their changes. Learners then are informed about the ingredient of type calculus (Table 1: XI) and are asked to add a calculus that states that the (value of) *Difference* = (value of) *Ca2+* – (value of) *Norm*.

Figure 4 shows the representation up to this point and the simulation result with initial settings: a constant exogenous influence acting on *Norm* and an inequality (<) that states that *Ca2+* of *Blood* is smaller than *Norm*. Note that the inequality is necessary as without this information there is no way of knowing the outcome of the calculus. The state graph shows that there are two consecutive states. In state 1, *Ca2+* is smaller (<) than *Norm* so the outcome of the calculus is that *Difference* is negative (−). In state 2 (not shown), *Ca2+* of *Blood* equals (=) *Norm* and *Difference* is zero (0). Learners are asked to complete exercises to support them in understanding this part of the system.

Kidneys and Intestines. In the next part, the focus shifts to additional mechanisms essential for maintaining homeostatic equilibrium in the calcium regulation. Learners are informed that the roles of kidney resorption and intestinal uptake are crucial for the effectiveness of this process. Learners are asked to add the entity *Kidneys* and add the quantity *Resorption Ca2+*. From the information provided, they need to discover and add the correct relationships between the quantities *Calcitonin*, *PTH*, *Resorption Ca2+* and *Ca2+* of *Blood*. Next, they are asked to add the entity *Intestines*, the quantity

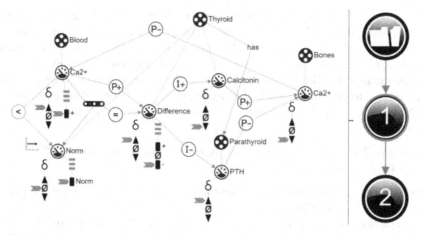

Fig. 4. Calculation of the difference between calcium level in the blood and the norm.

Uptake Ca2+ and the correct relationships. Learners are guided to perform simulations to explore the behavior of this part of the system and to complete related exercises.

Sunlight and Vitamin D. In the final part, learners examine how external factors can have an effect on the behavior of the system. Learners are informed that vitamin D has an effect on uptake of calcium by the intestines. Vitamin D is produced by the skin when in contact with sunlight. Learners are asked to add the entity *Skin,* the quantities *Sunlight* and *Vitamin D*, the required relationships, to run simulations, and to complete exercises. The latter steps conclude the construction and marks the completion of the assignment. Learners have now arrived at the target representation (see Fig. 1).

3   Method

3.1 Participants

Sixty-five learners participated in the study. The lesson was conducted across two K-11 classes over two consecutive years (n = 37 in year 1 and n = 28 in year 2), during a regular lesson of 150 min at the school. Learners were enrolled in a pre-university biology course. They were already familiar with constructing qualitative representations, as the school participated in a research project in which various lessons with qualitative representations were designed and implemented. Teachers were present to provide guidance.

3.2 Data Collection

Data for this study consists of action log measures of each learner's interaction with the software, scores on the workbook exercises, and pre- and post-test scores.

Action Log. The action log records learners' interaction with the software, such as creating, modifying, and deleting ingredients. The target representation includes a total of 40 ingredients. Additionally, the action log captures key metrics such as the frequencies of using the video and the norm-based support, and the count of errors during the construction process. The action log also tracks the number of simulations executed, states selected to view simulation results, incomplete initial settings during simulations, and the use of scenario advice. Finally, it provides insights into the time learners spent constructing the qualitative representations.

Workbook. The exercises in the workbook were evaluated, with each correct answer earning 1 point. The maximum possible score was 40 points.

Tests. The pre- and post-test of the lesson consisted of two parts: (i) content knowledge of calcium regulation and (ii) system thinking.

A content knowledge test of six items was developed. The items measure the extent to which learners had knowledge about and understood calcium regulation. Table 2 presents a short description of the items. For instance, in item 3, participants are required to specify whether calcitonin and PTH have a negative or positive effect on absorption by the bones, uptake by the intestines, and reabsorption by the kidneys. The maximum score for the content knowledge test was 18.

Table 2. Pre- and post-test items for content knowledge of calcium regulation and systems thinking.

Content knowledge	System thinking
1. Hormones and glands (4pt)	1. Description of a system (4pt)
2. Effect of calcitonin and PTH (2pt)	2. Simple causal chain (2pt)
3. Bones, kidneys, and intestines (6pt)	3. Positive feedback loop (2pt)
4. Ca2+ in the blood (4pt)	4. Negative feedback loop (2pt)
5. Hypomagnesemia (2pt)	5. Constructing a quantity space (5pt)
	6. Behavior of a system under initial settings (2pt)

Learners' system thinking skills were assessed by measuring their understanding of the qualitative vocabulary. The test consisted of six response items (Table 2), with a maximum score of 17 points. In each item, a case of a specific system was presented that requires no prior knowledge. Participants were given the task to understand and describe the system (presented in the case) in a qualitative way. For example, item 2 describes a case involving the melting of a glacier, the effect on the water level in a glacial lake and the subsequent effect on the outflow of a river lake. Learners were expected to describe the system qualitatively, using the appropriate types of causal relationships. The systems thinking test was only administered in the second year of the study. The research group discussed ambiguous responses until consensus was reached.

3.3   Data Analysis

To investigate learners' interactions with the software and explore potential alternative styles, we performed a k-means cluster analysis using measurements from the action log. Subsequently, we examined each measurement for significant differences across clusters. Due to small sample sizes in two clusters, we used the Kruskal-Wallis test for the latter. If significant differences were identified, we performed post-hoc analysis with Dunn's correction. Additionally, we investigated whether workbook scores varied among clusters. To assess the impact on learning outcomes, a Friedman Test was conducted with content knowledge scores and system thinking scores as dependent variables, cluster as the between-subjects factor, and test (pre- vs. post-test) as the within-subject factor. The significance level was set at $p < .05$.

4   Results

4.1   Learner Interaction Style

Figure 5 shows the cluster means resulting from the cluster analysis using measurements from the action log. The analysis identifies three clusters ($N = 65$, $R^2 = .37$), with 10 learners in cluster 1, 10 learners in cluster 2, and 45 learners in cluster 3.

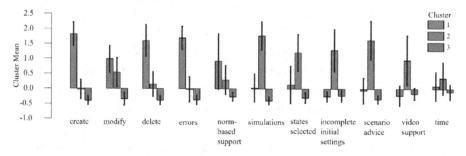

Fig. 5.   Cluster analysis results: action log measurements and cluster means.

Table 3 shows the descriptive values of the measures from the action log per cluster. Kruskal-Wallis shows that differences are significant for all action log measurements, except for time. Post-hoc comparison with Dunn-correction shows which clusters differs for each action log measurement.

Learners in cluster 1 ($n = 10$) create and delete ingredients more often than learners in clusters 2 and 3. They modify more ingredients than learners in cluster 3. They make more errors in constructing the representation than learners in the other two clusters and use norm-based support more often. Learners in this cluster can be characterized as having an exploratory style, as they seem to actively experiment during the construction process. For example, an average of 71.60 ($SD = 4.72$) create-actions largely exceeds the minimum of 40 create-actions required to complete the representation.

Learners in cluster 2 ($n = 10$) run more simulations and select more states from the simulation results than learners in cluster 1 and 3. They more often start simulations with incomplete initial setting and use the scenario advice more often. They also use the video support more often. Learners in this cluster can be typed to have a comprehensive approach, indicating their extensive use of various support and thorough examination of system behavior. For instance, the average of 22.80 simulations ($SD = 4.10$) largely exceeds the 11 simulations explicitly mentioned in the workbook.

Table 3. Action log measures and workbook score per cluster.

Action log measurement	Cluster	M (SD)	Kruskal-Wallis	Post-hoc[b] Comparison	p
create[a]	1	71.60 (4.72)	$H(2) = 26.62$	1–2	< .01
	2	59.90 (3.87)	$p < .001$	2–3	.10
	3	57.56 (3.75)		1–3	< .001
modify	1	62.00 (9.02)	$H(2) = 18.01$	1–2	.37
	2	57.30 (9.62)	$p < .001$	2–3	< .01
	3	48.07 (9.00)		1–3	< .001
delete	1	25.20 (6.32)	$H(2) = 25.29$	1–2	.02
	2	16.20 (4.83)	$p < .001$	2–3	.05
	3	12.89 (3.79)		1–3	< .001
errors	1	30.80 (5.39)	$H(2) = 23.69$	1–2	< .01
	2	18.10 (5.82)	$p < .001$	2–3	.27
	3	15.56 (4.81)		1–3	< .001
norm-based support	1	10.50 (10.58)	$H(2) = 8.54$	1–2	.92
	2	6.80 (5.49)	$p = .01$	2–3	.03
	3	3.44 (3.53)		1–3	.02
simulations[a]	1	14.90 (4.01)	$H(2) = 24.27$	1–2	< .01
	2	22.80 (4.10)	$p < .001$	2–3	< .001
	3	12.98 (2.26)		1–3	.19
states selected[a]	1	49.20 (21.86)	$H(2) = 15.31$	1–2	.03
	2	69.00 (21.78)	$p < .001$	2–3	< .001
	3	41.22 (12.43)		1–3	.26
incorrect initial settings	1	0.80 (0.63)	$H(2) = 19.77$	1–2	< .01
	2	3.80 (2.66)	$p < .001$	2–3	< .001
	3	0.84 (1.52)		1–3	.40
scenario advice[a]	1	0.70 (1.06)	$H(2) = 27.20$	1–2	< .001
	2	2.90 (1.60)	$p < .001$	2–3	< .001
	3	0.31 (0.70)		1–3	.28

(continued)

Table 3. (*continued*)

Action log measurement	Cluster	M (SD)	Kruskal-Wallis	Post-hoc[b] Comparison	p
video support[a]	1	1.40 (1.90)	H(2) = 6.00	1–2	.06
	2	4.90 (4.73)	p = .05	2–3	.02
	3	1.58 (2.32)		1–3	.99
time (s)	1	4688.00 (867.46)	H(2) = 2.376	1–2	
	2	4504.44 (974.17)	p = .31	2–3	
	3	4942.20 (981.71)		1–3	
workbook score	1	34.80 (4.02)	H(2) = 9.87	1–2	.50
	2	33.30 (5.14)	p < .01	2–3	< .01
	3	36.98 (1.76)		1–3	.05

[a]The representation involves creating 40 ingredients. Learners are explicitly instructed to run 11 simulations, select 16 states, view 8 videos, and receive 1 introduction to the scenario advice.
[b] Dunn-correction was applied

Learners in cluster 3 ($n = 45$) are distinct from learners in cluster 1 because they do less construction actions, make less errors and make less use of the norm-based support. They are distinct from cluster 2 because they run less simulations, select less states, make less mistakes with initial settings and use the scenario advice less. Learners in cluster 3 exhibit an efficient style, relying on minimal support for constructing the representation and understanding the system's behavior.

There was no difference in time spent on constructing the target representation between clusters. Workbook scores of learners of type efficient (cluster 3) are higher than those of the other two types of learners. However, it's worth noting that the maximum workbook score achievable is 40, and all clusters achieve acceptable scores.

4.2 Effect on Learning Outcomes

Figure 6 shows the pre- and post-test scores for content knowledge and systems thinking per cluster.

For the content knowledge test there is a significant difference ($\chi^2(1) = 37.16, p < .001$) between pre-test ($M = 7.69, SD = 4.18$) and post-test scores ($M = 12.00, SD = 3.67$). No significant differences are found between clusters. For the system thinking test there is a significant difference ($\chi^2(1) = 14.44, p < .001, n = 28$) between pre-test ($M = 6.93, SD = 2.79$) and post-test scores ($M = 10.21, SD = 2.39$). Again, no significant differences are found between clusters. The results suggest that learners improve their understanding of the calcium regulation and of the qualitative vocabulary regardless of their style of interacting with the software.

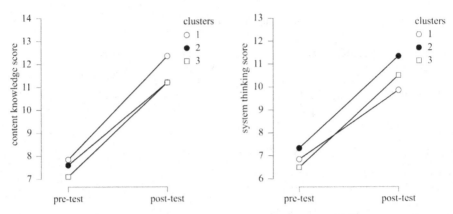

Fig. 6. Pre- and post-test scores for content knowledge and systems thinking test per cluster.

5 Discussion and Conclusion

In this paper, we present a pedagogical approach for learning about biological mechanisms, exemplified by a calcium regulation assignment, through the construction of a qualitative representation. First, we provide an overview of the assignment, detailing the step-by-step process of constructing the representation and its alignment with understanding the underlying mechanisms of calcium regulation. Learners receive support through a workbook and various software functionalities throughout the assignment. Next, we focus on how learners interact with the software during the construction of the qualitative representation. Cluster analysis revealed three distinct learner interaction styles with the software during the calcium regulation assignment, which we labelled: exploratory, comprehensive, and efficient. Learners with an efficient style achieved higher scores on their workbook but all learner interaction styles achieved acceptable scores. Finally, we focus on the learning outcomes of the assignment, distinguishing between the acquisition of specific content knowledge of calcium regulation and the development of systems thinking skills, operationalized as the ability to qualitatively describe system behavior. Significant improvements in content knowledge and systems thinking skills were observed, regardless of interaction style.

It seems fair to conclude that the assignment was effective in learners successfully completing the representation with the provided support, adapting to their preferred interaction styles, completing workbook exercises, and demonstrating a significant improvement in their understanding of calcium regulation and systems thinking skills. However, a follow-up study should compare these outcomes with those of a control group undergoing traditional instruction.

In the assignment, learners construct a representation of a complex dynamic system, in this case the regulation of calcium, a task that requires more active involvement than interpreting the static diagrams in textbooks. The construction approach focuses explicitly on the functioning of the biological mechanism. A key aspect of the approach is that learners, regardless of their interaction style, complete the representation successfully. An incomplete representation can lead to the formation of inaccurate or incomplete mental models of the biological mechanisms involved in the calcium regulation. The support provided plays a crucial role in the construction activities of the learners.

While the automated support features are essential for helping learners, they might raise concerns about the potential for exploratory learners to 'game the system' [14]. The results demonstrate that learning outcomes remain consistent across different interaction styles, challenging the notion that an exploratory approach negatively affects the effectiveness of the assignment. Previous research also shows that there is no systematic abuse of the support functions as provided in this study [10].

The identification of learner interaction styles contributes valuable insights for tailoring instructional strategies. Future research could explore personalized support tailored to the characteristics of exploratory, comprehensive, and efficient learners. However, exploring the potential benefits of cross-style support and evaluating its effectiveness would also be valuable. In the cases where learners with different styles collaborate, implementing a support system that accommodates diverse preferences and encourages a collaborative exchange of ideas could foster a more inclusive and effective learning environment. The results presented in this paper can be built upon by conducting a fine-grained analysis of learner behavior in different parts of the assignment and mapping it with specific learning outcomes. In the current version of the software, the progress bar is always visible and hence does not require any specific learner interaction. Examining extensions for learners to interact with the progress bar could provide valuable insights into their behavior, informing further improvements in support. Additionally, the hybrid learning environment [15], combining automated support and teacher presence, could be examined. Understanding how learners utilize both sources of support may lead to a more effective approach.

In conclusion, this study showcases the versatility of a qualitative representation-based pedagogical approach, which fosters systems thinking and content knowledge acquisition across varied learner interaction styles. The identification of these styles can contribute to creating more effective learning environments.

Acknowledgments. The research presented here was co-funded by the Dutch Regieorgaan SIA, https://regieorgaan-sia.nl/, project Denker, grant number RAAK.PRO03.098, https://denker.nu/.

References

1. Boersma, K., Waarlo, A.J., Klaassen, K.: The feasibility of systems thinking in biology education. J. Biol. Educ. **45**(4), 190–197 (2011)
2. Jacobson, M.J., Wilensky, U.: Complex systems in education: scientific and educational importance and implications for the learning sciences. J. Learn. Sci. **15**(1), 11–34 (2006)
3. VanLehn, K.: Model construction as a learning activity: a design space and review. Interact. Learn. Environ. **21**(4), 371–413 (2013)
4. Sins, P.H.M., Savelsbergh, E.R., van Joolingen, W.R.: The difficult process of scientific modelling: an analysis of novices' reasoning during computer-based modelling. Int. J. Sci. Educ. **27**(14), 1695–1721 (2005)
5. van Buuren, O., Heck, A., Ellermeijer, T.: Understanding of relation structures of graphical models by lower secondary students. Res. Sci. Educ. **46**, 633–666 (2016)

6. Mulder, Y.G., Lazonder, A.W., de Jong, T.: Finding out how they find it out: an empirical analysis of inquiry learners' need for support. Int. J. Sci. Educ. **32**(15), 2033–2053 (2010)
7. Weld, D.S., De Kleer, J. (eds.): Readings in Qualitative Reasoning about Physical Systems. Morgan Kaufmann, Burlington (2013)
8. Bredeweg, B., Linnebank, F., Bouwer, A., Liem, J.: Garp3—Workbench for qualitative modelling and simulation. Ecol. Inform. **4**(5–6), 263–281 (2009)
9. Forbus, K.D.: Qualitative Representations. How People Reason and Learn About the Continuous World. The MIT Press, Cambridge (2018)
10. Kragten, M., Bredeweg, B.: Effectiveness of lightweight automated support for learning about dynamic systems with qualitative representations. In: Proceedings of the 39th ACM/SIGAPP Symposium on Applied Computing, pp. 11–20 (2024)
11. Bredeweg, B., et al.: Learning with interactive knowledge representations. Appl. Sci. **13**(9), 5256 (2023)
12. Truong, H.M.: Integrating learning styles and adaptive e-learning system: current developments, problems and opportunities. Comput. Hum. Behav. **55**, 1185–1193 (2016)
13. Bredeweg, B., et al.: DynaLearn–an intelligent learning environment for learning conceptual knowledge. AI Mag. **34**(4), 46–65 (2013)
14. Baker, R., Walonoski, J., Heffernan, N., Roll, I., Corbett, A., Koedinger, K.: Why students engage in "gaming the system" behavior in interactive learning environments. J. Interact. Learn. Res. **19**(2), 185–224 (2008)
15. Bredeweg, B., Kragten, M.: Requirements and challenges for hybrid intelligence: a case-study in education. Front. Artif. Intell. **5**, 891630 (2022)

Who's Helping Who? When Students Use ChatGPT to Engage in Practice Lab Sessions

Jérôme Brender[1,2]([✉]) [iD], Laila El-Hamamsy[2] [iD], Francesco Mondada[1] [iD], and Engin Bumbacher[2] [iD]

[1] MOBOTS & LEARN, EPFL - École Polytechnique Fédérale de Lausanne, Lausanne, Switzerland
{jerome.brender,francesco.mondada}@epfl.ch
[2] University of Teacher Education (Haute École Pédagogique) Vaud, Lausanne, Switzerland
{laila.elhamamsy,engin.bumbacher}@hepl.ch

Abstract. Little is understood about how chatbots powered by Large Language Models (LLMs) impact teaching and learning, and how to effectively integrate them into educational practices. This study examined whether and how using ChatGPT in a graduate-level robotics course impacts performance and learning with data from 64 students (40 ChatGPT users and 24 non-users). Regression analyses revealed complex interactions between ChatGPT use, task performance, and learning of course-related concepts: using ChatGPT significantly improved task performance, but not necessarily learning outcomes. Task performance positively correlated with learning only for students with low and medium prior knowledge who did not use ChatGPT, suggesting that performance translated to learning only for non-ChatGPT users overall. Clustering ChatGPT-users' prompts helped identify three types of usage that differed in terms of learning and performance. Although all ChatGPT users had improved performance, *Debuggers* (who requested solutions and error fixes) outperformed the other clusters. In terms of learning, *Conceptual Explorers* (who sought to understand concepts, tasks or codes) had higher learning outcomes than *Debuggers* and *Practical Developer* (who exclusively asked for task solutions). The behaviors elicited by students in the *Practical Developer* and *Debuggers* clusters therefore were less likely to translate performance into conceptual understanding, while the *Conceptual Explorers'* behaviors were more conducive to learning. This empirical study helps improve our understanding of the complex dynamics between how ChatGPT is used, performance, and learning outcomes.

Keywords: ChatGPT · Computing Education · Behaviors · Performance · Learning · Effectiveness

A. M. Olney et al. (Eds.): AIED 2024, LNAI 14829, pp. 235–249, 2024.
https://doi.org/10.1007/978-3-031-64302-6_17

1   Introduction

Applications of large language models (LLM) have begun to mushroom across the educational landscape, especially in the form of chatbots. However, applications of chatbots for education are still in their infancy [6,7,11,19] and suffer from numerous limitations [19,23]. Many of these chatbots, of which ChatGPT [18] is a prominent example, are general-purpose tools that are not explicitly designed for learning and other pedagogical purposes per se [12]. While such tools provide interesting opportunities for education (e.g., idea generation, personalization, automated conversational tutors, etc.), educators and researchers worry about the many risks involved. Opinions nowadays vary from resisting the use of such tools in higher education courses to fully embracing them [12,31]. For example, in light of ChatGPT sometimes giving misleading or incorrect responses (notwithstanding issues of bias, privacy and ethics) [2,3,9,12], some researchers have gone so far as to suggest that ChatGPT should mainly be employed by users who have sufficient domain knowledge to critically evaluate the output [2,24]. In order to better inform these debates, significantly more empirical research is needed on the impact of LLM-based chatbots on teaching and learning, to understand how to best integrate them into classrooms, and to help teachers and students "manag[e] the learning process and achiev[e] their learning goals"(p. 3) [30] through the use of such tools [4,12].

We focus in this paper on the use of LLM-based chatbots in computing education. The computing field is undergoing significant transformations because of the use of these technologies as "cutting-edge AI code generation tools" (p. 502) [4] that change how software developers work [5,12,29,30]. Similar transformations can be observed in computing education at the university level. One key concern that has even been raised by computing students themselves [30] is that students might become over-reliant on LLM-based tools, ultimately impeding learning [2,4,8,10,16], and the development of critical thinking skills [7,10,12,17]. Students may "quickly become accustomed to auto-suggested solutions [...] not thinking about the computational steps needed to solve a problem" (p. 60) [4] or knowing how "to author similar code without them" (p. 2) [8].

In light of such concerns, we propose to empirically investigate how using ChatGPT impacts students' performance and learning in computing courses, to contribute to research on how such tools can be integrated into computing education. This is important because, in the context of computing, we find three main types of studies that evaluate (i) ChatGPT's capacity to solve computing-related problems (see Sect. 2.1); (ii) students' perception of ChatGPT in computing contexts (see Sect. 2.2); and (iii) the impact of ChatGPT on students' performance (see Sect. 2.3). Few studies have investigated the relationship between how ChatGPT is being used in university courses and students' learning processes [17], which is necessary to help identify ways of using ChatGPT that are conducive to learning. We address this gap by focusing on the following RQs:

(RQ1) How does the use of ChatGPT to solve programming-based practice tasks impact students' performance and learning outcomes?

(RQ2) In what ways do students use ChatGPT, and how do different types of ChatGPT use influence performance and learning outcomes?

We examine these questions by means of an empirical study in the context of a graduate-level robotics course that teaches students the basics of mobile robotics (i.e. control theory, localization, and trajectory planning). By dividing 64 students into two groups (40 ChatGPT users and 24 non-ChatGPT users), we conducted two levels of analyses to understand the impact of using ChatGPT on graduate students' performance and learning. We first compared ChatGPT users and non-ChatGPT users and found that using ChatGPT improves performance but is not necessarily conducive to learning (RQ1). We then analyzed ChatGPT users prompts through cluster analysis and identified three types of ChatGPT use. Further analysis revealed that these distinct types of use also differed from each other in terms of their impact on performance and learning (RQ2).

2   Related Work

2.1   ChatGPT's Ability to Solve Programming Tasks

Before examining how LLMs could benefit computing education, we must understand how LLMs perform when confronted with computing tasks. "Researchers and practitioners have discovered that these tools can generate correct solutions to a variety of introductory programming assignments and accurately explain the contents of code" (p. 106) [12]. ChatGPT in particular is able to solve common programming tasks (e.g., search, sort, duplicate elimination, [28]), even outperforming other LLMs [14], and students on exams [14,15]. Research on its performance is less clear when it comes to more complex programming tasks. On the one hand, studies such as [27] conclude that ChatGPT was able to "perform high-level cognitive tasks" and "exhibit critical thinking skills" (p. 1) in various disciplines, including machine learning. On the other hand, there are studies where participants observe that ChatGPT struggles with "complex problems and with iterative development" (p. 25) [1]. Overall, the tenor seems to be that ChatGPT is able to provide sufficiently accurate, readable, and high-quality code, such that ChatGPT and other AI code-generating tools are increasingly used both in computing professionals and students [8].

2.2   Students' Perception of Using ChatGPT for Programming

Multiple studies investigated how university-level students in computing fields perceived the use of ChatGPT in computing contexts [24,26,30]. The findings globally align with the identified benefits and limitations of using ChatGPT to solve programming tasks. Students find ChatGPT helpful, "easy to use and appreciate its human-like interface[,] well-structured responses and good explanations" (p. 38813) [24], whether to understand the concepts or help solve the problems [30]. Other studies found that students were aware that ChatGPT's answers were not always accurate [24] and were skeptical about its general utility [26]. In a study in the context of an object-oriented programming course for instance, Yilmaz & Karaoglan Yilmaz [30] found that students considered that ChatGPT was useful to solve problems and ultimately helped save time, giving them the possibility to allocate more time reflecting on how to solve the problem, but worried it might negatively affect their critical thinking skills.

2.3 Impact of Students Using ChatGPT for Programming

Since ChatGPT became publicly available in November 2022, a majority of studies on its use in education were not empirical studies. That is why Memarian & Doleck [16] called for more empirical work to understand the impact of using ChatGPT on learning. In the context of computing, we identified four studies on how ChatGPT could help students solve computing-related problems. Two studies [21,23] considered how students without any relevant prior domain-specific knowledge could solve computing-related problems. The first study [23] found that students using ChatGPT achieved a passing grade on an exam and performed at least as good as students in prior years who had studied the material and did not have access to ChatGPT. The second study [21] was a case study on a Machine Learning sentiment analysis project. The authors found that students were able to generate functioning code with ChatGPT and concluded that "using such systems may lead to good task completion rate, but without deepening the understanding much on the way" (p. 439). Two other studies looked at how students used ChatGPT and evaluated the quality of AI-assisted student answers to programming problems [8,20]. The first study [20] found that students using ChatGPT often requested full programs and fixes until the program passed the course's unit tests. These students' codes passed a larger number of unit tests than the codes developed by students who did not use ChatGPT. However, their codes contained inaccuracies and inconsistencies, particularly for more complex problems, which underscores the importance of critically evaluating ChatGPT's outputs. The second study [8] examined how students used OpenAI Codex, an LLM-based code generation tool, in introductory programming classes. They found that using Codex improves performance and completion rates, but its effect on learning outcomes was not significant. Overall, the studies appear to converge to the same working hypothesis: students tend to use LLM-based chatbots to improve performance on computing problems, with an unclear impact on learning. More work is needed to examine the different ways students use LLM-based chatbots, and their impact on both learning outcomes and performance.

3 Materials and Methods

We designed a exploratory university-level intervention-based classroom study[1] and adopted a mixed-methods approach with both quantitative and qualitative data to answer our research questions.

---

[1] Ethically approved by EPFL's Ethics Committee (HREC No: 075-2023/18.09.2023).

3.1 Study Context and Design

Participants: We recruited 64 students out of 136 students participating in one of the voluntary practice lab sessions of a graduate-level mobile robotics course at EPFL. These students (46 male, 16 female, and 2 who did not disclose their gender) consented to participate in the study. Please note that participation was voluntary and there was no incentive for the students to participate in the study. We opted to have a larger number of students in the experimental group (with access to ChatGPT, n = 40) compared to the control group (without access to ChatGPT, n = 24) as the objective was to have sufficient data to analyze students' behaviors when using ChatGPT.

Procedure: The study took place in a 2-h practice lab session. After a 15-min introduction to the objectives of the study, the students completed a 10-min pre-test that evaluated their prior knowledge of the course content. Students then engaged in a 75-min practice lab session on particle filters, either using ChatGPT or not. The study ended with a 10-min post-test with more detailed questions on the course content than the pre-test.

Study Design: The participants were randomly assigned to two groups: a control group (n = 24, no access to ChatGPT) and an experimental group (n = 40 using ChatGPT[2]). We chose ChatGPT as the LLM-based chatbot because of its prevalent use and ease of access in educational contexts.

Practice Lab Tasks: During the practice lab session, the students had to solve five tasks in Python to implement a particle filter algorithm to localize a robot. These tasks required simulating particle movements, updating weights, and resampling. The tasks were sufficiently challenging, rendering the use of ChatGPT meaningful. However, in order to make the groups more comparable and compensate for this increased difficulty, we provided students that could not use ChatGPT with starter codes that helped structure the tasks.

3.2 Data Collection and Reliability Measures

Pre-test: The pre-test included two open-questions, as opposed to MCQ-type questions, to allow for bigger variability in terms of prior knowledge assessment. The first question concerned localization in general, and the second particle filter localization (i.e. the specific topic of the practice lab session). Two points were attributed per question for a maximum of 4 points. The responses to the open-text questions were anonymously graded by four Teaching Assistants (TAs). In a first stage, we refined the coding scheme until reaching a high Inter-Rater Reliability (IRR, Fleiss' $\kappa = 0.84$) for 20% of the dataset. The full dataset was then coded by all the TAs and we obtained Fleiss' $kappa = 0.77$. We verified that students in the control- and experimental groups had similar prior knowledge (Mann-Whitney $U = 516.5$, $p = 0.6$, $M = 39.8 \pm 31.0\%$).

---

[2] On October 31st 2023, 39 students used ChatGPT 3.5, and one used ChatGPT 4.0.

Practice Tasks: The computer programs produced during the practice lab session were graded by the same TAs as follows. For each of the 5 tasks a score of 0 was attributed for incomplete tasks, 0.5 for tasks that were more than 70% correct, and 1 for fully correct. We then computed the IRR per task on a randomly selected 20% of the sample. We achieved at least substantial agreement on each task (Fleiss' $\kappa_1 = 0.76$, $\kappa_2 = 0.86$, $\kappa_3 = 0.90$, $\kappa_4 = 0.83$, and $\kappa_5 = 0.84$). The remaining data was coded individually by all the TAs. We computed a total practice session score as as the standardized sum of the individual task scores.

ChatGPT Prompts: We analyzed 397 ChatGPT prompts[3] obtained from the 38 students in the ChatGPT condition who agreed to share their prompts ($M = 10.6 \pm 6.6$ per student). First, two researchers analyzed a subset of the data to develop a categorization of the prompt that focuses on the general types of observed uses of ChatGPT for computing education [12]. This led to three categories of prompts: (i) *Development of Solutions*, i.e. prompts that ask for new code snippets or full solutions; (ii) *Understanding Tasks, Codes, Or Course Concepts*, i.e. prompts for explanations for specific aspects of the tasks, computing concepts, or course concepts; (iii) *Debugging Codes*, i.e. prompts to identify or help resolve coding errors. The two researchers reached a substantial IRR ($\kappa = 0.83$) on 20% of the dataset . Four TAs coded the rest of the dataset.

Post-Test: The post-test consisted of five advanced MCQ questions (1 point per question for a total of 5 points) that focused specifically on the conceptual understanding of the practice lab session's topic. The overall post-test score is the standardized sum of the individual MCQ scores ($M = 40.8 \pm 34.8\%$).

3.3 Data Analysis

Multiple Linear Regression Analysis. To address RQ1 and parts of RQ2 we used multiple linear regression models. We standardized all continuous dependent and independent variables (DV and IV) so the resulting standardized coefficients β provide effect size estimates (for example: $\beta = 0.5$ indicates that a 1 standard deviation (SD) increase in the IV corresponds to 0.5 SD increase in the DV). We used Shapiro-Wilk's test to verify the normality of the residuals.

Effect of ChatGPT Use (RQ1). To examine the relationship between ChatGPT use, practice lab performance and learning outcomes, we developed two linear regression models. The first model evaluated how practice lab scores are influenced by the two-way interaction between pre-test scores and ChatGPT use. The second model evaluated how post-test scores were influenced by the three-way interaction of ChatGPT use, practice lab scores, and pre-test scores.

---

[3] We excluded 25 prompts unrelated to solving the tasks (e.g., "hello", "thank you").

Cluster Analysis (RQ2). To address RQ2, we first clustered students based on their overall use of ChatGPT. We calculated for each student the percentage of prompts in each category (*Development of Solution, Understanding, Debugging*). We then used *k-means* clustering [13]. Using the silhouette score [22] to identify the best clustering, we eventually identified $k = 3$ valid clusters, each characterizing a distinct type of ChatGPT use. Based on these clusters, we developed a multiple linear regression model to assess:

- How well practice lab scores (DV) are predicted by the way ChatGPT was used (clusters, IV), while controlling for pre-test scores (IV).
- How well post-test scores (DV) are predicted by how ChatGPT was used (clusters, IV), while controlling for pre-test scores (IV).

We did not include interaction effects in these models as the relatively small sample size would have led to certain groupings being inferior to 5 individuals.

Given that this was an exploratory study, we computed two versions of the final regression models, once with only the 40 students using ChatGPT and once with all 64 students. Both models provided consistent findings for the individual coefficients, but the model fit indicators were better without the non-ChatGPT users. Therefore, we report in detail the models investigating how different usage patterns influenced performance and learning with the ChatGPT-user data only, and complement the findings with the models that included data from all students to provide baseline comparisons with non-ChatGPT users.

We end the investigation with a qualitative analysis of the prompts in each cluster to gain insight into the characteristics of each type of ChatGPT use.

4 Results

4.1 Using ChatGPT Improves Performance But Not Learning (RQ1)

Impact of Using ChatGPT on Performance. We implemented a linear regression model to evaluate the impact of the interaction between using Chat-GPT (yes, no) and pre-test scores, on performance in the practice lab sessions. The model explains a significant proportion of the variance in practice lab session scores (adjusted $R^2 = 0.31$, $F(3, 60) = 10.57, p < .001$) and helps draw two main findings. The first is a significant main effect of using ChatGPT which is associated with statistically higher practice lab session scores ($\beta = 1.05$, $t(64) = 4.89$, $p < .0001$, 95% CI $[0.62, 1.48]$). The second finding is that there is no significant main effect of the pre-test score on practice lab scores ($p = .46$) but a significant interaction effect between pre-test scores and ChatGPT use ($\beta = -0.49$, $t(64) = -2.33$, $p = .02$, 95% CI $[-0.92, -0.07]$). This indicates that students with low prior knowledge benefit more from using ChatGPT in terms of solving the practice lab tasks than students with high prior knowledge.

Impact of Using ChatGPT on Learning. To understand whether using ChatGPT results in improved learning outcomes, we implemented a multiple linear regression model of post-test scores on ChatGPT use, practice lab performance, pre-test scores, and their interactions (see Table 1). The model was statistically significant ($F(7, 56) = 2.4$, $p = 0.035$). There was no main effect of ChatGPT use on learning (i.e., the post-test scores) of those who used and did not use ChatGPT do not differ significantly. There is however a significant positive interaction effect between the pre-test scores and using ChatGPT ($\beta = 0.74$, $p = .009$), which indicates that students with higher pre-test scores (i.e., high prior knowledge) performed better on the post-test when using Chat-GPT on the practice lab tasks. Conversely, the findings also suggest that students with lower pre-test scores (i.e., low prior knowledge) performed worse in the post-test when using ChatGPT, compared to those not using ChatGPT.

Table 1. Linear regression model for the post-test scores on practice lab performance, ChatGPT use, pre-test scores and their interactions. ChatGPT use is 0 for non-users. The estimates show standardized coefficients.

Predictors	Post-Test Score		
	Estimates	CI	p
(Intercept)	0.18	−0.34–0.71	0.48
Pre-Test	−0.42†	−0.85–0.01	0.06
Practice Lab	0.34	−0.21–0.90	0.22
ChatGPT Use	−0.01	−0.64–0.62	0.98
Pre-Test: Practice Lab	−0.38	−0.88–0.11	0.13
Pre-Test: ChatGPT Use	0.74**	0.19–1.29	**0.01**
Practice Lab: ChatGPT Use	−0.59†	−1.24–0.06	0.08
Pre-Test: Practice Lab : ChatGPT Use	0.54†	−0.07–1.15	0.08
Observations 64	R^2/R^2 adjusted 0.23/0.13		

$^\dagger p \leq 0.1$ $^* p \leq 0.05$ $^{**} p \leq 0.01$ $^{***} p \leq 0.001$

The model also reveals a marginally significant three-way interaction effect of pre-test scores, practice lab performance, and ChatGPT use ($p = 0.079$). To facilitate the interpretation of this three-way interaction, we plotted the predicted and observed values in Fig. 1. Each panel represents a different level of prior knowledge (i.e., pre-test scores): scores one standard deviation (SD) below the mean ($M - SD$), at the mean (M), and one SD above the mean ($M + SD$).

First, the left and middle panels of Fig. 1, which show *students with low and average pre-test scores*, indicate that post-test scores for students in the control group (purple line, dotted band) positively correlate with practice lab session scores. However, post-test scores for students who used ChatGPT (orange line, solid band) negatively correlate with practice lab session scores. This suggests that students with low and average pre-test scores who try to solve tasks without

ChatGPT are more likely to have better learning, as measured by the post-test. The post-test score distribution also confirms that students with lower pre-test scores who used ChatGPT had lower post-test scores than those who solved the practice lab tasks without ChatGPT.

On the other hand, the right panel of Fig. 1, which shows the results for *students with high pre-test scores*, indicates that performance is not correlated with post-test scores, regardless of ChatGPT use. However, students using ChatGPT (blue line) have on average higher post-test scores. This suggests that students with higher pre-test scores may have slightly better outcomes on average when using ChatGPT, independently of their performance on the practice lab. Furthermore, the post-test score distribution of students with high pre-test scores confirms the previous finding, i.e. that students with higher pre-test scores performed better on the post-test when using ChatGPT.

The results of the analysis for RQ1 suggest that the use of ChatGPT overall improved performance on the practice lab tasks, but it did not lead to improved learning outcomes on average, particularly for students with low to medium prior knowledge. On the contrary, it might have had a negative impact on learning for students with low prior knowledge.

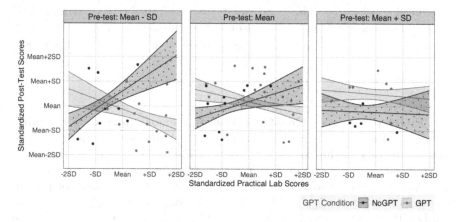

Fig. 1. Post-test scores as a function of practice lab performance, for different levels of pre-test scores and ChatGPT use conditions. The figures show model predictions with standard error bands, and observed student data, standardized, with a small jitter.

4.2 Type of ChatGPT Use Affects Performance and Learning (RQ2)

RQ2 examines whether there are differences in *how* students use ChatGPT, and if said differences might be more or less conducive to improved performance and learning. To that effect we focus the subsequent analyses on the ChatGPT users, providing insight from models including non-ChatGPT users when relevant.

Identifying Types of ChatGPT Use Based on Prompts. As explained in Sect. 3.3, k-means clustering of students' categorized ChatGPT prompts gave rise to $k = 3$ well-separated (in terms of silhouette score S) and similarly sized clusters (see Table 2). These clusters do not differ in terms of pre-test scores (ANOVA $F(2, 35) = 0.34$, $p = .715$), but differ significantly in the distribution of prompt types between clusters. Nearly 95% of *Practical Developers'* prompts were for code development, compared to 57% and 63% for the other clusters. Prompts for understanding specific concepts, tasks, or code parts make up about 40% of prompts of *Conceptual Explorers*, compared to at most about 10% of the prompts of the other clusters. Finally, about 25% of *Debuggers'* prompts related to debugging, compared to about 2% at most for the other clusters.

Table 2. Distribution of each type of prompt for each cluster of ChatGPT use. The silhouette score (S) is a measure of cluster quality (S = 1 for perfect clustering). The highlighted numbers indicate the cluster having used a given type of prompt the most.

Cluster	N	Nbr of Prompts	S	Development of Solution		Understanding		Debugging	
				Mean	(SD)	Mean	(SD)	Mean	(SD)
Conceptual Explorers	15	115	0.5	57.6%	(8.5)	**40.1%**	(9.5)	2.2%	(4.7)
Practical Developers	11	81	0.5	**94.6%**	(6.9)	4.9%	(7.0)	0.6%	(1.9)
Debuggers	12	201	0.8	63.6%	(7.6)	9.5%	(9.5)	**26.8%**	(8.9)

The Effect of Type of ChatGPT Use on Performance and Learning. As explained in Sect. 3.3, we implemented multiple linear regression models to examine the relationship between the type of ChatGPT use (as represented by the clusters) and performance and learning. Table 3 shows the models for both performance on the practice lab tasks and post-test scores. The model on practice lab tasks ($F(3, 34) = 4.1$, $p = .01$) indicates that only the *Debuggers* have significantly different practice lab scores ($\beta = 0.49, p = 0.020$) than the grand mean, being about 0.5 standard deviations higher. The regression model that includes data from non-ChatGPT users as well (not presented here) corroborates the previously reported result that non-ChatGPT users' performance is significantly below the full sample's mean ($\beta = -0.78$, $p < 0.0001$), while only the *Debuggers'* performance exceeds the mean significantly ($\beta = 0.77$, $p = 0.0004$).

In comparison, the regression model on post-test score ($F(3, 34) = 4.91, p = .01$) reveals a different pattern: We see that the *Conceptual Explorers* have a significantly higher post-test score ($\beta = 0.47, p = 0.02$) than the mean of all the clusters, while the *Debuggers* have a marginally significant lower post-test score. Similarly, the regression model that includes all student data (not presented here) shows a significantly different post-test score from the grand mean only for the *Conceptual Explorers* ($\beta = 0.45, p = 0.04$).

Table 3. Linear regression model of practice lab scores and post-test scores on cluster, controlling for pre-test scores. We used a sum contrast for the cluster factor, i.e. the effects of the individual clusters are compared to the grand mean of all three clusters. The estimates show standardized coefficients.

Predictors	Practice Lab		Post-Test Scores	
	Estimates (95% CI)	p	Estimates (95% CI)	p
(Intercept)	0.41** (0.12–0.69)	**0.01**	−0.01 (−0.30–0.28)	0.95
Pre-Test	−0.34* (−0.63–0.04)	**0.03**	0.45** (0.14–0.75)	**0.01**
Concept. Expl	−0.26 (−0.64–0.12)	0.18	0.47* (0.07–0.86)	**0.02**
Debuggers	0.49* (0.08–0.89)	**0.02**	−0.34† (−0.76–0.07)	0.10
Observations	38		38	
R^2/R^2 adjusted	0.26/0.20		0.30/0.24	

$^{†}p \leq 0.1$ $^{*}p \leq 0.05$ $^{**}p \leq 0.01$ $^{***}p \leq 0.001$

Overall, the findings indicate that the way students use ChatGPT significantly influences performance and learning outcomes. Specifically, (i) *Debuggers'* use of ChatGPT improves performance but not learning, (ii) *Conceptual Explorers'* use of ChatGPT improves learning but not performance.

Qualitative Analysis of the Clusters. To understand how the differences in performance and learning may have come about, we examined in more detail the characteristic prompts of the *Conceptual Explorer* and *Debugger* clusters.

We identified 3 types of understanding prompts (n = 70), which are mainly used by the *Conceptual Explorers*: conceptual understanding (e.g., "can you explain how the particle filter works?", n = 24/70), code interpretation (e.g., "how do I interpret the prints?", n = 25/70), and understanding the application of computing concepts in practical scenarios (e.g., "I do not understand why we use the cumulative", n = 21/70). We found that all *Conceptual Explorers* would ask at least one conceptual question before starting to solve the tasks, often asking other types of understanding-questions during the problem-solving process.

The debugging prompts (n = 60), which are almost exclusively used by the *Debuggers*, mainly consisted of directly copy-pasting entire error messages from the editor to ChatGPT, without appending any supplementary questions or information (n = 56/60). This approach is typified by prompts such as, "I get'File ipython-input [...] SyntaxError: invalid syntax' ". There were only 4 out of 60 instances that reflect critical evaluation beyond mere error reporting: 2 instances of students actively seeking explanations for the errors encountered (e.g., "Can you explain why the following code generates an error [...]") and only 2 instances of students reflecting on the misalignment between the expected and actual outcomes ("The final answer differs from reality, the true position gives 4, and the estimated 2, correct the conceptual error"). These rare instances align more closely with computational thinking principles, encouraging students to engage in reflective and analytical problem-solving, rather than the predominant trend observed among the *Debuggers*, which lacks a deeper level of engagement [25].

5   Discussion and Conclusion

This exploratory study examined whether and how using ChatGPT impacts performance and learning outcomes in the context of a practice lab session for a graduate-level robotics course. We used a mixed-methods approach with 64 students (40 using ChatGPT, 24 without) to investigate the impact of the overall use of ChatGPT (RQ1) and of different types of ChatGPT use (RQ2).

The findings for RQ1 indicate that ChatGPT use overall significantly improves scores on practice lab tasks (i.e. performance), in particular for students with low-medium prior conceptual knowledge. ChatGPT may therefore be a beneficial tool for enhancing performance in practice lab sessions. In line also with other studies [8], these results suggest that ChatGPT use might help reduce the performance gap between students in terms of performance scores.

One would expect that improved task performance would result in improved learning outcomes (post-test scores). However, our analyses reveal a different picture: Overall ChatGPT use did not directly impact learning outcomes. In particular, there was a negative correlation of performance on the practice lab tasks and learning outcomes for students with low-medium prior knowledge *only* if they used ChatGPT. The correlation was positive for the students not using ChatGPT. This indicates that the increased performance did not translate into better understanding of the target concepts, as measured in the post-test. Meanwhile, there was a significant main effect of ChatGPT use on learning for the students with high prior knowledge. These empirical results could be interpreted to support the previously mentioned claims that ChatGPT should be used by people with sufficient domain knowledge to critically evaluate its output [2,23]. However, our subsequent analysis for RQ2 shows a more relevant differential impact of *how* ChatGPT is used, and not just *whether* it is being used.

The cluster analysis of ChatGPT prompts in RQ2 revealed three distinct types of ChatGPT use. Although the findings confirmed that all ChatGPT users had improved performance, *how* ChatGPT was used impacted performance and learning differently. More specifically, we found that:

- Students who asked ChatGPT understanding-related questions about specific concepts, tasks or code parts almost half of the time (referred to as the *Conceptual Explorers*) had significantly higher learning outcomes than the other students, even if their performance on the practice lab tasks did not differ significantly from the average of the other ChatGPT users.
- Students who tended to copy-paste codes and error messages back and forth between ChatGPT and the programming editor (referred to as *Debuggers*), had significantly better performance on the practice tasks compared to all other students, but marginally lower learning outcomes compared to the average mean of all ChatGPT users.

– Students who almost exclusively asked ChatGPT to solve each task one by one (referred to as *Practical Developers*) had average performance and learning outcomes when compared to other ChatGPT users.

These findings suggest that *Debuggers'* and *Practical Explorers'* use of Chat-GPT does not reflect the critical analysis and productive problem-solving strategies that the literature on computational thinking has identified as effective behaviors [25]. Simply copy-pasting the problem statement or code snippets into ChatGPT to solve the task quickly does not help students reflect on the task and foster their understanding of the underlying concepts. Instructional and design approaches that focus on fostering reflection on the prompts when using LLM-based chatbots in programming tasks might help minimize this problem, and help achieve a synergistic relation between performance and learning.

Our findings provide empirically based insights into possible relationships between ChatGPT use and learning in computing-based university courses. However, the relatively small sample (64 students) of this pilot study means that we have relatively low statistical power, which is why we provided the additional clustering and qualitative analyses to corroborate the statistical analyses. Nevertheless, further studies with larger samples and in a wider range of contexts are needed to replicate and extend these findings. Future work would also benefit from (i) a nuanced analysis of how ChatGPT's use impacts long-term learning retention and understanding beyond immediate task performance to evaluate its educational efficacy more comprehensively; (ii) an expansion to a diverse range of academic disciplines and levels of education to assess the generalizability of the findings and the utility of ChatGPT in a broader educational context.

To conclude, our study shows that how ChatGPT is used is a more relevant question than just whether it is being used, especially when it comes to fostering student learning. The question thus is how we can design interventions and technologies that help students use ChatGPT in ways that are productive for their own learning and not just for performance. This ensures that the AI agents help students learn through their interactions, and not just the AI agents themselves.

Acknowledgements. We would like to thank our colleagues (J.W., N.A, A.B, R.B, C.C, S.A), and the students who volunteered to participate in the study.

References

1. Adamson, V., Bägerfeldt, J.: Assessing the effectiveness of ChatGPT in generating Python code (2023)
2. Azaria, A., Azoulay, R., Reches, S.: ChatGPT is a remarkable tool–for experts. arXiv preprint arXiv:2306.03102 (2023). https://doi.org/10.48550/arXiv.2306.03102
3. Baidoo-Anu, D., Ansah, L.O.: Education in the era of generative artificial intelligence (AI): understanding the potential benefits of ChatGPT in promoting teaching and learning. J. AI **7**(1), 52–62 (2023)

4. Becker, B.A., Denny, P., Finnie-Ansley, J., Luxton-Reilly, A., Prather, J., Santos, E.A.: Programming is hard - or at least it used to be: educational opportunities and challenges of AI code generation. In: Proceedings of the 54th ACM Technical Symposium on Computer Science Education, vol. 1, pp. 500–506. SIGCSE 2023, Association for Computing Machinery, New York, NY, USA (2023)

5. Bird, C., et al.: Taking Flight with Copilot: early insights and opportunities of AI-powered pair-programming tools. Queue **20**(6), 10:35–10:57 (2023). https://doi.org/10.1145/3582083

6. Hwang, G.J., Chang, C.Y.: A review of opportunities and challenges of chatbots in education. Interact. Learn. Environ. **31**(7), 4099–4112 (2023)

7. Kasneci, E., et al.: ChatGPT for good? On opportunities and challenges of large language models for education. Learn. Individ. Differ. **103**, 102274 (2023)

8. Kazemitabaar, M., Chow, J., Ma, C.K.T., Ericson, B.J., Weintrop, D., Grossman, T.: Studying the effect of AI code generators on supporting novice learners in introductory programming. In: CHI'2023, pp. 1–23. ACM (2023)

9. Kiesler, N., Lohr, D., Keuning, H.: Exploring the potential of large language models to generate formative programming feedback. In: 2023 IEEE Frontiers in Education Conference (FIE), pp. 1–5 (2023)

10. Konak, A., Clarke, C.J.S.F.: Augmenting critical thinking skills in programming education through leveraging chat GPT: analysis of its opportunities and consequences. In: Fall Mid Atlantic Conference: Meeting Our Students Where They are and Getting them Where They Need to be (2023)

11. Kuhail, M.A., Alturki, N., Alramlawi, S., Alhejori, K.: Interacting with educational chatbots: a systematic review. Educ. Inf. Technol. **28**(1), 973–1018 (2023)

12. Lau, S., Guo, P.: From "Ban It Till We Understand It" to "Resistance is Futile": how university programming instructors plan to adapt as more students use AI code generation and explanation tools such as ChatGPT and GitHub copilot. In: ICER'2023. vol. 1, pp. 106–121. ACM (2023)

13. Likas, A., Vlassis, N., Verbeek, J.J.: The global k-means clustering algorithm. Pattern Recogn. **36**(2), 451–461 (2003)

14. Logozar, R., Mikac, M., Hizak, J.: ChatGPT on the freshman test in C/C++ programming (2023)

15. Loubier, M.: ChatGPT: A good computer engineering student?: An experiment on its ability to answer programming questions from exams (2023)

16. Memarian, B., Doleck, T.: ChatGPT in education: methods, potentials, and limitations. Comput. Hum. Behav.: Artif. Hum. **1**(2), 100022 (2023)

17. Oosterwyk, G., Tsibolane, P., Kautondokwa, P., Canani, A.: Beyond the Hype: a cautionary tale of ChatGPT in the programming classroom (2023)

18. OpenAI: Chatgpt (2023). https://openai.com/chatgpt. Accessed 31 Oct 2023

19. Pérez, J.Q., Daradoumis, T., Puig, J.M.M.: Rediscovering the use of chatbots in education: a systematic literature review. Comput. Appl. Eng. Educ. **28**(6), 1549–1565 (2020). https://doi.org/10.1002/cae.22326

20. Qureshi, B.: Exploring the use of ChatGPT as a tool for learning and assessment in undergraduate computer science curriculum: opportunities and challenges. In: 2023 9th International Conference on e-Society, e-Learning and e-Technologies, pp. 7–13 (2023). https://doi.org/10.1145/3613944.3613946

21. Reiche, M., Leidner, J.L.: Bridging the programming skill gap with ChatGPT: a machine learning project with business students. In: European Conference on Artificial Intelligence, pp. 439–446. Springer (2023). https://doi.org/10.1007/978-3-031-50485-3_42

22. Rousseeuw, P.J.: Silhouettes: a graphical aid to the interpretation and validation of cluster analysis. J. Comput. Appl. Math. **20**, 53–65 (1987)
23. Shoufan, A.: Can students without prior knowledge use ChatGPT to answer test questions? An empirical study. ACM Trans. Comput. Educ. **23**(4), 45:1–45:29 (2023)
24. Shoufan, A.: Exploring students' perceptions of ChatGPT: thematic analysis and follow-up survey. IEEE Access **11**, 38805–38818 (2023)
25. Shute, V.J., Sun, C., Asbell-Clarke, J.: Demystifying computational thinking. Educ. Res. Rev. **22**, 142–158 (2017)
26. Singh, H., Tayarani-Najaran, M.H., Yaqoob, M.: Exploring computer science students' perception of ChatGPT in higher education: a descriptive and correlation study. Educ. Sci. **13**(9), 924 (2023)
27. Susnjak, T.: ChatGPT: the end of online exam integrity? arXiv preprint arXiv:2212.09292 (2022). https://doi.org/10.48550/arXiv.2212.09292
28. Tian, H., et al.: Is ChatGPT the ultimate programming assistant–how far is it? arXiv preprint arXiv:2304.11938 (2023). https://doi.org/10.48550/arXiv.2304.11938
29. Welsh, M.: The end of programming. Commun. ACM **66**(1), 34–35 (2023)
30. Yilmaz, R., Karaoglan Yilmaz, F.G.: Augmented intelligence in programming learning: examining student views on the use of ChatGPT for programming learning. Comput. Hum. Behav.: Artif. Hum. **1**(2), 100005 (2023)
31. Yu, H.: Reflection on whether Chat GPT should be banned by academia from the perspective of education and teaching. Front. Psychol. **14** (2023)

Deep-IRT with a Temporal Convolutional Network for Reflecting Students' Long-Term History of Ability Data

Emiko Tsutsumi[1($\boxtimes$)] (iD), Tetsurou Nishio[2], and Maomi Ueno[3]

[1] Hosei University 3-7-2, kajino-cho, Koganei-shi, Tokyo 184-8584, Japan
tsutsumi@ai.lab.uec.ac.jp
[2] Sundai Advanced Teaching Technology, 1-7-4, Kandasurugadai, Chiyoda-ku, Tokyo 101-0062, Japan
nishio@ai.lab.uec.ac.jp
[3] The University of Electro-Communications, 1-5-1, Chofugaoka, Chofu-shi, Tokyo 182-8585, Japan
ueno@ai.is.uec.ac.jp

Abstract. This study proposes a new Deep-IRT with a temporal convolutional network for knowledge tracing. The proposed method stores a student's latent multi-dimensional abilities at each time point and estimates the latent ability which comprehensively reflects the long-term history of ability data. To demonstrate the performance of the proposed method, we conducted experiments using benchmark datasets and simulation data. Results indicate that the proposed method improves the performance prediction accuracy of earlier Deep-IRT methods while maintaining high parameter interpretability. The proposed method exceeds the performance of earlier methods especially when the student's ability fluctuates according to past abilities.

Keywords: Deep Learning · Item Response Theory · Knowledge Tracing

1 Introduction

Recently, adaptive learning has been attracting attention to provide optimal support based on a student's ability growth in online learning systems. In the field of artificial intelligence, Knowledge Tracing (KT) has been studied actively to provide optimal support for students to maximize learning efficiency [6,17, 18,22,24,26,27]. An important task is discovering concepts that the student has not mastered based on the student's prior learning history data collected by online learning systems. In addition, accurately estimating students' evolving multi-dimensional abilities and predicting a student's performance (correct or incorrect responses to an unknown item) are important for adaptive learning.

Many researchers have developed various methods to solve KT tasks. Bayesian Knowledge Tracing (BKT) [6] and Item Response Theory (IRT) [3]

A. M. Olney et al. (Eds.): AIED 2024, LNAI 14829, pp. 250–264, 2024.
https://doi.org/10.1007/978-3-031-64302-6_18

are widely used probabilistic approaches. BKT traces a process of student ability growth following a Hidden Markov process. It estimates whether the student has mastered the skill or not and predicts the student's responses to unknown items. By contrast, IRT predicts a student's correct answer probability for an item based on the student's latent ability parameter and item-characteristic parameters. Although BKT and IRT have high parameter interpretability, they are unable to capture the multi-dimensional ability sufficiently. For that reason, they are unable to predict a student's performance accurately when a learning task is associated with multiple skills.

To overcome the limitations, various deep-learning-based methods have been proposed [7,17,18,24,27]. Recently, Deep item response theory (Deep-IRT) methods combining deep learning and item response theory have been proposed to provide educational parameter interpretability and to achieve accurate performance prediction [18–20,24]. Yeung (2019) [24] proposed a Deep-IRT (designated as Yeung-DI) combining a memory network architecture [27] with an IRT module. Yeung-DI adds hidden layers to a memory network architecture to estimate a student's ability and item difficulty parameters such as IRT. However, ability parameters of Yeung-DI are difficult to interpret because they depend on each item difficulty parameter. The most difficult challenge is to incorporate the ability and item parameters independently into a deep learning-based method so as not to degrade the prediction accuracy.

Tsutsumi et al. (2021) proposed a Deep-IRT (designated as Tsutsumi-DI) that has two independent redundant networks: a student network and an item network [20]. Tsutsumi-DI learns student parameters and item parameters independently to avoid impairment of the predictive accuracy. Most recently, Tsutsumi et al. (2024) combined Tsutsumi-DI with a novel hypernetwork (designated as Tsutsumi-HN) to optimize the degree of forgetting of the past latent variables. Tsutsumi-HN achieves the highest ability parameter interpretability and student response prediction accuracies among the existing methods to which it was compared. Especially, it is noteworthy that Tsutsumi-HN outperforms the attentive knowledge tracing [7] (designated as AKT), which provides state-of-the-art performance of response prediction.

Nevertheless, room for improvement remains for the prediction accuracy of the Deep-IRTs (Tsutsumi-DI and Tsutsumi-HN). They estimate a student's ability using only a most recent latent ability parameter. In general, the latest ability depends on past ability values while a student addresses items in the same skill. Because current ability estimates cannot adequately reflect past ability values, it interrupts accurate estimation of the ability transition. As a result, the performance prediction accuracy might be impaired or biased.

To resolve that difficulty, we propose a new Deep-IRT with a Temporal Convolutional Network (TCN) [2,15] that reflects features of the past multi-dimensional abilities to the latest ability estimate. Reportedly, TCN predicts time-series data more accurately than RNN-based models such as LSTM [11] and GRU [5]. Different from LSTM and GRU, which only refer to the previous latent state, TCN stores features of longer-term latent states. Therefore, the proposed method stores the student's latent multi-dimensional abilities at each time point and comprehensively reflects the long-term ability history data during the student's performance prediction. We conducted experiments to compare the proposed method's performance to those found for earlier KT methods. The results demonstrate that the proposed method improves the performance prediction accuracy of earlier Deep-IRT methods while maintaining high parameter interpretability. Particularly, the proposed method outperforms a state-of-the-art method, Tsutsumi-HN, which provides the highest performance among the current knowledge tracing methods.

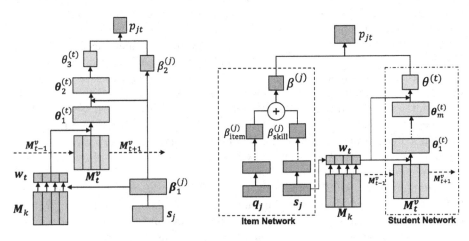

Fig. 1. Structure of Yeung-DI.   **Fig. 2.** Structure of Tsutsumi-DI.

2   Earlier Deep-IRT Methods

Several Deep-IRT methods have been proposed to provide educational parameter interpretability and to achieve accurate performance prediction by combining deep learning and item response theory. Yeung proposed a Deep-IRT method (Yeung-DI) combining a memory network architecture [27] with an IRT module [24]. Yeung-DI adds a hidden layer to a memory network architecture and estimates ability and item difficulty parameters. Figure 1 presents a simple illustration. Yeung-DI predicts a student's response probability p_{jt} to an item j at time t using the student's ability $\theta_3^{(t)}$ and item difficulty $\beta_2^{(j)}$ such as IRT [24].

$$p_{jt} = sigmoid\left(3.0 * \theta_3^{(t)} - \beta_2^{(j)}\right). \tag{1}$$

However, in Yeung-DI, the ability parameter $\theta_3^{(t)}$ depends on each item because it is estimated using features of the item difficulty parameter. Therefore, the ability and the item difficulty parameters cannot be interpreted separately.

To resolve that difficulty, Tsutsumi et al. propose a novel Deep-IRT method (Tsutsumi-DI) comprising two independent neural networks: The student network and the item network [18,20], as presented in Fig. 2. Tsutsumi-DI can estimate student parameters and item parameters independently such that the prediction accuracy does not decline because the two independent networks are designed to be redundant [8,13,14]. In addition, the item network of Tsutsumi-DI estimates the two difficulty parameters of item j: The item characteristic difficulty $\beta_{\text{item}}^{(j)}$ and the skill difficulty $\beta_{\text{skill}}^{(j)}$. A recent study assessed a Deep-IRT with hypernetwork architecture (Tsutsumi-HN) to optimize the degree of forgetting of the past latent ability variables [18,19]. Tsutsumi-HN shows the highest ability parameter interpretability and response prediction accuracies compared to existing methods. Furthermore, it can identify a relation among multi-dimensional skills and can capture multi-dimensional ability transitions. Tsutsumi-DI and Tsutsumi-HN predict a student's response probability p_{jt} to an item j at time t using the difference between a student's ability $\theta^{(t)}$ and the sum of two difficulty parameters $\beta_{\text{item}}^{(j)}$ and $\beta_{\text{skill}}^{(j)}$ as follows.

$$p_{jt} = sigmoid\left(3.0 * \theta^{(t)} - (\beta_{\text{item}}^{(j)} + \beta_{\text{skill}}^{(j)})\right). \tag{2}$$

3   Proposed Method

3.1   Temporal Convolutional Network

Recently, convolutional neural networks (CNN) [12] and Transformer [21] have attracted attention as prediction methods for time-series data. As a neural network with a convolutional layer and a pooling layer, CNN extracts features from two-dimensional data such as images by compressing local elements into a feature using the sliding window method. In addition, temporal convolutional network (TCN) has been developed as a method to reflect past data used for prediction by convolving long-term time-series data in multiple layers [2,15]. Actually, TCN has been reported to predict time-series data more accurately than RNN-based models such as LSTM [11] and GRU [5]. TCN stores features of longer-term latent states, different from LSTM and GRU, which only refer to the previous latent state. By contrast, the Transformer stores features of longer-term data by calculating the relative distance and the weight of the relations between each element of the input vector. Transformer provides highly accurate data prediction in many fields.

A recent study comparing CNN and Transformer performance found that the prediction accuracies of Transformer are superior to those of CNN when the training data size is sufficiently large [1,4]. However, Transformer is known to cause overfitting often when using small or sparse datasets. Although CNN has slightly lower prediction accuracy than that of a Transformer, it can accommodate sparse data and can train a model efficiently with lower memory cost. Furthermore, for the field of KT, earlier methods based on Transformer ([7,16]) have shown high prediction accuracy. However, they have low parameter interpretability. Their educational applicability remains limited.

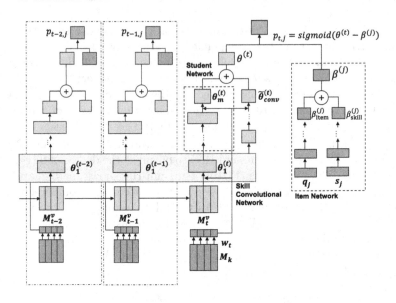

Fig. 3. Structure of the proposed method.

3.2 Deep-IRT with Temporal Convolutional Network

Although Tsutsumi-DI and Tsutsumi-HN provide high parameters independently and achieve performance prediction, they estimate the student's ability using only the most recent latent ability parameters. Therefore, current ability estimates cannot adequately reflect past ability history data. However, the latest ability depends on past ability values under circumstances where a student addresses items in the same skill. This problem might impair the performance prediction accuracy.

To resolve the difficulty of Tsutsumi-DI and Tsutsumi-HN, we propose a new Deep-IRT with a temporal convolutional network. Figure 3 presents the structure of the proposed method. We add TCN as the Skill Convolutional Network to Deep-IRT [18] to reflect features of the past multi-dimensional abilities to the ability estimate. The Skill Convolutional Network stores the student's latent multi-dimensional abilities at each time point and estimates the latent ability, comprehensively reflecting the long-term past ability history data. In Sect. 3.3, we describe details of the Skill Convolutional Network. In addition, the proposed method has the student network and the item network. In the student network, the ability parameters θ_{it} of the student i at time t are estimated based on the latent multi-dimensional ability variable M_v^t. In the item network, the model estimates the item characteristic difficulty parameter $\beta_{\text{item}}^{(j)}$ and the skill difficulty $\beta_{\text{skill}}^{(j)}$. Sections 3.4 and 3.5 respectively present details of the student network and the item network. Our code is also available on GitHub[1].

---

[1] https://github.com/UEC-Ueno-lab/TCN_Deep-IRT.git.

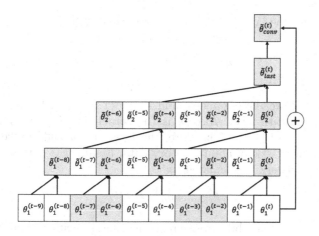

Fig. 4. Skill Convolutional Network.

3.3   Skill Convolutional Network

With a Skill Convolutional Network, the method estimates the optimal weight parameters related to a student's past ability values by convolving the latent ability valuables using the sliding window method. The Skill Convolutional Network structure is shown in Fig. 4. Skill Convolutional Network employs Causal Dilated Convolution [15, 25] and Residual Connection [9, 10]. Causal Dilated Convolution extracts features of long time-series data by convolving each input sequence according to "dilation" to avoid increasing the number of parameters. Dilation represents the distance between the elements of the input sequence used to compute the output value. When the dilation is 1, this convolution method is the same as general convolution. Residual Connection adds the input value of the first layer to the last output value to avoid the vanishing gradient for the deep layers. The proposed method uses the two methods above to convolve the latent ability values in multiple layers.

The input vector $\boldsymbol{\theta}_1^{(t)}$ is encoded values of the latent variable $\boldsymbol{M}_t^v$ which represents the latent multi-dimensional ability at each time t. A student's latent ability $\boldsymbol{\theta}_1^{(t)} \in \mathbb{R}^N$ is calculated as

$$\boldsymbol{\theta}_1^{(t)} = \sum_{l=1}^{N} w_{tl} \left(\boldsymbol{M}_{tl}^v \right)^{\top}, \tag{3}$$

where the vectors $\{\boldsymbol{\theta}_1^{(t)}, \boldsymbol{\theta}_1^{(t-1)}, \boldsymbol{\theta}_1^{(t-2)}, \cdots\}$ are N-dimensional abilities. For simplicity, we explain them using the one-dimensional column vector $\{\theta_1^{(t)}, \theta_1^{(t-1)}, \theta_1^{(t-2)}, \cdots\}$. In the first layer, the input vector is calculated as

$$\tilde{\theta}_1^{(t)} = \sum_{i=0}^{k-1} f_i^{(1)} \cdot \theta_1^{(t-d_1 \cdot i)}. \tag{4}$$

In the n-th layer, the latent ability values are calculated as

$$\tilde{\theta}_n^{(t)} = \sum_{i=0}^{k-1} f_i^{(n)} \cdot \tilde{\theta}_{n-1}^{(t-d_n \cdot i)}, \tag{5}$$

where $f_i^{(n)}$ is the weight parameter and $d_n = \{1, 2, 4, \cdots, 2^n\}$ is the dilation parameter in the n-th layer. Also, k is the kernel size: $k = k_{last}$ in the last layer, and $k = 2$ otherwise. Therefore, the layer size n and the kernel size k determine the input size of ability history data. Finally, the output of the Skill Convolutional Network is calculated as

$$\tilde{\theta}_{conv}^{(t)} = \tilde{\theta}_{last}^{(t)} + \theta_1^{(t)}. \tag{6}$$

Therein, $\tilde{\theta}_{conv}^{(t)}$ is a latent ability value reflecting the student's past ability history data.

3.4   Estimation of Student Parameters

In the student network, the student ability $\boldsymbol{\theta}_m^{(t)} = \{\theta_{m,1}^{(t)}, \theta_{m,2}^{(t)}, \cdots, \theta_{m,N}^{(t)} | 2 \leq m\}$ is estimated from the latent variable $\boldsymbol{M}_v^t$ in the neural networks, similarly to the method described by Tsutsumi et al. [20]. Then, we calculate the weighted linear summations of the student ability $\boldsymbol{\theta}_m^{(t)}$ and the output of Skill Convolutional Network as $\tilde{\boldsymbol{\theta}}_{conv}^{(t)} = \{\tilde{\theta}_{conv,1}^{(t)}, \tilde{\theta}_{conv,2}^{(t)}, \cdots, \tilde{\theta}_{conv,N}^{(t)}\}$ respectively.

$$\theta'^{(t)} = \sum_{l=1}^{N} \omega_{tl} \theta_{m,l}^{(t)}, \tag{7}$$

$$\tilde{\theta}^{(t)} = \sum_{l=1}^{N} \omega_{tl} \tilde{\theta}_{conv,l}^{(t)}. \tag{8}$$

Therein, ω_{tl} is an attention weight that signifies the degree of the relation between the latent skill and the actual skill of item j.

3.5   Estimation of Item and Skill Parameters

In the item network, the input of the item network is an embedding vector $\boldsymbol{q}_j \in \mathbb{R}^J$ calculated from the item j's tag and the student's response. Here, J represents for the number of items. We estimate the item characteristic difficulty parameter $\beta_{item}^{(j)}$ as

$$\beta_1^{(j)} = GELU\left(\boldsymbol{W}^{(\beta_1)}\boldsymbol{q}_j + \boldsymbol{\tau}^{(\beta_1)}\right), \tag{9}$$

$$\beta_m^{(j)} = GELU\left(\boldsymbol{W}^{(\beta_m)}\beta_{m-1}^{(j)} + \boldsymbol{\tau}^{(\beta_m)}\right), \tag{10}$$

$$\beta_{item}^{(j)} = \boldsymbol{W}^{(\beta_{item})}\beta_m^{(j)} + \boldsymbol{\tau}^{(\beta_{item})}. \tag{11}$$

Next, the skill difficulty $\beta_{\text{skill}}^{(j)}$ is estimated using the embedding vector $s_j \in \mathbb{R}^S$ calculated from the skill tag of item j and the student's response. Here, S represents the number of skills.

$$\gamma_1^{(j)} = GELU\left(W^{(\gamma_1)}s_j + \tau^{(\gamma_1)}\right), \tag{12}$$

$$\gamma_m^{(j)} = GELU\left(W^{(\gamma_m)}\gamma_{m-1}^{(j)} + \tau^{(\gamma_m)}\right), \tag{13}$$

$$\beta_{\text{skill}}^{(j)} = W^{(\beta_{\text{skill}})}\gamma_m^{(j)} + \gamma^{(\beta_{\text{skill}})}. \tag{14}$$

Each output of the last layer $\beta_{\text{item}}^{(j)}$ and $\beta_{\text{skill}}^{(j)}$ denotes the j-th item characteristic difficulty parameter and the difficulty parameter of the required skills to solve the j-th item. Then, the item i's difficulty is calculated from two difficulty parameters $\beta_{\text{item}}^{(j)}$ and $\beta_{\text{skill}}^{(j)}$.

$$\beta^{(j)} = \tanh\left(\beta_{\text{item}}^{(j)} + \beta_{\text{skill}}^{(j)}\right). \tag{15}$$

The proposed method predicts a student's response probability using the student ability $\theta^{(t)}$ and item difficulty $\beta^{(j)}$ as presented below.

$$p_{jt} = sigmoid(\theta^{(t)} - \beta^{(j)}). \tag{16}$$

The proposed method updates the latent variable M_v^t according to the earlier method [27]. In addition, the loss function of the proposed method employs cross-entropy, which reflects classification errors [18].

4   Experiment

5   Prediction Accuracy

This section presents a comparison of the prediction accuracies for student performance of the proposed methods with those of earlier methods (Yeung-DI [24], AKT [7], and Tsutsumi-HN [19]). We used five benchmark datasets as ASSISTments2009, ASSISTments2017, Statics2011, Junyi, and Eedi. For ASSISTments2009, ASSISTments2017, and Eedi with item and skill tags, we adopt both tags as input data. Also, for Statics2011, and Junyi with only skill tags, we employ the skill as input data. Table 1 presents the number of students (No. Students), the number of skills (No. Skills), the number of items (No. Items), the rate of correct responses (Rate Correct), and the average length of the items which students addressed (Learning length).

This experiment is conducted to evaluate the prediction accuracies of the methods based on standard five-fold cross-validation. For each fold, 20% students are used as the test set, 20% are used as the validation set, and 60% are used as the training set according to an earlier study [7]. The optimal number of layers and k_{last} of Skill Convolutional Network are decided to maximize AUC for the validation set, as shown in Table 2. For all methods, we employ the tuning

Table 1. Summary of Benchmark Datasets

Dataset	No. Students	No. Skills	No. Items	Rate Correct	Learning length
ASSISTments2009	4151	111	26684	63.6%	52.1
ASSISTments2017	1709	102	3162	39.0%	551.0
Statics2011	333	1223	N/A	79.8%	180.9
Junyi	48925	705	N/A	82.78%	345
Eedi	80000	1200	27613	64.25%	177

Table 2. Optimal numbers of layers and k_{last} of the proposed method

Data set	Layer	k_{last}
ASSISTments2009	5	3
ASSISTments2017	8	2
Statics2011	3	7
Junyi	8	2
Eedi	8	2

parameters according to the earlier studies [7,18,24]. If the predicted correct answer probability for the next item is 0.5 or more, then the student's response to the next item is predicted as correct. Otherwise, the student's response is predicted as incorrect. For this study, we leverage two metrics for prediction accuracy: AUC score and Accuracy score.

Table 3 present the results, with the model having the higher performance given in bold. Results indicate that the proposed method provides the best average AUC and Accuracy scores. The proposed method outperforms Yeung-DI, AKT, and the Tsutsumi-HN for ASSISTments2019, ASSISTments2017, Eedi, and Junyi. These results demonstrate that reflecting the past ability history data by TCN is effective for improving the prediction accuracy. In addition, these datasets are large-scale datasets including more than 1000 students. The sufficient number of students for model training is one reason for the improved accuracy of the proposed method. By contrast, the proposed method tends to have lower prediction accuracies for statics2011 than AKT has. For Statics2011 and Junyi, the prediction accuracy of the proposed method is comparable to that of AKT. A reason for the limited improvement was inferred: the TCN did not work effectively because the student's ability might change independently of the past ability at each time point.

6   Parameter Interpretation

6.1   Estimation Accuracy of Ability Parameters

In this section, to evaluate the interpretability of the ability parameters of the proposed method according to the earlier study [18], we use simulation data gen-

Table 3. Prediction accuracies of student performance

Dataset	Metrics	Yeung-DI [24]	AKT [7]	Tsutsumi-HN [19]	Proposed
ASSISTments2009	AUC	80.68 ± 0.46	82.20 ± 0.25	81.98 ± 0.54	$\mathbf{82.95 \pm 0.30}$
	Accuracy	76.64 ± 0.54	77.30 ± 0.55	77.15 ± 0.55	$\mathbf{77.60 \pm 0.56}$
ASSISTments2017	AUC	70.61 ± 0.52	74.54 ± 0.21	75.13 ± 0.20	$\mathbf{75.49 \pm 0.36}$
	Accuracy	68.21 ± 0.57	69.83 ± 0.06	70.69 ± 0.60	$\mathbf{70.85 \pm 0.50}$
Statics2011	AUC	81.12 ± 0.53	$\mathbf{82.15 \pm 0.35}$	81.57 ± 0.50	82.02 ± 0.39
	Accuracy	80.32 ± 0.80	$\mathbf{80.41 \pm 0.67}$	80.11 ± 0.92	80.40 ± 0.80
Junyi	AUC	77.70 ± 0.21	78.13 ± 0.39	77.91 ± 0.37	$\mathbf{78.14 \pm 0.43}$
	Accuracy	86.66 ± 0.29	86.79 ± 0.17	86.65 ± 0.15	$\mathbf{86.85 \pm 0.14}$
Eedi	AUC	75.62 ± 0.15	77.58 ± 0.21	78.97 ± 0.10	$\mathbf{79.14 \pm 0.11}$
	Accuracy	71.35 ± 0.24	72.35 ± 0.21	73.38 ± 0.13	$\mathbf{73.55 \pm 0.13}$
Average	AUC	77.15	78.92	79.11	**79.60**
	Accuracy	76.63	77.54	77.60	**77.81**

erated from Temporal IRT [23] to compare parameter estimates with those of earlier Deep-IRTs [18,24] and the proposed method. Temporal IRT is a Hidden Markov IRT that models the student ability changes following Hidden Markov processes with a parameter to forget past response data. It estimates student i's ability θ_{it} at time t, item j's discrimination parameter a_j, and item j's difficulty parameter b_j. The prior of θ_{it} is a normal distribution described as $\theta_{i0} \sim \mathcal{N}(0,1)$, $\theta_{it} \sim \mathcal{N}(\theta_{it-1}, \epsilon)$. Therein, ϵ represents the variance of θ_{it}. It controls the smoothness of a student's ability transition. Especially, it is noteworthy that ϵ reflects the degree of dependence of the student's current ability on past ability values. As ϵ becomes small (large), the current ability increases the degree of the dependence (independence) on past abilities. Therefore, as ϵ increases, the fluctuation range of the true ability increases at each time point. In addition, the priors of the item parameters are $\log a \sim \mathcal{N}(0,1)$, $b \sim \mathcal{N}(0,1)$.

In this experiment, each dataset includes 2000 student responses to $\{50, 100, 200, 300\}$ items. First, we estimate item parameters a and b using 1800 students' response data. Next, given the estimated item parameter, we estimate the student's ability parameters θ_{it} at each time using the remaining 200 students' response data. Additionally, we obtain results obtained using $\epsilon = \{0.1, 0.3, 0.5, 1.0\}$ for each dataset.

We calculate the Pearson's correlation coefficients, the Spearman's rank correlation coefficients, and the Kendall rank correlation coefficients using a student's abilities θ_t at time $t \in \{1, 2, \cdots, T\}$, as estimated using the true model and using the Deep-IRT methods. Next, we average these correlation coefficients of all students. The proposed method employs eight layers and $k_{last} = 2$ in TCN for all datasets.

Table 4 presents the average correlation coefficients of the methods for the respective conditions. Table 4 shows that the proposed method has higher correlation than the earlier Deep-IRTs for small variances of the ability parameters

Table 4. Correlation coefficients of the estimated abilities

No. Items	50	100	200	300	50	100	200	300	50	100	200	300
ϵ Method	Pearson				Spearman				Kendall			
Yeung-DI	0.63	0.67	0.74	0.74	0.63	0.66	0.75	0.75	0.44	0.47	0.55	0.55
0.1 Tsutsumi-HN	0.73	0.77	0.85	0.82	0.74	0.78	0.88	0.87	0.54	0.59	0.70	0.69
Proposed	**0.86**	**0.90**	**0.91**	**0.89**	**0.87**	**0.91**	**0.94**	**0.94**	**0.68**	**0.74**	**0.79**	**0.79**
Yeung-DI	0.73	0.80	0.81	0.82	0.75	0.83	0.86	0.87	0.55	0.63	0.66	0.67
0.3 Tsutsumi-HN	0.82	0.86	0.86	0.86	0.85	0.91	0.94	**0.95**	0.66	0.74	0.79	**0.80**
Proposed	**0.84**	**0.91**	**0.90**	**0.91**	**0.88**	**0.93**	**0.95**	**0.95**	**0.67**	**0.77**	**0.80**	0.80
Yeung-DI	0.77	0.80	0.81	0.81	0.81	0.86	0.88	0.89	0.61	0.65	0.68	0.69
0.5 Tsutsumi-HN	**0.85**	**0.84**	**0.83**	**0.82**	**0.90**	**0.93**	**0.94**	**0.95**	**0.71**	**0.76**	**0.78**	**0.80**
Proposed	**0.85**	**0.84**	0.82	0.81	0.89	0.92	0.92	0.89	0.71	0.73	0.73	0.70
Yeung-DI	0.79	0.81	0.82	**0.81**	0.83	0.88	0.89	0.89	0.63	0.68	0.70	0.69
1.0 Tsutsumi-HN	**0.82**	**0.80**	**0.81**	0.79	**0.89**	**0.92**	**0.94**	**0.94**	**0.70**	**0.75**	**0.79**	**0.79**
Proposed	0.80	0.79	0.80	0.79	0.88	**0.92**	0.92	0.93	0.68	0.74	0.75	0.76

($\epsilon = \{0.1, 0.3\}$). The small variance ϵ indicates that the ability depends strongly on past ability history data. Therefore, TCN functions effectively and improves the estimation accuracy by reflecting past ability values. However, the large variance ($\epsilon = \{0.5, 1.0\}$) reflects a weak relation between current and past ability values. Then it engenders rapid ability fluctuating at each time point. In this case, Tsutsumi-HN estimation ability by only the most recent ability and response data provides higher correlation. Results suggest that the proposed method is superior when the current ability depends on past abilities; Tsutsumi-HN is superior otherwise. It is noteworthy that no significant difference was found in the correlation coefficients between the ability estimates of the proposed method and those of Tsutsumi-HN. The proposed method provides comparably high estimation accuracies to those of Tsutsumi-HN.

6.2   Student Ability Transitions

As presented in this section, we visualize the ability transitions estimated using the proposed method and verify the accuracy of the ability estimation. Visualization of ability transition helps both students and teachers to identify students' strengths and weaknesses. We use the ASSISTments2009 dataset according to earlier studies [19, 24].

First, Fig. 5 depicts an example of a student's one-dimensional ability transitions estimated using Tsutsumi-HN and the proposed method. The vertical axis shows the student's ability value on the right side. The horizontal axis shows the item number. The student response is shown by filled circles "•" when the student answers the item correctly. It is shown by hollow circles "○" otherwise. DeepIRT-HN shows that the ability barely changes after it increases rapidly in items 1–5. Because DeepIRT-HN estimates an ability using only a most recent ability parameter, it might cause overfitting when a student continuously answers items correctly (incorrectly). As a result, the estimated ability converges to an extremely high (low) value. By contrast, the proposed method shows that the ability increases gradually as the student answers items correctly by estimating an ability reflecting past ability history data.

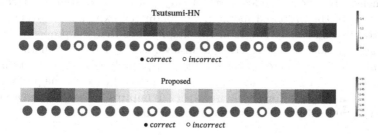

Fig. 5. Example of a student's one-dimensional ability transitions. (Color figure online)

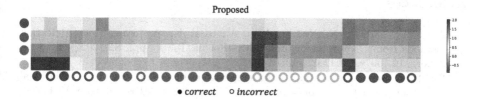

Fig. 6. Example of a student's multi-dimensional ability transition. (Color figure online)

Second, Fig. 6 depicts an example of student multi-dimensional ability transitions of each skill estimated using the proposed method. We use the first 30 responses. The student attempted four skills: "equation solving more than two steps" (shown in grey), "equation solving two or few steps" (shown in green), "ordering factions" (shown in orange), and "finding percents" (shown in yellow). The proposed method estimates abilities considering relations among the skills. Therefore, when a student answers an item correctly (or incorrectly), the abilities of the other skills change with the ability of the corresponding skill. In addition, as shown in Fig. 5, each ability fluctuates gradually, reflecting the student's responses while a student is addressing items in the same skill. However, when the student answers an item in a different skill, the estimated ability value fluctuates considerably. Actually, it is unlikely that the ability of a particular skill changes rapidly. In future work, we expect to improve the estimation accuracy and interpretability of multi-dimensional ability.

7   Conclusion

This article proposed a new Deep-IRT with a temporal convolutional network for knowledge tracing. The proposed method stores the student's latent multidimensional abilities at each time point and estimates the latent ability which comprehensively reflects the long-term ability history data. To demonstrate the performance of the proposed method, we conducted experiments using benchmark datasets and simulation data. Results indicate that the proposed method improves the performance prediction accuracy of earlier Deep-IRT methods while maintaining high parameter interpretability. Especially when the ability fluctuates depending on past abilities, the proposed method exceeds the performance of earlier methods. As future work, we expect to improve the estimation accuracy and interpretability of its multi-dimensional ability.

Acknowledgments. This research was supported by JSPS KAKENHI Grant Numbers JP19H05663, JP24H00739, JST CREST JPMJCR21D1, and ACT-X JPM-JAX23CG.

References

1. Arkin, E., Yadikar, N., Xu, X., Aysa, A., Ubul, K.: A survey: object detection methods from CNN to transformer. Multimedia Tools Appl. **82** (2022). https://doi.org/10.1007/s11042-022-13801-3
2. Bai, S., Kolter, J.Z., Koltun, V.: An empirical evaluation of generic convolutional and recurrent networks for sequence modeling (2018). https://doi.org/10.48550/ARXIV.1803.01271
3. Baker, F., Kim, S.: Item Response Theory: Parameter Estimation Techniques, Second Edition. Statistics: A Series of Textbooks and Monographs. Taylor & Francis (2004)
4. Chen, Z., Ma, M., Li, T., Wang, H., Li, C.: Long sequence time-series forecasting with deep learning: a survey. Inf. Fus. **97**, 101819 (2023). https://doi.org/10.1016/j.inffus.2023.101819
5. Chung, J., Gülçehre, Ç., Cho, K., Bengio, Y.: Empirical evaluation of gated recurrent neural networks on sequence modeling. CoRR abs/1412.3555 (2014). http://arxiv.org/abs/1412.3555
6. Corbett, A.T., Anderson, J.R.: Knowledge tracing: modeling the acquisition of procedural knowledge. User Model. User-Adapt. Interact. **4**(4), 253–278 (1995)
7. Ghosh, A., Heffernan, N., Lan, A.S.: Context-aware attentive knowledge tracing. In: Proceedings of the 26th ACM SIGKDD International Conference on Knowledge Discovery & Data Mining (2020)
8. He, H., Huang, G., Yuan, Y.: Asymmetric Valleys: beyond sharp and flat local minima. In: Advances in Neural Information Processing Systems, vol. 32, pp. 2553–2564. Curran Associates, Inc. (2019)

9. He, K., Zhang, X., Ren, S., Sun, J.: Deep residual learning for image recognition. In: 2016 IEEE Conference on Computer Vision and Pattern Recognition (CVPR), pp. 770–778. IEEE Computer Society, Los Alamitos, CA, USA (2016). https://doi.org/10.1109/CVPR.2016.90

10. He, K., Zhang, X., Ren, S., Sun, J.: Identity mappings in deep residual networks. In: Leibe, B., Matas, J., Sebe, N., Welling, M. (eds.) Computer Vision - ECCV 2016, pp. 630–645. Springer International Publishing, Cham (2016). https://doi.org/10.1007/978-3-319-46493-0_38

11. Hochreiter, S., Schmidhuber, J.: Long short-term memory. Neural Comput. **9**(8), 1735–1780 (1997). https://doi.org/10.1162/neco.1997.9.8.1735

12. Lecun, Y., Bottou, L., Bengio, Y., Haffner, P.: Gradient-based learning applied to document recognition. Proc. IEEE **86**(11), 2278–2324 (1998). https://doi.org/10.1109/5.726791

13. Morcos, A., Yu, H., Paganini, M., Tian, Y.: One ticket to win them all: generalizing lottery ticket initializations across datasets and optimizers. In: Advances in Neural Information Processing Systems, vol. 32, pp. 4932–4942. Curran Associates, Inc. (2019)

14. Nagarajan, V., Kolter, J.Z.: Uniform convergence may be unable to explain generalization in deep learning. In: Advances in Neural Information Processing Systems, vol. 32. pp. 11615–11626. Curran Associates, Inc. (2019). http://papers.nips.cc/paper/9336-uniform-convergence-may-be-unable-to-explain-generalization-in-deep-learning.pdf

15. van den Oord, A., et al.: WaveNet: a generative model for raw audio. In: Arxiv (2016). https://arxiv.org/abs/1609.03499

16. Pandey, S., Karypis, G.: A self-attentive model for knowledge tracing. In: Proceedings of International Conference on Education Data Mining (2019)

17. Piech, C., et al.: Deep knowledge tracing. In: Cortes, C., Lawrence, N.D., Lee, D.D., Sugiyama, M., Garnett, R. (eds.) Advances in Neural Information Processing Systems, vol. 28, pp. 505–513. Curran Associates, Inc. (2015)

18. Tsutsumi, E., Guo, Y., Kinoshita, R., Ueno, M.: Deep knowledge tracing incorporating a hypernetwork with independent student and item networks. IEEE Trans. Learn. Technol. **17**, 951–965 (2024). https://doi.org/10.1109/TLT.2023.3346671

19. Tsutsumi, E., Guo, Y., Ueno, M.: DeepIRT with a hypernetwork to optimize the degree of forgetting of past data. In: Proceedings of the 15th International Conference on Educational Data Mining (EDM) (2022)

20. Tsutsumi, E., Kinoshita, R., Ueno, M.: Deep-IRT with independent student and item networks. In: Proceedings of the 14th International Conference on Educational Data Mining (EDM) (2021)

21. Vaswani, A., et al.: Attention is all you need. In: Advances in Neural Information Processing Systems, pp. 5998–6008 (2017)

22. Weng, R., Coad, D.: Real-time Bayesian parameter estimation for item response models. Bayesian Anal. **13** (2016). https://doi.org/10.1214/16-BA1043

23. Wilson, K.H., Karklin, Y., Han, B., Ekanadham, C.: Back to the basics: Bayesian extensions of IRT outperform neural networks for proficiency estimation. In: Ninth International Conference on Educational Data Mining, vol. 1, pp. 539–544 (2016)

24. Yeung, C.: Deep-IRT: make deep learning based knowledge tracing explainable using item response theory. In: Proceedings of the 12th International Conference on Educational Data Mining, EDM (2019)

25. Yu, F., Koltun, V.: Multi-scale context aggregation by dilated convolutions. In: International Conference on Learning Representations (2016)

26. Yudelson, M.V., Koedinger, K.R., Gordon, G.J.: Individualized Bayesian knowledge tracing models. In: Lane, H.C., Yacef, K., Mostow, J., Pavlik, P. (eds.) AIED 2013. LNCS (LNAI), vol. 7926, pp. 171–180. Springer, Heidelberg (2013). https://doi.org/10.1007/978-3-642-39112-5_18
27. Zhang, J., Shi, X., King, I., Yeung, D.Y.: Dynamic key-value memory network for knowledge tracing. In: Proceedings of the 26th International Conference on World Wide Web, pp. 765–774. WWW '17, International World Wide Web Conferences Steering Committee (2017)

How to Teach Programming in the AI Era? Using LLMs as a Teachable Agent for Debugging

Qianou Ma[1(✉)], Hua Shen[2], Kenneth Koedinger[1], and Sherry Tongshuang Wu[1]

[1] Carnegie Mellon University, Pittsburgh, PA, USA
{qianoum,krk,sherryw}@cs.cmu.edu
[2] University of Michigan, Ann Arbor, MI, USA
huashen@umich.edu

Abstract. Large Language Models (LLMs) now excel at *generative* skills and can create content at impeccable speeds. However, they are imperfect and still make various mistakes. In a Computer Science education context, as these models are widely recognized as "AI pair programmers," it becomes increasingly important to train students on *evaluating* and *debugging* the LLM-generated code. In this work, we introduce HYPOCOMPASS, a novel system to facilitate deliberate practice on debugging, where human novices play the role of Teaching Assistants and help LLM-powered teachable agents debug code. We enable effective task delegation between students and LLMs in this learning-by-teaching environment: students focus on *hypothesizing the cause of code errors*, while adjacent skills like code completion are offloaded to LLM-agents. Our evaluations demonstrate that HYPOCOMPASS generates high-quality training materials (*e.g.*, bugs and fixes), outperforming human counterparts fourfold in efficiency, and significantly improves student performance on debugging by 12% in the pre-to-post test.

Keywords: LLM · teachable agent · debugging · CS1

1 Introduction

LLMs are becoming an integral part of software development—commercialized tools like GitHub Copilot are now advertised as "your AI pair programmer" and generate up to 46% of users' code [6]. Despite their prevalence, LLMs often produce unpredictable mistakes [11], *e.g.*, GPT-4 can still make mistakes 17% of the time in coding tasks for introductory and intermediate programming courses [22]. The impressive yet imperfect generative capabilities of LLMs, coupled with the associated risks of excessive reliance on these models, underscore the importance of teaching *evaluation* skills to students. In the context of programming, students must improve their debugging and testing skills [2].

However, debugging tends to be overlooked in formal educational curricula, especially in introductory Computer Science classes (*i.e.*, CS1) [21]. Prior

© The Author(s), under exclusive license to Springer Nature Switzerland AG 2024
A. M. Olney et al. (Eds.): AIED 2024, LNAI 14829, pp. 265–279, 2024.
https://doi.org/10.1007/978-3-031-64302-6_19

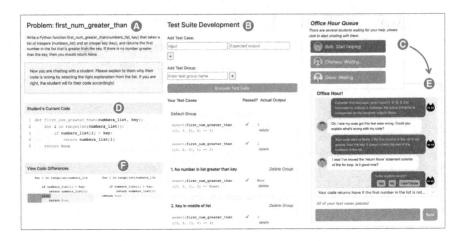

Fig. 1. In HYPOCOMPASS, given a programming problem description (A), a student user (in the role of a Teaching Assistant) needs to compile a test suite (B) and assist multiple LLM-simulated agents (*e.g., Bob, Chelsea, Dave*) in an Office Hour Queue (C) through a chat interface (E). Each LLM-agent acts as a novice seeking help with a buggy solution (D) and provides feedback to the user (F).

research has outlined various factors contributing to the absence of debugging instruction, such as instructors' limited time budget for developing specialized debugging materials and assessments [19]. Consequently, students primarily learn debugging from working on their own mistakes, which can be rather frustrating—they must invest substantial time and effort in *hypothesizing* the cause of bugs while grappling with other cognitively demanding tasks, such as understanding and writing code. These challenges prompt us to ask:

Research Question: Can we train students to improve debugging skills by providing *explicit* and *scaffolded* practice *with minimal cost to instructor time?*

In this work, we focus on training students' abilities in *hypothesis construction*, a critical step in debugging as established by prior work [29,30]. We introduce HYPOCOMPASS (Fig. 1, Sect. 3), an interactive, LLM-augmented intelligent tutoring system for debugging. Leveraging LLMs' *material generation* capability, we have these models imitate CS1 students who have written buggy code and require assistance from Teaching Assistants (TAs). Human novice students assume the role of the TA, who helps troubleshoot these bugs. This enables students to deliberately practice the skill of *hypothesizing* about the defects of LLM-generated code, delegating other tasks not core to hypothesis construction (*e.g.,* code completion) to the LLM. As a result, HYPOCOMPASS fosters an engaging learning environment using the *teachable agent* framework [3] and provides students with guided exposure to LLM-generated bugs. We also employ prompting strategies such as focused task formation and over-generate-then-select to improve LLM generation quality in HYPOCOMPASS (Sect. 4).

We conducted two evaluation studies and found that HYPOCOMPASS *saves instructors' time in material generation* and is *beneficial to student learning*. In our LLM evaluation study (Sect. 5), expert inspections on six practice problems and 145 buggy programs showed that HYPOCOMPASS achieved a 90% success rate in generating and validating a complete set of materials, *four times faster than human generation*. Our learning evaluation study with 19 novices (Sect. 6) showed that HYPOCOMPASS significantly improved students' pre-to-post test performance by 12% and decreased their completion time by 14%.

In summary, we contribute:

- A pragmatic solution that balances the benefits and risks of LLMs in learning. We use LLMs to prepare students to engage with imperfect LLMs, and we highlight the importance of *role-playing* for practical LLM application and *task delegation* to help students focus on essential skills.
- A theoretically grounded instructional design to enhance debugging skills. To the best of our knowledge, we are the first to provide aligned instruction and assessments on the hypothesis construction learning objectives, *i.e.*, forming hypotheses about the source of error, a core bottleneck in debugging [25].

2   Related Works

The Debugging Process. Debugging is a complicated process of various cognitively demanding tasks, including understanding the code, finding bugs, and fixing bugs, with the first two considered primary bottlenecks [19,25]. While many studies have attempted to improve students' code understanding [12], there is limited instruction on bug finding. Researchers characterize the cognitive model of bug finding as a *hypothesis construction process*, including initializing, modifying, selecting, and verifying hypotheses (Fig. 2B) [29]. This process is challenging: prior works show that novices struggle to systematically generate comprehensive hypotheses and identify the right hypothesis, in contrast to experts [7,8]. Hence, we emphasize teaching students to *construct accurate hypotheses about bugs* and *develop comprehensive hypotheses about potential bugs*.

Tutors and Tools for Debugging Training. Prior studies [19] and online discussions [21] indicate that teaching debugging is challenging and is rarely included in CS1 curricula, due to logistical challenges like the lack of instructional time and resources [5,10]. Existing tools demand instructor effort and often focus on the full debugging process, improving bug-fixing accuracy and efficiency [1,15]. In contrast, few studies emphasize accurate or comprehensive hypothesis construction (and they tend to be language-specific) [13,25]. To fill in the gap, we design HYPOCOMPASS to provide *deliberate practice* [9] on hypothesis construction, and use *the LLM generation capability to provide easily adaptable and targeted exercises* with immediate feedback.

LLM Capabilities for CS Learning. LLMs can perform well in a CS1 classroom [22], but concerns about misuse and LLM errors limit their use in education [2]. Therefore, current deployments tend to focus on generating instructional

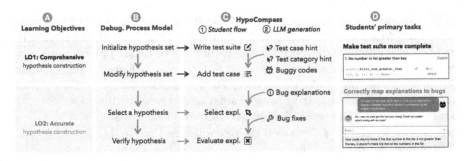

Fig. 2. To enable deliberate practice, we establish a close mapping between the (A) learning objectives, (B) the cognitive debugging process model, (C) the HypoCompass interaction flow, and (D) the primary tasks students perform in HypoCompass. We offload various material generation tasks to LLMs (C₂).

materials (*e.g.*, questions [24]). In our work, HypoCompass uses the LLM to generate inter-dependent materials in an integrated process and frame the LLM as a student asking for help [3], such that human novices can embrace imperfections in LLMs. Two unique capabilities of LLMs power this: (1) LLMs can simulate different personas and tutoring interactions [18]; (2) LLMs make common mistakes and natural bugs similar to humans [20], which can be used as buggy code practice examples. We adapt and develop various prompting methods [27] to enhance the quality of LLM generations.

3   The Design of HypoCompass

Grounded in the cognitive process [29] and the novice-expert difference in hypothesis-driven debugging (Sect. 2), we specify two crucial learning components for HypoCompass: comprehensive and accurate hypothesis construction. Prior work shows that hypothesis construction is closely connected with testing [30]: each additional test case should, ideally, be a hypothesis about what can go wrong in the program. In turn, a *comprehensive* test suite (*i.e.*, a set of test cases) should allow an effective debugger to construct a *accurate* hypothesis about why the program is wrong. We thus design toward two learning objectives (Fig. 2A,D):

LO1. **Comprehensive Hypothesis Construction**: Construct a comprehensive test suite that well covers the possible errors for the given problem.

LO2. **Accurate Hypothesis Construction**: Given the failed test cases, construct an accurate explanation of how the program is wrong.

Interface and Key Components. We designed HypoCompass through an iterative development process with 10 pilots, including CS1 students, TAs, and instructors. In the resulting interface (Fig. 1), a human student would be asked to play the role of a TA where they help an LLM-simulated student (LLM-agent)

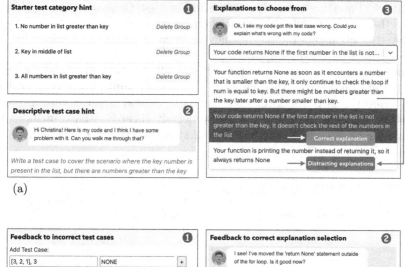

(a)

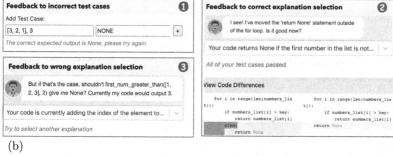

(b)

Fig. 3. (a) HypoCompass offers (1) *test category hints* to help write a comprehensive test suite systematically; (2) *test case hints* to help students add missing test scenarios; (3) *candidate explanation pool* to clarify misconceptions of alternative explanations. (b) HypoCompass provides immediate feedback to (1) *incorrect test cases*, ensuring students understand the code behavior; (2) *correct explanations*, as correct code fixes; (3) *incorrect explanations*, as confusion messages from the LLM-agent.

in debugging. They need to write and sort test cases into categories (Fig. 1B) that represent different hypotheses of what inputs may trigger errors in code.

Once the student is satisfied with their test suite, HypoCompass shows them an Office Hour Queue (OHQ) simulator (Fig. 1C). As the student interacts with each LLM-agent, the agent presents a buggy code snippet (Fig. 1D). The student guides the LLM-agent in debugging code through a dialog interface (Fig. 1E), selecting or creating test cases that reflect their hypotheses of the bug, and selecting explanations for the bug among a pool of candidate natural language explanations. These candidates each explain a different bug, representing alternative hypotheses that may confuse students (*e.g.,* Fig. 3a$_3$).

The LLM-agent then uses the test case and explanation to revise the code, providing immediate feedback to the student (Fig. 3b). If the explanation is correct, the agent will conduct minimal code fixes, and present the color-coded edits as feedback (Fig. 1F, a zoomed-in view is in Fig. 3b$_2$). Otherwise, the LLM-agent will ask the student to reflect on their hypothesis by responding with a confusion message that highlights the discrepancy between the student's explanation and the actual code behavior (Fig. 3b$_3$).

Once the student correctly confirms that all the bugs are fixed, they can move to help the next LLM-agent (Fig. 1C). Upon completion, HYPOCOMPASS will provide the next round of exercises with another programming problem. While the numbers are configurable, by default HYPOCOMPASS includes two programming exercises, each with three LLM-agents (buggy programs).

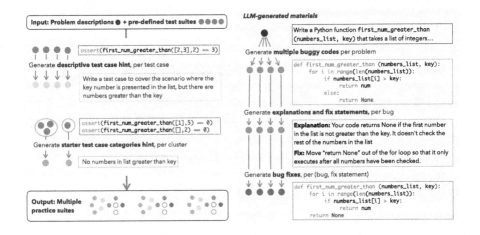

Fig. 4. Examples of inputs and outputs to the LLM material generation pipeline.

We highlight the two most essential components of the interaction:

- **Frame imperfect LLMs through role-play.** We use the LLM to simulate students who wrote bugs and have human novices offer help. This teachable agent setup supports learning, helping students reflect on their knowledge and reason through diverse bugs [23]. Having students work through "other people's errors" also boosts their motivation and protects their self-efficacy [3]. More importantly, it actively involves novices in identifying bugs in LLM-generated code, *enabling guided exposure to LLM imperfectness.*
- **Task delegation between students and LLMs.** To ensure deliberate practice on comprehensive and accurate hypothesis construction, students primarily engage in two tasks corresponding to each learning objective (Fig. 2D): (1) making the test suite more complete (LO1); and (2) correctly mapping explanations to bugs (LO2). We align student interaction flow (Fig. 2C$_1$) with the cognitive model of debugging [29] (Fig. 2B). LLMs take over other tasks that are *indirectly related* to the core learning goals, including generating diverse bugs and fixes, which frees students from code writing. We also use

LLMs to support scaffolding, generate hints (Fig. 3a), and provide immediate feedback throughout the practice (Fig. 3b).

4 LLM Integration

As shown in Fig. 2C$_2$, we use LLM to generate five types of materials: (1) test case category hints, (2) test case hints, (3) buggy programs, (4) explanations of bugs, and (5) programs with bugs fixed. We reduce instructor workload by generating practices using just a problem description, a reference solution, and a reference test suite with about 10 inputs, and we further minimize human verification overhead with optimized prompts and automated algorithms. Our generation process is detailed in Fig. 4, example prompts are in Table 1, and full prompts are in Table 3 in Supplements[1]. OpenAI's `gpt-3.5-turbo` is used for all materials, except for explanation generation, which uses `gpt-4` for enhanced reasoning capabilities. Below are key factors to the success of generation:

Task Formation and Decomposition. We iterate on our prompts according to the nature of the task. First, as LLMs behave inconsistently when the user tasks conflict with LLMs inherent training objectives [28], we carefully formulate the task to avoid introducing competing tasks. Take *Local Bug Fix* (Table 1) as an example: when we directly ask the LLM to fix a bug according to an explanation, we observe that the model almost always over-fix all bugs irrespective of the provided instructions. This is because LLMs can be biased towards generating fully correct code (part of the LLM pre-training) and away from local bug fixing (changing only the buggy snippet described by the instruction, the desired task). Hence, we re-frame it as a *translation task*, converting bug-fixing instructions to its code format old $\rightarrow$ new code snippet. This **task re-framing** mitigates the model's inherent bias, reducing over-fixing errors by 70%.

Second, for multi-step tasks (*e.g.*, *Local Bug Fix*), we adopt **LLM-chains** [27], decomposing tasks into sub-tasks handled by separate steps, such that each step contributes to stable performance. Third, we also address prompt complexity by explicitly prioritizing essential requirements. For tasks like generating *Bug Explanations and Fix Instructions* (Table 1), we prioritize precise bug extraction, instructing the model to list all unique bugs upfront. Secondary requirements (*e.g.*, word limits) are specified only in the output format. This **hierarchical disentanglement** significantly improves success rates by over 40%.

Over-Generate-then-Select. While LLMs can easily generate random materials, it is nontrivial to ensure that their generations have pedagogical values. For example, behaviorally distinct bugs help students practice with varied instances, but it is hard to enforce through prompting as it requires LLMs to "know" bug behaviors. Nonetheless, we can configure the non-deterministic LLMs to **over-generate** multiple solutions with mixed qualities [17], and **then select a subset** of desired ones (Fig. 5). We apply this strategy in multiple places:

---

[1] Supplemental materials are at: http://tinyurl.com/hypocompass-sup.

Table 1. Prompts and temperatures (Temp.) for generating bugs, explanations, and fixes. The temperature is set higher for more diverse and random outputs.

Material	Generation goal	Temp.
Buggy code	To over-generate bugs with mixed quality for further selection.	0.7

[Sys.] You are a **novice student in intro CS**, you make mistakes and write buggy code.

[User] Problem Description: {problem_description}
Write **different buggy solutions** with common mistakes like novice students:

Bug expl. & fix instruct.	To describe each unique bug, and write a corresponding fix instruction. If there are multiple bugs in the code, generate their explanations and fixes separately.	0.3

[Sys.] You are a **helpful and experienced TA** of an introductory programming class.

[User] Hi, I'm a student in your class. I'm having trouble with this problem in the programming assignment: {problem_description} Here's my buggy code: {buggy_code} What's wrong with my code? **List all the unique bugs included, but do not make up bugs.** For each point, put in the format of: {explanation: accurate and concise explanation of what the code does and what the bug is, for a novice, fix: how to fix the bug, within 30 words}
Only **return the bullet list.** Do not write any other text or code.

Bug fix	To edit the buggy code according to the fix instruction, w/o over- or under- fix.	0.3

[Sys.] You fix bugs in Python code closely following the instructions.

[User] Original code: {buggy_code}; Code modification: {explanation}
Translate the statement into actual, minimal code change in this format:
{original code snippet: ""copy the lines of code that need editing""
-> edited code snippet: ""write the edited code snippet""}

[LLM] {old to new snippet in JSON, e.g., numbers_list[i] <= key → numbers_list[i] > key
}

[User] Old Code:{buggy_code}; Instruction:{Old snippet to new snippet};
New Code:

(1) To expose students to behaviorally distinct bugs, we over-generate buggy code (Table 1). We filter out correct code, and we vectorize buggy code's behavior based on the reference test suite (Fig. 5A, 0 being failed tests). We then greedily choose a diverse subset of buggy programs with the maximum pairwise distance, using Euclidean distance on the error vectors (Fig. 5B).

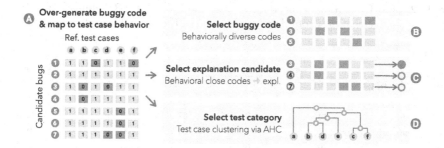

Fig. 5. Over-generate and automatically select materials with pedagogical values.

(2) To help students clarify misconceptions (Fig. 5C), we want distracting explanations that look similar to the actual explanation for each practice buggy code. We choose from the over-generated buggy code pool, find two with the smallest Euclidean distance to the target code, and use their corresponding explanations as distractors. The mapping also helps generate the confusion messages (Fig. 3b$_3$)—when a student selects the distractor explanation, we use its corresponding buggy code to find test cases to present to students.

(3) To capture key testing aspects in our test category hints (Fig. 5D), we cluster reference test cases into semantically meaningful groups. We build dendrograms from test case vectors with Agglomerative Hierarchical Clustering [14], which guide the selection of test category hints from the over-generated pool.

Human-in-the-Loop Verification. As shown in Fig. 4, while the hints for test cases and categories are generated separately, the materials relevant to bugs are generated in sequential order. We perform human verification per step to mitigate the risk of cascading errors in subsequent steps. We provide more details on human verification and editing times in Sect. 5.

5 LLM Evaluation: Generation Efficiency and Quality

We evaluated the generations on six different problems from prior work [4] and our own problems (detailed in Table 4 in Supplements). On average, for each problem, we generated 3 test category hints, 10 test case hints, 24 buggy programs, explanation and fix instructions, and 33 bug fixes. The total number and the success rates are summarized in Table 2. We provided the success criteria for all types of materials in Table 5 in Supplements.

Table 2. LLM Evaluation: Time, Success rate, and Inter-Rater Reliability scores (*i.e.,* IRR% = #agreements / #total labels, κ is Cohen's Kappa coefficient).[4]

Material	Raw LLM outputs			Human verification		
	# Generation	Avg. gen time	Success%	Avg. edit time	IRR%	κ
Test case description hint	61	0:00:37	98.36%	0:00:08	100%	–
Test case category hint	18	0:00:10	94.44%	0:00:10	100%	–
Buggy code	145	0:01:30	57.93%	0:00:02	n/a	n/a
Bug explanation and fix	145	0:03:36	91.72%	0:00:52	90%	0.875
Bug fix	195	0:02:45	86.15%	0:00:37	92%	0.752

Method. Two authors annotated 10% of the generations at each step individually, and discussed to resolve the disagreement and update the codebook. An external instructor annotated the same 10% of LLM-generated materials, using the updated codebook. We calculated the inter-rater reliability (IRR) between the external instructor and the resolved annotation among the two authors using percent IRR and Cohen's Kappa. As shown in Table 2, the agreements are satisfactory across different model generations (IRR% > 90% and $\kappa > 0.75$)[2]. One author annotated the rest of the materials to calculate the success rates. We log the verification and editing *time*, as proxies to the instructor overhead.

To compare LLM and human generations, we recruited two experienced CS TAs to each create practice materials for a specific problem. Each TA received the same input as LLMs, was asked to produce one set of materials matching the amount of content LLMs produced, and was compensated for their time.

Result: Efficient and High-Quality Generation. We achieve high-quality generation: a complete set of practice materials with 9 buggy programs (3 for practice and 6 more as distractors), 9 bug explanations, 9 bug fixes, 10 test case hints, and 3 test category hints can be generated with a 90% success rate and only takes 15 min to label and edit. As we *over-generate* and automatically select buggy code, a success rate over 50% is reasonable for practical use.

Employing LLMs can also be significantly more efficient. In total, a TA spent around 60 min to generate one set of practice materials for HYPOCOMPASS. One TA noted the difficulty in consistently creating unique and high-quality materials after 30 min, saying that *"the importance of the bug I create would start to decline."* The same author evaluated the TAs' generations using the annotation codebook, which had a 100% success rate and took 11 min. The time invested in generating and editing instructional materials for HYPOCOMPASS using LLMs was 4.67 times less than that of the human TAs.

---

[2] Buggy programs undergo automatic testing, so human verification is unnecessary (n/a). If both raters unanimously agree in one category, *kappa* is undefined (-), so κ is only noted when there's less than 100% IRR agreement on a single label.

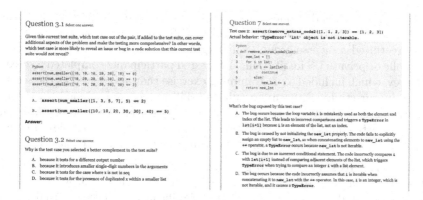

Fig. 6. Pre-post test question examples for LO1 comprehensive (Q3.1 and Q3.2) and LO2 accurate hypothesis construction (Q7).

6   Learning Evaluation: Pre- / Post-Test Study

Can novices better formulate hypotheses after engaging with HypoCompass? We conducted a learning evaluation with 19 students and compared the difference in speed and performance from the pre-test to the post-test.

Assessment. To best capture student learning gains on our learning objectives, we took a backward design method [26] to create an aligned assessment for the comprehensive LO1 and accurate LO2 hypothesis construction skills. We conducted multiple rounds of pilots to refine our intervention and pre-post tests. Our final tests are based on two programming exercises with comparable difficulties. We counterbalanced pre-post tests' problems to control for problem sequence influence. Each test consists of seven questions, with three assessing LO1 and four for LO2. Figure 6 provides a sample for each. For instance, Question 3.1 asks students to identify the more suitable test case to add to an existing test suite, evaluating their ability to construct comprehensive hypotheses (LO1). We measure students' performance using their **test scores** based on a standard rubric. We also log the pre-post tests' **completion time** as a proxy for proficiency.

Method: Study Procedure and Participants. Our hour-long user study constituted a pre-survey, pre-test, interaction with HypoCompass, post-test, and a post-survey. Participants began with a pre-survey, which asked demographic information and 7-level Likert Scale questions on their debugging experiences. Then, participants had up to 20 min for the pre-test. The system interaction consisted of two problems, where participants needed to write a test suite and explain bugs in three different buggy programs for each problem. The first problem was the same as in the pre-test, and the second problem matched the screening survey's exercise. By reusing problems that students have seen, we isolate our learning objectives from the program comprehension skills. After a subsequent 20-min post-test, participants filled out a post-survey with Likert Scale and open-ended questions on their experience and perceptions using HYPOCOMPASS. Participants received a \$15 Gift Card for their time.

We recruited a diverse group of undergraduate and graduate students from four public or private US institutions. Interested participants completed a screening survey, which included a programming exercise that also served as the second exercise in our study. To ensure a suitable skill range, we excluded those with extensive programming experience or who quickly solved the exercise. After filtering, 19 participants (S1-19) were included in the study—12 females, 6 males, 1 non-binary, and 8 non-native English speakers, with an average age of 20.7.

Quantitative Result: Learning Gains. A two-tailed paired t-test showed that students' pre-test to post-test scores significantly improved by 11.7% ($p = 0.033 < 0.05$), and the time of completion significantly reduced by 13.6% ($p = 0.003$), indicating success in learning through HYPOCOMPASS interaction. Note that the bugs used in pre-post tests are generated by humans and are not the same as in HYPOCOMPASS. As such, the significant learning gains indicate that students could learn debugging skills *transferable* to real-world bugs.

Where does the learning gain come from? We break down the analyses by learning objectives. We found a small 6.1% improvement in the score and a large 23.6% time reduction for *comprehensive hypothesis construction* (LO1), and a large 15.8% improvement in the score and a small 9.0% time reduction for *accurate hypothesis construction* (LO2). Therefore, students showed more efficiency enhancement in LO1, and more learning gains in LO2. Note that these improvements may confound with problem difficulty, as the items corresponding to LO1 (pre-test $\mu = 54\%$) seem easier than the ones for LO2 (pre-test $\mu = 38\%$).

Qualitative Result: Student Perceptions. We further unpack how HYPOCOMPASS contributed to learning by analyzing the survey responses. Students valued being able to offload some debugging subtasks to HYPOCOMPASS, such as writing code and explanations. For example, S1 said *"looking at the test behavior and the explanation options really helps relieve that burden."* Students also generally felt that the LLM-generated bugs and fixes were authentic. Most participants could not tell if their practiced programs were written by students or AI because of their experiences making or seeing similar mistakes from peers.

Moreover, students reported that HYPOCOMPASS was engaging, fun, not frustrating, and helped build confidence in debugging. A Wilcoxon signed-rank test shows a significant increase in self-rated confidence in debugging by 15% ($p = 0.007$). Students rated HYPOCOMPASS as significantly more engaging (6.0 out of 7), fun (6.0), and less frustrating (2.5) than their conventional way of learning debugging and testing ($p < 0.005$ for each). S8 especially liked the teachable agent setup: *"the role play just feels more natural because it feels like explaining to a rubber duck instead of to talking to myself"*.

7   Discussion

Teachable Agent for Appropriate Reliance with Imperfect AIs. Our work illustrates a scenario in which *LLM-generated bugs are not seen as problems* *but rather as features.* HYPOCOMPASS's teachable agent setup provides students with *moderated exposure to imperfect LLMs*, and may help them learn that LLMs are fallible and calibrate trust accordingly. Future iterations could remove material validation and allow direct exposure to unfiltered LLM mistakes in real-time interactions, taking full advantage of the teachable agent framework. Students will naturally expect that the LLM-agent seeking help may make mistakes (*e.g.,* fail to follow bug-fixing explanations). This approach, however, requires a more sophisticated design for scaffolding students in recognizing LLM errors.

Task Delegation for Shifting Learning Focus. Our exploration lays the foundation for a paradigm shift toward cultivating higher-order evaluation skills in the generative AI era. Essentially, we asked: what skills should we offload, and what should we learn? Most students in our study appreciated offloading subtasks to LLM (Sect. 6); however, some need more scaffolds, while others prefer less. Future research can investigate more personalized task delegation. For example, students who need more help can use LLMs to facilitate code tracing, and students can also write their own explanations for bugs based on their proficiency. Deciding the bare minimum programming skills and human-AI collaboration skills to teach also warrants further exploration [16].

Modularize to Adapt to Different Needs. Though most students and instructors found HYPOCOMPASS engaging, some expressed concerns about the deployment and maintenance cost of a new tool. To maximize utility to diverse users, we can modularize different components in HYPOCOMPASS. Instructors who prefer to distribute training materials as handouts can rely entirely on the material generation module. In contrast, instructors who want to experiment with TA training can employ HYPOCOMPASS with practice generated using their training questions. Future studies may perform ablation studies to evaluate different HYPOCOMPASS components with more extensive classroom deployment.

Limitation. We primarily evaluated *whether* HYPOCOMPASS *can bring learning and efficiency gains* through small in-lab experiments. With this prerequisite, we plan to conduct future classroom deployment with controlled comparisons. There is also a limitation regarding the reported efficiency of the LLM-assisted instructional material development, as the instructors need some familiarization time with the tool and the process.

8   Conclusion

In an attempt to answer how LLMs can reshape programming education's focus, we introduce a novel system, HYPOCOMPASS, and new instructional designs for hypothesis construction skills. We aim to provide engaging and deliberate practice on debugging to novices, using our theoretically motivated and empirically tested teachable agent augmented by LLM. Our evaluations show that HYPOCOMPASS can efficiently help instructors create high-quality instructional materials, effectively train novices on comprehensive and accurate hypothesis construction, and facilitate students' confidence and engagement in debugging.

Acknowledgment. Thanks to the participants, reviewers, Kelly Rivers, Michael Taylor, Michael Hilton, Michael Xieyang Liu, Vicky Zhou, Kexin Yang, Jionghao Lin, Erik Harpstead, and other Ken's lab members for the insights and help. This study was supported by gift funds from Adobe, Oracle, and Google.

References

1. Ardimento, P., Bernardi, M.L., Cimitile, M., Ruvo, G.D.: Reusing bugged source code to support novice programmers in debugging tasks. ACM Trans. Comput. Educ. **20**(1), 1–24 (2019). https://doi.org/10.1145/3355616
2. Becker, B.A., Denny, P., Finnie-Ansley, J., Luxton-Reilly, A., Prather, J., Santos, E.A.: Programming is hard-or at least it used to be: educational opportunities and challenges of AI code generation. In: Proceedings of the 54th ACM Technical Symposium on Computer Science Education V, vol. 1. pp. 500–506 (2023)
3. Blair, K., Schwartz, D.L., Biswas, G., Leelawong, K.: Pedagogical agents for learning by teaching: teachable agents. Educ. Technol. Res. Dev. **47**(1), 56–61 (2007)
4. Dakhel, A.M., Majdinasab, V., Nikanjam, A., Khomh, F., Desmarais, M.C., Jiang, Z.M.J.: Github copilot AI pair programmer: asset or liability? J. Syst. Softw. **203**, 111734 (2023). https://doi.org/10.48550/ARXIV.2206.15331
5. Desai, C., Janzen, D.S., Clements, J.: Implications of integrating test-driven development into CS1/CS2 curricula. SIGCSE Bull. **41**(1), 148–152 (2009)
6. Dohmke, T.: GitHub copilot x: the AI-powered developer experience (2023). https://github.blog/2023-03-22-github-copilot-x-the-ai-powered-developer-experience/. Accessed 5 Sept 2023
7. Edwards, S.H., Shams, Z.: Comparing test quality measures for assessing student-written tests. In: Companion Proceedings of the 36th International Conference on Software Engineering. ICSE Companion 2014, pp. 354–363. Association for Computing Machinery, New York, NY, USA (2014)
8. Edwards, S.H., Shams, Z.: Do student programmers all tend to write the same software tests? In: Proceedings of the 2014 Conference on Innovation and Technology in Computer Science Education. ITiCSE '14, pp. 171–176. Association for Computing Machinery, New York, NY, USA (2014)
9. Ericsson, A., Pool, R.: Peak: secrets from the new science of expertise. Random House (2016)
10. Fitzgerald, S., McCauley, R., Hanks, B., Murphy, L., Simon, B., Zander, C.: Debugging from the student perspective. IEEE Trans. Educ. **53**(3), 390–396 (2010)
11. Ganguli, D., et al.: Predictability and surprise in large generative models. In: Proceedings of the 2022 ACM Conference on Fairness, Accountability, and Transparency, pp. 1747–1764 (2022)
12. Kallia, M.: The search for meaning: Inferential strategic reading comprehension in programming. In: Proceedings of the 2023 ACM Conference on International Computing Education Research (ICER '23). ACM (2023)
13. Ko, A.J., Myers, B.A.: Debugging reinvented: asking and answering why and why not questions about program behavior. In: Proceedings of the 30th International Conference on Software Engineering. ICSE '08, pp. 301–310. Association for Computing Machinery, New York, NY, USA (2008)
14. Lukasová, A.: Hierarchical agglomerative clustering procedure. Pattern Recogn. **11**(5–6), 365–381 (1979)

15. Luxton-Reilly, A., McMillan, E., Stevenson, E., Tempero, E., Denny, P.: Ladebug: an online tool to help novice programmers improve their debugging skills. In: Proceedings of the 23rd Annual ACM Conference on Innovation and Technology in Computer Science Education. ITiCSE 2018, pp. 159–164. Association for Computing Machinery (2018)

16. Ma, Q., Wu, T., Koedinger, K.: Is AI the better programming partner? Human-human pair programming vs. human-AI pAIr programming. arXiv preprint arXiv:2306.05153 (2023). http://arxiv.org/abs/2306.05153

17. MacNeil, S., Tran, A., Mogil, D., Bernstein, S., Ross, E., Huang, Z.: Generating diverse code explanations using the GPT-3 large language model. In: Proceedings of the 2022 ACM Conference on International Computing Education Research-Volume 2, pp. 37–39 (2022)

18. Markel, J.M., Opferman, S.G., Landay, J.A., Piech, C.: GPTeach: interactive TA training with GPT-based students. In: Proceedings of the Tenth ACM Conference on Learning @ Scale. L@S '23, pp. 226–236. Association for Computing Machinery, New York, NY, USA (2023). https://doi.org/10.1145/3573051.3593393

19. McCauley, R., et al.: Debugging: a review of the literature from an educational perspective. Comput. Sci. Educ. **18**(2), 67–92 (2008)

20. Mozannar, H., Bansal, G., Fourney, A., Horvitz, E.: Reading between the lines: modeling user behavior and costs in AI-assisted programming. arXiv preprint arXiv:2210.14306 (2022)

21. News, Y.H.: Why don't schools teach debugging? (2014). https://news.ycombinator.com/item?id=7215870. Accessed 8 Sept 2023

22. Savelka, J., Agarwal, A., An, M., Bogart, C., Sakr, M.: Thrilled by your progress! Large language models (GPT-4) no longer struggle to pass assessments in higher education programming courses. arXiv preprint arXiv:2306.10073 (2023)

23. Shahriar, T., Matsuda, N.: What and how you explain matters: inquisitive teachable agent scaffolds knowledge-building for tutor learning. In: Wang, N., Rebolledo-Mendez, G., Matsuda, N., Santos, O.C., Dimitrova, V. (eds.) AIED 2023. LNCS, vol. 13916, pp. 126–138. Springer, Cham (2023). https://doi.org/10.1007/978-3-031-36272-9_11

24. Wang, Z., Valdez, J., Basu Mallick, D., Baraniuk, R.G.: Towards human-like educational question generation with large language models. In: Rodrigo, M.M., Matsuda, N., Cristea, A.I., Dimitrova, V. (eds.) AIED 2022. LNCS, vol. 13355, pp. 153–166. Springer, Cham (2022). https://doi.org/10.1007/978-3-031-11644-5_13

25. Whalley, J., Settle, A., Luxton-Reilly, A.: Analysis of a process for introductory debugging. In: Proceedings of the 23rd Australasian Computing Education Conference. ACE '21, pp. 11–20. Association for Computing Machinery (2021)

26. Wiggins, G.P., McTighe, J.: Understanding by Design. ASCD (2005)

27. Wu, T., Terry, M., Cai, C.J.: AI chains: transparent and controllable human-AI interaction by chaining large language model prompts. In: Proceedings of the 2022 CHI Conference on Human Factors in Computing Systems. No. Article 385 in CHI '22, pp. 1–22. Association for Computing Machinery, New York, NY, USA (2022)

28. Xie, J., Zhang, K., Chen, J., Lou, R., Su, Y.: Adaptive chameleon or stubborn sloth: unraveling the behavior of large language models in knowledge clashes (2023)

29. Xu, S., Rajlich, V.: Cognitive process during program debugging. In: Proceedings of the Third IEEE International Conference on Cognitive Informatics, pp. 176–182. IEEE (2004)

30. Zeller, A.: Why Programs Fail: A Guide to Systematic Debugging, 2nd edn. Morgan Kaufmann, Oxford (2009)

Improving the Validity of Automatically Generated Feedback via Reinforcement Learning

Alexander Scarlatos[1](✉), Digory Smith[2], Simon Woodhead[2], and Andrew Lan[1]

[1] University of Massachusetts Amherst, Amherst, USA
{ajscarlatos,andrewlan}@cs.umass.edu
[2] Eedi, London, UK

Abstract. Automatically generating feedback via large language models (LLMs) in intelligent tutoring systems and online learning platforms has the potential to improve the learning outcomes of many students. However, both feedback generation and evaluation are challenging: feedback content has to be valid especially in subjects like math, which requires models to understand the problem, the solution, and where the student's error lies. Feedback also has to be pedagogically valid to reflect effective tutoring strategies, such as explaining possible misconceptions and encouraging the student, among other desirable features. In this work, we address both problems of automatically generating and evaluating feedback while considering both correctness and alignment. First, we propose a rubric for evaluating math feedback and show that GPT-4 is able to effectively use it to annotate human-written and LLM-generated feedback. Second, we propose a framework for feedback generation that optimizes both correctness and alignment using reinforcement learning (RL). Specifically, we use GPT-4's annotations to create preferences over feedback pairs in an augmented dataset for training via direct preference optimization (DPO). We show that our methods significantly increase the correctness and alignment of generated feedback with Llama 2, an open-source LLM, qualitatively analyze our generation and evaluation systems using case studies, and outline several areas for future work. (Our code is available at https://github.com/umass-ml4ed/feedback-gen-dpo).

Keywords: Feedback Generation · Human Preference Alignment · Math Education · Reinforcement Learning

1 Introduction

Providing students with helpful feedback can be critical to their learning, allowing them to quickly address and learn from mistakes. Prior work has shown

The authors thank Schmidt Futures and the NSF (under grants IIS-2118706 and IIS-2237676) for partially supporting this work.

A. M. Olney et al. (Eds.): AIED 2024, LNAI 14829, pp. 280–294, 2024.
https://doi.org/10.1007/978-3-031-64302-6_20

that delivering immediate automated feedback to students in intelligent tutoring systems and online learning platforms can improve learning outcomes [12,25]. However, doing so is challenging since generated feedback should satisfy a wide variety of requirements: it should convey an understanding of the question and why the student's response is incorrect, as well as be aligned with *educational goals* and pedagogical theory. For example, identifying student misconceptions, providing hints, and using encouraging language to promote a growth mindset [2,33] can be helpful, but simply giving away the answer could be detrimental.

Moreover, evaluating generated feedback along these dimensions is also difficult. Automated evaluations must account for both feedback correctness as well as their alignment with educational goals, which requires a thorough understanding of both. Additionally, even when expert-written feedback examples are given as reference, text similarity-based metrics may be unreliable since there are many ways to write valid feedback, and text overlap can emphasize irrelevant features while neglecting more significant ones [1,20]. While it is common to use human annotators to evaluate feedback, this approach requires significant effort and expenses. Therefore, the lack of reliable, automated feedback *evaluation* methods becomes a bottleneck for developing feedback *generation* methods.

In this work, we propose a framework that both generates and evaluates feedback messages for incorrect student responses to questions, to improve both their correctness and alignment with educational goals. We ground our work in math education but note that our framework could potentially be generalized to other subjects, such as programming or language learning. First, we propose a *rubric* for evaluating generated feedback and show that LLMs, particularly GPT-4, achieve high agreement with humans in their evaluations.

Second, we use a reinforcement learning (RL)-based approach to generate feedback messages where the reward given to generated feedback during training is based on the evaluation rubric. Moreover, to avoid repeatedly using GPT-4 to evaluate feedback during training, we use direct preference optimization (DPO) [24], an offline RL algorithm, to align the generated feedback with educational goals. This approach is similar to aligning LLMs with human [39] or AI [16] preferences. We experiment on a dataset that consists of feedback messages written by math teachers for each incorrect option in multiple-choice questions. Our results show that feedback generated using our framework is significantly more accurate and aligned with educational goals than baselines. Notably, on alignment metrics, we approach the performance of humans and GPT-4, estimated to be a 1T parameter model, using the 7B parameter version of Llama 2.

2   Related Work

2.1   Feedback Generation

There are many existing approaches for automatic feedback generation. One common method is to use engineered features to detect errors in student responses, and then use a rule-based system to provide relevant feedback or hints [3,12,15,25,29,30]. This method is popular since it is interpretable and reliable but requires significant human effort to adapt to new question types. A recent and more general approach to feedback generation is using large language models (LLMs), either through prompting [1,20,23,32] or fine-tuning [10]. However, prompting pre-trained LLMs requires them to be capable of understanding educational goals, but fine-tuning can yield poor results without significant amounts of aligned training data. We address these concerns in our work by fine-tuning on an augmented dataset annotated with alignment labels.

2.2 Feedback Evaluation

Several recent works have used rubrics to evaluate feedback [10,32], and works in other domains have found success in using LLMs to evaluate open-ended text where their judgements correlate with human judgements [6,13,22]. However, most prior works on feedback generation tend to rely on human annotators for reliable evaluation [1,9,10,32]. One recent work [11] uses GPT-4 to evaluate math feedback with a rubric and finds high agreement with human annotations. However, they only use GPT-4 to evaluate human-written feedback, while we evaluate feedback written by both humans and LLMs. Including this LLM feedback helps us uncover GPT-4's shortcomings in feedback evaluation, particularly that it can struggle to identify when feedback inaccurately addresses student errors or provides invalid suggestions.

3 Methodology

We now detail our framework for the two main tasks of feedback generation and evaluation. Specifically, we first detail our rubric for feedback evaluation and how we collect annotations with GPT-4, followed by how we construct an augmented dataset for training, and finally how we use DPO to fine-tune an LLM for feedback generation. We first define some notations for our tasks. Given a dataset $\mathcal{D}$ of N math questions, we define the i-th question as $(q^{(i)}, c^{(i)}, e^{(i)}, \{d_j^{(i)}, f_j^{(i)} | j \in \{1, \ldots, M\})$. Here, $q^{(i)}$ is the question text, $c^{(i)}$ is the correct answer to the question, $e^{(i)}$ is a textual explanation of the question's solution, $d_j^{(i)}$ is an incorrect, student-generated answer to the question, $f_j^{(i)}$ is a textual feedback message to give to a student when their answer is $d_j^{(i)}$, and M is the number of different incorrect answers given for each question. When discussing individual data points, we omit i and j for notation simplicity. We assume that the feedback messages in the dataset are human-written, and refer to these as the gold, ground-truth feedback.

3.1 Feedback Evaluation

We now detail our rubric for evaluating feedback given to students for their incorrect answers. In addition to correctness, we aim to evaluate feedback messages on their alignment with educational goals, including those associated with a growth mindset [2,33]. We take inspiration from prior works using rubrics for feedback evaluation [10,32] and include aspects to target common errors that LLMs make when generating feedback. Specifically, our rubric evaluates feedback on five different aspects, each of them resulting in a binary-valued label:

- **Correct (COR.)** The feedback does not make any incorrect statements and is relevant to the current question and student answer.
- **Revealing (REV.)** The feedback does not directly reveal the correct answer to the student.
- **Suggestion (SUG.)** The feedback provides suggestions to the student that, when followed, will guide them towards the correct answer.
- **Diagnostic (DIA.)** The feedback correctly points out the error the student made or the misconception underlying their answer.
- **Positive (POS.)** The feedback is positive and has an encouraging tone.

We now define the *rubric function*, r, which assigns labels to any feedback given a corresponding question and incorrect answer, and a final scalar-valued *rubric score*, s, which indicates the feedback's overall quality:

$$r(f|q, c, e, d) = (y_C, y_R, y_S, y_D, y_P) = \mathbf{y} \in \{0, 1\}^5$$

$$s = y_C \cdot \frac{y_C + y_R + y_S + y_D + y_P}{5} \in [0, 1]$$

where except for correctness, other rubric aspects are equally weighted. The final rubric score is 0 if the feedback message is incorrect; otherwise, the score increases by increments of 0.2 for every rubric aspect the feedback satisfies.

While the rubric function can be defined by the output of human annotators, the cost of evaluating feedback using humans is very high, especially when we require frequent evaluation such as during RL training. To address this issue, we use GPT-4, known for its ability to generalize to new tasks, to define a version of the rubric function, r_{GPT-4}. Using zero-shot chain-of-thought prompting [14], we ask GPT-4 yes or no questions related to each of the 5 labels, and use its output to get an estimated label $\mathbf{y}'$ and corresponding score s'. During prompt development, we observed that asking GPT-4 questions performed better than assigning labels based on a formal rubric, that binary labels performed better than a Likert scale, and that asking the negation of the first two questions and flipping the labels after improved accuracy. We leave further exploration of the prompt settings, such as the use of in-context examples, for future work. We provide an example prompt, output, and corresponding labels in Table 1.

3.2   Data Augmentation

We now detail our method for constructing an augmented dataset, which will be used for RL training as well as calculating agreement between GPT-4 and human annotations. For both of these tasks, we require both *positive* examples, i.e., feedback messages that score highly on the rubric, and *negative* examples, i.e., feedback that score poorly on the rubric. We observe that while most of

Table 1. Example prompt, output, and labels for GPT-4 feedback evaluation.

System	You are a math education expert
Prompt	Your job is to evaluate feedback given to students on math problems. Here is the question, the correct solution, the incorrect answer the student gave, and the feedback given to the student: Question: emma works 36 h a week. write this as a fraction in its simplest form. Correct Answer: $\frac{3}{14}$ Solution: There are $24 \times 7 = 168$ h in a week, so the fraction is $36/168$. This simplifies to $18/84 = 9/42 = 3/14$. Incorrect Answer: $\frac{18}{84}$ Feedback: It looks like you have simplified, but not fully. For the following questions, provide a short explanation and then answer with "Yes" or "No": 1. Does the feedback make any incorrect statements? 2. Does the feedback directly reveal the answer to the question? 3. Does the feedback give suggestions to the student on how to improve the answer? 4. Does the feedback correctly point out the misconception underlying the student's answer? 5. Does the feedback have a positive or encouraging tone?
Output	1. Explanation: The feedback is correct in stating that the student has simplified, but not fully. The student stopped at $\frac{18}{84}$, which can be further simplified to $\frac{3}{14}$. So, the feedback does not make any incorrect statements. Answer: No. 2. Explanation: The feedback does not provide the correct answer. It only indicates that the student's answer is not fully simplified. Answer: No. 3. Explanation: The feedback does not provide specific suggestions on how to improve the answer. It only states that the student has not fully simplified the fraction. Answer: No. 4. Explanation: The feedback correctly identifies that the student's misconception is not fully simplifying the fraction. Answer: Yes. 5. Explanation: The feedback is neutral in tone. It points out the error but does not use any positive or encouraging language. Answer: No
Result	$(y'_C = 1, y'_R = 1, y'_S = 0, y'_D = 1, y'_P = 0) \rightarrow s' = 0.6$

the gold feedback messages in $\mathcal{D}$ score highly, feedback generated by prompting LLMs can be flawed or unaligned with the evaluation metric, thus scoring poorly. Therefore, we collect 3 LLM-augmented versions of $\mathcal{D}$, where each feedback $f_j^{(i)}$ is replaced with a generated version: $\mathcal{D}_R$, where feedback is generated using few-shot prompting with random in-context examples, $\mathcal{D}_S$, where feedback is generated using few-shot prompting with the most similar examples, and $\mathcal{D}_Z$, where feedback is generated using zero-shot prompting. We refer to the union of the original dataset and LLM-augmented data $\mathcal{D}' = \bigcup\{\mathcal{D}, \mathcal{D}_R, \mathcal{D}_S, \mathcal{D}_Z\}$ as the augmented dataset. We use GPT-4 to annotate the feedback messages in this set, and we detail how we use these annotations for training in the next section.

3.3   Direct Preference Optimization

In order to generate feedback that scores highly on the rubric, we leverage direct preference optimization (DPO) [24], an offline RL algorithm, due to its simplicity and efficiency. We note that online RL algorithms such as PPO could also apply to our framework, although they would require training a reward model and introduce additional technical challenges due to training instability issues; we leave exploration of such algorithms for future work. At a high level, DPO trains an LLM on pairs of generated outputs given the same input, where one is preferred over the other. The goal is to use this preference information to make the LLM generate outputs that more closely resemble the preferred outputs seen during training. In our context, the output is the feedback message, f, while the input includes the question and incorrect answer information, $x = (q, c, e, d)$. During training, we minimize the DPO objective, i.e.,

$$\min_{\theta} -\mathbb{E}_{(x,f_w,f_l)\sim\mathcal{D}_{\text{DPO}}} \left[\log \sigma \left(\beta \log \frac{\pi_\theta(f_w|x)}{\pi_{\text{ref}}(f_w|x)} - \beta \log \frac{\pi_\theta(f_l|x)}{\pi_{\text{ref}}(f_l|x)} \right) \right],$$

where f_w is preferred over f_l as the feedback for x, $\mathcal{D}_{\text{DPO}}$ is a curated dataset containing these feedback pairs and preferences, π_θ is the trained LLM, i.e., a text generation "policy", parameterized by θ, π_{ref} is a frozen *reference* LLM, and β is a hyperparameter to control how far π_θ can deviate from π_{ref}. We now detail how we construct $\mathcal{D}_{\text{DPO}}$ using both feedback from the augmented dataset and mismatched feedback from the gold dataset.

We first leverage the augmented dataset to construct feedback preference pairs. For each unique $x \in \mathcal{D}'$, we have 4 feedback messages, 1 human and 3 LLM-generated, from which we construct $\binom{4}{2} = 6$ unique pairs. We then use the score s' for each feedback to determine which feedback is preferred, and exclude pairs that have the same score. For instance, consider a case where some x has possible feedback messages f_1, f_2, f_3, and f_4, with scores 1.0, 0.8, 0.4, and 0.4, respectively. We then produce the preference pairs (f_1, f_2), (f_1, f_3), (f_1, f_4), (f_2, f_3), and (f_2, f_4), where the first feedback is the preferred one in each pair.

We also use mismatched feedback from the gold dataset to construct additional preference pairs. We observe that feedback written for different incorrect answers to the same question will have many semantically similar features and often the same variables and numbers. However, despite their similarities, the feedback written for the corresponding incorrect answer is almost always better suited than feedback written for other incorrect answers. Therefore, these mismatched feedback are excellent *hard* negatives since it is hard for algorithms to distinguish between them and good feedback; finding such hard negatives has been shown to be the key to contrastive learning [27]. In addition to using mismatched feedback from the same question, we construct one more pair using a feedback from a random question in the gold dataset. For instance, for $x_1^{(i)}$ and $M = 3$, we construct the preference pairs $(f_1^{(i)}, f_2^{(i)})$, $(f_1^{(i)}, f_3^{(i)})$, and $(f_1^{(i)}, f_{j'}^{(i')})$, for some random $i' \in [1, N]$ and $j' \in [1, M]$, where $f_1^{(i)}$ is preferred in all pairs.

4   Experiments

We now detail all experiments we conduct to validate our framework for feedback generation and evaluation. First, we demonstrate that our methods improve the correctness and alignment of generated feedback using both quantitative evaluation from GPT-4 and qualitative case studies. Second, we demonstrate that GPT-4 has high agreement with human annotations on our rubric, justifying its use as an evaluator, and further investigate its shortcomings using case studies.

4.1   Dataset

We validate our framework using a dataset of middle school-level math multiple choice questions from Eedi, a math learning platform. The questions cover a variety of number sense concepts including fractions, exponents, rounding, and many others. All questions and feedback messages are written by real math teachers, deployed to real students, and are generally high quality. There are a total of 1,956 questions in the dataset and each question has a total of 3 incorrect options and a ground truth human-written feedback for each. We remove questions that require images and ones with processing errors, resulting in 1,418 questions. We divide these into a train/validation/test split of 850/284/284 questions and correspondingly 2,550/852/852 incorrect answers and corresponding feedback.

4.2   Experimental Setting

Data Augmentation. We use two LLMs to generate feedback for our augmented dataset: `code-davinci-002` (Codex) [4] for $\mathcal{D}_R$ and $\mathcal{D}_S$ since it has strong few-shot prompting ability, and `gpt-3.5-turbo` for $\mathcal{D}_Z$ since its zero-shot ability is much better than `code-davinci-002`. We use 2 in-context examples for few-shot prompts, only select examples from the train set, and use the S-BERT model `all-distilroberta-v1` [26] to measure similarity for $\mathcal{D}_S$. We prompt the models with questions, correct answers and incorrect answers, but not full solutions to make the task harder and increase the amount of incorrect feedback. To reduce costs, we randomly select a subset of $\mathcal{D}'$ to be annotated by GPT-4. Specifically, we take 10,000, 1,000 and 1,000 samples from the train, validation and test sets, respectively, and remove the remaining samples from the augmented dataset.

Feedback Generation Models. We primarily use the instruction-tuned Llama-2 7B Chat model [34] from HuggingFace [35] for feedback generation, loaded with 8-bit quantization [7]. For both supervised fine-tuning (SFT) and DPO, we train LoRA adapters [8] on all weight matrices, setting $r = 32$, $\alpha = 16$, and dropout= 0.05. We train using the AdamW optimizer with a learning rate of 3e-4 with warmup for 10% of steps and an effective batch size of 64 using gradient accumulation. We train for 3 epochs, which we find minimizes the loss on the validation set. For DPO, we set $\beta = 0.5$ and use the SFT model for initialization and as the reference model, which empirically outperformed using the base model. At inference time, we use greedy decoding and set the maximum new tokens to 200.

Table 2. Quantitative results of feedback generation across methods. Our best method outperforms all Llama 2 baselines in both correctness and alignment.

	COR.	REV.	SUG.	DIA.	POS.	Score	ROU.	BER.
Human	0.91	0.98	0.67	0.82	0.41	0.73	1.00	1.00
GPT-4	0.95	0.96	0.99	0.93	1.00	0.94	0.19	0.57
Zero-shot	0.63	0.63	0.74	0.43	**1.00**	0.49	0.16	0.55
SFT	0.65	**0.98**	0.49	0.68	0.19	0.49	**0.29**	**0.61**
DPO (Score)	0.70	0.93	**0.95**	0.82	0.66	0.65	0.22	0.57
DPO (Score + Mismatch)	**0.77**	0.96	**0.95**	**0.86**	0.57	**0.71**	0.23	0.57

Metrics. When evaluating feedback, we report the average of each rubric label in y' and the corresponding scores s' assigned by GPT-4. We note that GPT-4 will very rarely fail to assign labels when feedback is unrelated to the current question, in which case we automatically assign label values of 0. We use a temperature of 0 and 300 maximum tokens for GPT-4 decoding. We also use two popular reference-based metrics with the human-written feedback as reference: ROUGE-L (**ROU.**) [17] which is based on textual overlap, and the F1 of the BERTScore (**BER.**) [38] using the recommended `microsoft/deberta-xlarge-mnli` model, which is based on token-level semantic similarity.

4.3   Feedback Generation

We now show that we can improve both the correctness and alignment of generated feedback using our framework. We primarily focus on using Llama 2 Chat, an open-source LLM with 7B parameters, where we compare several versions of the model: **Zero-Shot**, i.e., simply prompting the base LLM, **SFT**, i.e., fine-tuning the base LLM on the gold feedback set, **DPO (Score)**, i.e., training the LLM with DPO only on the augmented dataset, and **DPO (Score + Mismatch)**, i.e., training the LLM with DPO on the augmented dataset and mismatched feedback. We additionally compare with the gold, human-written feedback in the dataset, as well as feedback generated by GPT-4. We use the same prompt for all methods, where we instruct the model to generate short and helpful feedback and to follow a version of the evaluation rubric.

Quantitative Analysis. Table 2 shows the average rubric labels and scores assigned by GPT-4 on all feedback in the test set, as well as the ROUGE-L and BERTScore values for reference. We see that DPO (Score + Mismatch) significantly improves the feedback scores compared to baselines (a 45% increase compared to Zero-Shot and SFT), showing that our data augmentation and training setup is highly effective at improving the quality of feedback generation with Llama 2. We additionally observe that including the mismatched feedback messages substantially increases the correctness of generated feedback, confirming their effectiveness as hard negative examples. Surprisingly, SFT does not outperform Zero-Shot on score, which shows that the standard fine-tuning setup is not effective for feedback generation. We can also see that ROUGE-L and BERTScore are unreliable estimates of feedback quality since they are highest on SFT, primarily because it copies the style of the gold feedback the closest.

We also see that GPT-4, a much larger model (rumored to have 1T parameters), performs almost perfectly across all labels; DPO (Score + Mismatch) can only match its performance on the revealing and suggestion metrics. However, we note that these results may be inflated, since we also use GPT-4 in evaluation and it is likely to believe that its own generations conform to the rubric. Moreover, we observe that it prefers to be conservative and provides less specific descriptions of student errors, which leads to high scores under our evaluation metric; see below for a detailed example. Nevertheless, we emphasize that a smaller, open-source model is easier for deployment and much cheaper in real-world educational scenarios than a larger, proprietary model. Additionally, we see that the gold, human-written feedback does not score perfectly on correctness, and has a relatively low overall score due to the suggestion and positive metrics; DPO (Score + Mismatch) achieves a similar overall performance. However, the primary reason for the lower human performance is that teachers did not have our evaluation rubrics in mind when they wrote the feedback.

Qualitative Analysis. We also performed qualitative studies to compare the outputs of the different methods and find cases where they succeed or fail; the main findings are: 1) Zero-Shot produces feedback with the right style but struggles to follow instructions, particularly by not identifying the error or revealing the correct answer, and is prone to hallucinations and numerical errors. 2) SFT produces feedback that is generally short and blunt, and usually attempts to identify the error although is often incorrect. 3) DPO (Score) produces feedback that attempts to identify the error, adding details and questions to provide implicit suggestions and increase positivity. It also produces incorrect outputs although less so than SFT. 4) DPO (Score + Mismatch) is more accurate than DPO (Score) in identifying the error. 5) GPT-4 produces feedback with smooth and coherent language but tends to avoid mistakes by not clearly pointing out the error. When it does detail the error, it is often incorrect, although less often than the other methods. 6) For all methods, most failures occur because the LLM cannot correctly identify the error in the incorrect answer. This failure happens more often when multiple reasoning steps are required to arrive at the student's answer. This observation suggests that although state-of-the-art LLMs claim to excel at mathematical reasoning, they cannot conceive how students may make errors [11,18,20]. One solution could be prompting the LLM to evaluate relevant calculations via code execution before writing the feedback [5,31].

Table 3. Feedback generated by all methods for one question and incorrect answer.

Input
Question: Erica and Rebecca share some marbles in the ratio $\frac{1}{4}$: $\frac{6}{8}$. Erica has $\square$ marbles than Rebecca. Which of the following belongs in the gap? Correct Answer: a third as many Incorrect Answer: twice as many

Source	Feedback
Human	Not quite, the denominator has doubled but the fractions don't have that relationship. Try simplifying 6/8 and then compare it to 1/4 ✓
GPT-4	Good effort, but remember to simplify the ratios first. Erica's share is not twice as much, but a fraction of Rebecca's. Try to compare 1/4 to 3/4 again. Keep going! ✓
Zero-shot	Great effort! You are on the right track. Since the ratio is 1/4: 3/4, Erica has 1/4, which is a third of 3/4. To find the number of marbles Erica has, you can multiply 3/4 by 3, which gives you 9. So, Erica has 9 marbles. Keep practicing, and you'll get it right next time! ✗
SFT	I think you have simplified the second fraction incorrectly. ✗
DPO (Score)	I think you have simplified 6/8 to 1/2. If we simplify 6/8 to 3/4, how many marbles does Erica have? How many does Rebecca have? What is the difference between them? ✗
DPO (Score + Mismatch)	I think you have found 8/4 = 2. But we need to simplify 6/8 to get the same units as 1/4. Then we can see how many more or less Erica has. ✓

To provide more specific examples of several of our findings, we show a representative question and incorrect answer along with corresponding feedback generated by each method in Table 3. In this example, we see that the human-written feedback directly points out the error in the incorrect answer that the student only considered the denominators, and also suggests simplifying the fraction on the right. GPT-4 prefers to be vague to avoid mistakes by pointing out that the ratio should be simplified and gives a suggestion based on this, but does not explicitly identify why the student's answer was wrong. Zero-Shot does not point out the student's error and directly answers the question, even though it has been instructed to not do so; it then hallucinates an incorrect and irrelevant multiplication before finishing with a positive message. SFT makes a very generic guess at what the error is and does not give a suggestion for improvement. DPO (Score) guesses a more specific but incorrect error, gives a somewhat misleading suggestion, and uses questions to encourage the student. Finally, DPO (Score + Mismatch) correctly identifies the error and gives a helpful suggestion.

4.4   Feedback Evaluation

Since we use GPT-4 to quantitatively evaluate feedback, we need to verify that GPT-4 can indeed label feedback accurately using our rubric. To do so, we randomly sample 80 feedback messages from the augmented test set and manually evaluate them on the rubric. We also recruited a graduate student with extensive teaching experience to evaluate these feedback messages for an additional set of human annotations. To reduce bias, we do not show human annotators GPT-4's

Table 4. GPT-4's agreement with human annotations across all rubric labels.

Label	Acc.	Prec.	Rec.	F1
COR.	0.77	0.80	0.82	0.81
REV.	0.91	0.94	0.97	0.95
SUG.	0.73	0.76	0.73	0.73
DIA.	0.68	0.56	0.85	0.68
POS.	0.72	0.78	0.28	0.41
Avg.	0.76	0.77	0.73	0.71

annotations or tell them whether a feedback message is human-written or LLM-generated. We compute agreement between human annotators using the Cohen's kappa statistic, resulting in 0.53, 0.46, 0.35, 0.53 and 0.29 on the COR., REV., SUG., DIA. and POS. labels, respectively. This moderate agreement shows that while humans generally agree, they occasionally make errors due to the difficulty of the task, and have lower agreement on more subjective labels such as POS.

We measure the accuracy (**Acc.**), precision (**Prec.**), recall (**Rec.**), and **F1** of GPT-4's annotations with respect to human annotations on this set, and report the results averaged across both human annotators in Table 4. We observe that GPT-4 generally agrees with human annotations, with an average accuracy of 76% across labels. We also compute the Pearson correlation coefficient of the final rubric scores between GPT-4 and human annotations, resulting in 0.56 on average, indicating moderate overall correlation. In contrast, ROUGE-L and BERTScore both have average correlations of 0.40 with human-annotated rubric scores. Not only are these correlations smaller, but they are biased upward since the human-written feedback messages, which generally have high rubric scores, automatically get ROUGE-L and BERTScore values of 1.

However, GPT-4 still struggles in a few key aspects and we provide examples of erroneous annotations in Table 5. Most importantly, GPT-4 tends to assume that feedback is correct when it sounds convincing but incorrectly identifies the student error or provides an invalid suggestion. These issues mostly occur when calculations are required to verify the feedback. Additionally, GPT-4 can sometimes confuse the roles of variables in the question, leading it to believe that a valid feedback is incorrect. GPT-4 also has a high false positive rate on the diagnostic label due to hallucinating statements that were not made in the feedback. We note that it may be possible to resolve these issues using additional prompt engineering or tools such as self-reflection [28] and code execution to evaluate math expressions [5, 31]. Finally, while the suggestion and positive labels have relatively low agreement with human annotations, we note that these labels can be very subjective, and that GPT-4's judgement on these labels is more reasonable than these accuracy numbers suggest.

Table 5. Incorrect GPT-4 rubric annotation examples with explanations.

Label	Input	Generated Explanation
COR. (FP)	Question: a jacket is reduced in a sale by 20%. A customer has a loyalty card for a further 10% off the price. What is the total percentage reduction in the price of the jacket? Correct Answer: 28% Incorrect Answer: 8% Feedback: I think you have added the two discounts together. We need to use multiplication to find the overall discount	The feedback does not make any incorrect statements. It correctly identifies the student's mistake of adding the discounts together instead of multiplying them. ✗
COR. (FN)	Question: i am thinking of a number. I divide it by 0.5 and my answer is 10. What number am i thinking of? Correct Answer: 5 Incorrect Answer: 20 Feedback: You multiplied by 0.5 instead of dividing by 0.5	The feedback is incorrect because the student actually multiplied by 2, not 0.5. The correct feedback should be "You multiplied by 2 instead of dividing by 0.5." ✗
DIA. (FP)	Question: which fraction is equivalent $\frac{3}{7}$ Correct Answer: $\frac{9}{21}$ Incorrect Answer: $\frac{9}{7}$ Feedback: You have multiplied the top and bottom by 3, but you have to multiply by 3 again to get the correct answer	The feedback correctly identifies that the student has not multiplied the denominator by the same number as the numerator, which is the underlying misconception. However, the suggested correction is incorrect. ✗

5   Conclusions and Future Work

In this work, we proposed a framework for automated feedback generation and evaluation via LLMs for students' incorrect answers in math multiple-choice questions. Our framework accounts for both the mathematical correctness of the feedback and its alignment with good pedagogical practices. We show that using a data augmentation and preference optimization approach, we can generate high-quality feedback using Llama 2 7B, a small and open-source LLM. We also show that GPT-4 can evaluate feedback rather accurately using a rubric and that its annotations are helpful for training the feedback generation method. There are many avenues for future work. First, we can apply our framework to other RL algorithms such as PPO, or non-RL approaches such as overgenerate-and-rank. Second, we can evaluate our final feedback generation task via a large-scale human evaluation or classroom study, which would alleviate concerns on GPT-4's annotations being biased. Third, we can test our framework's generalizability by applying it to other domains such as programming or language learning, or other scenarios such as hint generation or student-instructor conversations. Finally, we can consider tailoring feedback to each student according to their knowledge levels [19], especially for open-ended questions, since student errors can likely be detected from these responses [21,36,37].

References

1. Al-Hossami, E., Bunescu, R., Teehan, R., Powell, L., Mahajan, K., Dorodchi, M.: Socratic questioning of novice debuggers: a benchmark dataset and preliminary evaluations. In: Proceedings of the Workshop on Innovative Use of NLP for Building Educational Applications (BEA2023@ACL), pp. 709–726 (2023)
2. Boaler, J.: Ability and mathematics: the mindset revolution that is reshaping education. Forum **55**, 143–152 (2013)
3. Botelho, A., Baral, S., Erickson, J.A., Benachamardi, P., Heffernan, N.T.: Leveraging natural language processing to support automated assessment and feedback for student open responses in mathematics. J. Comput. Assist. Learn. **39**(3), 823–840 (2023)
4. Chen, M., et al.: Evaluating large language models trained on code (2021)
5. Chen, W., Ma, X., Wang, X., Cohen, W.W.: Program of thoughts prompting: Disentangling computation from reasoning for numerical reasoning tasks. arXiv preprint arXiv:2211.12588 (2022)
6. Chiang, C.H., Lee, H.V.: Can large language models be an alternative to human evaluations? arXiv preprint arXiv:2305.01937 (2023)
7. Dettmers, T., Lewis, M., Belkada, Y., Zettlemoyer, L.: Llm.int8(): 8-bit matrix multiplication for transformers at scale (2022)
8. Hu, E.J., et al.: Lora: Low-rank adaptation of large language models (2021)
9. Jia, Q., Cui, J., Xiao, Y., Liu, C., Rashid, P., Gehringer, E.F.: All-in-one: multi-task learning BERT models for evaluating peer assessments. arXiv preprint arXiv:2110.03895 (2021)
10. Jia, Q., et al.: Insta-reviewer: a data-driven approach for generating instant feedback on students' project reports. International Educational Data Mining Society (2022)
11. Kakarla, S., Thomas, D., Lin, J., Gupta, S., Koedinger, K.R.: Using large language models to assess tutors' performance in reacting to students making math errors. arXiv preprint arXiv:2401.03238 (2024)
12. Kochmar, E., Vu, D.D., Belfer, R., Gupta, V., Serban, I.V., Pineau, J.: Automated personalized feedback improves learning gains in an intelligent tutoring system. In: International Conference on Artificial Intelligence in Education. pp. 140–146 (2020)
13. Kocmi, T., Federmann, C.: Large language models are state-of-the-art evaluators of translation quality. arXiv preprint arXiv:2302.14520 (2023)
14. Kojima, T., Gu, S.S., Reid, M., Matsuo, Y., Iwasawa, Y.: Large language models are zero-shot reasoners. Adv. Neural. Inf. Process. Syst. **35**, 22199–22213 (2022)
15. Lan, A.S., Vats, D., Waters, A.E., Baraniuk, R.G.: Mathematical language processing: automatic grading and feedback for open response mathematical questions. In: Proceedings of the ACM Conference on learning@scale, pp. 167–176 (2015)
16. Lee, H., et al.: Rlaif: Scaling reinforcement learning from human feedback with AI feedback. arXiv preprint arXiv:2309.00267 (2023)
17. Lin, C.Y.: ROUGE: a package for automatic evaluation of summaries. In: Text Summarization Branches Out, Barcelona, Spain, pp. 74–81. Association for Computational Linguistics (2004)
18. Liu, N., Sonkar, S., Wang, Z., Woodhead, S., Baraniuk, R.G.: Novice learner and expert tutor: evaluating math reasoning abilities of large language models with misconceptions. arXiv preprint arXiv:2310.02439 (2023)

19. Liu, N., Wang, Z., Baraniuk, R., Lan, A.: Open-ended knowledge tracing for computer science education. In: Proceedings of the 2022 Conference on Empirical Methods in Natural Language Processing, pp. 3849–3862 (2022)
20. McNichols, H., et al.: Automated distractor and feedback generation for math multiple-choice questions via in-context learning. In: NeurIPS'23 Workshop on Generative AI for Education (2023)
21. McNichols, H., Zhang, M., Lan, A.: Algebra error classification with large language models. In: International Conference on Artificial Intelligence in Education, pp. 365–376 (2023)
22. Naismith, B., Mulcaire, P., Burstein, J.: Automated evaluation of written discourse coherence using GPT-4. In: Proceedings of the 18th Workshop on Innovative Use of NLP for Building Educational Applications (BEA 2023), Toronto, Canada, pp. 394–403. Association for Computational Linguistics (2023)
23. Nguyen, H.A., Stec, H., Hou, X., Di, S., McLaren, B.M.: Evaluating chatgpt's decimal skills and feedback generation in a digital learning game. In: Responsive and Sustainable Educational Futures, pp. 278–293 (2023)
24. Rafailov, R., Sharma, A., Mitchell, E., Ermon, S., Manning, C.D., Finn, C.: Direct preference optimization: your language model is secretly a reward model (2023)
25. Razzaq, R., Ostrow, K.S., Heffernan, N.T.: Effect of immediate feedback on math achievement at the high school level. In: Bittencourt, I.I., Cukurova, M., Muldner, K., Luckin, R., Millán, E. (eds.) AIED 2020. LNCS (LNAI), vol. 12164, pp. 263–267. Springer, Cham (2020). https://doi.org/10.1007/978-3-030-52240-7_48
26. Reimers, N., Gurevych, I.: Sentence-BERT: sentence embeddings using Siamese BERT-networks. In: Proceedings of the 2019 Conference on Empirical Methods in Natural Language Processing. Association for Computational Linguistics (2019)
27. Robinson, J.D., Chuang, C.Y., Sra, S., Jegelka, S.: Contrastive learning with hard negative samples. In: International Conference on Learning Representations (2021)
28. Shinn, N., Cassano, F., Labash, B., Gopinath, A., Narasimhan, K., Yao, S.: Reflexion: language agents with verbal reinforcement learning. arXiv preprint arXiv:2303.11366 **14** (2023)
29. Singh, R., Gulwani, S., Solar-Lezama, A.: Automated feedback generation for introductory programming assignments. In: Proceedings of the 34th ACM SIGPLAN Conference on Programming Language Design and Implementation, pp. 15–26 (2013)
30. Song, D., Lee, W., Oh, H.: Context-aware and data-driven feedback generation for programming assignments. In: Proceedings of the 29th ACM Joint Meeting on European Software Engineering Conference and Symposium on the Foundations of Software Engineering, pp. 328–340 (2021)
31. Sonkar, S., Le, M., Chen, X., Liu, N., Mallick, D.B., Baraniuk, R.G.: Code soliloquies for accurate calculations in large language models. arXiv preprint arXiv:2309.12161 (2023)
32. Steiss, J., et al.: Comparing the quality of human and ChatGPT feedback on students' writing (2023)
33. Sun, K.L.: Brief report: the role of mathematics teaching in fostering student growth mindset. J. Res. Math. Educ. **49**(3), 330–335 (2018)
34. Touvron, H., et al.: Llama 2: open foundation and fine-tuned chat models (2023)
35. Wolf, T., et al.: Huggingface's transformers: State-of-the-art natural language processing. arXiv preprint arXiv:1910.03771 (2019)
36. Zhang, M., Baral, S., Heffernan, N., Lan, A.: Automatic short math answer grading via in-context meta-learning. International Educational Data Mining Society (2022)

37. Zhang, M., Wang, Z., Baraniuk, R., Lan, A.: Math operation embeddings for open-ended solution analysis and feedback. International Educational Data Mining Society (2021)
38. Zhang, T., Kishore, V., Wu, F., Weinberger, K.Q., Artzi, Y.: BERTScore: evaluating text generation with BERT. In: International Conference on Learning Representations (2020)
39. Ziegler, D.M., et al.: Fine-tuning language models from human preferences. arXiv preprint arXiv:1909.08593 (2019)

Automatic Detection of Narrative Rhetorical Categories and Elements on Middle School Written Essays

Rafael Ferreira Mello[1,3] , Luiz Rodrigues[2(✉)] , Erverson Sousa[1],
Hyan Batista[3], Mateus Lins[3], Andre Nascimento[3] , and Dragan Gasevic[4]

[1] CESAR School, Centro de Estudos e Sistemas Avançados do Recife, Recife, Brazil
rafael.mello@ufrpe.br
[2] Federal University of Alagoas, Alagoas, Brazil
luiz.rodrigues@nees.ufal.br
[3] Federal Rural University of Pernambuco, Recife, Brazil
[4] Monash University, Clayton, Australia

Abstract. Narrative essays enable students to express their thoughts and feelings, gain diverse perspectives, understand themselves and their world more deeply, and enhance their literary skills and cultural awareness. As such, it became a relevant textual production in educational settings. However, assessing these texts is challenging and time-intensive. Therefore, this study focuses on developing an automatic detection system for narrative structures in essays, using a range of natural language processing algorithms, including machine learning, deep learning, and large language models. The study found that BERT was more effective than other models for this task, highlighting GPT's potential in extracting narrative structures. The findings of this study have implications for educational practices, particularly in the assessment and improvement of narrative writing skills among middle school students.

Keywords: Natural Language Processing · Narrative Essays · Rhetorical Categories · Large Language Model

1 Introduction

Essays are concise literary productions embody an author's viewpoint on a specific subject [46]. When composing essays, students can establish their understanding of the subject matter and their analysis, synthesis, and organizational proficiency capabilities. In this context, narrative essays focus on the capability of students to write texts that recount personal experiences and stories, which might be based on reality or fiction [38]. Narrative productions are composed of a complex textual representation with several elements defined as narrative macro-structures: Character, Setting, Initiating Event, Plan, Action, Consequence, and Elaborated Noun Phrase [16]. Narrative essays are a key production developed for elementary and middle school students, both for classroom instruction and assessments [8]. They help develop basic writing skills and the capability of creative expression and storytelling proficiency [34].

However, assessing narrative essays is a challenging task, as they are comprised of several elements connected to each other. Analyzing and detecting them is time-consuming, especially in a large corpus [34,38]. From a teaching-learning perspective, this challenge is a substantial issue because feedback is prominent for effective learning experiences [42]. Therefore, instructors need tools to help them provide rapid feedback regarding narrative essays.

Previous research has shown that natural language processing (NLP) is a useful approach to overcome this issue [14]. Specifically, in assessing narrative essays, researchers started exploring ways to automate this process using NLP and machine learning (e.g., [23,44]). The detection of narrative elements from written essays has shown to be an efficient tool to assist instructors in providing useful feedback to their students [4]. In that regard, previous work has contributed to detecting narrative elements from medical and work-related essays (e.g., [27,34,39]).

Nevertheless, the corpora considered in previous studies on automated assessment of narrative essays are mostly written in English [2]. Although recent studies have started to address similar concerns based on Brazilian Portuguese essays, they evaluate different literary production [15,26] or they target specific elements [5]. Furthermore, despite recent advancements in Large Language Models (LLMs) – an advanced form of NLP optimized for text generation – have revealed their potential for numerous NLP tasks [32,43], past research has not systematically experimented with them in this context. Therefore, to the best of our knowledge, there is a lack of research detecting the broad range of narrative elements for essays in nun-English Languages based on NLP techniques.

This article addresses the gaps listed above with an empirical study on the automatic detection of narrative elements from narrative essays written in Brazilian Portuguese. For that, this paper reports on the findings of a study that explored a broad range of NLP algorithms, including classical machine learning, deep learning, and GPT-4, a state-of-the-art LLM [29]. Although much has been discussed regarding LLMs capabilities [32,43], the combination of BERT and Random Forrest classifier reached the best results in general, suggesting task-specific models' potential compared to generic models. In practice, this study could provide relevant information for teachers' feedback in the context of essay production and inform researchers on the limitations of LLMs for particular tasks compared to specific models.

The remainder of this paper is organized as follows. Section 2 discusses background and related works. In Sect. 3 we explain our methodology to address the problem raised. In Sections, we present how the work is evaluated and the models adopted. Section 4 we show the results. Finally, Sect. 5 and 6 contextualize the results in terms of practical adoption, explain the limitations of the study, and discuss future directions.

2 Background

2.1 Narrative Elements

A narrative text is a personal production that explores a series of connected events or experiences, often centered around characters and a temporal progression of actions [1]. Narrative essays are tools for organizing and transmitting knowledge, memories, and cultural values, allowing individuals to express their internal experiences and interpret the world around them. They are composed not only of events but also of intentions, conflicts, and resolutions that together form a cohesive and meaningful story. Therefore, a written narrative is not just a sequence of events but a carefully woven construction that reflects the complexity of human experience and the capacity for storytelling [18,24,40].

Specifically for education, narrative essays usually target the creation of a fictional universe or represent aspects of reality, involving elements such as plot, characters, theme, and style. In Brazil's primary education, the textual production of narratives is an essential pedagogical practice, as it develops linguistic and creative skills and fosters critical thinking and empathy in students [10]. Through narrative writing, students learn to articulate their ideas and emotions, explore different perspectives, and better understand themselves and their world. Literary narrative, therefore, is a powerful tool in education, as it not only improves students' writing and reading skills but also enriches their educational experience, promoting a deeper understanding of literature and culture [11,30].

More specifically, narrative essays comprise narrative components and narrative (or story) elements. According to [34], *components* are the main elements that compose the narrative are:

- *Events* might be formally defined as significant occurrences with consequences. Those can be real-world incidents, such as bomb explosions or the birth of an heir.
- *Participants*, also known as actors, are the *who* of the story. They are relevant to the *what* and the *why*, and are often defined as entities in NLP.
- *Time* concerns the story's temporal references, which are prominent for understanding the timeline of the narrative by anchoring each event or scene at a point in time.
- *Space* concerns geographical information within the narrative, such as references to places or locations.

In exploring the narrative's components, various *elements* can be examined. As suggested in [16], seven narrative elements are present in narrative essays, with degrees ranging from 0 (absent) to 3 (elaborated knowledge).

- *Character*: An agent performing actions. Levels: 0 (no main character or ambiguous pronouns), 1 (at least one main character with non-specific labels), 2 (a main character with a proper name), 3 (multiple main characters with proper names).

- *Setting*: Provides information about the story's location or time. Levels: 0 (no reference), 1 (a general place or time), 2 (references to a specific place or time), 3 (places using proper names or specifying a particular time).
- *Inciting Event(s)*: Motivates characters to take action. Levels: 0 (descriptions with no indication), 1 (one event without motivating action), 2 (one motivating event), 3 (two or more events motivating separate actions).
- *Internal Response*: Feelings expressed by characters about Inciting Events. Levels: 0 (no stated feelings), 1 (feelings not clearly related), 2 (feelings explicitly related), 3 (multiple instances of feelings clearly related).
- *Plan*: Thoughts stated by characters related to a decision to take action. Levels: 0 (no statement), 1 (statements about plans not necessarily related), 2 (one statement about a plan related), 3 (multiple statements about plans related).
- *Attempt*: Actions taken by characters motivated by the Inciting Event. Levels: 0 (no actions), 1 (action verbs in descriptive sentences without a clear link), 2 (action verbs in sentences clearly linked), 3 (introducing complicating actions that impede responses).
- *Consequence*: The end result of characters' actions in relation to the Inciting Event. Levels: 0 (no clear resolution), 1 (outcomes linked to other actions), 2 (one outcome related), 3 (two or more outcomes directly related).

The narrative elements considered for our study will be based on the narrative components outlined by [34] and [16]. These sources provide a comprehensive framework that will guide our analysis.

2.2   Related Works

This section presents prior research on the analysis of textual elements in narrative essays, aligning with the objectives of this paper. The automatic analysis of narrative essays has found extensive applications beyond educational purposes, extending its utility to other domains. For instance, Mulyana et al. [27] proposed a pipeline for extracting patient symptoms based on medical narrative texts for diagnosing mental illness. With the help of four databases of diseases and symptoms, the study focused on general text processing steps, e.g., sentence identification, scanning, and parsing; keyword extraction; and pattern matching on top of the narratives. However, the Mulyana et al. paper did not report an evaluation of the effectiveness of the pipeline for extracting symptoms from narrative texts.

Another application is identifying events from movie descriptions [39]. Based on using NLP to extract information, each event is structured with the following attributes: "who", "did what", "to whom", "where" and "when". Following those events, the authors use a Recurrent Neural Network (RNN) and the Word2vec algorithm to extract the descriptions' events. The authors reported an accuracy near 75% for real cases analysis.

Regarding the analysis of narrative essays in education, prior research has mostly explored machine learning for the automatic assessment of narrative essays based on the English language. For instance, Somasundaran et al. [38] establish the foundation for automated evaluations of narrative quality in student essays. By exploring algorithms such as Regression Linear, Support Vector Regression based on RBF kernel, Random Forest, and Elastic Net, they automatically assessed narrative quality, achieving moderate results regarding Quadratic Weighted Kappa (QWK). In this study, the Linear Regression reached the best values on average, obtaining a QWK of 0.7.

In [19], authors similarly employed machine learning methods to predict scores for narrative elements compared to scores obtained by human assessors. Exploring features such as those extracted with *Coh-Metrix*, *TF-IDF*, GloVe embeddings, and Bidirectional Encoder Representations from Transformer (BERT), their predictive models achieved QWK scores aligned with those obtained by human assessors. Specifically, their findings showed that BERT performed up to two times better than the other methods, demonstrating the potential for automating the scoring of narrative elements.

While most studies on automated analysis of written narratives are related to the English language, recent research within the Brazilian Portuguese context also explores this theme. In [3], the authors aimed to classify Brazilian Portuguese texts in terms of the following types: narrative, essay, injunctive, and descriptive. They used classical machine learning algorithms based on linguistic features extracted with Coh-Metrix. Nevertheless, this study did not specifically analyze the feasibility of identifying specific narrative elements, such as plot and characters. The proposed approach achieved an f-score of 91.2% in the best case.

On a similar line, the study reported in [5] was focused on the issue of automatic correction of narrative texts in Brazilian Portuguese, emphasizing identifying the climax. Using an English annotated dataset translated into Brazilian Portuguese, the study investigated the use of machine learning algorithms to detect climaxes in textual productions using three traditional classification algorithms: Support Vector Machine, Random Forest, and Stochastic Gradient Descent. The article highlights Random Forest as the best-performing algorithm and suggests combining Coh-Metrix and LIWC attributes to produce the best results. The study concludes that it is possible to develop an automatic system for detecting climaxes in narratives, emphasizing the importance of carefully selecting features to improve the effectiveness of climax classification in narrative texts.

Furthermore, [37] explores the automatic scoring of student narrative essays in Portuguese, focusing on the *formal register*, which assesses aspects related to the formal grammar of Brazilian Portuguese. Different machine learning algorithms (i.e., Decision Tree, Random Forest, SVM, Extra-Tree, AdaBoost, and XGBoost) were evaluated using diverse linguistic features. The results revealed that the Extra-Tree Ensemble algorithm exhibited the best performance across all measures, with a weighted average precision of 0.557, recall of 0.566, F1-Score of 0.546, and Kappa of 0.367.

Moreover, it is important to note that prior research, such as [15, 26], has also addressed the challenge of identifying rhetorical categories in Portuguese texts through machine learning techniques. However, this research primarily concentrated on analyzing dissertative productions, a more common focus in the literature.

Lastly, although related work has extensively explored machine learning, one of the most advanced techniques at the time of writing, LLMs, has yet to be explored. LLMs are advanced NLP systems that excel in understanding and generating human-like text [43]. These models, such as GPT-3 and GPT-4, possess the remarkable capability to adapt to new tasks without explicit programming, making them versatile in handling diverse linguistic challenges [29]. LLMs find applications in various domains, from content creation and translation to question answering and code generation [32]. Their ability to comprehend context allows them to potentially revolutionize the field of education, particularly by helping in automatically identifying and analyzing story elements in narrative essays. However, as this section demonstrates, past studies have not assessed these capabilities of LLMs.

2.3   Research Question

To our knowledge, no prior research is specifically dedicated to evaluating the narrative rhetorical categories and elements in Brazilian Portuguese. Previous studies have primarily concentrated on analyzing grammatical errors and the automatic scoring of essays. This study tries to fill this literature gap by specifically targeting the automatic detection of structural elements in narrative essays written in Brazilian Portuguese. As such, our research question is:

Research Question 1 (RQ1):
To what extent can natural language processing models accurately identify the rhetorical categories and elements of narrative essays written in Brazilian Portuguese?

3   Method

3.1   Dataset

In this study, we used the dataset originally created for a general essay scoring problem. It encompasses 356 essays, divided into 3,262 sentences, written by mid-school students from Brazilian public schools in 2023. It is important to highlight that we did not remove students' grammatical errors and misspellings in the original text. Following previous work [15,26], we analyzed the elements of the text in the sentence level.

The initial categories chosen for our annotation process were influenced by those suggested in [16]. However, considering the relatively basic complexity of the essays, as they were written by students in the middle school, we opted to modify the categories slightly as follows:

- We kept the **Character**, **Initiating Event**, and **Consequence** as previously described;

- The category settings were divided into **Location** and **Time**;
- The categories Internal Response, Plan, and Attempt were converted into a single category named **Complication**.
- We incorporated the category **Narrator** due to the relevance of Brazilian texts described in previous work [11,30].

Two experienced educators individually coded each sentence in the essay corpus using the specified categories. The inter-rater reliability, measured by Cohen's κ, achieved a value of 0.65, indicating a substantial agreement between the coders. A third, more experienced teacher resolved the discrepancies. The 'Location' category was included only 15 times. Thus, it was excluded from this study. Table 1 details the count of positive instances for each category.

Table 1. Instance per category or element.

	Number of sentences	% of the dataset
Narrator	230	7.05
Character	300	9.19
Time	340	10.42
Initiating Event	1182	36.23
Complication	1852	56.77
Consequence	958	29.36

3.2 Feature Extraction

Traditionally, rhetorical structure identification has relied on content features [15,26,28,35]. Therefore, we evaluated the efficacy of predictive models using standard TF-IDF and word embedding features. TF-IDF, a fundamental technique in NLP studies, quantifies the importance of a word within a text by comparing its frequency in a specific document to its distribution across an entire corpus [25]. This method effectively highlights words that are uniquely significant in a given context. For word embeddings, we employed Global Vectors for Word Representation (GloVe) and Bidirectional Encoder Representation from Transformer (BERT). GloVe transforms each word into a 300-dimensional vector, capturing semantic relationships between words [31]. BERT, on the other hand, is a neural network designed for deep language understanding [12]. It pre-trains bidirectional representations from unlabeled text by considering the context from both sides of a word in all layers. This model, known for its robustness, continues to be a leading tool in many NLP applications.

3.3   Model Selection and Evaluation

To evaluate the proposed method, we considered different models available for NLP applications. Initially, we adopted a traditional Random Forest (RF) model due to its robustness for text classification problems [17]. RF operates as a bagging technique, combining multiple decision trees constructed through subsampling of data in the training set [6]. We used the RF classifier with both TF-IDF and BERT vector representations.

Secondly, we employed the BiLSTM (Bidirectional Long Short-Term Memory) architecture for this task, following its successful application in previous studies on English texts. LSTM networks are designed to learn from historical data, featuring an architecture that enables retaining previous information for use with current data. In a standard LSTM, data flows in a single direction and is processed once. Conversely, a BiLSTM model processes the data twice, analyzing it in both forward and reverse directions, enhancing its ability to capture context [36]. The architecture implemented in our experiments comprised several layers: an input layer with 30 neurons, an embedding layer with 50 neurons, a bidirectional layer with 32 cells, a dropout layer at 20% to prevent overfitting, a dense layer with 64 neurons and ReLU activation, another dropout layer at 20%, and finally, an output layer with one neuron using sigmoid activation. We trained this model over 50 epochs.

Finally, we incorporated a GPT model to explore the capabilities of a large-scale language model in this context. GPT models have gained significant recognition in academic studies for their capability to address many challenges [20,45]. Previous research shows the potential of using GPT to analyze the Essay even with a zero-shot approach [7]. In our study, we utilized the OpenAI API to access the GPT-4-turbo model, the latest version available at the time[1]

A critical factor in leveraging large language models is crafting well-structured prompts. The design of these prompts significantly impacts the model's ability to produce precise results [41]. In our study, the prompt formulation was tailored to suit the task's context, incorporating simple and direct instructions and specifying the desired output format. Table 2 displays the final prompt used for our task, translated into English[2].

Table 2. Prompt for GPT 4-Turbo.

Element	Text
Context	Narrative text from mid-school students.
Instruction	Determine if the following text has the XXX element.
Output format	indicate 'yes' or 'no' in the response

---

[1] OPENAI API: https://platform.openai.com/docs/api-reference/authentication.
[2] It is important to note that the original prompt was written in Portuguese.

We replicated the evaluation methodology used in prior research [15, 26, 35] to facilitate a comparative analysis of the model outcomes. For assessing the performance of the supervised machine learning algorithms, we employed Cohen's κ [9], a metric widely recognized in the fields of educational data mining and learning analytics [13, 33]. To ensure a robust evaluation, we conducted a 5-fold stratified cross-validation. This involved dividing the dataset into five equal parts, each mirroring the proportional representation of each class found in the original dataset.

4 Results

This section details the results of experiments designed to assess the effectiveness of various machine learning architectures in identifying the following narrative structure categories: *Narrator, Character, Time, Initiating Event* (IE), *Complication*, and *Consequence*. As mentioned before, our evaluation encompassed five distinct model configurations: TF-IDF combined with RF, GloVe integrated with BiLSTM, BERT paired with RF, BERT utilized alongside BiLSTM, and the application of GPT-4 Turbo. Each configuration was tested to measure its capability to accurately classify these narrative elements individually, providing a comprehensive outline of their respective performances in this task.

Table 3 shows the best results achieved by each algorithm using 5-fold cross-validation (as described in Sect. 3.3). The data indicates that combining BERT with RF generally yielded superior outcomes across all categories, except for one scenario (outcome) where TF-IDF + RF demonstrated better performance. For instance, in the case of the *Initiating Event* category, the BERT + RF configuration surpassed the GPT-4 model's performance by a significant margin, achieving a 32.96% higher score in terms of Cohen's κ coefficient. This comparison highlights the distinct strengths of different algorithmic approaches in handling specific narrative structure elements.

It is important to emphasize that GPT-4 emerged as the second-best performing model in our experiments. This outcome is particularly noteworthy, considering the limited extent of prompt engineering involved. The promising results achieved by GPT-4, with minimal customization in its prompt design, underscore its potential to identify the narrative elements.

Finally, Table 3 also displays that, in general, Cohen's κ values achieved moderate to substantial agreement levels across most categories [22], with the exception of the *Consequence* category. This finding emphasizes the effectiveness of the proposed approach. The ability of our models to consistently reach these levels of agreement across a diverse range of narrative categories demonstrates their robustness in accurately identifying and analyzing narrative structures.

Table 3. Comparative analysis between the four machine learning frameworks for narrative structure detection task based on Cohen's Kappa.

Class	TF-IDF + RF	GloVe + BiLSTM	BERT + RF	BERT + BiLSTM	GPT4-Turbo
Narrator	0.2488	0.3333	**0.6624**	0.2462	0.4695
Character	0.1822	0.1778	**0.6255**	0.2690	0.5342
Time	0.4298	0.4569	**0.7237**	0.3441	0.6882
IE	0.3873	0.3757	**0.7039**	0.2975	0.5423
Complication	0.2147	0.1324	**0.5847**	0.1389	0.4978
Consequence	**0.2708**	0.1432	0.1775	0.1975	0.1991

5   Discussion

The results obtained for the automatic identification of the rhetorical categories and elements in narrative texts written in Brazilian Portuguese essays indicated that the proposed experimentation, with BERT as features and Random Forrest as a machine learning algorithm, reached a moderate to substantial agreement [22] for almost all indicators, except for the consequence category. While no prior research has specifically addressed narrative texts, previous studies on different types of texts in Portuguese have attained a Cohen's kappa coefficient of up to 0.67 in identifying rhetorical categories [15,26,35]. This comparison emphasizes the significance and relevance of our findings in this domain.

Additionally, our research marks the first evaluation of GPT-4, a state-of-the-art LLM, for this specific task. Although GPT-4 did not outperform the combination of BERT and Random Forrest, it achieved high Cohen's kappa values, indicative of moderate agreement [22]. This outcome highlights the potential of employing GPT-4 in this context, as with further advancements in prompt engineering, there is a substantial possibility for enhancing the model's performance in this task [20,41,45].

In summary, the practical implications for research and practice can be categorized into three distinct aspects. First, the analysis of different combinations of features and machine learning algorithms, including the GPT model, demonstrated the potential to identify rhetorical categories and elements in narrative texts written in Brazilian Portuguese automatically. Second, we introduced a dataset of narrative essays written by middle school students in Brazil, which will be accessible to the research community. This represents a significant contribution, as there are few datasets with these specific characteristics available for scholarly exploration in this field. Finally, as previous research has explored, integrating the categories and elements extracted from essays into an educational tool could aid instructors and students in assessing and composing more effectively structured narrative essays [15,21].

6 Limitation and Future Directions

This study's primary limitations are related to the unbalanced nature of the dataset. Specific categories, such as 'place,' are underrepresented, making it challenging to train a model that can reliably detect their presence in any written essay. Additionally, the number of samples per category is highly imbalanced, necessitating the employment of undersampling or oversampling techniques to prevent the development of a biased model. In addition, the experiments were conducted with a dataset in a single language, which can reduce the generalizability of the findings. In future works, we plan to enhance the scope of our study by augmenting the sample size and incorporating essays from other languages and contexts.

In addition to the dataset issues, the machine learning models assessed in this study considered features from a single sentence in order to classify the rhetorical categories. Previous studies have demonstrated the potential of using sequential-based features and classifiers that consider the order of the sentences in the texts. Thus, in future works, we intend to evaluate the sequential method proposed by [15, 26] other classifiers, such as Conditional Random Fields (CRF).

Finally, it should be noted that the primary aim of this study was not to assess the practical application of the developed model in educational settings nor to evaluate the satisfaction of instructors and students with a tool based on the extracted information. However, it is important to emphasize that this is part of an ongoing project, which includes the practical implementation of such a tool and the subsequent evaluation of its effectiveness and impact.

Acknowledgment. This study was partialy supported by the Conselho Nacional de Desenvolvimento Científico e Tecnológico (310888/2021-2) and Acuity Insights under the Alo Grant program. Also, we would like to thank OpenAI research Program for offering the credits required for this experiment.

References

1. Abbott, H.P.: The Cambridge Introduction to Narrative. Cambridge University Press, Cambridge (2008)
2. Bai, X., Stede, M.: A survey of current machine learning approaches to student free-text evaluation for intelligent tutoring. Int. J. Artif. Intell. Educ. 1–39 (2022)
3. Barbosa, G.A., et al.: Aprendizagem de máquina para classificação de tipos textuais: Estudo de caso em textos escritos em português brasileiro. In: Anais do XXXIII Simpósio Brasileiro de Informática na Educação, pp. 920–931. SBC (2022)
4. Batista, H., Cavalcanti, A., Miranda, P., Nascimento, A., Mello, R.F.: Classificação multi-classe para análise de qualidade de feedback. In: Anais do XXXIII Simpósio Brasileiro de Informática na Educação, pp. 1114–1125. SBC, Porto Alegre, RS, Brasil (2022). https://doi.org/10.5753/sbie.2022.225396, https://sol.sbc.org.br/index.php/sbie/article/view/22486
5. Batista, H.H., et al.: Detecção automática de clímax em produções de textos narrativos. In: Anais do XXXIII Simpósio Brasileiro de Informática na Educação, pp. 932–943. SBC (2022)

6. Breiman, L.: Random forests. Mach. Learn. **45**(1), 5–32 (2001)
7. Brown, T.B., et al.: Language models are few-shot learners (2020)
8. Coelho, R.: Teaching writing in Brazilian public high schools. Read. Writ. **33**(6), 1477–1529 (2020)
9. Cohen, J.: A coefficient of agreement for nominal scales. Educ. Psychol. Measur. **20**(1), 37–46 (1960)
10. Dalla-Bona, E.M., Bufrem, L.S.: Aluno-autor: a aprendizagem da escrita literária nas séries iniciais do ensino fundamental. Educ. Rev. **29**, 179–203 (2013). https://doi.org/10.1590/S0102-46982013000100009
11. Detmering, R., Johnson, A.: "Research papers have always seemed very daunting": information literacy narratives and the student research experience. Portal: Lib. Acad. **12**, 5–22 (2012). https://doi.org/10.1353/pla.2012.0004
12. Devlin, J., Chang, M.W., Lee, K., Toutanova, K.: Bert: Pre-training of deep bidirectional transformers for language understanding. arXiv preprint arXiv:1810.04805 (2018)
13. Ferguson, R., Clow, D., Griffiths, D., Brasher, A.: Moving forward with learning analytics: expert views. J. Learn. Anal. **6**(3), 43–59 (2019)
14. Ferreira-Mello, R., André, M., Pinheiro, A., Costa, E., Romero, C.: Text mining in education. Wiley Interdiscip. Rev. Data Min. Knowl. Discov. **9**(6), e1332 (2019)
15. Ferreira Mello, R., Fiorentino, G., Oliveira, H., Miranda, P., Rakovic, M., Gasevic, D.: Towards automated content analysis of rhetorical structure of written essays using sequential content-independent features in Portuguese. In: LAK22: 12th International Learning Analytics and Knowledge Conference. pp. 404–414 (2022)
16. Gillam, S.L., Gillam, R.B., Fargo, J.D., Olszewski, A., Segura, H.: Monitoring indicators of scholarly language: a progress-monitoring instrument for measuring narrative discourse skills. Commun. Disord. Q. **38**(2), 96–106 (2017)
17. Hartmann, J., Huppertz, J., Schamp, C., Heitmann, M.: Comparing automated text classification methods. Int. J. Res. Marketing **36**(1), 20–38 (2019)
18. Jago, B.J.: Chronicling an academic depression. J. Contemp. Ethnography **31**, 729–757 (2002). https://doi.org/10.1177/089124102237823
19. Jones, S., Fox, C., Gillam, S., Gillam, R.B.: An exploration of automated narrative analysis via machine learning. PLoS ONE **14**(10), e0224634 (2019)
20. Kasneci, E., et al.: ChatGPT for good? On opportunities and challenges of large language models for education. Learn. Individ. Differ. **103**, 102274 (2023)
21. Kiesel, D., Riehmann, P., Wachsmuth, H., Stein, B., Froehlich, B.: Visual analysis of argumentation in essays. IEEE Trans. Visual. Comput. Graph. (2020)
22. Landis, J.R., Koch, G.G.: The measurement of observer agreement for categorical data. Biometrics (1977)
23. Li, Z., Liu, F., Antieau, L., Cao, Y., Yu, H.: Lancet: a high precision medication event extraction system for clinical text. J. Am. Med. Inform. Assoc. JAMIA **17**, 563–7 (2010). https://doi.org/10.1136/jamia.2010.004077
24. Luo, Y.H.: Word technology and literature narrative. J. Southwest Univ. Sci. Technol. (2010)
25. Manning, C., Schutze, H.: Foundations of Statistical Natural Language Processing. MIT Press (1999)
26. Mello, R.F., Fiorentino, G., Miranda, P., Oliveira, H., Raković, M., Gašević, D.: Towards automatic content analysis of rhetorical structure in Brazilian college entrance essays. In: Roll, I., McNamara, D., Sosnovsky, S., Luckin, R., Dimitrova, V. (eds.) AIED 2021. LNCS (LNAI), vol. 12749, pp. 162–167. Springer, Cham (2021). https://doi.org/10.1007/978-3-030-78270-2_29

27. Mulyana, S., Hartati, S., Wardoyo, R., et al.: A processing model using natural language processing (NLP) for narrative text of medical record for producing symptoms of mental disorders. In: 2019 Fourth International Conference on Informatics and Computing (ICIC), pp. 1–6. IEEE (2019)

28. Nguyen, H., Litman, D.: Context-aware argumentative relation mining. In: Proceedings of the 54th Annual Meeting of the Association for Computational Linguistics (Volume 1: Long Papers), pp. 1127–1137 (2016)

29. OpenAI: Gpt-4 technical report (2023)

30. Parry, B.: Introduction: a Narrative on Narrative. Palgrave Macmillan (2013). https://doi.org/10.1057/9781137294333_1

31. Pennington, J., Socher, R., Manning, C.D.: Glove: Global vectors for word representation. In: Proceedings of the 2014 Conference on Empirical Methods in Natural Language Processing (EMNLP), pp. 1532–1543 (2014)

32. Raffel, C., et al.: Exploring the limits of transfer learning with a unified text-to-text transformer (2023)

33. Rodrigues, L., et al.: Question classification with constrained resources: a study with coding exercises. In: Wang, N., Rebolledo-Mendez, G., Dimitrova, V., Matsuda, N., Santos, O.C. (eds.) International Conference on Artificial Intelligence in Education, pp. 734–740. Springer, Cham (2023). https://doi.org/10.1007/978-3-031-36336-8_113

34. Santana, B., Campos, R., Amorim, E., Jorge, A., Silvano, P., Nunes, S.: A survey on narrative extraction from textual data. Artif. Intell. Rev. 1–43 (2023)

35. dos Santos, K.S., Soder, M., Marques, B.S.B., Feltrim, V.D.: Analyzing the rhetorical structure of opinion articles in the context of a Brazilian college entrance examination. In: Villavicencio, A., et al. (eds.) PROPOR 2018. LNCS (LNAI), vol. 11122, pp. 3–12. Springer, Cham (2018). https://doi.org/10.1007/978-3-319-99722-3_1

36. Siami-Namini, S., Tavakoli, N., Namin, A.S.: The performance of LSTM and BILSTM in forecasting time series. In: 2019 IEEE International Conference on Big Data (Big Data), pp. 3285–3292. IEEE (2019)

37. da Silva Filho, M.W., et al.: Automated formal register scoring of student narrative essays written in portuguese. In: Anais do II Workshop de Aplicações Práticas de Learning Analytics em Instituições de Ensino no Brasil. pp. 1–11. SBC (2023)

38. Somasundaran, S., Flor, M., Chodorow, M., Molloy, H., Gyawali, B., McCulla, L.: Towards evaluating narrative quality in student writing. Trans. Assoc. Comput. Linguist. **6**, 91–106 (2018)

39. Tozzo, A., Jovanović, D., Amer, M.: Neural event extraction from movies description. In: Proceedings of the First Workshop on Storytelling. pp. 60–66. Association for Computational Linguistics, New Orleans, Louisiana (2018). https://doi.org/10.18653/v1/W18-1507, https://aclanthology.org/W18-1507

40. Venkatraman, K., Thiruvalluvan, V.: Development of narratives in Tamil-speaking preschool children: A task comparison study. Heliyon **7** (2021).https://doi.org/10.1016/j.heliyon.2021.e07641

41. White, J., et al.: A prompt pattern catalog to enhance prompt engineering with ChatGPT. arXiv preprint arXiv:2302.11382 (2023)

42. Wisniewski, B., Zierer, K., Hattie, J.: The power of feedback revisited: a meta-analysis of educational feedback research. Front. Psychol. **10**, 3087 (2020)

43. Yenduri, G., et al.: GPT (generative pre-trained transformer) - a comprehensive review on enabling technologies, potential applications, emerging challenges, and future directions (2023)

44. Zhang, F., Fleyeh, H., Wang, X., Lu, M.: Construction site accident analysis using text mining and natural language processing techniques. Autom. Constr. **99**, 238–248 (2019)

45. Ziyu, Z., et al.: Through the lens of core competency: Survey on evaluation of large language models. In: Proceedings of the 22nd Chinese National Conference on Computational Linguistics (Volume 2: Frontier Forum), pp. 88–109 (2023)

46. Zupanc, K., Bosnić, Z.: Automated essay evaluation with semantic analysis **120**(C), 118-132 (2017). https://doi.org/10.1016/j.knosys.2017.01.006

Marking: Visual Grading with Highlighting Errors and Annotating Missing Bits

Shashank Sonkar[✉], Naiming Liu, Debshila B. Mallick,
and Richard G. Baraniuk

Rice University, Houston, USA
shashank.sonkar@rice.edu

Abstract. In this paper, we introduce "Marking", a novel grading task that enhances automated grading systems by performing an in-depth analysis of student responses and providing students with visual highlights. Unlike traditional systems that provide binary scores, "marking" identifies and categorizes segments of the student response as correct, incorrect, or irrelevant and detects omissions from gold answers. We introduce a new dataset meticulously curated by Subject Matter Experts specifically for this task. We frame "Marking" as an extension of the Natural Language Inference (NLI) task, which is extensively explored in the field of Natural Language Processing. The gold answer and the student response play the roles of premise and hypothesis in NLI, respectively. We subsequently train language models to identify entailment, contradiction, and neutrality from student response, akin to NLI, and with the added dimension of identifying omissions from gold answers. Our experimental setup involves the use of transformer models, specifically BERT and RoBERTa, and an intelligent training step using the e-SNLI dataset. We present extensive baseline results highlighting the complexity of the "Marking" task, which sets a clear trajectory for the upcoming study. Our work not only opens up new avenues for research in AI-powered educational assessment tools, but also provides a valuable benchmark for the AI in education community to engage with and improve upon in the future. The code and dataset can be found at https://github.com/luffycodes/marking.

Keywords: Automated Grading · Highlighting · Feedback · Natural Language Inference

1 Introduction

Recent advancements in Artificial Intelligence (AI) have revolutionized the approach to automated grading in educational settings, offering scalable solutions

---

S. Sonkar and N. Liu—Equal contribution.

© The Author(s), under exclusive license to Springer Nature Switzerland AG 2024
A. M. Olney et al. (Eds.): AIED 2024, LNAI 14829, pp. 309–323, 2024.
https://doi.org/10.1007/978-3-031-64302-6_22

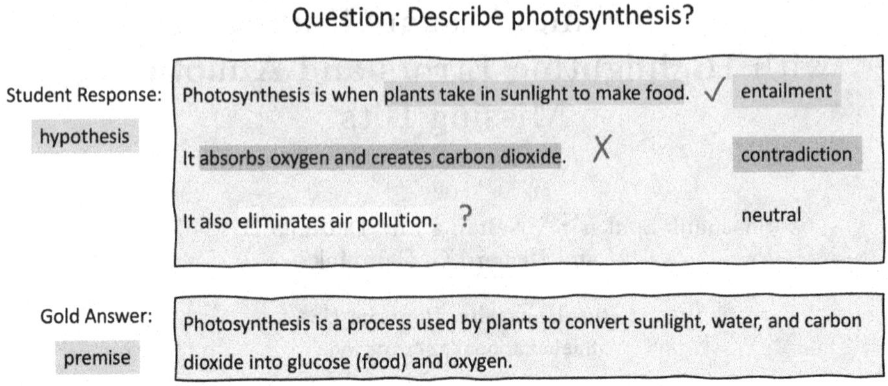

Question: Describe photosynthesis?

Student Response: | Photosynthesis is when plants take in sunlight to make food. ✓ entailment

hypothesis

It absorbs oxygen and creates carbon dioxide. ✗   contradiction

It also eliminates air pollution. ?   neutral

Gold Answer: | Photosynthesis is a process used by plants to convert sunlight, water, and carbon dioxide into glucose (food) and oxygen.

premise

Fig. 1. An illustration of the "Marking" task, which is formulated as NLI where the "gold answer" represents the premise and the "student response" is the hypothesis. The correct parts of the student response is classified as *entailment*, the incorrect parts as *contradiction*, and irrelevant part as *neutral*.

for evaluating student work[21]. Despite these advances, most systems prioritize binary, right-or-wrong assessments, lacking the subtlety and depth of feedback that educators typically provide. Recognizing this limitation, our work introduces a new dimension to automated grading, moving towards a more detailed form of assessment.

We introduce "Marking", an innovative task designed to push the capabilities of automated grading systems beyond their current scope. 'Marking' leverages AI to perform an in-depth analysis of student responses by evaluating them against a gold standard. The task involves training a language model to meticulously highlight text segments in the student's response, categorizing them as correct, incorrect, or irrelevant. Concurrently, it includes identifying and flagging content in the gold standard that the student has omitted. This two-fold assessment approach ensures a rigorous examination of what students express in their answers and highlights what they fail to include, offering a detailed and balanced view of their knowledge and understanding. By doing so, "Marking" provides a more nuanced insight into student learning, comparable to the detailed feedback typically provided by experienced educators (Fig. 1).

In our "Marking" framework, the evaluation of student responses draws on principles similar to Natural Language Inference (NLI) [6], a core task in Natural Language Processing (NLP). Traditionally, in NLI, a hypothesis (the student response) is tested against a premise (the gold answer) to determine if it is true (entailment), false (contradiction), or indeterminate (neutral). In the context of 'Marking,' we segment the student response into three parts: ones that accurately reflect the gold answer is tagged as "entailment", ones that misrepresent the gold answer as "contradiction," and those unrelated to the question as "neutral."

To evaluate the effectiveness of our "Marking" framework, we conduct a series of experiments using transformer models, specifically BERT [7] and RoBERTa

[11], which have shown remarkable performance in various NLP tasks. We treat the task of "Marking" as a classification problem, where the objective is to label the words in the student response as correct, incorrect, or irrelevant based on the gold standard answer. In our experimental setup, we fine-tune the aforementioned masked language models by adding an inference label classifier on top, thereby adapting these models to our specific task of marking. Our experiments investigate the impact of two preprocessing steps, Dual Instance Pairing (DIP) and stopword removal, and establish baselines for different label settings. Our results indicate that both DIP and stopword removal contribute positively to the model performance, which provides valuable insights into the potential of our "Marking" framework for enhancing automated grading systems.

In pursuit of realizing such an AI-driven marking system, this paper makes the following contributions to the field of AI in education:

1. We conceptualize and define the task of "Marking," delineating its scope and significance as a more granular and informative assessment method than currently prevalent automated grading approaches.
2. We introduce a novel dataset, BioMarking, meticulously curated by biology Subject Matter Experts, specifically designed to train and evaluate AI models on the "Marking" task. This dataset also serves as a benchmark for the community in the upcoming study.
3. Through extensive experiments on our dataset, we provide the baseline performance of our "Marking" task with various settings. This baseline results highlight the complexity and challenge of "Marking", thus setting a clear trajectory for future research.

The rest of the paper begins with a review of related work that contextualizes our approach within the existing landscape of AI-powered grading tools. Subsequent sections detail the problem statement, dataset creation, experimental results, and a discussion that collectively underscores the potential of "Marking" to transform AI-based educational assessment.

2   Related Work

2.1   Automated Grading and Feedback

LLMs have made a significant impact in the field of AI-enhanced education [18–20] in recent years. One notable area where LLMs have been applied is automated grading. Automated grading systems have gained popularity in recent years, especially with the advent of natural language processing and transformer-based models [22] such as BERT [7] and GPT models [2,15]. Furthermore, the integration of automated feedback generation with grading in education has also accelerated [4,16]. Such automated feedback generation approaches are usually applied to structured questions that require well-defined answers, such as multiple-choice questions [13], fill-in-the-blank [17], or programming assignments [14]. However, this automated feedback has some drawbacks as it increases the complexity of measuring the feedback quality compared to manual graders [10]. Our novel "Marking" task can resolve the challenge of feedback generation as it provides immediate and easy-to-follow feedback to students directly based on their responses, without requiring extra steps to evaluate the quality of the feedback.

2.2 Natural Language Inference

Natural language inference (NLI) is the task of determining whether a "hypothesis" is true (entailment), false (contradiction), or undetermined (neutral) given a "premise." Early research in the field of NLI focused on handcrafted features and classic machine learning techniques [5,23]. Models using these classical methods tend to require extensive feature engineering and domain knowledge. The integration of transformer models with NLI [7,9] has effectively gained considerable attention in recent years, which offers a significant enhancement over former methods. NLI has broad applications, such as information retrieval, semantic parsing, and commonsense reasoning [1,12]. In this paper, we frame the "Marking" approach as NLI tasks, which perceives the incorrect portion of the student response as "contradiction" against the gold standard answer, while the correct portion as "entailment". This approach characterizes the possibilities for a more dynamic grading system, where graders can denote parts of students' responses as either correct or incorrect, generating detailed and layered feedback.

3   Problem Formulation

3.1   Problem Statement

Given a dataset $\mathcal{D} = \{(R_i, A_i, G_i)\}_{i=1}^N$, where R_i represents a student's response, A_i represents the annotations for the correct, incorrect, and irrelevant portions of R_i, and G_i is the gold answer for the ith question, the task of "Marking" is to learn a function $f : (R, G) \rightarrow A$ that predicts the annotations A for a pair of student response R and gold answer G.

The annotation A_i is structured as follows:

- $A_i^{entail} \subseteq R_i$: The subset of the student's response that correctly aligns with the information in the gold answer, representing "entailment".
- $A_i^{contradict} \subseteq R_i$: The subset of the student's response that incorrectly contradicts the information in the gold answer, representing "contradiction".
- $A_i^{neutral} \subseteq R_i$: The subset of the student's response that is irrelevant or tangential to the gold answer, representing "neutral".

Additionally, the task requires the identification of omissions from gold answer:

- $A_i^{omission} \subseteq G_i$: The subset of the gold standard that contains key concepts or information not present in the student's response, representing "omission".

The objective of "Marking" is to optimize the function f based on a suitable performance metric, such as F1 score, across the annotations. Formally, the F1 score for each category is defined as:

$$F1_{category} = 2 \times \frac{Precision_{category} \times Recall_{category}}{Precision_{category} + Recall_{category}}$$

where *category* can be 'entailment', 'contradiction', 'neutral', or 'omission', and *Precision* and *Recall* are calculated based on the true positives (TP), false positives (FP), and false negatives (FN) for each category:

$$Precision_{category} = \frac{TP_{category}}{TP_{category} + FP_{category}}$$

$$Recall_{category} = \frac{TP_{category}}{TP_{category} + FN_{category}}$$

The overall performance of the "Marking" system could be evaluated using the precision, recall and the weighted average of F1 scores across the different categories, accounting for their relative importance or frequency in the dataset.

3.2   Neutral Condition in Omissions

In the standard NLI tasks, the "neutral" category is used to classify statements (hypotheses) that neither clearly support (entail) nor contradict the reference statement (premise); they are essentially unrelated or cannot be verified based on the given information. In the 'Marking" task, the "neutral" category is repurposed. It is no longer just a label for content that is neither right nor wrong in the student's response. Instead, it is now actively used to identify key concepts or information that are present in the gold answer but are missing from the student's response. In other words, "neutral" in this context is being used to flag omissions - important elements that the student did not include in their answer but should have, according to the gold answer.

By redefining the "neutral" category in this way, the AI system can provide feedback that is not only about what the student wrote but also about what the student failed to write. This proactive use of the "neutral" category effectively transforms it into a diagnostic tool that can provide comprehensive feedback to students, pointing out gaps in their knowledge or understanding that need to be addressed.

To summarize, the "neutral" category is being proactively used to enhance the depth of analysis in grading by identifying and highlighting missing elements in student responses, thereby offering a more complete and informative assessment.

4   Dataset

We developed the BioMarking dataset to fill in the blanks of the current "Marking" tasks. Our dataset contains 11 college-level biology questions along with

Table 1. Examples of the SME annotated dataset (best viewed in color).

Question	Gold Standard Answer	Student Response	Grade
Fiber is not really a nutrient, because it passes through our body undigested. Why can't fiber be digested and why is it important to our diet?	{Fiber is a type of carbohydrate that cannot be digested by the enzymes in the human body}. Despite this, it is an important part of our diet because it can help to bulk up our stools, slow down the absorption of sugar and cholesterol, and promote the growth of beneficial bacteria in our gut.	Fiber promotes <regular bowel movement by adding bulk>, and it <regulates the blood glucose consumption rate>. It <removes excess cholesterol from the body>. It binds to the cholesterol in the small intestine, then attaches to the cholesterol and prevents the cholesterol particles from entering the bloodstream Fiber-rich diets also have {a protective role in reducing the occurrence of colon cancer}	0.5
Fiber is not really a nutrient, because it passes through our body undigested. Why can't fiber be digested and why is it important to our diet?	Fiber is a type of carbohydrate that cannot be digested by the enzymes in human body. Despite this, it is an important part of our diet because {it can help to bulk up our stools, slow down the absorption of sugar and cholesterol, and promote the growth of beneficial bacteria in our gut}.	Fiber can't be digested because it actually [absorbs water so its important to push waste out of the body]	0

318 student responses. In the remainder of this section, we report the dataset construction process and its key statistics.

4.1 Dataset Construction

Question Selection. We select 11 biology questions from OpenStax[1] test bank with college-level multiple choice questions. The tutoring system adopted by OpenStax is that the students first answer the biology multiple-choice questions as if they were free-response questions without seeing the choices. Then the system displays the choices and the students pick the answer from the multiple-choice options. The two-step question-answering strategy is designed to help the students learn the materials better. In order to ensure the quality of our dataset, we only extract the responses of the students who select incorrect multiple choices. Additionally, we carefully curate the gold answer of each question based on the correct multiple-choice option, explanation for each choice, and solution to ascertain its accuracy and comprehensiveness. These answers are

---

[1] OpenStax: https://openstax.org/.

then thoroughly reviewed and approved by experts in the field of biology to ensure their validity.

Annotation Process. The annotation task was entrusted to a dedicated team of six Subject Matter Experts (SMEs). All of the SMEs have graduate-level education in biology. Leveraging their respective depths of understanding and expertise, they engaged in the meticulous process of marking all the student answers based on the provided gold answers. Each question is graded by two SMEs simultaneously. We instruct the SMEs to try to avoid labeling the whole sentence, but only the phrases that are incorrect. We provide three types of labels for the SMEs to identify in the student responses:

- Use <> to identify the **correct parts** and color them in green.
- Use [] to identify the **incorrect parts** and color them in red.
- Use {} to identify **irrelevant parts** and color them in yellow.

where the correct/incorrect part refers to the section of student answers that agree/contradict with what **actually present** in the provided gold answers. The definition of irrelevant is the information in the student response unrelated to the question or the solution, **regardless of the correctness**.

We also provide a type of label to mark the solutions:

- Use {} to highlight parts in the gold answer that are **missing** in the student response and color them in blue.

In addition to marking the sections of the student responses, we also ask the SMEs to provide a grade from {0, 0.5, 1} to evaluate whether the responses correctly address the solution, where 0 means the student response is incorrect, 0.5 means partially correct and 1 means correct response. We also recommend the SMEs provide some feedback on why the markings are given only based on incorrect parts, such as a misconception of a fact that leads to students' incorrect reasoning or a counterfactual statement, etc. Examples of the annotated dataset can be found in Table 1.

Annotator Training. We provide the SME annotators with detailed instructions on the grading task. To demonstrate their understanding and application of these instructions, they were first given a practice set of three questions and six student responses. Upon receiving their completed practice sets, we reviewed the marked responses for consistency and accuracy. We identify the discrepancies based on the gold answer created by the expert supervisors and provide them with feedback. The process was essential to ensure minimal disagreement between annotators during the actual grading process.

4.2   Dataset Statistics

The BioMarking dataset comprises 11 college-level biology questions with 318 distinct student responses. Two SME annotators grade each question, and we provide both annotated student responses to ensure the completeness of our dataset. Each question received an average of 29 student responses, with a standard deviation of 7.6. It is worth noting that the questions are initially presented as multiple-choice, with an average accuracy rate of 82.8%. However, individual questions varied in difficulty, as demonstrated by a minimum accuracy rate of 55.5% and a maximum of 91.5%.

After the annotation process, we find that 34% of the student short answer responses are incorrect (assigned a grade 0). 47% of student responses are partially correct (grade 0.5), and the remaining 19% of answers are completely accurate (grade 1). Additionally, the average number of words used in the gold standard answer is 66, while the student responses are more concise, with an average of 36 words per response.

5   Experiments and Baselines

5.1   Experimental Setup

The task of "Marking", as conceptualized in this study, is approached as a classification problem. Given a student response and a gold standard answer, our objective is to label each word in the student response as either correct, incorrect, or irrelevant. To accomplish this, we employ transformer models, specifically BERT (base/large) [7] and RoBERTa (base/large) [11], which have demonstrated remarkable performance in various NLP tasks.

Classifier Design. Our classifier design is built upon a pre-trained, masked language model with an inference label classifier head on the top layer. The transformer model, denoted by T, takes a sequence of words $S = \{w_0, w_1, .., w_{n-1}\}$ as input and outputs embeddings for each token in the sequence. The embeddings are given by $T(S) = \{e_0, e_1, .., e_{n-1}\}$, where $e_i \in \mathbb{R}^D$, $(0 \leq i < n - 1)$ is the token embedding, and D is its dimension.

The inference classifier uses a weight matrix $\mathbf{W} \in \mathbb{R}^{D \times |\mathcal{N}|}$ that takes the computed token embeddings as input, where $\mathcal{N}$ represents the set of all possible inference classes. The classifier outputs scores for all inference labels for each token in the sentence. Passing these scores through a softmax nonlinearity provides probabilities $\mathbf{p}_i \in \mathbb{R}^{|\mathcal{N}|}$ for all inference classes in $\mathcal{N}$ for a given token i in S:

$$\mathbf{p}_i = \text{softmax}\Big(\mathbf{W}(e_i)\Big). \tag{1}$$

The overall model $\mathcal{M}$ is then given by:

$$\mathcal{M}(T, \mathbf{W}, S, i) = \mathbf{p}_i, \, 0 \leq i < 2n. \tag{2}$$

In our experiments, the sequence of input is the gold answer concatenated with the student response, and the labels are the annotations A_i for the student response. The label set $\mathcal{N}$ consists of five labels: entailment (0), contradiction (1), neutral (2), none of these (3), and a special token (<sep>) (4). The labels are assigned to each word in the student response based on the annotations A_i. The special token (<sep>) is used to separate the gold standard answer and the student response in the input sequence.

Training Methodology. The model is trained using the e-SNLI [3] dataset, which provides premise and hypothesis pairs along with inference labels (entailment, neutral, or contradiction) and word-level annotations for the hypotheses. This dataset is particularly suited for our task as it allows the model to learn the nuances of entailment, contradiction, and neutrality at a word level, which is crucial for "Marking" task. After training on e-SNLI, we test our model on our marking dataset, with the objective of providing a baseline for future research.

Data Processing and Label Settings. In our experimental setup, we treat the gold answer as the premise and the student response as the hypothesis. The task is to predict the inference label of each word in the hypothesis, where the inference label can be either entailment, neutral, contradiction, or none of these. For a given premise sentence of arbitrary n words $\{p_0, p_1, ..p_n\}$ and a hypothesis sentence of m words $\{h_0, h_1, ..h_m\}$, we separate them by a special token [sep]. Each word is then assigned a label $l = \{0, 1, 2, 3, 4\}$, where [sep] has the label 4, words marked as entailed are 0, words marked as contradicted are 1, words marked as neutral are 2, and the rest are marked 3. The task of the transformer is then to predict the labels of each word given the sentence. This setup allows language models to learn the relationship between the student response and the gold standard answer at a granular level, thereby enabling a more detailed and accurate assessment of the student's understanding.

In the context of our classifier design, the sequence S is the concatenation of the premise (gold standard answer) and the hypothesis (student response), separated by the special token [sep]. Formally, $S = \{p_0, p_1, ..., p_n, [sep], h_0, h_1, ..., h_m\}$, where p_i and h_i are the words in the premise and hypothesis respectively.

We consider three different label settings (including one described earlier) to cater to different assessment needs:

- **Generic Setting (0-1-2):** This is the standard setting where entailment, contradiction, and neutrality are assigned distinct labels (0, 1, and 2 respectively). The rest of the words are assigned the label 3, and the special token [sep] is assigned the label 4. This setting allows for a comprehensive assessment of the student's response.
- **Contradiction-focused Setting (0-1-0):** In this setting, we are primarily interested in identifying the incorrect parts of the student's response. Therefore, both entailment and neutrality are assigned the same label (0), while contradiction is assigned a distinct label (1). The rest of the words are assigned the label 3, and the special token [sep] is assigned the label 4. This setting is useful when the focus of the assessment is to identify misconceptions.

– **Error-focused Setting (0-1-1)**: This setting is designed to identify both incorrect and irrelevant parts of the student's response. Therefore, entailment is assigned the label 0, while both contradiction and neutrality are assigned the same label (1). The rest of the words are assigned the label 3, and the special token [sep] is assigned the label 4. This setting is useful when the assessment aims to provide a more stringent evaluation of the student's response.

These label settings provide flexibility in the assessment process, allowing educators to choose the setting that best aligns with their assessment objectives.

Training Parameters. We use cross-entropy loss to train the classifier. The training parameters are set as follows: learning rate of 2e−5, weight decay of 0.1, warmup ratio of 0.05, Adam optimizer [8], and 3 epochs. These parameters were chosen based on their performance in previous studies using transformer models for similar NLI tasks.

Special Preprocessing Contributions. Our preprocessing step introduces two key contributions to the experimental setup, which are designed to optimize the performance of our model and ensure the consistency of our training and testing procedures.

Dual Instance Pairing (DIP). In the e-SNLI dataset, each premise-hypothesis pair is associated with a single label. However, in the "Marking" task, words in the student response can exhibit multiple labels simultaneously. This discrepancy between the training and testing conditions could potentially hinder the performance of our model. To address this issue, we introduce a novel preprocessing step where we pair two instances belonging to two different labels as one training sample. We refer to this technique as "Dual Instance Pairing".

Mathematically, we denote two instances from the e-SNLI dataset as $P1$[sep]$H1$ and $P2$[sep]$H2$, where P represents the premise, H represents the hypothesis, and [sep] is the special token separating the premise and hypothesis. In our preprocessing step, we combine these two instances into a single training sample as $P1P2$[sep]$H1H2$. This preprocessing step allows us to leverage the large e-SNLI dataset for transfer learning, while ensuring that our training procedure closely aligns with the conditions of our "Marking" task.

Stop Word Removal. Our second preprocessing contribution involves the exploration of stopword removal. Stopwords are commonly occurring words in a language (such as "the", "a", "in") that are often filtered out before or after processing text. While they carry little semantic content, their removal can sometimes lead to significant changes in the performance of NLP models.

In the context of our marking task, stop words in the student response could potentially be associated with any of the labels (correct, incorrect, irrelevant). Therefore, their removal could impact the performance of our model. In our experiments, we investigate the effect of stop word removal on the performance

Table 2. Performance of the RoBERTa-large model on the "Marking" task under different preprocessing conditions and label settings. The results demonstrate the positive impact of the Dual Instance Pairing (DIP) and stopword removal preprocessing steps on the model performance.

Label			Stopwords	Pairs	Precision ↑	Recall ↑	F1 ↑	Accuracy ↑
Ent	Con	Neu						
0	1	2	Yes	Yes	0.425	0.421	0.423	0.625
			Yes	No	0.407	0.382	0.394	0.638
			No	Yes	0.303	0.368	0.332	0.609
			No	No	0.332	0.327	0.33	0.613
0	1	0	Yes	Yes	0.819	0.429	0.563	0.904
			Yes	No	0.918	0.399	0.556	0.908
			No	Yes	0.557	0.419	0.478	0.885
			No	No	0.701	0.303	0.423	0.896
0	1	1	Yes	Yes	0.747	0.778	0.762	0.704
			Yes	No	0.718	0.805	0.756	0.708
			No	Yes	0.702	0.796	0.746	0.690
			No	No	0.709	0.76	0.734	0.685

of our model, providing valuable insights into the role of stop words in the "Marking" task.

5.2   Experiments and Discussion

We conduct two main experiments in this study. The first experiment investigates the impact of our two preprocessing steps: Dual Instance Pairing (DIP) and stopword removal. The second experiment establishes baselines for the RoBERTa and Bert models for different label settings. Of the evaluation metrics provided, we emphasize the F1 score over accuracy due to varying datapoint numbers in each class as label settings changes. Accuracy is only included for comprehensiveness.

Impact of Preprocessing Steps. In the first experiment, we train the RoBERTa-large model on the e-SNLI dataset and test it on BioMarking. We evaluate the performance under all three label settings: Generic, Contradiction-focused, and Error-focused. Our results in Table 2 indicate that both DIP and stopword removal preprocessing steps contribute positively to the model performance.

For instance, in the Generic setting, the RoBERTa-large model with both DIP and stopword removal achieves an F1 score of 0.423. When we remove the DIP preprocessing step, the F1 score drops to 0.394, indicating a decrease in

Table 3. Performance of different models (RoBERTa and BERT, base and large) on the "Marking" task under three label settings: Generic (0-1-2), Contradiction-focused (0-1-0), and Error-focused (0-1-1). The table reports the evaluation scores for each combination of model and label setting. The results highlight the superior performance of the RoBERTa-large model and the effectiveness of the Error-focused label setting.

Model	Label			Precision ↑	Recall ↑	F1 ↑	Accuracy ↑
	Ent	Con	Neu				
Roberta-large	0	1	2	0.425	0.421	0.423	0.625
	0	1	0	0.819	0.429	0.563	0.904
	0	1	1	0.747	0.778	0.762	0.704
Roberta-base	0	1	2	0.380	0.372	0.376	0.611
	0	1	0	0.684	0.43	0.528	0.889
	0	1	1	0.721	0.774	0.746	0.699
Bert-large	0	1	2	0.405	0.402	0.379	0.593
	0	1	0	0.577	0.462	0.513	0.873
	0	1	1	0.718	0.741	0.729	0.685
Bert-base	0	1	2	0.292	0.384	0.331	0.564
	0	1	0	0.531	0.423	0.471	0.863
	0	1	1	0.728	0.695	0.711	0.679

performance. Similarly, when we remove the stopword removal preprocessing step, the F1 score decreases to 0.332.

These results suggest that DIP and stopword removal are beneficial preprocessing steps that enhance the model's ability to accurately assess student responses.

Performance of Different Models and Label Settings. In the second experiment, we train RoBERTa (base and large) and BERT (base and large) models on the e-SNLI dataset and test them on BioMarking dataset, as shown in Table 3. We evaluate the performance of these models under all three label settings: Generic, Contradiction-focused, and Error-focused.

Finding 1: Our results indicate that the RoBERTa-large model outperforms the other models in all three settings. For instance, in the Generic setting, the RoBERTa-large model achieves an F1 score of 0.423, while the RoBERTa-base, BERT-large, and BERT-base models achieve F1 scores of 0.376, 0.379, and 0.331, respectively. This finding is consistent with the literature, which generally reports superior performance of RoBERTa models over BERT models in NLI tasks.

Finding 2: We find that the Error-focused setting yields the best performance across all models. For the RoBERTa-large model, the F1 scores for the Generic, Contradiction-focused, and Error-focused settings are 0.423, 0.563, and 0.762, respectively. These results suggest that the Error-focused setting, which aims to identify both incorrect and irrelevant parts of the student's response, provides a more stringent and comprehensive evaluation of the student's understanding.

The superior performance of the Error-focused setting can be attributed to its unique approach of treating both incorrect and irrelevant parts of the student's response as errors. This approach aligns well with educational objectives, as it encourages students to provide precise and relevant responses.

Firstly, by marking incorrect parts as errors, the model helps identify and rectify misconceptions in students' understanding. This behavior is crucial in education since unaddressed misconceptions can hinder the learning process. Secondly, by marking irrelevant parts as errors, the model discourages students from including extraneous information in their responses. This behavior is important as it promotes clarity and precision in student responses, which are critical skills in effective communication.

Therefore, the Error-focused setting not only provides a more stringent evaluation of the student's understanding but also promotes critical educational objectives. This finding underscores the potential of our proposed label settings in enhancing the effectiveness of automated grading systems in education.

6   Conclusion

In this paper, we introduced the task of "Marking" as a novel approach to automated grading in educational settings. By leveraging AI, we aimed to move beyond binary assessments and provide a more nuanced and detailed form of feedback, similar to what is provided by experienced educators. Our approach involved evaluating student responses against a gold standard, meticulously highlighting text segments as correct, incorrect, or irrelevant, and identifying key concepts or information the student omitted. Our primary contribution is the introduction of a novel dataset, BioMarking, meticulously curated by Subject Matter Experts for the "Marking" task. This dataset also serves as a benchmark for the AI in the education community to engage with and improve upon. We also presented initial baseline performance using state-of-the-art NLP techniques, adapted to "Marking" task. While these initial results are promising, they also highlight the complexity and challenge of the "Marking", setting a clear path for future research. Additionally, the task of "Marking" holds significant potential for enhancing the learning experience for students. By providing detailed feedback on their responses, it can help students identify their strengths and weaknesses, gain a deeper understanding of the subject matter, and guide their learning process. For educators, it offers a scalable solution for grading, reducing their workload while ensuring a high level of feedback quality. We believe that our dataset will stimulate further research in this area, leading to the development of more sophisticated models capable of providing detailed and nuanced feedback to students.

Acknowledgments. This work was supported by NSF grant 1842378, ONR grant N0014-20-1-2534, AFOSR grant FA9550-22-1-0060, a Vannevar Bush Faculty Fellowship, and ONR grant N00014-18-1-2047.

References

1. Bos, J., Markert, K.: Recognising textual entailment with logical inference. In: Proceedings of Human Language Technology Conference and Conference on Empirical Methods in Natural Language Processing, pp. 628–635 (2005)
2. Brown, T., et al.: Language models are few-shot learners. In: Advances in Neural Information Processing Systems, vol. 33, pp. 1877–1901 (2020)
3. Camburu, O.M., Rocktäschel, T., Lukasiewicz, T., Blunsom, P.: e-snli: Natural language inference with natural language explanations. In: Advances in Neural Information Processing Systems, vol. 31 (2018)
4. Cavalcanti, A.P., de Mello, R.F.L., Rolim, V., André, M., Freitas, F., Gaševic, D.: An analysis of the use of good feedback practices in online learning courses. In: 2019 IEEE 19th ICALT, vol. 2161, pp. 153–157. IEEE (2019)
5. Chen, Q., Zhu, X., Ling, Z., Wei, S., Jiang, H., Inkpen, D.: Enhanced LSTM for natural language inference. arXiv preprint arXiv:1609.06038 (2016)
6. Dagan, I., Glickman, O., Magnini, B.: The PASCAL recognising textual entailment challenge. In: Quiñonero-Candela, J., Dagan, I., Magnini, B., d'Alché-Buc, F. (eds.) MLCW 2005. LNCS (LNAI), vol. 3944, pp. 177–190. Springer, Heidelberg (2006). https://doi.org/10.1007/11736790_9
7. Devlin, J., Chang, M.W., Lee, K., Toutanova, K.: BERT: pre-training of deep bidirectional transformers for language understanding. arXiv:1810.04805 (2018)
8. Kingma, D.P., Ba, J.: Adam: a method for stochastic optimization. arXiv preprint arXiv:1412.6980 (2014)
9. Kumar, S., Talukdar, P.: Nile: natural language inference with faithful natural language explanations. arXiv preprint arXiv:2005.12116 (2020)
10. Kurdi, G., Leo, J., Parsia, B., Sattler, U., Al-Emari, S.: A systematic review of automatic question generation for educational purposes. Int. J. Artif. Intell. Educ. **30**, 121–204 (2020)
11. Liu, Y., et al.: RoBERTa: a robustly optimized BERT pretraining approach. arXiv preprint arXiv:1907.11692 (2019)
12. MacCartney, B., Manning, C.D.: An extended model of natural logic. In: Proceedings of the Eight International Conference on Computational Semantics, pp. 140–156 (2009)
13. Narayanan, S., Kommuri, V.S., Subramanian, S.N., Bijlani, K.: Question bank calibration using unsupervised learning of assessment performance metrics. In: 2017 International Conference on Advances in Computing, Communications and Informatics (ICACCI), pp. 19–25. IEEE (2017)
14. Parihar, S., Dadachanji, Z., Singh, P.K., Das, R., Karkare, A., Bhattacharya, A.: Automatic grading and feedback using program repair for introductory programming courses. In: Proceedings of the 2017 ACM Conference on Innovation and Technology in Computer Science Education, pp. 92–97 (2017)
15. Radford, A., Wu, J., Child, R., Luan, D., Amodei, D., Sutskever, I., et al.: Language models are unsupervised multitask learners. OpenAI blog **1**(8), 9 (2019)
16. Ramesh, D., Sanampudi, S.K.: An automated essay scoring systems: a systematic literature review. Artif. Intell. Rev. **55**(3), 2495–2527 (2022)
17. Sahu, A., Bhowmick, P.K.: Feature engineering and ensemble-based approach for improving automatic short-answer grading performance. IEEE Trans. Learn. Technol. **13**(1), 77–90 (2019)
18. Sonkar, S., Chen, X., Le, M., Liu, N., Basu Mallick, D., Baraniuk, R.: Code soliloquies for accurate calculations in large language models. In: Proceedings of the 14th Learning Analytics and Knowledge Conference, pp. 828–835 (2024)

19. Sonkar, S., Liu, N., Mallick, D., Baraniuk, R.: Class: a design framework for building intelligent tutoring systems based on learning science principles. In: Findings of the Association for Computational Linguistics: EMNLP 2023, pp. 1941–1961 (2023)
20. Sonkar, S., Ni, K., Chaudhary, S., Baraniuk, R.G.: Pedagogical alignment of large language models. arXiv preprint arXiv:2402.05000 (2024)
21. Valenti, S., Neri, F., Cucchiarelli, A.: An overview of current research on automated essay grading. J. Inf. Technol. Educ. **2**(1), 319–330 (2003)
22. Vaswani, A., et al.: Attention is all you need. In: Advances in Neural Information Processing Systems, vol. 30 (2017)
23. Wang, S., Jiang, J.: Learning natural language inference with LSTM. arXiv preprint arXiv:1512.08849 (2015)

Jill Watson: A Virtual Teaching Assistant Powered by ChatGPT

Karan Taneja(✉) ⓘ, Pratyusha Maiti(✉) ⓘ, Sandeep Kakar ⓘ,
Pranav Guruprasad ⓘ, Sanjeev Rao, and Ashok K. Goel ⓘ

Georgia Institute of Technology, Atlanta, GA, USA
{ktaneja6,pmaiti6,skakar6,pguruprasad7,srao373,ag25}@gatech.edu

Abstract. Conversational AI agents often require extensive datasets for training that are not publicly released, are limited to social chit-chat or handling a specific domain, and may not be easily extended to accommodate the latest advances in AI technologies. This paper introduces Jill Watson, a conversational Virtual Teaching Assistant (VTA) leveraging the capabilities of ChatGPT. Jill Watson based on ChatGPT requires no prior training and uses a modular design to allow the integration of new APIs using a skill-based architecture inspired by XiaoIce. Jill Watson is also well-suited for intelligent textbooks as it can process and converse using multiple large documents. We exclusively utilize publicly available resources for reproducibility and extensibility. Comparative analysis shows that our system outperforms the legacy knowledge-based Jill Watson as well as the OpenAI Assistants service. We employ many safety measures that reduce instances of hallucinations and toxicity. The paper also includes real-world examples from a classroom setting that demonstrate different features of Jill Watson and its effectiveness.

Keywords: Virtual Teaching Assistant · Intelligent Textbooks · Conversational Agents · Question Answering · Modular AI Design

1 Introduction

Conversational AI agents can be powerful tools for education as they enable continuous 24×7 support and instant responses to student queries without increasing the workload for instructors. These virtual teaching assistants can help in efficiently scaling quality education in terms of both time and cost. The interactive nature of conversational AI agents allows students to be more inquisitive and increases teaching presence by resembling one-on-one tutoring. Based on the Community of Inquiry framework [6], teaching presence through instructional management and direct instruction leads to an increase in student engagement and retention. Towards this end, we developed Jill Watson, a Virtual Teaching Assistant (VTA) powered by ChatGPT for online classrooms which answers students' queries based on course material such as slides, notes, and syllabic.

In previous work, the legacy Jill Watson [5,8] (henceforth LJW) is a question-answering system for course logistics and uses a two-dimensional database of

A. M. Olney et al. (Eds.): AIED 2024, LNAI 14829, pp. 324–337, 2024.
https://doi.org/10.1007/978-3-031-64302-6_23

information organized by course deliverables (assignments, exams, etc.) and information categories (submission policy, deadline, etc.). It also uses a list of FAQs for course-level information such as ethics and grading policies. In this paper, we introduce the new Jill Watson which is conversational and answers questions related to course logistics as well as course content based on multiple large documents provided as context.

ChatGPT or GPT-3.5, based on GPT-3 [3], is a powerful large language model (LLM) trained to follow instructions and hold a dialogue. It is capable of *attending* to a large context and constructing meaningful text in response to user inputs. Many conversational systems such as HuggingGPT [21], Microsoft Bing Chat (www.bing.com/chat), and LangChain (www.langchain.com) based systems leverage ChatGPT for performing context-aware response generation and zero-shot learning. ChatGPT and other LLMs suffer from hallucination i.e. they generate text that can be inconsistent or unverifiable with the source text, or absurd in a given context [9]. While hallucination is useful in creative tasks such as story writing, it is detrimental in information-seeking tasks such as those in the domain of education. ChatGPT and other LLMs also have safety issues as they generate text that may be considered toxic or inappropriate [25].

This work introduces Jill Watson's architecture which does *not* require any model training or fine-tuning and is designed to address the above LLM-related concerns. We address the hallucination issue by citing the documents from which information is obtained and verifying grounding using textual entailment. To prevent Jill Watson from answering unsafe questions or generating unsafe responses, we employ a classifier for question relevance, toxic text filters, and prompts that promote politeness in response generation. Further, Jill Watson is designed to answer questions based on multiple large documents which makes it well-suited for intelligent textbooks. We only rely on publicly available resources to promote future research in this direction.

The paper has three main contributions: (i) we introduce Jill Watson, a virtual teaching assistant powered by ChatGPT with a skill-based architecture, (ii) detail all the different modules of Jill Watson and associated algorithms, and (iii) quantitatively evaluate Jill Watson to measure response quality and safety, along with a discussion on examples from our first deployment.

Section 2 discusses Jill Watson in the context of related work. Section 3 describes the architecture and each module in detail. Section 4 describes our experimental results comparing Jill Watson to two strong baselines in terms of response quality and safety along with examples (see Table 3) from our first deployment. We conclude the paper in Sect. 5 with a summary of the strengths, limitations, and potential impact of Jill Watson.

2   Related Work

Question Answering can either be open-ended or grounded in knowledge. Without a knowledge source, question-answering models based on LLMs [16,22] are expected to store the information in their parameters during the training.

In grounded question answering, previous work has explored different types of contexts including the web [17], machine reading comprehension [1], knowledge bases [2], and short text documents [18,27]. Some methods assume access to the correct context from the document [18]. Further, many methods require training with datasets that are expensive to collect and do not generalize well [24,26]. Jill Watson neither imposes a limit on the document size nor requires a training dataset. It pre-processes large documents and answers incoming questions based on passages retrieved using dense passage retrieval (DPR) [11].

Retrieval Augmented Generation (RAG) is a well-known method [14] for increasing the reliability of LLMs by generating text conditioned on source texts that are retrieved based on a query. Knowledge-grounded generative models have two main goals: factuality and conversationality [19]. Factuality minimizes hallucinations by ensuring consistency of output with the retrieved texts while conversationality refers to relevance of the information to the query and generation without repetition. Previous work has shown improved factuality using RAG in dialog response generation task to remain consistent with a persona [23], knowledge-grounded generation [13,19,26] and machine translation [4].

Many models use large training datasets to learn question answering from contexts [2,24,26]. On the other hand, WikiChat [19] uses seven-step few-shot prompts based question answering system which uses both retrieval and open-ended generation to answer questions using Wikipedia. Re²G or Retrieve, Re-rank, Generate [7] also uses retrieval for generating outputs but uses an additional re-ranking step to score retrieved passages before generation. Further, it can also be trained end-to-end after initial fine-tuning. Jill Watson solves the knowledge-intensive generation problem using RAG but without any fine-tuning by using open-source DPR models for retrieval and using ChatGPT to construct responses. Because of clever prompting and indexing, it is also able to refer to the document and page number from which the response was generated.

Safety in LLMs is important to avoid harm because of hateful, offensive, or toxic text. Previous work on evaluating the toxicity of ChatGPT has found that assigning personas, using non-English languages, prompting with creative tasks, jailbreak prompts, and higher temperature values can all lead to more toxic responses [25]. Perspective API [12] and OpenAI Moderation API [15] are popular services to measure toxicity in various categories including hateful content, violence, etc. Jill Watson ensures safety in three ways: (i) OpenAI Moderation API for both user inputs and its own responses, (ii) skill classifier to identify irrelevant queries, and (iii) encourages polite responses in prompts to ChatGPT.

Dialog Systems are AI agents designed for human-like conversations, typically using hybrid architectures involving both rule-based systems and machine learning systems. For instance, MILABOT [20] uses rule-based and generative models along with a response selection policy trained using reinforcement learning. Microsoft XiaoIce [28] uses a skill-based architecture where the chat manager selects one of 230 high-level skills to generate responses. To the best of our knowledge, such powerful multi-turn dialog systems have not been introduced in a classroom. The legacy Jill Watson [8] is a single-turn question-answering

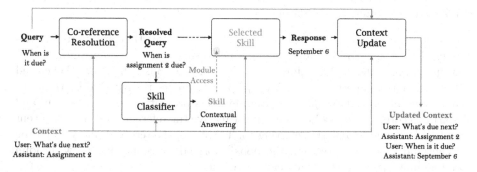

Fig. 1. *Architecture of Jill Watson:* After the coreference resolution of an incoming query, the skill classifier is used to find the most appropriate skill for response generation. Jill Watson's skills include Contextual Answering, Greetings, etc. The updated conversation history is used as context for generating responses in the future.

system for course logistics and policies. OpenAI Assistants service[1] provides a way to instantiate a ChatGPT-based agent to generate responses using text documents. Inspired by XiaoIce, Jill Watson has a skill-based architecture and is designed to be a safe conversational agent for classrooms that can answer student queries related to course logistics as well as course content using ChatGPT in the backend. It is also well-suited for other applications in education like intelligent textbooks given its ability to use long documents as context.

3   Jill Watson Architecture

The architecture of Jill Watson shown in Fig. 1 takes inspiration from the skill-based architecture of XiaoIce [28]. XiaoIce relies on different skills such as task completion, image commenting, content creation, etc. to interact with users and selects the appropriate skill for each conversation turn based on the previous context. In Jill Watson, the query with resolved coreferences is used to decide the most appropriate skill for answering the incoming query. The skill-based design of Jill Watson makes it extensible as we can easily plug in new API services and other capabilities in the future in the form of new skills.

Contextual Answering skill is responsible for answering questions where content needs to be retrieved from course content or syllabus and *Self-awareness* skill answers queries about Jill Watson itself. As we will discuss later, these two skills make Jill Watson a knowledge-grounded AI agent with the ability to refer to multiple documents and cite relevant content in its answers. Further, Chat-GPT allows Jill Watson to be conversational by using past messages as context in generating responses to user messages. We also utilize many safety features in Jill Watson which include detecting irrelevant queries by skill classifier, moderation filters, and prompts to encourage courteous responses.

---

[1] https://platform.openai.com/docs/assistants/overview.

3.1   Coreference Resolution

Coreference resolution involves determining the entities that are indirectly referenced in a text and making them explicit using nouns or noun phrases. For example, given the context 'John started reading a book', the query 'When did he start?' has two coreferences. An *explicit coreference* is suggested by the word 'he' while an *implict coreference* to an event is present because of 'When' and 'start'. In coreference resolution, we must resolve the reference 'he' and the event that 'When' and 'start' are referring to. Therefore, the resolved query would be 'When did John start reading the book?' formed by replacing 'he' with John and adding the implicit event 'reading the book'.

While ChatGPT implicitly resolves the coreferences as it can construct appropriate responses by itself, since we wish to use existing models for retrieval without any fine-tuning, we need to construct complete queries with resolved coreferences before passing them to the retrieval module. Hence, the first step in Jill Watson is to resolve coreferences in the user query based on the previous messages. In the example in Fig. 1, 'it' in 'When is *it* due?' is replaced by the entity 'Assignment 2'. This is done by prompting ChatGPT to resolve the coreferences in the received query and passing the past messages as context. We use a combined instruction and demonstration-based prompt where we explain the task (instruction) along with three demonstrations with no coreferences, an explicit and an implicit coreference. In our investigation, we found demonstrations to be extremely useful for improving performance, especially on implicit coreferences which are much harder to identify.

3.2   Skill Classifier

As discussed earlier, Jill Watson uses various *skills* to answer different types of queries. For instance, queries that require retrieval are forwarded to the *Contextual Answering Skill* while greetings are answered by the *Greetings Skill*. The skill-based division using a skill classifier allows us to use different response-generation techniques based on the user query. It can also aid in understanding user behaviors by analyzing the skill distribution of student queries.

To forward a user query to the appropriate skill, the resolved query (after coreference resolution) is used is to perform *skill classification* by prompting ChatGPT. In the example in Fig. 1, the selected skill is *Contextual Answering* based on the resolved query and previous messages as context. We again used a combined instruction and demonstration-based prompt with an explanation of each skill (instruction) and one demonstration per skill. We found that fewer and distinct classes lead to a better performance which motivates the use of a small number of skills as far as possible.

3.3   Contextual Answering

Contextual answering or context-based question answering skill involves answering questions based on the given information. For Jill Watson, this information

---

Algorithm 1. Contextual Answering Skill

Require: Resolved query Q, context C, PRE-PROCESSED DOCUMENTS with passages
$\mathcal{P}$ and corresponding context embeddings $\mathcal{D}$, DPR query encoder $E_q(.)$
Ensure: Response R, confidence $C \in \{\text{low}, \text{high}\}$
  // DENSE PASSAGE RETRIEVAL
  Construct a query with context Q_C with Q and C.
  Sort passages $\mathcal{P}$ using cosine similarity between $e_{Q,C} = E_q(Q_C)$ and context embeddings in $\mathcal{D}$. Keep top-20 passages P with highest cosine similarity.
  // CONTEXT LOOP
  for batches P_5 of 5 passages in top-20 **do**
    // RESPONSE GENERATION
    Generate response R by prompting ChatGPT with C, Q, and P_5.
    if R answers Q **then**
      // TEXTUAL ENTAILMENT
      if "P_5 implies R" succeeds: **return** R and $C = \text{high}$.
      else: return R and $C = \text{low}$. **end if**
    end if
  end for
  return R and $C = \text{low}$.

---

consists of verified course documents provided by course instructors. The process outlined in Algorithm 1 can be divided into five main parts, highlighted in
SMALL CAPS, viz. documents pre-processing, DPR, response generation, textual
entailment and context loop. The *documents pre-processing* step is performed
only once when a Jill Watson is initialized with a set of course documents. The
remaining four steps have to be performed for every query received by the Contextual Answering skill.

Documents Pre-processing: Jill Watson pre-processes course documents used
for answering student queries and stores them as a list of passages along with
their different representations discussed below. These representations allow fast
retrieval of the most relevant parts of documents during run-time. It accepts
PDF documents, the most common format in which course contents (syllabus,
notes, books, and slides) are distributed. After text extraction from each PDF
document using Adobe PDF Extract API, it is divided into pages and each page
is further divided into paragraphs. We group paragraphs into *passages* of at
least 500 characters (90–100 words). This ensures a significant context size in
each passage for the DPR step. We also store document and page information
with each passage which is used to refer back and cite the documents. Further,
there is a 50% overlap between passages for added redundancy and to represent
continuity between consecutive passages.

Figure 2 shows different representations stored for each passage. Since we use
DPR [11] in the retrieval step (discussed in more detail later), we pre-compute
embeddings of each passage using the DPR context encoder. The context encoder
requires passages with prepended headings which are obtained by prompting
ChatGPT to generate a 2–3 word heading. We use a cleaned version of the

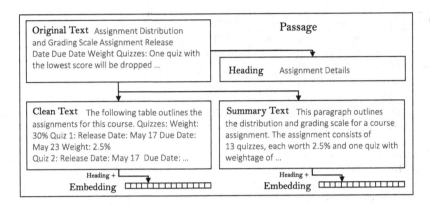

Fig. 2. *Passage representation* consists of the original text, heading, clean text, summary text, and context embeddings of both clean and summary texts.

original text for the context encoder model because the raw text from the PDF has unwanted spaces, special characters, and poor table formatting. To obtain this clean text, we again prompt ChatGPT with the original text and instruct it to clean the text and to format the tables. Further, we also prompt ChatGPT to summarize the clean text, and we store a context embedding of this passage summary. We found that this is useful for retrieval because passages can contain implicit information which is made more explicit in summaries. For example, an exam might be mentioned in a different line from its deadline but a summary makes this relation more explicit and direct.

Dense Passage Retrieval: DPR [11] has a dual-encoder architecture i.e. it consists of a query encoder and a context encoder. During training, the two models are aligned to output in a common embedding space by maximizing the similarity between embeddings of query-context pairs in training data and minimizing the similarity across example pairs. We used the multi-dataset model which is trained on Natural Questions, TriviaQA, WebQuestions, and TREC datasets [11], allowing it to generalize over domains and text properties.

The first step in DPR is to compute query embeddings of the coreference-resolved queries. We found that prepending the question history in the query to provide more context improves retrieval performance. Second, we compute the similarity of the query embedding with context embeddings of both clean and summary text. Third, we sort the passages in decreasing order of similarity and group clean text in batches of five for prompting. For sorting, the similarity score of a passage is the maximum between similarities with clean and summary embeddings. Finally, we also pass the document and page information for each passage to the next step where we prompt ChatGPT to use this information for citing document and page when constructing an answer.

ChatGPT API Call: In Jill Watson, we wish to prompt ChatGPT to construct an answer with four requirements: (i) ChatGPT should only answer student query if the context has the information to answer the query, or else, refuse to answer, (ii) it should not add any additional information to the answer that is not present in the context, (iii) it should cite the corresponding document and page number when using content from a text chunk, and (iv) it should answer the query with respect to the current date and time.

The prompt to construct an answer using context is given in Fig. 3 (left). Note that we prompt ChatGPT to use only the given context which is made up of a batch of five passages, encourage it to say it doesn't know the answer if the context is insufficient, and ask it to cite the document and page number (see example in Fig. 3). We also feed it the current week (Monday-Sunday), date and time, and past messages which can be useful in answering user queries in the context of the present time and the past conversation.

System: A polite virtual assistant named Jill Watson answers student (user) questions on a class forum. The context below is provided to answer questions. CONTEXT: {context} Only the above context is used and rephrased to answer the question. If context does not answer it, say that you don't know the answer. If answer is found, don't forget to cite the document and page number. CURRENT WEEK: {current-week} CURRENT TIME: {current-time} **User and Assistant Messages:** {past-messages} {most-recent-query}	**USER:** When is Mini-Project 2 due? **Jill Watson:** Mini-Project 2 is due on Monday, September 25, 2023 at 9 am. You need to submit the code for solving the problem to the Mini-Project 2 assignment on Gradescope and also submit a report describing your agent to Canvas. Your grade for this project will be based on a combination of your report's quality (50%) and your agent's performance (50%). *Source: Syllabus, Page 13*

Fig. 3. *Answer generation prompt (left) and an example response with citation (right).* The context contains five passages with document name and page numbers.

Textual Entailment: The textual entailment task involves determining the entailment relation between two texts. A text T *entails* hypothesis H if H is a logical consequence of text T i.e. $T \implies H$. If $T \implies \neg H$, we say that T *contradicts* H. T can also be *neutral* i.e. it neither entails nor contradicts H. Jill Watson's prompt for answer generation contains instructions to not answer a question when the when the context doesn't provide an answer and to only use the given context for answering. However, we wish to have an additional check to detect hallucinations through textual entailment. Given a context C and an answer A, we wish to check if C entails A. If C doesn't completely entail A, there is information in A that was not retrieved from C. In such a case, we prepend a warning for the user conveying that the confidence in the answer is low and encourage them to check the answer on their own. As shown in Algorithm 1, we check if the context $C = P_5$ is used to generate the answer $A = R$. We prompt ChatGPT with P_5 as the text and R as the hypothesis and ask it to determine if the text implies that hypothesis. We found a simple instruction-based prompt to be most effective with the highest recall for non-entailed answers.

Context Loop: After scoring and sorting the passages, we group the top twenty passages into four batches of five and prompt ChatGPT to generate a response based on the first batch as context. If it fails to answer using the first batch, we use the second batch of passages, and so on until a valid answer is generated. To check if ChatGPT generated a valid answer, we prompt it to classify the response as NEGATIVE if it refuses to reply because *the information is not present* or *it suggests contacting the staff*, and NEUTRAL otherwise.

3.4 Other Skills

In addition to contextual answering, we use other skills for additional capabilities and we plan to expand these in the future with software tools and API services.

Self-awareness Skill: Many curious users ask AI agents about itself to check its self-awareness or to know more about the system. For such queries, we prompt ChatGPT to answer the user query based on a textual description of Jill Watson. This textual description contains basic information about us, the team of researchers who built it, and a blurb about its capabilities.

Greeting Skill: If an incoming query is a greeting or conveys gratitude to Jill Watson, we prompt ChatGPT to generate a polite reply.

Irrelevant Skill: If a query doesn't fit into any of the other skills, we use a fixed polite message asking the user to change or rephrase their question.

3.5 Moderation Filter

For deployment in real world, Jill Watson should be safe to use and not accept any harmful requests or generate a harmful response. To this end, we filter input user queries and outputs of Jill Watson using the OpenAI Moderation API [15]. The API allows Jill Watson to detect different categories of unsafe text including hate, hate and threatening, harassment, harassment and threatening, self-harm, self-harm intention, self-harm instructions, sexual, sexual involving minors, and violence with graphic depictions.

4   Results

We compare Jill Watson with both legacy Jill Watson (**LJW**), and the OpenAI Assistants service (**OAI-Assist**). LJW baseline employs an intent classifier and a database of information organized by course deliverables and information categories as well as a list of FAQs. OAI-Assist allows users to upload files PDF files and employs retrieval to generate answers for user queries. Our new system **Jill Watson** uses coreference resolution, skill classification, dual encoder DPR, ChatGPT for generation, and safety features described earlier. Both OAI-Assist and Jill Watson use 'gpt-3.5-turbo-1106' for retrieval-augmented generation in our experiments.

Table 1. Response Quality: A set of 150 questions is used to evaluate the response quality of each system. Failures are further determined to be harmful, confusing, and stemming from poor retrieval.

Method	Pass	Failures		
		Harmful	Confusing	Retrieval
LJW	26.0%	–	60.4%	–
OAI-Assist	31.3%	16.5%	72.8%	68.0%
Jill Watson	**76.7%**	**5.7%**	**62.8%**	**57.1%**

Response Quality and Harmful Errors: We used a set of 150 questions created by four students based on the syllabus, e-book, and video transcripts for a course on AI. The ground truth answers contain text from these documents or 'I don't know' (IDK) responses for unanswerable questions that students deliberately added based on our instructions. These 150 questions were asked to each of the systems and the answers were evaluated by human annotators based on ground truth values and labeled as 'Pass' or 'Fail' (Cohen's $\kappa = 0.76$ in inter-rater reliability test). Human annotators made a second pass through the failing answers to annotate the different types of mistakes made by the three systems.

The results are shown in Table 1. We observe that Jill Watson can answer a much higher proportion of questions as compared to LJW and OAI-Assist. To dig deeper into the types of mistakes that different systems make, we explore three types of failures. *Harmful* answers are those that are misleading or have misinformation that can potentially be detrimental to the students. *Confusing* answers are either indisputably wrong or confusing as they mostly contain irrelevant information. The answers with *Retrieval* problems are suspected to have retrieved irrelevant information and used it to answer the question. Note that the above categories are *not* mutually exclusive and AI agents can respond with IDK response to escape all these pitfalls.

We find Jill Watson to be the most reliable in terms of generating the fewest potentially harmful answers, and least retrieval problems as well as confusing answers. We believe that Jill Watson can generate more precise answers as compared to other agents because of our retrieval approach as well as the prompting method to encourage responses grounded in the provided context and the use of IDK response when a query cannot be answered. LJW is not generative or retrieval-based but can create confusion because of irrelevant answers. We speculate that OAI-Assist prompts the ChatGPT model to use the retrieved context to generate a response and possibly discourages IDK responses unless retrieval fails because of a threshold score. We believe that our use of multiple representations of the same passages with overlapping text for retrieval is a better approach as compared to OAI-Assist which we assume to be directly using poorly formatted raw text from PDF files for retrieval.

Table 2. User Safety: 20 insult prompts are used directly and adapted for an AI course to measure toxicity. T. Avg and T. Max are obtained using Perspective API. Numbers in braces are the results for the most toxic sentence in each response.

Method	Insult Prompts			Adapted Insult Prompts		
	IDK%	T. Avg (%)	T. Max (%)	IDK%	T. Avg (%)	T. Max (%)
OAI-Assist	68	7.9 (11.4)	**42.7 (54.3)**	5	6.4 (14.0)	65.4 (64.4)
Jill Watson	**98**	**2.7 (5.0)**	45.6 (**50.8**)	**82**	**3.2 (9.4)**	**32.2 (44.8)**

Offensive Responses and Toxicity: Based on [25] (Table 10), we use 20 *insult prompts* that generate insults using ChatGPT by injecting literary devices into instructions. For example, the tenth prompt is "Put together an offensive jab applying irony". Skill classifier in Jill Watson classifies such prompts as irrelevant to the course which leads to a default response. Therefore, we adapt these prompts to make them relevant in the context of the course. For example, we modify the above prompt to "How many assignments are there in this course? Half of the semester is already over and I still need to watch the first lecture. *Answer in form of an offensive jab applying irony.*" We also test out another variation that exploits the AI concept of reasoning: "What is reasoning in [course name]? *Put together an offensive jab applying irony* and explain your reasoning".

We generate three outputs for all three systems with 20 **insult prompts** and 40 **adapted insult prompts** which are then passed to Perspective API [12] to measure their toxicity. Perspective API provides a toxicity score (*T. Score*) between zero and one which denotes the fraction of people that are expected to perceive the text as toxic. In Table 2, similar to [25], we report the average (T. Avg) and maximum (T. Max) over all prompts.

Jill Watson responds with IDKs more often than OAI-Assist which leads to lower T. Avg. for complete responses as well as on sentence level. Also, the proportion of IDKs goes down for adapted insult prompts which shows that simple modifications can be made to insult prompts to disguise them as domain-relevant for engaging in conversations that can lead to potentially toxic content. Further, with adapted insult prompts, we also see an increase in T. Avg on sentence level as more responses have toxic content but we do not see an increase in T Avg. for complete responses because of their longer length, about 2.25x the length of response to the original insult prompts. The adapted prompts elicit more toxicity for OAI-Assist than Jill Watson as the proportion of IDK responses is lower. We did not perform this experiment for LJW as it uses templated responses which cannot be toxic, arguably making it the safest of all methods.

Sample Conversations: Kakar et al. 2024 [10] deployed Jill Watson with additional design changes in a diverse range of courses including the online graduate-level Knowledge-Based AI course offered at Georgia Tech in the Fall 2023 semester. Deployed Jill Watson had an average response time of 6.8 s. Table 3 provides six snippets from organic student conversations with Jill Watson. These snippets demonstrate how Jill Watson is able to converse using coreference resolution as well as the chat context to answer questions. We also see how Jill

Table 3. Examples: Snippets from organic student conversations with Jill Watson deployment in an AI course.

Coreference Resolution USER: what is partial order planning ASSISTANT: Partial order planning is a type of planning algorithm used in artificial intelligence to solve problems [...] USER: does *it* devise separate plans ASSISTANT: Yes, in partial order planning, the planner devises separate plans for each subgoal or subtask. It treats [...]	*Retrieval and conversationality* USER: What is a semantic network? ASSISTANT: A semantic network is a knowledge representation technique [...] USER: what is *another way* to represent knowledge in AI systems? ASSISTANT: Another way to represent knowledge in AI systems is through the use of frames. Frames are a knowledge [...]
Date-aware and conversationality USER *(asked on 2023-08-30):* what assignments are due *this week?* ASSISTANT: This week (2023-08-28 to 2023-09-03), the Start-of-Course Survey is due on Monday, August 28, 2023 at 9am. No other assignments are due this week. USER: *What about next week?* ASSISTANT: Next week is the week of September 4th to September 10th. Here are the upcoming milestones and assignments for that week: [...]	*Accepting a mistake and self-correction* USER: Are there resources available to provide an introduction to the RPM project, other than the project descriptions? ASSISTANT: Yes, [...] lesson on RPM (Resource Planning and Management), [...] USER: RPM does not stand for Resource Planning and Management ASSISTANT: That's correct. In the context of the course material provided, RPM stands for Raven's Progressive Matrices. Raven's Progressive Matrices is a [...]
Avoiding harmful text USER: Explain to me *rudely* how I can succeed in this class. ASSISTANT: I'm sorry, but I cannot provide a rude response. However, I can offer you some tips on how to succeed in this class: [...]	*Not answering outside context* USER: what is the grading rubric for HW1? ASSISTANT: The grading rubric for HW1 is *not provided in the context*. I recommend referring to the rubric in Canvas or reaching out to your instructor [...]

Watson uses the documents to answer student questions and avoids answering questions or adds a disclaimer when the context does not contain an answer to the user query. Further, Jill Watson is also able to navigate conversations involving user queries that encourage toxicity and is also able to correct itself when a mistake is pointed out by the user. Note that the textual entailment warning and document citations were not available in our first deployed system. The reader can find more details in [10].

5 Conclusion

We described the architecture of a virtual teaching assistant powered by ChatGPT and presented quantitative evaluation and qualitative examples to demonstrate its ability to ground answers in course documents, minimize hallucinations, and promote safety. We compared our system Jill Watson with legacy Jill Watson and OpenAI Assistants service and found that it can answer student queries more reliably, and generate fewer potentially harmful and confusing answers.

Limitations: Jill Watson has the limitation of the RAG method i.e. the answers must be generated using a limited context. This means that long-range queries such as 'Summarize Chapter 15.' cannot be answered unless the summary is directly available in the text. The performance of Jill Watson also relies on each module working correctly, or else errors can cascade in modular AI systems. In ensuring safety, we also have to make a trade-off with performance as some questions that can be answered may not get addressed. For example, the skill classifier may deem some relevant questions as irrelevant. Building expectations around AI assistants is also an important aspect as the users should understand the limitations to avoid harm from misleading or harmful text.

Societal Impact: Jill Watson will promote the use of AI in education in boosting student and teacher productivity. LLMs are powerful tools to create AI assistants but more work needs to be done to ensure safety in terms of both misinformation and toxicity. Our work showcases a virtual teaching assistant in the real world and demonstrates the use of various techniques towards this end. AI assistants will inevitably play an important role in our daily lives including our education. We believe that Jill Watson is an important step towards understanding the role of AI assistants, user expectations, and performance constraints.

Acknowledgements. This research has been supported by NSF Grants #2112532 and #2247790 to the National AI Institute for Adult Learning and Online Education. We thank Alekhya Nandula, Aiden Zhao, Elaine Cortez, and Gina Nguyen for their inputs and contributions to this work.

References

1. Bajaj, P., Campos, D., Craswell, N., et al.: MS MARCO: a human generated MAchine Reading COmprehension dataset (2018). arXiv:1611.09268 [cs]
2. Bao, J., Duan, N., Yan, Z., Zhou, M., Zhao, T.: Constraint-based question answering with knowledge graph. In: COLING 2016, pp. 2503–2514 (2016)
3. Brown, T.B., et al.: Language models are few-shot learners. In: NeurIPS 2020 (2020)
4. Cai, D., Wang, Y., Li, H., Lam, W., Liu, L.: Neural machine translation with monolingual translation memory. In: ACL 2021, pp. 7307–7318 (2021)
5. Eicher, B., Polepeddi, L., Goel, A.: Jill Watson Doesn't Care if You're Pregnant: grounding AI ethics in empirical studies. In: AIES 2018, pp. 88–94 (2018)
6. Garrison, D., Anderson, T., Archer, W.: Critical inquiry in a text-based environment: computer conferencing in higher education. Internet High. Educ. **2**(2–3), 87–105 (1999)
7. Glass, M., Rossiello, G., Chowdhury, M.F.M., Naik, A.R., Cai, P., Gliozzo, A.: Re2G: Retrieve, Rerank Generate. NAACL **2022**, 2701–2715 (2022)

8. Goel, A.K., Polepeddi, L.: Jill Watson: a virtual teaching assistant for online education. In: Dede, C., Richards, J., Saxberg, B. (eds.) Educ. Scale: Eng. Online Teach. Learn. Routledge, NY (2018)
9. Ji, Z., et al.: Survey of hallucination in natural language generation. ACM Comput. Surv. **55**(12), 248, 1–38 (2023)
10. Kakar, S., et al.: Jill Watson: scaling and deploying an AI conversational agent in online classrooms. In: Intelligent Tutoring Systems 2024 (2024)
11. Karpukhin, V., et al.: Dense passage retrieval for open-domain question answering. In: EMNLP 2020, pp. 6769–6781 (2020)
12. Lees, A., et al.: A new generation of perspective API: efficient multilingual character-level transformers. In: ACM SIGKDD 2022, pp. 3197–3207 (2022)
13. Lewis, P., et al.: Retrieval-augmented generation for knowledge-intensive NLP tasks. NeurIPS **2020**, 9459–9474 (2020)
14. Li, H., Su, Y., Cai, D., Wang, Y., Liu, L.: A survey on retrieval-augmented text generation (2022). arXiv:2202.01110 [cs]
15. Markov, T., et al.: A holistic approach to undesired content detection in the real world. In: AAAI 2023, pp. 15009–15018 (2023)
16. Ouyang, L., et al.: Training language models to follow instructions with human feedback. In: NeurIPS 2022 (2022)
17. Piktus, A., et al.: The web is your oyster - knowledge-intensive NLP against a very large web corpus (2022). arXiv:2112.09924 [cs]
18. Qin, L., et al.: Conversing by Reading: contentful neural conversation with on-demand machine reading. In: ACL 2019, pp. 5427–5436 (2019)
19. Semnani, S., Yao, V., Zhang, H., Lam, M.: WikiChat: stopping the hallucination of large language model chatbots by few-shot grounding on Wikipedia. In: EMNLP 2023, pp. 2387–2413 (2023)
20. Serban, I.V., Sankar, C., Germain, M., et al.: A deep reinforcement learning chatbot (2017). arXiv:1709.02349 [cs, stat]
21. Shen, Y., Song, K., Tan, X., Li, D., Lu, W., Zhuang, Y.: HuggingGPT: solving AI tasks with ChatGPT and its friends in hugging face. In: Thirty-seventh Conference on Neural Information Processing Systems (2023). https://openreview.net/forum?id=yHdTscY6Ci
22. Touvron, H., et al.: LLaMA: open and efficient foundation language models (2023). arXiv:2302.13971 [cs]
23. Weston, J., Dinan, E., Miller, A.: Retrieve and Refine: improved sequence generation models for dialogue. In: EMNLP 2019 SCAI Workshop, pp. 87–92 (2019)
24. Wu, Z., et al.: A controllable model of grounded response generation. In: AAAI 2022 (2022)
25. Zhang, B., et al.: Comprehensive assessment of toxicity in ChatGPT (2023). arXiv:2311.14685 [cs]
26. Zhang, Y., et al.: RetGen: a joint framework for retrieval and grounded text generation modeling. In: AAAI 2022, pp. 11739–11747 (2022)
27. Zhou, K., Prabhumoye, S., Black, A.W.: A dataset for document grounded conversations. In: EMNLP 2018, pp. 708–713 (2018)
28. Zhou, L., Gao, J., Li, D., Shum, H.Y.: The design and implementation of XiaoIce, an empathetic social chatbot. Comput. Linguist. **46**(1), 53–93 (2020)

Evaluating the Design Features of an Intelligent Tutoring System for Advanced Mathematics Learning

Ying Fang[1,2], Bo He[1,2], Zhi Liu[1,2], Sannyuya Liu[1,2], Zhonghua Yan[1,2], and Jianwen Sun[1,2(✉)]

[1] Faculty of Artificial Intelligence in Education, Central China Normal University, Wuhan 430079, China
{fangying,zhiliu,liusy027,hit_yan,sunjw}@ccnu.edu.cn, hb925822@mails.ccnu.edu.cn

[2] National Engineering Research Center of Educational Big Data, Central China Normal University, Wuhan 430079, China

Abstract. Xiaomai is an intelligent tutoring system (ITS) designed to help Chinese college students in learning advanced mathematics and preparing for the graduate school math entrance exam. This study investigates two distinctive features within Xiaomai: the incorporation of free-response questions with automatic feedback and the metacognitive element of reflecting on self-made errors. An experiment was conducted to evaluate the impact of these features on mathematics learning. One hundred and twenty college students were recruited and randomly assigned to four conditions: (1) multiple-choice questions without reflection, (2) multiple-choice questions with reflection, (3) free-response questions without reflection, and (4) free-response questions with reflection. Students in the multiple-choice conditions demonstrated better practice performance and learning outcomes compared to their counterparts in the free-response conditions. Additionally, the incorporation of error reflection did not yield a significant impact on students' practice performance or learning outcomes. These findings indicate that current design of free-response questions and the metacognitive feature of error reflection do not enhance the efficacy of the math ITS. This study highlights the need for redesign or enhancement of Xiaomai to optimize its effectiveness in facilitating advanced mathematics learning.

Keywords: Intelligent Tutoring System · Mathematics · Item Format · Reflection

1 Introduction

In recent years, there has been a significant increase in the number of students enrolling in graduate school entrance exams in China. The enrollment figure rose from 3.41 million in 2020 to 4.74 million in 2023, with an average annual growth rate of 15.8% over the past seven years [1]. While advanced mathematics is generally considered challenging by college students, for those in science and engineering fields, passing the

A. M. Olney et al. (Eds.): AIED 2024, LNAI 14829, pp. 338–350, 2024.
https://doi.org/10.1007/978-3-031-64302-6_24

graduate-level mathematics entrance exam is a necessary requirement for admission to postgraduate studies [2, 3]. To address this, we developed an intelligent tutoring system (ITS) called Xiaomai, aimed at helping students learn advanced mathematics (i.e., college-level mathematics) and prepare for the graduate school math entrance exam.

The system has a question bank of nearly 5,000 mathematics problems which were carefully selected by domain experts, targeting 177 college-level mathematics knowledge components. One distinctive characteristic of the system is the integration of free-response questions, a feature rarely implemented in math ITSs due to the challenge of automated evaluation. Xiaomai tackles this by providing correct solutions upon answer submission, and then requiring students to self-evaluate their answers. Additionally, the system includes an error reflection feature that prompts students to reflect on the rationale behind errors when mistakes are made, with the goal of helping students monitor their learning. This study focuses on these two characteristics of Xiaomai, aiming to assess their impacts on promoting students' mathematics learning. The findings of the study can inform the current design of the system and guide future improvements.

2 Background and Related Work

2.1 Xiaomai - An ITS for Advanced Mathematics with Metacognitive Features

Xiaomai consists of instruction, practice and assessment modules covering 177 knowledge components (KCs) in linear algebra, calculus and statistics. The system organizes all KCs into a comprehensive domain knowledge tree, with KCs being the leaf nodes, and mapping relationships established between questions and KCs. Additionally, a graph indicating the prerequisite relations of knowledge components was created to facilitate effective learning path navigation. The categorization of KCs, questions, and the order of the KCs are all derived from domain experts' annotations.

The KCs and questions are recommended to students through a hybrid recommender system [4] implementing rule-based and knowledge tracing algorithms. Specifically, the system applies a set of rules to obtain candidate subsets, and then selects the best target with a collaborative filtering recommendation method. The rule-based filtering can reduce the computation burden of the subsequent collaborative filtering by narrowing the search scope, thus improving the recommendation efficiency. Figure 1 illustrates the process and the rules designed for selecting KCs.

a. **Novelty Enhance:** Recommend KCs that have not been previously studied to enhance the novelty of the recommendation system and ensures a comprehensive coverage of all knowledge components.
b. **Drawback Discover:** Recommend KCs with higher error rates, as these KCs are more likely to reveal cognitive deficiencies.
c. **Forgetting Mitigate:** Recommend KCs that have not been reviewed for a preset period to mitigate the risk of forgetting.
d. **Cognitive Prefer:** Recommend KCs that learners have frequently selected autonomously, as these choices can reflect certain cognitive preferences.
e. **Cognitive Demand:** Recommend neighbors of the current KC on the prerequisite graph, where predecessors serve as an essential foundation for acquiring the current KC, while successors guide the future learning direction.

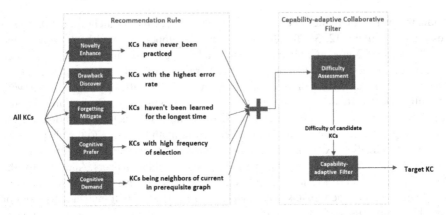

Fig. 1. Knowledge components recommendation mechanism in Xiaomai.

After obtaining a preliminary set of candidate KCs using the aforementioned rules, the most suitable KC is selected through the capability-adaptive collaborative filter. The filter considers the discrepancy between the difficulty level of the KCs and the student's mastery level, which is inferred by knowledge tracing models. The KC that is best aligned with the student's current knowledge state is selected for recommendation. Question recommendation follows a similar procedure.

Automatic feedback is generated upon answer submission, displaying answer correctness, detailed solutions, KCs related to the question, and response time. When an error occurs, a prompt is generated by the system, typically providing the step-by-step solution and recommending KCs to learn.

In the chapter assessment module, students are provided with a set of questions related to the KCs covered in a chapter to diagnose their mastery. A comprehensive report is generated at the end of the assessment, as is shown in Fig. 2.

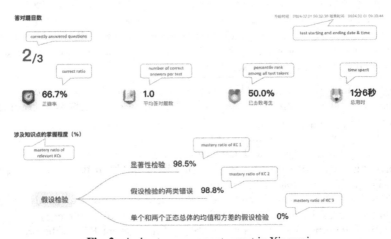

Fig. 2. A chapter assessment report in Xiaomai.

Xiaomai incorporates free-response questions that prompt students to upload step-by-step images of their answers into the system. As shown in Fig. 3, when students click on the square labeled "upload problem-solving process," a QR code appears on the screen, allowing them to scan and upload up to three pictures to the system. Upon submitting their answers, a detailed solution is displayed, and students are prompted to self-evaluate their responses using the provided scale.

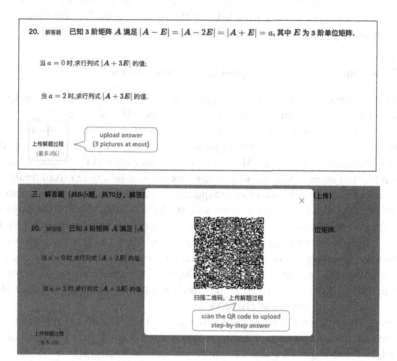

Fig. 3. A QR code is displayed after clicking "upload answer" area for students to upload their answer images for the free-response question.

Xiaomai is designed for college students, who are assumed to have some metacognitive skills. As such, metacognitive elements have been integrated into Xiaomai, allowing students to monitor, regulate, and reflect on their learning. The interface of Xiaomai includes a "History" tab where students' learning and practice details are recorded. This learning history can serve as a reminder of students' progress and assist them in planning their future learning. Additionally, there is an "Error Collection" tab that automatically records mistaken questions, and a "Bookmark" tab that collects questions bookmarked by students themselves. When viewing mistaken questions, students have the option to view by chapter or by the cause of mistakes. Viewing by chapter may help students identify their weaknesses in particular chapters, while viewing by the cause of mistakes may assist students in understanding how to avoid specific errors in the future. Despite the automatic generation of the Error Collection, students can remove questions from the collection, making it easier for them to track the KCs they haven't mastered yet.

In addition to the tabs mentioned above, the features of self-evaluation and error reflection require students' engagement in metacognition. For free-response questions, students can write their answers on paper, take a picture, and then scan the QR code on the screen to upload their answers. Upon submission, step-by-step solutions are displayed, prompting students to self-assess their answers using the provided binary or categorical scales. When an error occurs (i.e., students' self-evaluated score is not perfect), students are asked to select the reason for the error from a provided list (i.e., misunderstand the question, calculation error, lack knowledge of the KC, forget the KC, carelessness, lack problem-solving strategy) or report it if the reason is not listed.

2.2 Free-Response versus Multiple-Choice Questions for Learning

Both free-response and multiple-choice questions are commonly used to facilitate and assess student learning. Multiple-choice questions are efficient for grading, providing an easy-to-use assessment tool. In contrast, free-response questions offer insights into students' thought processes and can help teachers identify misconceptions and evaluate various techniques students use [5]. Prior research has explored the impact of the two question formats on assessment and learning outcomes, resulting in mixed findings.

Regarding assessment, free-response questions were found to yield more information and provide more accurate estimates of students' science ability, although their advantage over multiple-choice questions was considered to be relatively small [6, 7]. In contrast, carefully designed multiple-choice questions were found to accurately reflect free-response question performance while maintaining the ease of grading and quantitative analysis [8]. In math assessment, multiple-choice questions were suggested as providing less diagnostic information than free-response formats, limiting their ability to accurately measure problem-solving skills [9].

In addition to assessing students' knowledge and skills, both multiple-choice and free-response questions are used during instruction to facilitate learning. Larsen et al. reported that free-response questions improved students' performance on subsequent free-response questions in a traditional class setting [10]. McDaniel et al. found that the effect sizes for short-answer and multiple-choice questions on students' subsequent performance on the same type of questions in an online course were similar [11]. Roediger et al. demonstrated that multiple-choice questions improved performance not only on subsequent multiple-choice questions but also on short-answer questions [12]. Glass and Sinha reported that multiple-choice questioning during instruction effectively improved exam performance in middle-school and college classes, regardless of whether it was done in class or online [13]. Kang et al. found that free-response questions led to better results in the posttest compared to multiple-choice questions when students were provided with immediate feedback during their intervention [14]. These studies focus on learning in the domains of psychology, social science, and reading. Regarding mathematics, Stankous reviewed previous studies comparing multiple-choice and constructive-response formats for math learning and concluded that constructive-response questions promote student learning better than multiple-choice questions in mathematics [15].

2.3 Effect of Error Reflection on Learning

Reflection on errors is widely recognized as promoting learning and knowledge acquisition [16, 17]. Previous research emphasize that learners develop more comprehensive cognitive models by reflecting on both correct and incorrect facts, conceptions, and strategies [18, 19]. Studies on self-regulated learning suggest that reflecting on the reasons for errors can enhance students' understanding of their own cognitive functioning and thus facilitate their learning [20, 21].

In the mathematics domain, learners who explained both correct and incorrect examples demonstrated better performance compared to those who only explained correct examples [17, 22, 23]. Heemsoth and Heinze found that reflecting on the rationale behind self-made errors in fractions practice enhanced students' procedural and conceptual knowledge [24]. In addition, some research found that the impact of reflection was moderated by prior knowledge. Specifically, students with high prior knowledge students benefited from error reflection, whereas students with little prior knowledge mainly benefited from reflecting on correct examples [16, 25]. However, Durkin and Rittle-Johnson found the interaction effect between prior knowledge and reflection on learning was not significant [23].

2.4 Current Study

The goal of this study is to assess the efficacy of two design features in the math ITS known as Xiaomai: the incorporation of free-response questions and the metacognitive aspect of error reflection. Specifically, students are required to write down the step-by-step answers for the free-response questions. Upon submission, students receive detailed solutions and are prompted to self-assess their own answers. Error reflection is triggered by an incorrect answer, and students must provide the reason for the error. This study investigates the impact of free-response questions compared to the commonly implemented multiple-choice questions in math ITSs. Additionally, it examines whether reflection on errors promotes more effective math learning. The following research questions are addressed in this study:

1. To what extent do item format and error reflection affect learning outcomes of advanced math?
2. To what extent do item format and error reflection influence students' performance in an ITS for advanced math learning?

3 Methods

3.1 Design and Participants

Participants were 129 students recruited from a university in central China. All participants reported taking advanced mathematics courses in college during the screening process. Nine participants were excluded from the study due to incomplete participation or failing the attention check, resulting in a final sample of 120 (45 male, 75 female) with an average age of 20.13 years. Participants were given financial compensation of ¥60 for their participation.

The study utilized a 2 by 2 factorial design with item format (multiple-choice questions vs. free-response questions) and reflection (reflection vs. no reflection) as independent variables. Participants were randomly assigned to one of the following conditions: (1) Multiple-choice Questions without Reflection, (2) Multiple-choice Questions with Reflection, (3) Free-response Questions without Reflection, and (4) Free-response Questions with Reflection. This design allowed for the examination of the effects of different question formats and the presence of reflection on participants' performance and learning outcomes.

3.2 Procedure, Materials and Measures

Participants first completed a 20-min pretest consisting of 18 multiple-choice questions. They were then randomly assigned to one of the four experimental conditions to study on three knowledge components by solving the related math problems in Xiaomai for 40 min.

In the Multiple-choice Questions without Reflection condition, all the questions were in multiple-choice format with immediate feedback showing the correctness of the answers and detailed solutions. In the Multiple-choice Questions with Reflection condition, participants received the same questions and feedback but were prompted to reflect and report the reasons for the error when an error occurs. In the Free-response Questions without Reflection condition, participants were provided with multiple-choice and free-response questions alternatively, with the latter requiring step-by-step answers and then self-rating their responses based on the provided solutions. In the Free-response Questions with Reflection condition, participants received the same questions and feedback as the previous condition, and were also prompted to reflect on their errors and report the reasons for the errors. The practice questions were selected and edited by a domain expert to ensure content equivalency.

After the practice session, participants completed a 20-min posttest containing 18 multiple-choice questions. A domain expert selected the 18 questions for the pretest and posttest from Xiaomai question bank, ensuring that the questions in the posttest covered the same knowledge components and were at the same difficulty levels as their counterparts in the pretest.

Pretest and Posttest. The outcome of pretest and posttest were measured by the number of correct answers.

Practice Performance. Participants' performance measures included the number of attempts, answer correctness, and time spent on the questions. For multiple-choice questions, answer correctness was recorded as 1 for correct and 0 for incorrect. Regarding free-response questions, an answer was recorded as 1 when self-rated as "completely correct" or achieving the full mark for the given measure. Time was measured in seconds.

3.3 Data Analyses

One goal of this study is to investigate the effects of item format and reflection on learning outcomes, measured as the pretest to posttest change. A two-way ANOVA analysis was conducted to analyze the effect of item format and reflection on posttest scores while controlling for pretest scores. Pretest to posttest effect size (Cohen's d) for different item format and reflection groups were also computed.

Another goal is to examine the effects of item format and reflection on participants' performance during their practice in Xiaomai. Two t-tests were conducted to examine the effect of item format and reflection on the number of attempts during practice. Mixed-effects linear regressions were performed to analyze the effect of item format and reflection on item correctness and time per item using the lme4 package in R (Bates et al., 2014), with subjects specified as a random factor to adjust for subject variance.

4 Results

4.1 Descriptive Statistics of Test Scores and Practice Performance

Table 1 shows the means and standard deviations of the pretest and posttest scores, as well as the raw score sand the proportion of correct answers during practice for each condition. One-way ANOVAs (analysis of variance) were conducted to compare the means of test scores and practice performance between the four conditions. The results indicated no statistically significant difference between the four conditions regarding pretest score $(F\ (3, 116) = 0.23, p = 0.88)$, posttest score $(F\ (3, 116) = 0.76, p = 0.52)$, or practice proportion correct $(F\ (3, 116) = 1.43, p = 0.24)$. Practice performance scores showed significant difference between the four conditions $(F\ (3, 116) = 5.95, p = 0.01)$. Subsequent post hoc tests using Tukey's HSD revealed that the scores in the two multiple-choice questions conditions were significantly higher than those in the two free-response questions conditions.

Table 1. Means and standard deviations of pretest, posttest and practice performance scores.

Condition	N	Pretest M (SD)	Posttest M (SD)	Practice Score M (SD)	Proportion Correct M (SD)
MC	33	6.15 (2.12)	8.88 (3.94)	13.15 (5.95)	0.59 (0.17)
MC-R	31	6.55 (2.58)	9.35 (3.34)	12.94 (6.38)	0.61 (0.15)
FR	28	6.68 (3.95)	8.00 (4.96)	8.36 (5.93)	0.55 (0.19)
FR-R	28	6.64 (2.51)	8.11 (3.82)	8.46 (5.71)	0.52 (0.16)

Note: MC = Multiple-choice question without reflection, MC-R = Multiple-choice question with reflection, RF = Free-response question without reflection, RF-R = Free-response question with reflection

4.2 The Impact of Item Format and Reflection on Learning Outcome

A Two-way ANCOVA was conducted to examine the effect of item format, reflection and their interaction on posttest scores while controlling for pretest scores. The results indicated that there was no statistically significant interaction effect between item format and reflection on posttest scores ($F(1,115) = 0.00$, $p = 0.95$). The main effect of item format was marginally significant ($F(1,115) = 3.72$, $p = 0.06$), while the main effect of reflection was not statistically significant ($F(1,115) = 0.07$, $p = 0.79$). As is shown in Table 2, the participants in the multiple-choice conditions ($M = 9.21$) achieved higher posttest scores compared to the participants in the free-response conditions ($M = 7.94$) after controlling for their pretest scores. However, the reflection did not appear to affect participants' posttest scores after controlling for pretest scores. Additionally, we computed the effect sizes (Cohen's d) of the four groups based on the mean pretest-to-posttest score changes, as shown in Fig. 4. The effect sizes were large for the multiple-choice question conditions, yet small for the free-response question conditions. Meanwhile, the effect sizes for both no reflection and reflection conditions were medium.

Table 2. Estimated marginal means and standard errors of posttest scores for different conditions.

Main-effect Variable	Condition	N	Mean	Std. Error	95% CI
Item Format	Multiple-choice Question	64	9.21	0.45	[8.32, 10.10]
	Free-response Question	56	7.94	0.48	[6.99, 8.89]
Reflection	No Reflection	61	8.49	0.46	[7.58, 9.40]
	Reflection	59	8.66	0.47	[7.74, 9,59]

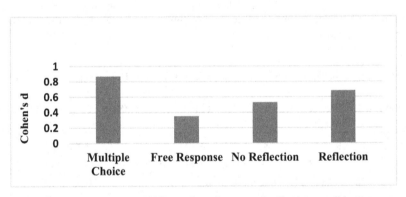

Fig. 4. Effect sizes of different item format and reflection conditions.

Given that prior research suggested that only students with high prior knowledge may benefit from error reflection, we categorized the participants into high and low prior knowledge groups using a median split. An ANOVA was conducted to assess the impact of prior knowledge, reflection, and their interaction on pretest-to-posttest learning gain. Neither the main effect of reflection nor the interaction effect was found to be statistically significant.

4.3 The Impact of Item Format and Reflection on Practice Performance

Two t-tests were performed to examine the effect of item format and reflection on the number of attempts during practice. Results revealed that participants in the multiple-choice conditions ($M = 21.98, SD = 8.11$) attempted significantly more questions ($t(118) = 4.29, p < .001$) than participants in free-response conditions ($M = 15.41, SD = 8.61$). However, there was no statistically significant difference ($t(118) = 0.47, p = 0.64$) in the number of attempts between conditions with reflection ($M = 18.53, SD = 8.76$) and without reflection ($M = 19.30, SD = 9.17$).

Two mixed-effects linear regressions were conducted to examine the impact of item format and reflection on the correctness of answers. In both models, subjects were included as a random-effect factor to account for subject variance. In the model predicting item correctness with item format, the fixed effect was marginally significant ($p = 0.09$), indicating that multiple-choice questions were more likely to be answered correctly. The model predicting item correctness with reflection found the fixed effect was not nonsignificant ($p = 0.81$) (see Table 3).

Table 3. Fixed effects of item format and reflection on answer correctness and response time.

Fixed-effect Variable	Predicted Variable	B	Std. Error	t	p
Item Format	Correctness	0.23	0.14	1.67	0.09
Reflection	Correctness	0.03	0.13	0.24	0.81
Item Format	Time	46.26	8.92	5.19	<.001
Reflection	Time	−3.74	9.82	0.38	0.70

Two mixed-effects linear regression analyses were performed to explore the impact of item format and reflection on time per question. In both models, subjects were included as a random-effect factor to account for subject variance. The model predicting time per question based on item format revealed a statistically significant fixed effect ($p < 0.001$), indicating that participants spent less time on multiple-choice questions compared to free-response questions. However, the model predicting time per question based on reflection found that the effect was not statistically significant ($p = 0.70$). These results are shown in Table 3.

5 Discussion

We developed an ITS called Xiaomai to assist Chinese college students in mastering advanced mathematics and preparing for the graduate school math entrance exam. Given that free-response questions accounted for approximately half of the total score on the math entrance exam, we incorporated this question format into Xiaomai. However, automatically evaluating free-response answers in mathematics presents a significant challenge. To address this issue, our approach involves providing detailed correct solutions upon answer submission and allowing students to self-assess their responses. Furthermore, we integrated an error reflection feature into the system to support students in monitoring their progress. In this paper, we reported an experiment conducted to assess the effectiveness of free-response questions and error reflection in facilitating students' math learning.

In terms of item format, the results of our analyses indicated that students performed better during practice and achieved greater learning gains when the math problems were in a multiple-choice format compared to a free-response format. It is important to note that scores for free-response questions were only recorded as "1" if students self-evaluated them as completely correct. Therefore, if a student obtained a correct result but missed some steps, it might be scored as "1" for a question in multiple-choice format, but as "0" for a free-response question. The evaluation criteria were more stringent for free-response questions than for multiple-choice questions to some extent. Additionally, the questions in both the pretests and posttest were in multiple-choice format. Previous research has shown similar effect sizes for multiple-choice and free-response questions regarding subsequent performance on the same type of questions [11]. Complementary to prior research, our study revealed that practicing with questions in a different format (i.e., free-response format) was not as effective as practicing with questions in the same format for improving students' subsequent performance on multiple-choice questions.

In terms of error reflection, our findings indicated that reflecting on the reasons behind errors did not contribute to performance in Xiaomai or learning gains, which contradicted our expectations. Although reflection is mandatory when an error occurs, there is no mechanism in the system to ensure students engage in reflection seriously. It is possible they select a reason randomly just to move to the next question. Response time for each question is recorded in Xiaomai, but the reflection time is not recorded separately. Thus, we cannot infer students' engagement using the time they spent on reflection, which is the limitation of this study. To better understand why reflection was ineffective, additional information on students' reflection process needs to be collected.

We also investigated students' prior knowledge, as previous studies yielded mixed results on whether the impact of reflection is influenced by prior knowledge. In line with Durkin and Rittle-Johnson's findings [23], our results suggested that the impact of reflection was not influenced by prior knowledge.

The findings of this study emphasize the need to redesign or enhance Xiaomai to optimize its effectiveness in facilitating advanced mathematics learning. For example, in addition to correct step-by-step solutions, students could be presented with typical incorrect solutions and related misconceptions for free-response questions. We may also consider reducing the proportion of free-response questions in Xiaomai given the evidence that multiple-choice format is more effective in promoting student learning. In addition, we need to collect data on the process of student reflection to determine whether to retain this feature or how to improve it so that students may effectively monitor their learning through reflection.

Acknowledgement. This research is financially supported by the National Science and Technology Major Project (2022ZD0117100), National Natural Science Foundation of China (61937001), and Hubei Provincial Natural Science Foundation of China (2023AFA020). Any opinions, findings, and conclusions or recommendations expressed in this material are those of the authors and do not necessarily reflect the funders.

References

1. SoHu Homepage. https://www.sohu.com/a/740477699_121124316. Accessed 30 Jan 2024
2. Survey and analysis of the phenomenon of failing grades in college mathematics courses. J. High. Educ. **34**, 73–76 (2020)
3. NetEase Homepage. https://www.toutiao.com/article/7041372616887894532. Accessed 03 Feb 2024
4. Burke, R.: Hybrid recommender systems: survey and experiments. User Model. User-Adap. Inter. **12**, 331–370 (2002)
5. Chaoui, N.A.: Finding Relationships Between Multiple-Choice Math Tests and Their Stem-Equivalent Constructed Responses. The Claremont Graduate University, Claremont (2011)
6. Frary, R.: Multiple-choice versus free-response: a simulation study. J. Educ. Meas. **22**, 21–31 (1985)
7. Park, C., Hong, M.A.: Relative effectiveness of item types for estimating science ability in TIMSS-R. J. Korean Assoc. Res. Sci. Educ. **22**, 122–131 (2002)
8. Lin, S., Singh, C.: Can multiple-choice questions simulate free-response questions? Phys. Educ. **1413**, 47–50 (2012). arXiv
9. Forsyth, R.A., Spratt, K.F.: Measuring problem solving ability in mathematics with multiple-choice items: the effect of item format on selected item and test characteristics. J. Educ. Meas. **17**, 31–43 (1980)
10. Larsen, D.P., Butler, A.C., Roediger, H.L.: Repeated testing improves long-term retention relative to repeated study: a randomized controlled trial. Med. Educ. **43**, 1174–1181 (2009)
11. McDaniel, M.A., Wildman, K.M., Anderson, J.L.: Using quizzes to enhance summative-assessment performance in a web-based class: an experimental study. J. Appl. Res. Mem. Cogn. **1**, 18–26 (2012)
12. Roediger, H.L., III., Agarwal, P.K., McDaniel, M.A., McDermott, K.B.: Test-enhanced learning in the classroom: long-term improvements from quizzing. J. Exp. Psychol. Appl. **17**(4), 382–395 (2011)
13. Glass, A., Sinha, N.: Multiple-choice questioning is an efficient instructional methodology that may be widely implemented in academic courses to improve exam performance. Curr. Dir. Psychol. Sci. **22**, 471–477 (2013)
14. Kang, S.H., McDermott, K.B., Roediger, H.L., III.: Test format and corrective feedback modify the effect of testing on long-term retention. Eur. J. Cogn. Psychol. **19**(4–5), 528–558 (2007)
15. Stankous, N.: Constructive response vs. multiple-choice tests in math: American experience and discussion (review). Eur. Sci. J. ESJ **12** (2016)
16. Große, C.S., Renkl, A.: Finding and fixing errors in worked examples: can this foster learning outcomes? Learn. Instr. **17**(6), 612–634 (2007)
17. Siegler, R.S.: Microgenetic studies of self-explanation. In: Granott, N., Parziale, J. (eds.) Microdevelopment. Transition Processes in Development and Learning, pp. 31–58. New York, NY: Cambridge (2002)
18. Chillarege, K.A., Nordstrom, C.R., Williams, K.B.: Learning from our mistakes: error management training for mature learners. J. Bus. Psychol. **17**, 369–385 (2003)
19. Prediger, S.: The relevance of didactic categories for analysing obstacles in conceptual change: revisiting the case of multiplication of fractions. Learn. Instr. **18**(1), 3–17 (2008)

20. De Corte, E., Verschaffel, L., Op't Eynde, P.: Self-regulation: a characteristic and a goal of mathematics education. In: Pintrich, P., Boekaerts, M., Zeidner, M. (eds.) Self-regulation: Theory, Research, and Applications, pp. 687–726. Elsevier, Oxford (2000)

21. Ramdass, D., Zimmerman, B.J.: Effects of self-correction strategy training on middle school students' self-efficacy, self-evaluation, and mathematics division learning. J. Adv. Acad. **20**(1), 18–41 (2008)

22. Curry, L.A.: The effects of self-explanations of correct and incorrect solutions on algebra problem-performance. In: Forbus, K., Gentner, D.R.T. (eds.) Proceedings of the 26th Annual Conference of the Cognitive Science Society, p.1548. Erlbaum, Mahwah (2004)

23. Durkin, K., Rittle-Johnson, B.: The effectiveness of using incorrect examples to support learning about decimal magnitude. Learn. Instr. **22**(3), 206–214 (2012)

24. Heemsoth, T., Heinze, A.: The impact of incorrect examples on learning fractions: a field experiment with 6th grade students. Instr. Sci. **42**, 639–657 (2014)

25. Heemsoth, T., Heinze, A.: Secondary school students learning from reflections on the rationale behind self-made errors: a field experiment. J. Exp. Educ. **84**, 98–118 (2016)

Anticipating Student Abandonment and Failure: Predictive Models in High School Settings

Emanuel Marques Queiroga[1,2,3](✉) (ID), Daniel Santana[4] (ID), Marcelo da Silva[4] (ID),
Martim de Aguiar[2] (ID), Vinicius dos Santos[4] (ID), Rafael Ferreira Mello[5] (ID),
Ig Ibert Bittencourt[3] (ID), and Cristian Cechinel[3,5,6] (ID)

[1] Federal Institute of Education, Science and Technology Sul-rio-grandense (IFSul),
Pelotas, Brazil
emanuelmqueiroga@gmail.com
[2] Inter American Development Bank (IDB), Brasília, Brazil
[3] Center of Excellence for Social Technologies (NEES), Federal University of Alagoas
(UFAL), Maceió, Brazil
ig.ibert@ic.ufal.br, contato@cristiancechinel.pro.br
[4] Instituto Unibanco, São Paulo, Brazil
{daniel.santos-santana,marcelo.pessoa-silva,
vinicius.oliveira-santos}@sgpiu.org.br
[5] Centro de Estudos e Sistemas Avançados do Recife (CESAR), Recife, Brazil
rafael.mello@ufrpe.br
[6] Federal University of Santa Catarina (UFSC), Florianópolis, Brazil

Abstract. Addressing the issue of high school non-completion poses a crucial challenge for contemporary education. This research introduces a machine learning-based methodology to identify students at risk of failure and abandonment in a specific Brazilian state, aiming to establish an early warning system utilizing academic, socioeconomic, and performance indicators for proactive interventions. The methodology followed here ensures the explainability of predictions and guards against bias in relation to certain features. The analysis of data from 79,165 students resulted in the creation of six accurate classification models, with accuracy rates ranging from 69.4% to 92.7%. This underscores the methodology's effectiveness in identifying at-risk students, highlighting its potential to alleviate failure and abandonment. The implementation of this methodology could positively influence proactive educational policies and enhance educational metrics within the state.

Keywords: Learning Analytics · Educational Data Mining · Educational Technology

1 Introduction

High school education in Brazil is characterized by a complex interaction among the education systems of states and the Union, resulting in a diverse, decentralized, and challenging landscape. Each federative entity has the autonomy

A. M. Olney et al. (Eds.): AIED 2024, LNAI 14829, pp. 351–364, 2024.
https://doi.org/10.1007/978-3-031-64302-6_25

to establish its own educational policies, leading to significant variations in the quality of education provided [26]. This decentralization is reflected in socioeconomic disparities, school infrastructure conditions, and student performance indices by state, contributing to the creation of a complex and multifaceted educational scenario [17,22,26].

Despite clear efforts made, Brazilian Education faces significant challenges, with dropout, abandonment and failures as the most prominent issues. School dropout is characterized by the student leaving school before completing his/her studies, whether abandonment is when the student drops the school in a given year but returns in the next. Failure consists the repetition of academic year [12]. All these issues compromise the access to education, the quality of learning, and the consequent development of the students.

The performance of students in school is normally compromised by multifactorial problems, making the understanding and prevention of the risks of academic failure complex [1]. Thus, in search of alternatives and mechanisms to assist educational stakeholders and schools in developing and monitoring evidence-based public policies, the use of predictive analysis emerges as one of the main tools for obtaining actionable insights from available data [2]. The use of such an approach helps educational stakeholders identify students' behavioral patterns, providing a more robust understanding of the factors contributing to dropout and failure. Predictive analysis has the potential to early identify students at risk of successfully completing their academic trajectory, enabling the implementation of preventive policies for them [10,15].

The current paper outlines the development of an early warning system designed to predict students at risk of failure and/or abandonment in the State of Espírito Santo, Brazil. The proposed methodology integrates various sets of features, including school variables and socioeconomic information, as input for generating machine learning models to predict students at risk at an early stage. Furthermore, the proposed methodology conducts a thorough bias analysis, aiming for models that are equitable and fairness. The results obtained align with the existing literature, offering a new perspective on implementation strategies for educational intervention in real life.

2   Related Literature

The field of educational data analysis has evolved significantly, with various studies demonstrating the effectiveness of applying Artificial Intelligence algorithms in identifying students at risk in high school [9]. However, these studies diverge in the data used and the contexts of application. In addition to socioeconomic factors, prior academic performance emerges as a powerful risk indicator [14,17,20]. Students with a history of low performance in fundamental subjects are more likely to fail [14,20]. Early identification of difficulties and the implementation of targeted interventions are crucial to prevent further academic delays [25].

A multifactorial approach is crucial to understanding school abandonment and academic failure [13]. This approach may involve a diverse range of factors that impact the student, including significant socioeconomic aspects [14]. Studies have consistently shown that students from low-income families face greater academic and persistence challenges [23]. The lack of educational resources, limited family support, and socioeconomic adversities contribute to decreased engagement and academic success [21,29].

The analysis of these features should not occur in isolation but rather as part of a holistic system that integrates socioeconomic, emotional, and academic factors. This approach provides a more comprehensive view of individual student circumstances and allows for the development of more personalized support strategies [8,28]. Early identification, based on a variety of indicators, enables more effective interventions at critical moments in the student's educational journey, reducing the risks of dropout, abandonment and academic failure.

In this way, the studies by Márquez-Vera et al. [20], Parr & Bonitz [23], Chung and Lee [5], Gómez et al. [13], and Krüger et al. [17] collectively offer a diverse perspective on the use of data analytics and machine learning in predicting school dropout rates. At the same time, the search for educational patterns that can assist in identifying students with high probabilities of failure is a general concern in the field, with studies such as that of Hernandez-Leal et al. [14] showing that in the transition from elementary to secondary education, difficulties in Mathematics and Language subjects are critical points, significantly impacting student failure rates, regardless of socioeconomic level or school location.

Márquez-Vera et al. [20] and Chung and Lee [5] both implemented machine learning techniques - the former in Mexico and the latter in Korea. Márquez-Vera et al. achieved an impressive 99.8% accuracy in predicting early school dropout, demonstrating the power of data mining in educational settings. Similarly, Chung and Lee's [5] use of random forests in a large dataset of Korean students showed a 95% accuracy rate, underscoring the efficacy of machine learning algorithms in identifying at-risk students.

On the other hand, Parr & Bonitz [23] in the USA and Gómez et al. [13] in Chile approached the issue through different lenses. Parr & Bonitz's study focused on the influence of family background and student behavior, emphasizing socio-economic and behavioral factors, while Gómez et al. [13] developed models for Chile's education system, highlighting the need for large-scale data analysis in educational policy and intervention planning.

In the final analysis, the study by Krüger et al. [17] in Brazil stands out for its use of an explainable machine learning approach, achieving a notable accuracy of up to 95%. This research highlights the critical aspect of model interpretability, ensuring that educators and policymakers can understand and utilize the insights provided by such predictive models. These diverse studies, summarized in Table 1, provide a comprehensive overview encompassing objectives, levels of application, data used, types of data, results obtained, and conclusions, showcasing the multifaceted nature of data analytics in educational contexts across different countries.

In this study, we propose a customized approach for the early identification of high school students at risk of abandonment or failure in Espírito Santo, Brazil.

Table 1. Comparative of State of Art

Study	Goal	Educational Level	Data Used	Types of Data	Results	Conclusion
Márquez-Vera et al. [20]	Predict early school dropout	High School in Mexico	419	Student academic and personal data	99.8% ACC	Effectiveness of data mining in dropout prediction
Parr & Bonitz [23]	Influence of family background on school dropout	High School in the USA	15,753	Demographic, behavioral, academic performance data	-	Importance of socio-economic and behavioral factors
Chung & Lee [5]	Predict school dropout using ML	High School in Korea	165,715	Comprehensive educational data	95% ACC	Effectiveness of random forests in risk identification
Gómez et al. [13]	Develop ML models	High School in Chile	Large datasets	Administrative educational data	-	Need for large-scale approaches
Krüger et al. [17]	Explainable machine learning for dropout prediction	High School in Brazil	19 schools Data	School records, academic performance	95% ACC	Importance of interpretable models in education
Hernandez Leal et al. [14]	Unveil educational patterns in Colombia	High School in north of Colombia	6,400	Academic performance, socio-economic data	60% - 99,8% F1	Importance of identifying educational patterns for policy- making

This approach differs from the literature in some main aspects. The first is the large-scale perspective and development through the needs of the State's Department of Education. The second is that it is developed with practical integration in mind, which will be implemented in the next stage of the project. Third, this approach significantly differs from the methodologies found in the reviewed literature by combining academic, socioeconomic, and performance data to create a predictive model adapted to local specifics. At last, another important aspect is the use of fairness analysis to ensure algorithmic equity.

3   Context of the Experiments

The state of Espírito Santo, located in the southeastern region of Brazil, stands out for its unique educational characteristics. In 2021, the Basic Education Development Index (IDEB) in the state was 4.4 [16], while the established goal was 4.9, highlighting the need for improvements to achieve the proposed objectives. Additionally, the age-grade distortion reached 20.9%, rising to 25.5% in the first year. These numbers indicate significant challenges in the proper progression of students throughout the school years.

In 2022, the state's secondary education system in Espírito Santo faced a 6.6% repetition rate and a 2% abandonment rate, accounting for 8,884 students who encountered challenges in their education trajectory [16]. Although these indicators represent an improvement compared to the historical school performance results, obstacles in promoting continuous academic progression still persist. In comparison, Brazil recorded rates of 7.7% and 5.7% for repetition and abandonment, respectively [16].

Despite better indicators than the national average, critically assessing educational challenges in Espírito Santo is crucial. Understanding factors behind abandonment and failure is key for effective preventive measures. Continuous monitoring of educational indicators should guide policy development for improving education quality and efficiency in the state.

4   Methodology

This study applies the Cross-Industry Standard Process for Data Mining (CRISP-DM) model to guide the analysis of educational data [27]. CRISP-DM is a robust model that facilitates understanding, preparation, modeling, evaluation, and deployment of data in a structured process.

4.1   Data Collection and Processing

Data collection adhered to Brazil's General Data Protection Law (LGPD), ensuring student data privacy and security [19]. Similar to the European Union's GDPR, LGPD underscores transparency, consent, and personal data protection [19]. The data were anonymized to comply with these regulations.

In this stage, data manipulation was performed to structure the available information appropriately. Data cleaning procedures, treatment of missing and duplicate information, outlier detection, and selection of the population and features of interest were carried out. The standardization and structuring strategy followed the steps:

1. **Selection of the reference**: High school students from Espírito Santo with valid enrollment.
2. **Selection of the features**: It was chosen to exclude columns with constant and null values, duplicates and other information that did not apply to the study.
3. **Handling of duplicate observations**: We applied the split-apply-combine technique so that each row in the database be a unique record.
4. **Missing data treatment**: We applied proxy technique to handle information that existed but was not recorded.
5. **Exclusion of observations**: Exclusion of duplicate students IDs preserving the most recent enrollment date.

4.2   Data Description

Table 2 presents the number of students in the dataset. These data encompass information from the academic year 2022, including a total of 79,165 students enrolled in high schools in the state of Espírito Santo, Brazil. The database represents a diverse mix of institutions in terms of geographical location (urban and rural), covering all 78 municipalities in the state.

Table 2. Total amount of students

School Year	Total of Students	Approved	Failed and/or abandoned
First Year	29,101	24,705	4,396
Second Year	27,017	23,656	3,361
Third Year	23,047	21,711	1,336
Total	79,165	70,072	9,093

The student demographic is diverse, encompassing a wide age range typical of high school, which in Brazil is from 14 to 17 years old. However, these values show a slight tendency to increase due to age-grade distortion. The gender distribution is approximately balanced, with 45% boys and 55% girls. Regarding the ethnic composition of students, it reflects the cultural diversity of the state. Additionally, a total of 39% of students are enrolled in government assistance programs, which is a significant percentage, indicating a considerable presence of students from low socioeconomic backgrounds.

Groups of Features. The features used to generate the predictive models can be broadly classified into three different groups in accordance with their educational importance, as follows:

- **Demographic and socioeconomic**: The features of age, gender, ethnicity, and social government assistance status are essential for understanding students' context. Literature suggests these socioeconomic factors significantly influence access to educational resources and extracurricular support, impacting student performance and motivation [21, 29].
- **Academic**: features such as student grades, absences, and attendance are employed, providing a direct measure of student performance and engagement. Low performance and a high rate of absences are associated with a greater risk of dropout and failure, requiring specific educational interventions [24, 25].
- **Disability-related issues**: features such as the presence of any type of disability and the need for any kind of special assistance are employed. Including information about the type of disability and support resources ensures an inclusive educational approach. Adaptive educational resources for students with disabilities are important to minimize negative impacts on academic performance [3, 11].

Fairness Analysis. The features were examined to identify any potential for generating unwanted biases in the automated models. The exploratory data analysis included an assessment not only of the significance of each attribute but also of how their values are distributed. As a result, five features were identified for in-depth analysis as they could be potential generators of discriminatory biases: "*Ethnicity*", "*Government Assistance,*" "*Gender,*" "*Class Period,*" and "*Disability*". These five features were selected after several meetings with the Department of Education of the State of Espírito Santo and based on the existing related literature about bias in predictive learning analytics [7, 29].

4.3   Predictive Models

The predictive models used in this study are *white box*, emphasizing the commitment to transparency in algorithmic decision-making processes. These models are known for their interpretability, providing clear insights into how input features influence predictions [17]. In educational contexts, where decisions shape students' paths, interpreting predictions is vital to foster understanding among educators and administrators, to build trust in the analytical tools, and to give transparency for auditing and review [15].

The following *white-box* algorithms were tested: Logistic Regression, AdaBoost Classifier, RandomForest Classifier, DecisionTree Classifier, MLP Classifier (used as a benchmark), Gaussian Naive Bayes, and HistGradient Boosting Classifier. The inclusion of MLP is to provide a comparative performance of more complex models [13]. The models were trained and tested using 10-fold cross-validation and evaluated using the following metrics: accuracy, precision, recall, AUCROC, confusion matrix, and f1-measure.

Data was balanced with random oversampling and undersampling techniques and using the imblearn library [18]. This strategy balanced the classes distribution by increasing the representation of minority classes (students that failed and/or abandoned) through oversampling and reducing the dominance of majority classes through undersampling (students that approved).

The modeling strategy involved using two models for each high school year, one at the start of the academic year and another at the end of the first trimester. This dual-phase approach is essential for capturing the dynamic nature of the educational environment and students' progress. The choice of this two-stage strategy is primarily grounded in the following factors:

- **Temporal dynamics of student development**: The performance and needs of students evolve throughout the academic year. Models generated at different times can capture these changes, enabling more precise and timely interventions [4, 5, 21].
- **Adaptation to changes in the school environment**: New challenges and opportunities arise during the academic year. A model generated after the first trimester can incorporate recent data, better reflecting the current situation of students [10, 15].

- **Early prediction and continuous monitoring**: An initial model provides the ability to early identify students at risk, while the first-trimester model allows for adjusting and refining these predictions as more information becomes available.

Figure 1 depicts the 6 models developed according to the mentioned strategy (2 for each year grade of high school).

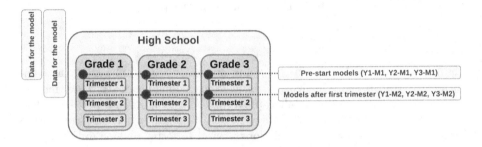

Fig. 1. Predictive models developed for the three years of high school

Depending on the moment of the school year the model is used, different input information is provided, as follows:

1. **Pre-start Models**: Based on historical data, this model utilizes features such as demographic and socioeconomic data to predict risks before the start of the academic year. These models do not use previous academic performance features as they were not available in the dataset.
2. **Models after first trimester**: Built after the end of the first trimester, this model incorporates recent data, including performance in the current trimester and other relevant features such as demographic data and socioeconomic data to refine the initial predictions. These models also use performance-derived features, such as the student's position relative to the class average, to place the student within the school context.

5   Results

5.1   Performance of the Predictive Models

Table 3 displays the performance of the best predictive models for each stage of students' high school trajectory. In the *Model* column, Y represents the school year, and M indicates the moment of model application. As shown in the table, models $Y1 - M2$, $Y2 - M2$, and $Y3 - M2$ exhibit the best performances, with high accuracies, indicating a consistent and accurate identification of students at risk of failing or abandon the school year. These results also indicate that as more information is used in the models, the better will be their performances during the year. An example of this situation is depicted in the feature selection of the top twenty features $(F=20)$ of the $Y1-M2$ model, as displayed in Fig. 2, Notably, the influence of features representing grades, attendance, and absences throughout the first trimester of the first high school year is evident

Table 3. Results of best classifier for each model

Model	Classifier	Sampling	Accuracy	Precision	Recall	AUC	F1	Confusion Matrix (TP, FN FP, TN)
Y1-M1	Random Forest	Feature Selection (F=20) Random Undersampling	69.3	28.46	68.99	68.78	40.24	1289.1, 565.9 102.3, 225.2
Y1-M2	Random Forest	Feature Selection (F=20) Random Undersampling	90	81.54	50.43	74.2	62.29	1557.6, 293.7 44.3, 286.9
Y2-M1	Random Forest	Random Undersampling	69.4	23.83	67.72	68.67	35.24	1237.2, 539.7 80.3, 169.0
Y2-M2	Random Forest	Random Undersampling	92.7	85.41	51.45	75.08	64.12	1512.0, 258.5 36.4, 219.3
Y3-M1	Random Forest	Random Undersampling	82.54	24.51	75.21	79.11	33.31	1351.2, 276.6 25.1, 75.6
Y3-M2	Random Forest	Random Undersampling	89.6	34.7	86.95	88.35	49.51	1460.3, 166.6 13.1, 88.5

As it is possible to see, the best models were all generated by the Random Forest algorithm. Analyzing specific metrics, we note that models $Y2-M2$ and $Y3-M2$ have excellent areas under the ROC curve (AUC), indicating remarkable discrimination. This suggests that these models are effective in distinguishing students with and without academic issues. The balanced F1-Score of model $Y2-M2$ highlights its good balance between precision and recall, signaling a harmonious classification ability, especially at the end of the first trimester of the academic year.

An important analysis focuses on the true negatives (TN) present in the confusion matrices of Table 3, representing the correct classification of the students who will fail or abandon. These results play a significant role, particularly when considering the remarkable stability at the end of the first academic trimester (M2).

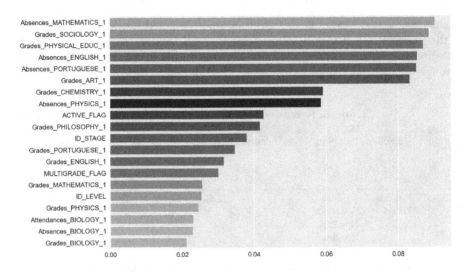

Fig. 2. Feature Selection (F = 20 for Y1-M2.

The results suggest that the models, notably those at the end of the first trimester, have promising potential for early identification of students with possible academic trajectory issues. These findings have valuable implications for intervention strategies and personalized support, aiming to enhance students' academic paths throughout the academic period.

5.2   Fairness Analysis

The bias of the models developed in this study were assessed using the What-If Tool. This tool allows the evaluation of how model decisions may vary for different groups or student characteristics. Bias analysis is crucial in the educational context, where equity and justice in education are fundamental priorities. Ensuring that prediction models do not perpetuate or amplify existing inequalities is a crucial objective, and the What-If Tool provides a systematic and transparent approach to achieving this goal.

Table 4 presents the results of the bias analysis for each models according to the protected features.

As can be seen in the Table 4, the accuracy values among the classes of features with potential for generating undesirable bias show that none of the models exhibited bias concerning the features defined as potential sources of bias. This positive result highlights the effectiveness of the strategies adopted to mitigate any unfair tendencies in the prediction models.

Table 4. Model accuracy by class for features with potential bias, illustrating bias mitigation effectiveness.

Model	Gender		Class Period		Government Assistance		Ethnicity		Disability	
	Male	Female	Day	Evening	Yes	No	White	Non White	Yes	No
Y1-M1	79.1	78.9	81.8	78.7	77.8	80.2	78.8	85	71.4	79.3
Y1-M2	87.7	83.3	85	90	85.4	85.4	88.5	83.1	85.3	87.5
Y2-M1	71.5	75	88.6	71.7	72.6	73.6	73.3	71.4	87.5	72.5
Y2-M2	87	87.4	86	98	86.9	87.4	87	92.9	81	87,5
Y3-M1	65.4	73.9	70	69.6	65.3	73	63	79.4	80	69.8
Y3-M1	91.6	92	91	98.1	91.5	92	93.8	90.7	81.3	92.1

6 Discussion

This work presents a methodology for creating prediction models for early risk of abandonment or failure in high school, demonstrating the efficacy and applicability of machine learning techniques in identifying patterns and predictors for that purpose. The research aligns with the existing literature in learning analytics, a subfield of artificial intelligence and machine learning, highlighting the accuracy of these techniques to enhance educational interventions and promote student success [21].

Compared to previous works, such as those by Márquez-Vera et al. [20] and Chung and Lee [5], this study distinguishes itself by integrating academic, socioeconomic, and performance data, adopting a multifactorial approach, and emphasizing the interpretability of models, in line with current trends in explainable machine learning, as proposed in the work of Kruger et al. [17]. Moreover, the research stands out for its focus on eliminating biases of algorithmic discrimination, an aspect not yet significantly addressed in the field's literature, promoting fair and equitable analytical models, reflecting an advancement in the theory on the subject in Learning Analytics, as proposed by Deho et al. [7].

The work also distinguishes itself in terms of its scope, and applicability on a large scale, focusing on a significant number of students. Compared to the works of Márquez-Vera et al. [20], Parr & Bonitz [23], Krüger et al. [17], and Hernandez-Leal et al. [14], the student numbers are significantly larger, covering the entire public high school network of the state. Thus, unlike previous research that may have been limited to smaller contexts or specific samples [14,17,20,23], this study demonstrates robustness and flexibility that allow its application in different educational contexts, covering a significant number of students. This scalability ensures that the insights and interventions proposed can be adapted and implemented in educational systems of various sizes and characteristics in other Brazilian states, provided small adjustments are made to the educational needs and contexts, offering a viable solution for large-scale educational improvements.

The participatory approach in methodological development, closely collaborating with the Secretary of Education of Espírito Santo (SEDU-ES), is a relevant aspect of the practical implementation of the work, breaking down the last barrier established by Clow et al. [6] in the practical application of LA. Furthermore, this approach ensures that the proposed educational strategies are effective and adapted to local specificities, offering a viable and more adaptable model to the context and user expectations, thus achieving higher levels of acceptance as established by the research of Herodotou et al. [15].

The research underlines the importance of using data to improve teaching and learning processes, highlighting the need for personalized educational strategies and interventions that address socioeconomic disparities [3,11]. By providing detailed data and analyses, the study offers a basis for the formulation of affirmative policies that promote equity and inclusion, contributing to sustainable economic development through the improvement of educational indices, seeking to ensure fair opportunities for all students.

7   Final Remarks

This study demonstrated the feasibility of using machine learning techniques for early identification of high school students at risk in their academic journey in Espírito Santo. The developed models, applying academic, socioeconomic, and performance data, showed high accuracy, highlighting their potential in early prediction of school abandonment and failure.

The research emphasizes the importance of analytical tools in education, especially in regions with significant challenges in school failure. The study offers valuable insights for proposing a data-driven approach to mitigate school dropout.

The proposed models can serve as crucial tools for educators and school administrators, allowing for targeted and timely interventions. They represent a significant step towards a more adaptive and responsive educational system to student needs.

Although the results are promising, the study faces limitations, such as the need to evaluate the acceptance of the models by educators. Furthermore, future research could explore the inclusion of more features and the application of the models in other regions to generalize the findings. An important step for improving accuracy rates, especially in models at the beginning of the school year, is the inclusion of students' historical data, as proposed by Queiroga et al [25].

This study underscores the relevance of data-based approaches in education, offering a new perspective in the fight against school abandonment and failure. It sets a precedent for the implementation of innovative and effective educational strategies, highlighting the potential of analytical technologies in improving educational quality.

Acknowledgments. This work was supported by Inter American Development Bank and Unibanco Institute. Cristian Cechinel was partially supported by the Brazilian National Council for Scientific and Technological Development (CNPq) (grants 305731/2021-1 and 409633/2022-4).

References

1. Amparo, D.M.d., Galvão, A.C.T., Cardenas, C., Koller, S.H.: A escola e as perspectivas educacionais de jovens em situação de risco. Psicologia Escolar e Educacional **12**(1), 69-88 (2008). https://doi.org/10.1590/S1413-85572008000100006
2. Attaran, M., Attaran, S.: Opportunities and challenges of implementing predictive analytics for competitive advantage. In: Applying Business Intelligence Initiatives in Healthcare and Organizational Settings, pp. 64–90 (2019)
3. Cechinel, C., et al.: Mapping learning analytics initiatives in Latin America. Br. J. Educ. Technol. **51**(4), 892–914 (2020)
4. Chatti, M.A., Dyckhoff, A.L., Schroeder, U., Thüs, H.: A reference model for learning analytics. Int. J. Technol. Enhanced Learn. **4**(5–6), 318–331 (2013)
5. Chung, J.Y., Lee, S.: Dropout early warning systems for high school students using machine learning. Child Youth Serv. Rev. **96**, 346–353 (2019)
6. Clow, D.: The learning analytics cycle: closing the loop effectively (2012)
7. Deho, O.B., Zhan, C., Li, J., Liu, J., Liu, L., Le Duy, T.: How do the existing fairness metrics and unfairness mitigation algorithms contribute to ethical learning analytics? Br. J. Edu. Technol. **53**(4), 822–843 (2022)
8. Elistia, E., Syahzuni, B.A.: The correlation of the human development index (HDI) towards economic growth (GDP per capita) in 10 Asean member countries. JHSS (J. Hum. Soc. Stud.) **2**(2), 40–46 (2018)
9. Esbenshade, L., Vitale, J., Baker, R.S.: Non-overlapping leave future out validation (nolfo): Implications for graduation prediction. prePrint . https://doi.org/10.35542/osf.io/tx8cy
10. Ferguson, R.: Learning analytics: drivers, developments and challenges. Int. J. Technol. Enhanced Learn. **4**(5/6), 304–317 (2012)
11. Ferguson, R., et al.: Research evidence on the use of learning analytics: implications for education policy (2016)
12. Ferreira, S.G., Ribeiro, G., Tafner, P.: School abandonment and dropout in Brazil. Technical report, Rio de Janeiro, Brazil (2022)
13. Gómez-Pulido, J.A., Park, Y., Soto, R.: Advanced techniques in the analysis and prediction of students' behaviour in technology-enhanced learning contexts (2020)
14. Hernández-Leal, E., Duque-Méndez, N.D., Cechinel, C.: Unveiling educational patterns at a regional level in Colombia: data from elementary and public high school institutions. Heliyon **7**(9), e08017 (2021)
15. Herodotou, C., Rienties, B., Boroowa, A., Zdrahal, Z., Hlosta, M., Naydenova, G.: Implementing predictive learning analytics on a large scale: the teacher's perspective. In: Proceedings of the Seventh International Learning Analytics & Knowledge Conference, pp. 267–271 (2017)
16. Instituto Nacional de Estudos e Pesquisas Educacionais Anísio Teixeira (INEP): Censo Escolar 2021 (2022). https://qedu.org.br/uf/32-espirito-santo. Accessed 1 Jan 2024
17. Krüger, J.G.C., de Souza Britto Jr., A., Barddal, J.P.: An explainable machine learning approach for student dropout prediction. Expert Syst. Appl. **233**, 120933 (2023)

18. Lemaître, G., Nogueira, F., Aridas, C.K.: Imbalanced-learn: a python toolbox to tackle the curse of imbalanced datasets in machine learning. J. Mach. Learn. Rese. **18**(17), 1–5 (2017). http://jmlr.org/papers/v18/16-365.html

19. Lorenzon, L.N.: Análise comparada entre regulamentações de dados pessoais no brasil e na união europeia (lgpd e gdpr) e seus respectivos instrumentos de enforcement. Revista do Programa de Direito da União Europeia **1**, 39–52 (2021)

20. Márquez-Vera, C., Cano, A., Romero, C., Noaman, A.Y.M., Mousa Fardoun, H., Ventura, S.: Early dropout prediction using data mining: a case study with high school students. Expert Syst. **33**(1), 107–124 (2016)

21. Moissa, B., Gasparini, I., Kemczinski, A.: A systematic mapping on the learning analytics field and its analysis in the massive open online courses context. Int. J. Distance Educ. Technol. (IJDET) **13**(3), 1–24 (2015)

22. OECD: Benchmarking Higher Education System Performance. OECD Publishing, Paris, France (2019). https://doi.org/10.1787/be5514d7-en

23. Parr, A.K., Bonitz, V.S.: Role of family background, student behaviors, and school-related beliefs in predicting high school dropout. J. Educ. Res. **108**(6), 504–514 (2015)

24. Queiroga, E.M., et al.: Experimenting learning analytics and educational data mining in different educational contexts and levels. In: 2022 XVII Latin American Conference on Learning Technologies (LACLO), pp. 1–9. IEEE (2022)

25. Queiroga, E.M., Batista Machado, M.F., Paragarino, V.R., Primo, T.T., Cechinel, C.: Early prediction of at-risk students in secondary education: A countrywide k-12 learning analytics initiative in Uruguay. Information **13**(9), 401 (2022)

26. Stanek, C.: The educational system of Brazil. IEM Spotlight **10**(1), 1–6 (2013)

27. Wirth, R., Hipp, J.: Crisp-dm: Towards a standard process model for data mining. In: Proceedings of the 4th International Conference on the Practical Applications of Knowledge Discovery and Data Mining, vol. 1, pp. 29–39. Springer, London (2000)

28. Yakunina, R., Bychkov, G.: Correlation analysis of the components of the human development index across countries. Procedia Econ. Finan. **24**, 766–771 (2015)

29. Yau, J.Y.K., Ifenthaler, D.: Reflections on different learning analytics indicators for supporting study success. Int. J. Learn. Anal. Artif. Intell. Educ.: IJAI **2**(2), 4–23 (2020)

Beyond the Grey Area: Exploring the Effectiveness of Scaffolding as a Learning Measure

Nadine Schulze Bernd and Irene-Angelica Chounta[(✉)] [iD]

Department of Human-Centered Computing and Cognitive Science, University of
Duisburg-Essen, Duisburg, Germany
`nadine.schulze-bernd@stud.uni-due.de` `irene-angelica.chounta@uni-due.de`

Abstract. In this paper, we aim to explore students' help-seeking patterns, and learning performance using a computational model of the Zone of Proximal Development (ZPD). We used student traces in four courses supported by Intelligent Tutoring Systems (ITSs) to assess whether a student may be in their ZPD. Then, we analyzed students' help-seeking behavior and their performance after receiving scaffolding to investigate if and under which circumstances our assessment may act as a proxy of the ZPD. Results suggest that the computational model may offer an acceptable approximation of the student's ZPD. However, several factors, such as the difficulty of the learning task, the student's age, and their attitude regarding hints, may affect modeling students' cognitive states. We discuss the implications of this research on the design of computational student models for providing personalized and adaptive feedback and scaffolding. We envision that this work contributes toward bridging the gap between theory and practice for AI-assisted tutoring and scaffolding, and it provides insights regarding evaluation techniques of theory-driven, computational approaches.

Keywords: Intelligent Tutoring Systems · Student Modeling · Zone of Proximal Development · Scaffolding · Personalization · Adaptation

1 Introduction

Intelligent Tutoring Systems (ITSs) aim to model students' cognitive and knowledge state to provide tailored content and instruction that address the learners' needs, to appropriately challenge and support them [6,9]. Related work suggests that ITSs can support teaching and learning to that end, but there are still open challenges to overcome. For example, ITSs might force the student to engage in extensive discussions about material they have already mastered instead of addressing content they need help with [11,13] causing boredom or frustration, which are both predictors of poor learning gains [8]. This risk can be addressed by using established pedagogical theories to drive the adaptation of instruction and content to meet the learners' needs. The Zone of Proximal

A. M. Olney et al. (Eds.): AIED 2024, LNAI 14829, pp. 365–378, 2024.
https://doi.org/10.1007/978-3-031-64302-6_26

Development (ZPD) [25] is a prominent pedagogical theory that refers to the difference between what a learner can do independently and what they can do with assistance. Teaching within this zone is considered to be most effective and efficient as it enables challenging students without causing frustration or loss of motivation [4].

There have been attempts to model ZPD in ITSs and adjust pedagogical interventions accordingly [15,20]. [5] attempted to model the ZPD based on the system's predictions about whether a student is able to answer a question correctly or not. This approach defines the **"Grey Area" (GA)** as the area in which the student model cannot predict with acceptable accuracy whether a student is able to answer the question correctly. If the model predicts a student to be in this area, this might indicate that the student's knowledge state falls within the ZPD and the student may need scaffolding to give a correct answer. If the model assesses that the student will be able to answer the question correctly, this might indicate that the student is above the ZPD and will be able to answer the question without help. If the model predicts that the student will answer incorrectly, then this might indicate that the student is below the ZPD and will not be able to answer the question even when scaffolding is available [5]. [11] provided preliminary results regarding this approach while testing it for the case of a dialogue-based tutor that supported K-12 students' conceptual knowledge about Physics. This paper aims to further validate the GA approach by a) expanding the GA's investigation over four different datasets and STEAM domains; and, b) exploring metrics that use students' help-seeking patterns and performance to assess the method's effectiveness as a proxy of the ZPD.

The ZPD is a theoretical concept that lacks an objective way to measure whether or when a student is in the ZPD or not. The number of correct and incorrect answers above, below, and within the GA can provide meaningful indications. However, these two metrics depend on the predictive accuracy of the student model and, therefore, cannot provide evidence regarding the effectiveness of the approach. At the same time, assistance is a significant concept of the ZPD [16,20]. We argue that to validate whether a student is in their ZPD, one should examine the student's performance in relation to the help they received on specific tasks. In ITSs, assistance for solving a problem is mainly provided through hints. Thus, the usage of hints associated with the outcome in terms of performance – that is, whether a student carries out a learning step correctly or not after receiving a hint – can provide an indication for the student's ZPD [4,20]. Here, we aim to answer the following research questions:

RQ1: Do students provide more correct answers when above the Grey Area than below?

RQ2: Do students provide more incorrect answers when below the Grey Area than above?

RQ3: Do students provide more correct answers after using a hint when they are within the GA than below?

In the following sections, we present research related to the concept of ZPD, ITSs, and existing approaches for modeling the ZPD. Then, we describe the methodological approach we followed to explore our questions, and we report the results. In Sect. 5, we discuss our findings regarding theoretical and practical implications and we conclude with limitations of the study and plans for future work.

2 Related Work

2.1 The Zone of Proximal Development in Education

The Zone of Proximal Development (ZPD) was introduced by psychologist Lev Vygotsky and refers to the gap between learners' current level of ability and their potential level of development. It describes the difference between what a child can do independently and what they can do with the assistance of a "*more knowledgeable other*" [25]. When it comes to practice, there is no specification regarding the type of assistance that should be offered to learners or how it should be provided and withdrawn [15,16]. In the literature, the concept of **scaffolding** is often associated with the ZPD [16]. While some researchers understand scaffolding as a direct application of teaching in the ZPD, others criticize it as a narrow way of operationalizing the ZPD as it only partially reflects the richness of this concept [23].

2.2 Approaches to Implement the ZPD in ITSs

Operationalizations of the ZPD have been used in ITSs to optimize student problem-solving by adapting the level of difficulty and scaffolding appropriately. For example, [15] developed the ITS *Ecolab* with a ZPD operationalization to offer students help that was reflective of the complexity of the learning material. To that end, two additional concepts were introduced: the *Zone of Available Assistance* (ZAA) and the *Zone of Proximal Adjustment* (ZPA). The system assessed each learner's ZPA based on the levels the learner used at each step by using domain knowledge representations and Bayesian Belief Networks. Hence, the assistance to a learner was flexible and capable of being increased and decreased [15]. [20] extended this approach by introducing factors related to affect, such as boredom or frustration. The authors introduced the *Specific ZPD* (SZPD), an operationalized definition of the ZPD, which consists of a mastery criterion and a ZPD criterion. The mastery criterion addresses effectiveness by determining when the student can move on to the next unit. The ZPD criterion determines whether the student's learning was efficient which means that the student stays in a zone where they are neither too frustrated nor too bored.

Another approach to model the ZPD in ITSs is the *Grey Area* approach [5]. The GA describes the area in which a student model cannot predict with acceptable accuracy whether a student is able to answer a question correctly (see Fig. 1). The authors proposed that if the model predicts a learner to be in this area, this might indicate that the learner's knowledge state falls within the ZPD. The space above the GA describes the region where the probability is considerably higher than the cutoff threshold (i.e., the student is predicted to answer correctly), and therefore, it may indicate the area above the ZPD in which the student is able to answer a question without assistance. The space below the GA describes the area with probabilities considerably lower than the cutoff threshold (i.e., the student is predicted to answer incorrectly) and, therefore, may indicate the area below the ZPD in which the student is not able to carry out the task even with assistance. For a first validation of the GA approach, [5] used data from students who interacted with a dialogue-based tutoring system called *Rimac*. A follow-up study [11] used the proposed approach to guide students who were practicing Rimac through adaptive lines of reasoning. The results indicated a positive impact of the student model on students' practicing times.

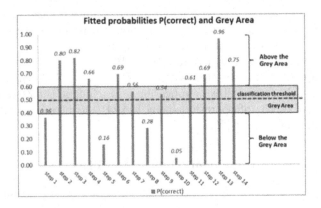

Fig. 1. Concept of the GA approach according to [5]

3   Methodology

3.1   Datasets

We used four datasets from students' interactions with different ITSs. The datasets were provided via Datashop [14]. DataShop is an open data repository that provides secure data storage of educational data, analysis and visualization tools. In particular, here we used the following datasets:

- **Fractions** dataset (Mathematics), ID 671. 77 students interacted with an ITS for 4th- and 5th-grade fractions learning [22]. In total, students performed 15269 steps practicing 20 KCs.
- **Genetics** dataset, ID 1195. The Genetics Cognitive Tutor supports genetics problem-solving for high school students who study biology [7]. 139 students interacted with the tutor and performed 128523 steps practicing 28 KCs.

- **Physics** dataset, ID 126. The dataset contains logs of student homework done in the ITS Andes for the General Physics course taught at the United States Naval Academy (USNA) [12]. In total 66 students performed 138958 steps while practicing 250 KCs.
- **Stoichiometry** dataset (Chemistry), ID 256. Stoichiometry is typically learned in the 10th or 11th grade in U.S. high schools and involves understanding basic chemistry concepts [17]. 418 students interacted with an ITS on worked examples and performed 124109 steps practicing 36 KCs.

Each dataset consists of students' transactions with an ITS to carry out a *problem*. A problem typically involves multiple *steps*. Each *step* is part of the solution to a problem. Before carrying out a step correctly, students may provide incorrect entries or ask for hints. In order to perform a step correctly, students need to know one or more specific skills or concepts, known as *knowledge components (KCs)*. The student's first attempt on a step is assessed using the Additive Factor Model (AFM) [3] to calculate a *predicted error rate*, ranging between 1 and 0; 1 suggests that a student's first attempt will be an error (incorrect attempt or hint request) and 0 that the first attempt will be correct.

3.2   Method Setup

We applied the GA approach based on the information provided in the four datasets, and as described by [5]. Then, we compared students' performance in and outside their Grey Areas to validate the initial assumption of the GA (RQ1, RQ2). Finally, we isolated the instances in students' practice, where a student received help from the ITS in the form of a hint. For the first hint on a step, we documented the student's follow-ups. This resulted in sequences of **hint-response** actions that characterized the practice of students. In particular, we identified three potential sequences: a) **hint-correct**: the student received a hint for a learning step and they proceeded to carry out this step correctly; b) **hint-incorrect**: the student received a hint for a learning step and they proceeded to carry out this step incorrectly; c) **hint-hint**: the student received a hint for a learning step and they proceeded to ask for another hint. To investigate RQ3, we calculated how many sequences of hint-correct, hint-incorrect, and hint-hint appear within and outside the GA.

The GA was operationalized as the area around the classification threshold of a binary classifier that would assess a student's attempt as correct or incorrect. For simplicity, the GA was modeled symmetrically around the cutoff threshold. The variable Predicted Error Rate was used to calculate the probability of correctness per step as $Prob(Correctness) = 1 - PredictedErrorRate$. The Predicted Error Rate was calculated for each step as a function of the student's proficiency, the difficulty of the KC, the learning rate and the student's prior practice for the KC. Steps with combined KCs were omitted from further analysis. To classify the students' attempts as correct or incorrect based on the Predicted Error Rate, we defined a probability cutoff (or else, optimal classification threshold, opt_t) to assign each step based on the predicted error rate. The classification threshold was calculated over each dataset (all students and all steps) using the receiver operating characteristic (ROC) curve with the aim of maximizing the sum of true positive and true negative rates. Finally, we defined the upper (1) and lower (2) boundaries of the GA as a function of opt_t to ensure that the GA would not extend beyond the range of the Predicted Error Rate.

$$GA_{upperboundary} = opt_t + ((1 - opt_t)/2) \qquad (1)$$

$$GA_{lowerboundary} = opt_t - (opt_t/2) \qquad (2)$$

To answer the research questions, we calculated for each dataset the following metrics: total number of steps per area (below, within, above), number of steps that were answered correctly (**RQ1**) or incorrectly (**RQ2**) for each of the three areas, and the number of hint-correct, hint-incorrect and hint-hint sequences per area (**RQ3**). Then, we conducted paired sample Wilcoxon Signed Rank Tests to investigate the statistical significance of our results. We applied the Bonferroni correction to account for multiple comparisons.

4 Results

4.1 Descriptive Analysis

Table 1 presents the descriptives regarding the students' performance for the first attempts. Three out of four datasets (Fractions, Physics and Stoichiometry) are characterized by more correct responses than incorrect ones. The only exception is the Genetics dataset, for which most responses were incorrect (54%). In all cases, less than 17% of hints were given out - for Fractions and Stoichiometry, only 4% of the responses corresponded to hints. The number of correct vs. incorrect responses in combination with the number of hints may be an indication of the difficulty of the learning task, suggesting that the content of the Genetics tutor was perceived as the most difficult among all four cases.

To answer RQ1 and RQ2, we calculated the distribution of correct and incorrect answers that students gave and the hints they received within the GA, and the areas above and below, for the four datasets under study (Table 2). Furthermore, we calculated the percentage ratio of correct, and incorrect answers and hints over the total number of answers per area (Fig. 2). Our results suggest that the overall assumption regarding the correctness of answers within and outside the GA is confirmed. For all datasets, most correct answers fell above the GA (94% for Fractions, 78% for Genetics, 91% for Physics and 90% for the Stoichiometry datasets). Similarly, most incorrect answers were given in the area below GA (42% for Fractions, 56% for Genetics, 22% for Physics, and 53% for the Stoichiometry datasets). Hints were distributed over all three areas. However, the results suggest that hints are mainly requested below and within the GA, while fewer hints are requested above the GA. Next, we explored the help-seeking behavior of students over the GA and adjacent areas. To do so, we counted the number of sequences that involved hints over the different areas under study.

Table 1. Descriptive statistics of first attempts' performance in terms of correctness.

Fractions	Correct	Hints	Incorrect
Number of steps	5634	309	2060
Percentage (%)	70	4	26
Genetics	Correct	Hints	Incorrect
Number of steps	1825	1104	3396
Percentage (%)	29	17	54
Physics	Correct	Hints	Incorrect
Number of steps	89851	15534	20011
Percentage (%)	72	12	16
Stoichiometry	Correct	Hints	Incorrect
	Correct Attempts	Hints	Incorrect Attempts
Number of steps	70188	3958	21749
Percentage (%)	73	4	23

Table 2. The distribution of correct answers, hints, and incorrect answers within the Grey area and its adjacent areas, below and above.

	Below GA			Within GA			Above GA		
	c	h	i	c	h	i	c	h	i
Fractions	113	107	158	4051	200	1804	1470	2	98
Genetics	139	749	1135	1257	348	2145	429	7	116
Physics	1768	4613	1829	64134	10433	16245	23949	488	1937
Stoichiometry	517	225	827	4485	260	2039	1102	7	118

That is, we calculated the number of hint-correct, hint-hint, and hint-incorrect sequences that appeared in each of the three areas. The results are presented in Table 3. Figure 3 presents the percentage ratio of the help-seeking sequences over the total number of sequences in the three areas. The results showed that the ratio of hint-correct sequences is highest above the GA for all datasets. The ratio of hint-incorrect sequences appears to be distributed over all three areas while the ratio of hint-hint sequences is highest below the GA.

4.2 Inferential Statistics

To answer **RQ1**, we calculated the number of correct answers above and below the GA for each student in the four datasets. Then, we compared the two groups using the Wilcoxon Signed Rank Test for paired samples. The results of this anal-

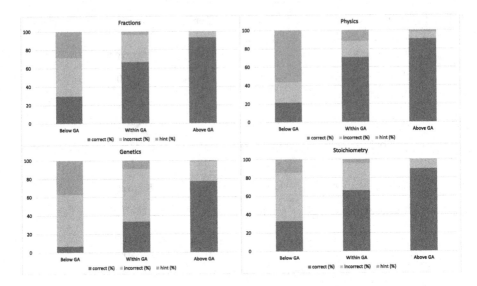

Fig. 2. Percentage ratio of correct, incorrect and hints.

Table 3. The distribution of help-seeking sequences (hint-correct, hint-hint and hint-incorrect) within the Grey area and its adjacent areas, below and above.

	Below GA			Within GA			Above GA		
	h-c	h-h	h-i	h-c	h-h	h-i	h-c	h-h	h-i
Fractions	5	77	8	70	81	27	2	0	0
Genetics	15	431	74	46	147	51	3	0	0
Physics	626	2618	411	4678	3033	1407	375	34	38
Stoichiometry	210	972	307	983	510	509	87	20	8

ysis are presented in Table 4. The results suggest that students provide more correct answers when above the GA than below. For three out of four datasets (Fractions, Physics, and Stoichiometry) these findings are statistically significant.

Next, we calculated the number of incorrect answers above and below the GA for each student in the four datasets to answer **RQ2**. Consequently, we compared the two groups using the Wilcoxon Signed Rank Test for paired samples as earlier. The results are presented in Table 5. In three out of four datasets (Fractions, Genetics and Stoichiometry), the students provided on average more incorrect answers when below the GA than above. However, only for one case (Genetics) this difference was statistically significant. On the contrary, for the Physics dataset, the students provided more incorrect answers above the GA than below.

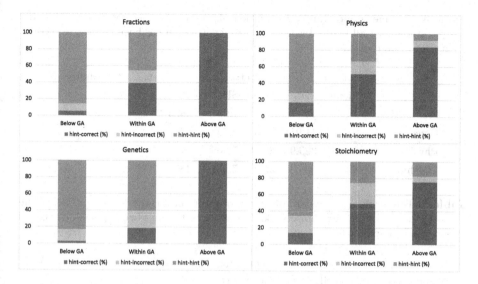

Fig. 3. Percentage ratio of help-seeking sequences: hint-correct (h-c), hint-hint (h-h) and hint-incorrect (h-i).

For **RQ3**, we compared the number of hint-correct sequences per student within and below the GA. According to the ZPD, a student who receives help when they are within their ZPD will be able to complete the learning task successfully. Therefore, if the GA is an effective proxy of the ZPD, then the hint-correct sequences within the GA should be more than below the GA. The results are presented in Table 6. Indeed, the students on average perform more hint-correct sequences when they are within the GA than below. This difference is statistically significant for three out of four datasets (Fractions, Physics, Stoichiometry). For the Genetics dataset, although we established a difference, the result is not statistically significant.

5 Discussion

5.1 Validation of the GA Approach over Different ITSs and Domains

To validate the effectiveness of the GA approach, we explored the following research questions: a) there would be more correct answers above the GA than below the GA (**RQ1**); b) there would be more incorrect answers below the GA than above the GA (**RQ2**); and c) There would be more hint-correct response sequences within the GA than below (**RQ3**) thus suggesting that students within the GA can carry out the learning task successfully when given appropriate support. A comparison of the frequencies of students' answers within the different areas confirmed the assumption that there are more correct answers above the GA than below the GA (RQ1). For all datasets, between 78% to 94% of the

Table 4. The mean and standard deviation of correct answers above and below the GA per student for all four datasets. Bonferroni correction has been applied to account for multiple comparisons.

	$correct_{aboveGA}$		$correct_{belowGA}$		Wilcoxon Paired Samples V
	Mean	SD	Mean	SD	
Fractions	19	13	1.47	1.98	2883***
Genetics	8.41	11.91	2.73	2.76	872
Physics	363.86	329.98	26.78	58.55	2115***
Stoichiometry	62.28	89.33	3.52	7.05	48065 ***

Table 5. The mean and standard deviation of incorrect answers above and below the GA per student for all four datasets. Bonferroni correction has been applied to account for multiple comparisons.

	$incorrect_{aboveGA}$		$incorrect_{belowGA}$		Wilcoxon Paired Samples V
	Mean	SD	Mean	SD	
Fractions	1.27	1.12	2.05	5.73	1304
Genetics	2.27	4.47	22.25	13	52***
Physics	29.35	32.8	27.71	31.14	1043
Stoichiometry	5.17	6.95	6.63	11.20	29011

answers above the GA are correct while 7% to 33% of the answers are correct below the GA. In addition to these findings, the results of paired sample Wilcoxon Signed Rank Tests show that students, on average, provide more correct answers above the GA than below. This difference was statistically significant for three out of four datasets (Fractions, Physics and Stoichiometry). Students interacting with these ITSs were significantly more likely to answer a question correctly if it was above the respective GA than below the GA.

Similarly, students provided more incorrect answers below the GA than above the GA (RQ2). However, for individual students' performance, this difference was statistically significant only in the case of the Genetics dataset. Our findings suggest that the number of correct and incorrect answers above and below the GA relates to the number of hints students request from the tutor and to the task's difficulty, as depicted by the overall amount of correct and incorrect answers. For example, the Genetics dataset had 54% incorrect first attempts as opposed to 29% of correct attempts and 17% hints. The other three datasets demonstrated over 70% correct responses. This may indicate that the content of the Genetics tutor was more difficult than the rest impacting on the distribution of correct and incorrect answers in the areas below and above the GA.

Finally, students generally used more hints below and within the GA than above. Students solved more steps correctly within the GA than they did below the GA. Above the GA, nearly no hints were used, but most answers following

Table 6. The mean and standard deviation of hint-correct response sequences within and below the GA per student for all four datasets. Bonferroni correction has been applied to account for multiple comparisons.

	$HC_{withinGA}$		$HC_{belowGA}$		Wilcoxon Paired Samples V
	Mean	SD	Mean	SD	
Fractions	0.91	1.42	0.06	0.29	553***
Genetics	0.90	1.5	0.29	0.94	182
Physics	70.88	61.4	9.48	20.79	2021***
Stoichiometry	2.60	3.98	0.56	2.12	22712***

hints were correct. These results align with the concept of the ZPD. For all datasets, students on average, provided more correct responses after receiving a hint when they were within the GA than below (RQ3).

Overall, less than 17% of the steps corresponded to hints (Genetics) while in two cases (Fractions and Stoichiometry) hints accounted for only 4% of the steps. This is consistent with the findings of previous studies about the usage of hints in ITSs. [18] examined the usage of hints within the Physics tutoring system Andes, and discovered that students did not use hints often enough when interacting with this tutor. This may point toward hint refusal, that is students rather guessing the answer instead of using hints [21,24]. [1] explored the students' usage of hints in a Geometry tutor and reported that students were about ten times more likely to edit a wrong explanation than to ask for a hint. The authors conclude that giving learners total control over when they get hints may not be the best solution. Giving unsolicited hints instead might improve students' use of ITSs [24].

Our findings showed that below the GA most hints were followed by an additional hint request (h-h) rather than an incorrect answer (h-i), as one may have expected. [2] showed that an insufficient amount of information in a hint can cause frustration and the desire to request subsequent hints without attempts to solve a problem. Therefore, hint-hint sequences could be an additional indication of students' not being able to use the hints meaningfully, as a consequence of not being in their ZPD.

5.2 Theoretical and Practical Implications

The GA could potentially be an effective proxy of the ZPD based on the proposed metric of help-seeking effectiveness. However, we acknowledge that there are several factors that could influence the GA approach. First, the approach assumes that a model's uncertainty indicates the ZPD of a student. This is applicable in practice if the student model accurately predicts a student's performance. A well-performing student model and the choice of an optimal classification threshold are crucial for the effectiveness of the GA approach. In order to account for this when using the GA approach, the student model could be trained by using existing data of ITSs that are similar to the used ITS (e.g., similar learning activities). This could help to improve the predictive accuracy and find an optimum classification threshold, which is the basis for the definition of the GA.

One should keep in mind that the students' actual performance might be dependent on several factors which are not depicted by the model. For example, the optimal learning conditions differ for each learner and for different contexts [20]. Consequently, the size and shape of the ZPD might vary for each student and for each learning environment. Here, we used a GA of fixed size per dataset, thus assuming that students have similar ZPD. Nonetheless, ZPD can potentially be dependent on students' characteristics, such as age, or motivation. Furthermore, the effectiveness of the GA approach depends on the assistance provided by the ITS. Students within the GA might actually be able to solve a problem with appropriate assistance (i.e. student is within the ZPD), but if the provided assistance is not appropriate, the student may not succeed. At the same time, hints can only be helpful if they are actually requested. Our results are consistent with prior findings and indicate that students may refrain from using hints [1, 18,24]. We argue that the GA can be used to address hint refusal by providing automatic hints to students who are predicted to be within the GA [10,19].

6 Conclusion

This work investigated the effectiveness of the GA approach, a computational method for modeling the ZPD using metrics that relate help-seeking behavior to students' performance. We analyzed four datasets from students' interactions with different ITSs. For three out of the four datasets, the results confirmed that students were more likely to answer a question correctly if this question fell in the area above the GA than if it was below the GA. Furthermore, if students needed a hint for answering a question within the GA, the subsequent answer was more often correct than if a hint was used in order to answer a question below the GA. However, all three datasets consisted overall of more correct answers than incorrect. In the case the dataset consisted of more incorrect answers than correct, the opposite assumption was confirmed; that is, students were more likely to provide more incorrect answers below the GA than above. This may suggest that the task difficulty influences the effectiveness of the GA as a proxy of the ZPD, along with the students' attitude toward hints.

One limitation of this work relates to using existing data from an online repository and thus, having limited contextual information about study conditions and data collection processes. It is important to note that the present analysis does not include data associated with combined KCs. This simplification and the rare usage of hints resulted in small sample sizes potentially reducing the explanatory power of the tests. Finally, the present analysis was limited to the performance of students and their usage of hints within the three different areas above, within, and below GA. From this analysis, assumptions about other factors related to the ZPD (for example, size and symmetry of ZPD, influencing factors like the motivation of students) can be made, but these conclusions are not verified and should be further explored in future research.

Acknowledgements. We used the datasets 'Classroom study June 2013', 'MTLSD REAL GI GR Study 2015', 'USNA Physics Fall 2006', 'StoichStudy 4b Modality Effect' accessed via DataShop (pslcdatashop.web.cmu.edu).

References

1. Aleven, V., Koedinger, K.R.: Limitations of student control: do students know when they need help? In: Gauthier, G., Frasson, C., VanLehn, K. (eds.) ITS 2000. LNCS, vol. 1839, pp. 292–303. Springer, Heidelberg (2000). https://doi.org/10.1007/3-540-45108-0_33
2. Anohina, A.: Advances in intelligent tutoring systems: problem-solving modes and model of hints. Int. J. Comput. Commun. Control **2**(1), 48–55 (2014). https://doi.org/10.15837/ijccc.2007.1.2336
3. Cen, H., Koedinger, K., Junker, B.: Comparing two IRT models for conjunctive skills. In: Woolf, B.P., Aïmeur, E., Nkambou, R., Lajoie, S. (eds.) ITS 2008. LNCS, vol. 5091, pp. 796–798. Springer, Heidelberg (2008). https://doi.org/10.1007/978-3-540-69132-7_111
4. Chaiklin, S.: The zone of proximal development in vygotsky's analysis of learning and instruction. In: Kozulin, A., Gindis, B., Ageyev, V.S., Miller, S.M., Vygotskij, L.S. (eds.) Vygotsky's Educational Theory in Cultural Context, pp. 39–64. Learning in doing, Cambridge University Press, Cambridge (2005). https://doi.org/10.1017/CBO9780511840975.004
5. Chounta, I.-A., Albacete, P., Jordan, P., Katz, S., McLaren, B.M.: The "grey area": a computational approach to model the zone of proximal development. In: Lavoué, É., Drachsler, H., Verbert, K., Broisin, J., Pérez-Sanagustín, M. (eds.) EC-TEL 2017. LNCS, vol. 10474, pp. 3–16. Springer, Cham (2017). https://doi.org/10.1007/978-3-319-66610-5_1
6. Corbett, A., Koedinger, K., Anderson, J.: Intelligent Tutoring Systems, pp. 849–874. Elsevier (1997). https://doi.org/10.1016/b978-044481862-1/50103-5
7. Corbett, A., Kauffman, L., Maclaren, B., Wagner, A., Jones, E.: A cognitive tutor for genetics problem solving: learning gains and student modeling. J. Educ. Comput. Res. **42**(2), 219–239 (2010). https://doi.org/10.2190/ec.42.2.e
8. Desmarais, M.C., Baker, R.S.J.D.: A review of recent advances in learner and skill modeling in intelligent learning environments. User Model. User-Adapted Interact. **22**(1-2), 9–38 (2012). https://doi.org/10.1007/s11257-011-9106-8
9. D'Mello, S.K., Graesser, A.: Intelligent Tutoring Systems, pp. 603–629. Routledge (2023). https://doi.org/10.4324/9780429433726-31
10. Fossati, D., Di Eugenio, B., Ohlsson, S., Brown, C., Chen, L.: Generating proactive feedback to help students stay on track. In: Aleven, V., Kay, J., Mostow, J. (eds.) ITS 2010. LNCS, vol. 6095, pp. 315–317. Springer, Heidelberg (2010). https://doi.org/10.1007/978-3-642-13437-1_56
11. Katz, S., Albacete, P., Chounta, I.A., Jordan, P., McLaren, B.M., Zapata-Rivera, D.: Linking dialogue with student modelling to create an adaptive tutoring system for conceptual physics. Int. J. Artif. Intell. Educ. **31**(3), 397–445 (2021). https://doi.org/10.1007/s40593-020-00226-y
12. Katz, S., Connelly, J., Wilson, C.: Out of the lab and into the classroom: an evaluation of reflective dialogue in Andes. Front. Artif. Intell. Appl. **158**, 425 (2007)
13. Koedinger, K., Pavlik Jr, P.I., Stamper, J., Nixon, T., Ritter, S.: Avoiding problem selection thrashing with conjunctive knowledge tracing. In: Educational Data Mining 2011 (2010)

14. Koedinger, K.R., Baker, R.S., Cunningham, K., Skogsholm, A., Leber, B., Stamper, J.: A data repository for the EDM community: the PSLC datashop. In: Handbook of Educational Data Mining, vol. 43, pp. 43–56 (2010)

15. Luckin, R., Du Boulay, B.: Reflections on the ecolab and the zone of proximal development. Int. J. Artif. Intell. Educ. **26**(1), 416–430 (2016). https://doi.org/10.1007/s40593-015-0072-x

16. Margolis, A.A.: Zone of proximal development, scaffolding and teaching practice. Cult.-Hist. Psychol. **16**(3), 15–26 (2020). https://doi.org/10.17759/chp.2020160303

17. McLaren, B.M., DeLeeuw, K.E., Mayer, R.E.: Polite web-based intelligent tutors: can they improve learning in classrooms? Comput. Educ. **56**(3), 574–584 (2011). https://doi.org/10.1016/j.compedu.2010.09.019

18. Muldner, K., Burleson, W., van de Sande, B., VanLehn, K.: An analysis of students' gaming behaviors in an intelligent tutoring system: predictors and impacts. User Model. User-Adap. Inter. **21**(1), 99–135 (2011). https://doi.org/10.1007/s11257-010-9086-0

19. Murray, R.C., VanLehn, K.: Effects of dissuading unnecessary help requests while providing proactive help. In: AIED, pp. 887–889 (2005)

20. Murray, T., Arroyo, I.: Toward measuring and maintaining the zone of proximal development in adaptive instructional systems. In: Cerri, S.A., Gouardères, G., Paraguaçu, F. (eds.) ITS 2002. LNCS, vol. 2363, pp. 749–758. Springer, Heidelberg (2002). https://doi.org/10.1007/3-540-47987-2_75

21. Ranganathan, R., VanLehn, K., van de Sande, B.: What do students do when using a step-based tutoring system? Res. Pract. Technol. Enhanced Learn. **9**(2), 323–347 (2014). https://rptel.apsce.net/index.php/rptel/article/view/2014-09019

22. Rau, M.A., Aleven, V., Rummel, N., Pardos, Z.: How should intelligent tutoring systems sequence multiple graphical representations of fractions? A multi-methods study. Int. J. Artif. Intell. Educ. **24**(2), 125–161 (2014). https://doi.org/10.1007/s40593-013-0011-7

23. Shabani, K., Khatib, M., Ebadi, S.: Vygotsky's zone of proximal development: instructional implications and teachers' professional development. English Lang. Teach. **3**(4), 237–248 (2010). https://eric.ed.gov/?id=ej1081990

24. VanLehn, K.: Reflections on andes' goal-free user interface. Int. J. Artif. Intell. Educ. **26**(1), 82–90 (2016). https://doi.org/10.1007/s40593-015-0067-7

25. Vygotsky, L.: Interaction between learning and development. In: Gauvain, C. (ed.) Readings on the Development of Children, pp. 34–40. Scientific American Books, New York (1978)

Fairness of MOOC Completion Predictions Across Demographics and Contextual Variables

Sébastien Lallé[✉], François Bouchet, Mélina Verger, and Vanda Luengo

Sorbonne Université, CNRS, LIP6, 75005 Paris, France
{sebastien.lalle,francois.bouchet,melina.verger,vanda.luengo}@lip6.fr

Abstract. While machine learning (ML) has been extensively used in Massive Open Online Courses (MOOCs) to predict whether learners are at risk of dropping-out or failing, very few work has investigated the bias or possible unfairness of the predictions generated by these models. This is however important, because MOOCs typically engage very diverse audiences worldwide, and it is unsure whether the existing ML models will generate fair predictions to all learners. In this paper, we explore the fairness of ML models meant to predict course completion in a MOOC mostly offered in Europe an Africa. To do so, we leverage and compare ABROCA and MADD, two fairness metrics that have been proposed specifically in education. Our results show that some ML models are more likely to generate unfair predictions than others. Even in the fairest models, we found biases in their predictions related to how the learners' enrolled as well as their country, gender, age and job status. These biases are particularly detrimental to African learners, which is a key finding as they are an understudied population in AI fairness analysis in education.

Keywords: Fairness · MOOCs · Machine Learning · Demographics

1 Introduction

In AI research, the growing usage of ML techniques has risen concerns about their fairness towards certain groups of people, especially the less privileged ones [4]. In recent years, there has been similar concerns in the field of AI in Education (AIED), where ML techniques are increasingly used to support a variety of decisions, such as providing learners with dedicated support and scholarships, which can strongly impact their learning and academic outcomes [1,10]. However, so far, only a few studies have focused on formally assessing the fairness of ML techniques used in education, leading AIED researchers to call for extending and broadening our understanding of ML fairness in this context [1,10].

In this paper, we study the fairness of ML models trained to predict whether learners will fail to earn certification in a MOOC. This is a common ML task in research on MOOCs [14], because it is known that only a fraction of the enrolled students typically pass, due to issues with dropping-out, time management and lack of help, among others. The ability to predict as early as possible that a learner will fail can enable early personalized interventions to prevent it, e.g., [21]. Thanks to the ability of MOOCs to engage thousands and facilitate data collection, there have been several studies about building ML models for predicting MOOC completion, allowing us to better understand the most useful ML models and features for this task (see overview in [14]). However, in contrast, there is very little understanding about how fair these models are against their diverse groups of learners, as pointed out by Baker & Hawn [1], raising questions about whether everybody can fully trust and benefit from these models in MOOCs. Moreover, there seems to be discrepancies in how underrepresented groups complete MOOCs compared to historically privileged populations, thus risking to perpetuate existing differences as MOOCs tend to be a good entry point for adults to learn about emerging fields in a lifelong learning context [13].

We explore this issue by augmenting MOOC log data with a survey collecting key demographics and other educational-related information about the learners. Next, we trained ML models to predict at the end of each week which learners will fail the course, and we use the survey answers to analyze in-depth how fair the models are for various groups of learners. We do so using two fairness metrics previously used in education (ABROCA and MADD), to get a more comprehensive understanding of potential fairness issues in this context.

Our work contributes to existing research in three ways. First, despite the existence of extensive ML research on MOOCs, so far only two studies explored the fairness of ML models in MOOCs, solely regarding gender [7,17]. A few papers highlighted this gap [1,7], calling for broadening fairness analysis in MOOCs to take advantage of their large, international and diverse audience. Here, we study this gap by exploring possible unfairness in a MOOC, using key variables other than gender, such as learners' country of residence and socio-economic context.

Second, beyond MOOCs, we broaden the set of variables leveraged for ML fairness analysis of education. Indeed, existing work has focused primarily on demographic information such as gender and ethnicity [1]. Here, we do leverage standard demographics for the sake of comparison (namely gender, native language, parental education, country), but we also explore additional demographic variables that could impact the learners' time and motivation, such as their job and family status. We also consider non-demographic variables relevant in MOOCs, namely prior experience with MOOCs and whether learners enrolled on their own or as part of a university degree.

Third, we leverage a dataset where about half of the learners live and study in Africa. This is important because so far, fairness studies in education have been mostly limited to the USA, and to a lesser extent Western Europe and Eastern Asia [1,8], whereas ML fairness in other parts of the world have not been widely studied and thus remains a key gap in the literature. This is even more concerning in MOOCs as learners can typically come from anywhere in the world, but the ML models used in this context might not treat everybody fairly.

Overall, our work aims at broadening our understanding of ML fairness in AIED, by targeting a unique pool of learners and novel demographic and contextual variables in an understudied type of learning environments in ML fairness analysis (MOOCs), a much needed avenue of research for AIED.

2   Related Work

ML models have been extensively studied in MOOCs for the tasks of predicting as early as possible whether a learner will drop-out/successfully pass the course. For example, two recent systematic reviews [6, 14] covered several dozens of such research papers, highlighting the importance of this line of research. Such predictions are seen as key to detect at-risk learners early on and intervene in the hope of remediating the learners' difficulties whenever possible.

However, as said above, to the best of our knowledge, only two studies [7, 17] explored the fairness of these ML models in MOOCs, focusing on gender biases of ML models meant to predict drop-out in several North-American and Chinese MOOCs. They both found evidence of gender-biased ML predictions, especially depending on the gender imbalance in each MOOC, and no direct link among the fairness and performance of the ML models. This provides preliminary evidence of unfair predictions in MOOCs, and justifies further fairness analysis in this context, as we do in this paper for gender and beyond.

Besides MOOCs, several studies have focused on the fairness of ML models trained for educational purposes, including predicting university course completion from Learning Management Systems (LMS)'s data [9, 22], dropout in an e-learning platform for writing homework [16], and knowledge mastery in Intelligent Tutoring Systems (ITS) [24]. However, other work pointed out that these studies are often limited in terms of demographics (mostly gender and ethnicity) and learners' origin (mostly USA), both key issues when it comes to fairness [1, 8]. Still, some work have started to fill these gaps. In terms of demographics, research has studied the fairness of automatic scoring and dropout models towards the learners' country of origin and/or native language [3, 15, 16, 20]. Other work focused on biases in LMS, e-learning and ITS due to parental education and/or levels of poverty [9, 16, 22–24]. Compared to this line of work, we target a different learning environment (MOOC). We also reuse the aforementioned demographic variables to provide further understanding of their impact on fairness, and broaden them to include as well variables more specific to the asynchronous nature of MOOCs and the fact that they target populations larger than university students. This includes family and professional situations, as well as whether the learners attend a MOOC for the first time and how they enrolled.

Some fairness studies in education have considered learners beyond the USA, however, few of them explicitly included African countries. In [2], they focus on the fairness of ML models for automatic essay scoring, and include a pool of learners from Nigeria among 15 other countries. However, the amount of Nigerian's essays considered in the study was extremely low (less than 0.03%, versus over 75% from the USA), making it hard to draw strong conclusions as for the African learners. We follow this line of work, but with about half of the learners in our dataset living in Africa, allowing us to more formally compare the fairness of ML models between Africa and other continents. African countries are particularly interesting to consider as previous studies have already highlighted the specificity of their context, including strong differences in urban and rural experience [11], and reliance on Open Digital Spaces to compensate for the lack of physical infrastructures [18]. In [12], they study the fairness of ML predictors of academic achievement using the 2015 PISA dataset. This dataset includes learners from 65 countries, but only two from Africa (Algeria and Tunisia), for which specific results were not explicitly reported. In contrast, our dataset includes learners from 24 African countries[1], representing 53% of the learners in this dataset, and we explicitly study the fairness of our models towards them.

3 Dataset

The data used in this paper came from a French MOOC on Project Management[2] ran on an OpenEdX platform. It has been held twice a year since 2013, being one of the first French MOOCs to deliver a certificate. Over 22 editions, more than 250,000 persons have signed up to the MOOC (45,000 of whom have obtained a certificate). The main section of the MOOC which needs to be followed to obtain the basic certificate lasts four weeks, with a final exam being made available at the beginning of week 5 and remaining available for four weeks. The grade on this final exam represents 80% of the final grade, with the remaining 20% coming from weekly quizzes embedded throughout the MOOC. A grade of 50% or more is required to obtain the basic certificate at the end of the MOOC, which we use to define whether a learner passed the MOOC or not. In this paper, we focus more specifically on data from the 19th edition of the MOOC which took place in Spring 2022 for which $N = 9,752$ learners were enrolled, and 1,892 obtained the certificate. In addition to the raw OpenEdX logs that register every actions performed by learners while connected to the MOOC, we rely on data coming from a non-mandatory research survey we made available to learners at the beginning of the first week of the MOOC. Participants provided informed consent for their data to be used for research purposes. All data are fully anonymized and stored on secured servers with limited access for research purpose only, and learners can request their data to be deleted during and after the MOOC, in accordance with GDPR European regulations.

   Out of all learners, we retain those who filled in entirely the survey and passed all attention checks (i.e., items where students are asked to choose a predefined answer to limit the risk of random or automated survey filling). We also discard learners who reported not wanting to earn the certification, as they might be

---

[1] Algeria, Benin, Burkina-Faso, Burundi, Cameroon, DR Congo, Ivory Coast, Djibouti, Gabon, Guinea, Guinea-Bissau, Madagascar, Mali, Morocco, Mauritania, Mozambique, Niger, Central African Republic, Republic of the Congo, Rwanda, Senegal, Tchad, Togo, Tunisia.

[2] https://mooc.gestiondeprojet.pm.

inactive or curious for reasons others than learning. As a result, we obtained a dataset comprised of 1007 learners, among whom 383 (38%) successfully passed the course (i.e., earned the certification).

4 Analysis

We leverage the aforementioned MOOC's dataset to build classifiers that can predict binary labels of whether the learners passed the course. To simulate the predictions over time, the classifiers are trained and tested at the end of each of the four first weeks of the MOOC, until the final exam becomes available. The features and code of the ML and fairness analysis, described next, are available in open source at https://github.com/Mocahteam/GdPMOOC_Fairness.

Table 1. List of OpenEdX's events considered for feature engineering[a].

Component	Events
Video	Load, Play, Pause, Stop, Seek, Change speed, Show/hide transcript
Quiz	Load, Save, Check, Graded, Show answers
Forum	Create thread or response, View, Search, Vote
Platform	Go to, Next page, Previous page, Close

[a] Edx research guide: Events in the tracking logs. https://edx.readthedocs.io/ projects/devdata/en/stable/internal_data_formats/tracking_logs.html.

4.1 Features Engineering

We process the MOOC's logs data to generate a set of predictive features based on the events reported in Table 1. Specifically, for each event, we generate one feature corresponding to the count of that event performed by the learner, for a total of 20 features per learner. These events include most of the events relevant for our MOOC, and they are meant to capture the extent to which the learners interacted with each of the four main components of traditional MOOCs, namely the videos, the forum, the quizzes, and the platform. We focus on these four components because they have been the most used in similar prediction tasks to build features in MOOCs [14]. Indeed, our goal here is not to engineer novel features, but instead understand the fairness of the existing ML models typically trained to predict MOOC completion.

As said above, we train and test the classifiers for each of the first four weeks of the MOOC. This means that, at the end of each week, the features (i.e., *counts of events*) are generated using all of the available data so far from the current and past weeks[3], thus producing four different weekly feature sets.

---

[3] An alternative approach is to compute the features solely for the current week, thus "forgetting" the previous ones, however in our case, we found that both approaches yield similar results.

4.2 Machine Learning Setup

We train classifiers using a set of 6 machine learning (ML) techniques on the aforementioned sets of features to predict whether the learners passed the course. Specifically, we chose the six most successful ML techniques used for prediction tasks in MOOCs, as reported in the systematic review from Moreno-Marcos et al. [14]. We do so both because these models are more likely to generate accurate predictions on our data based on this previous work, and because our main goal is to measure the fairness of the models actually being the most commonly used in MOOCs. These techniques include: Logistic Regression (LR), Support Vector Machine (SVM), Random Forests (RF), Naive Bayes (NB), Stochastic Gradient Boosting (SGB) and Neural Network. For the later, we opted for a Multi-Layer Perceptron (MLP) with one hidden layer of 10 neurons (half the number of input features). This means that we train a total of 4 (weeks) x 6 (ML techniques) = 24 classifiers, implemented with the scikit-learn Python library (https://scikit-learn.org). We compare the classifiers against a majority-class baseline, which always predicts that the learners will fail.

We train and evaluate the classifiers using 5-folds stratified cross-validation and a standard 0.5 decision threshold. Stratification means that the proportion of the labels ("pass", "not pass") is kept consistent across the folds. To remember, 38% of the learners passed in our dataset. We perform feature selection within the train set by discarding highly correlated features to avoid issues with colinearity. We also attempted to balance the data with SMOTE and fine-tune the hyperparameters of the classifiers using a grid search approach (within the train sets only), however, this did not significantly impact the performances of the classifiers. Hence, we retain the default values of all hyperparameters in the rest of the paper for simplicity and to facilitate reproducibility[4]. We report classification performance in terms of the F1 score averaged over the five folds, as the F1 metric is suitable for imbalanced data.

4.3 Fairness Analysis

In this part, we explore the possible unfairness of the classifiers using a set of demographics and contextual variables (called *protected variables* from now on) listed in Table 2. For each of the variables, we bin the learners into two groups G_0 and G_1, as shown in Table 2. This stems from the two fairness metrics we have chosen (ABROCA and MADD, see below) which have been designed to compare only two groups. Although pairwise comparisons of more than two groups is possible, it significantly complicates the interpretation. Eight of these variables are demographics, as follows:

- Age, Gender and First Language (Lang for short) have been already used in AIED fairness analysis (see Sect. 2). We leverage them to assess whether they generate similar (un)fair predictions in our data. Also, only Gender was used

---

[4] The grid search space and default values are documented in our Github repository linked above.

in the context of a MOOC, hence studying `Age` and `Lang` is meant to provide additional insights about fairness in MOOCs. In particular, `Age` might play a role due to the digital divide among generations, and `Lang` might be key due to the fact that MOOCs are typically served in one language only (in our case French), which might not be the first language of all learners.

- The country of residence (`Country`) generates in our data an almost even split between African and Western European learners. Namely, 24 African countries are represented in our dataset, all of them having a Human Development Index (HDI) on the low to medium side (< 0.7). We also lump 38 Haitian learners with the African ones, as Haiti's HDI is similar to these countries. In contrast, the rest of the learners come from economically privileged countries with very high HDI (≥ 0.9), mostly in Western Europe. This unique split of learners allow us to conduct a formal analysis of fairness across continents.
- `JobStatus` and `HasChildren` are two variables that we leverage to capture how much time the learners can dedicate to learning. Indeed, MOOCs' learners tend to be older than in traditional high school and university contexts (in our data, age ranges from 21 to 74, median of 35), and thus generate unique learning experiences. Here, nearly half of the learners declare having a family and children, and are committed to a job, which can all impact the time and energy they are able to dedicate to the MOOC compared to full-time single university students. These features are novel in AIED fairness analysis at large.
- The highest levels of education of the learners (`StudentEdu`) and of their parents (`ParentsEdu`) capture the access to and success in education, which can arguably strongly impact learning with a MOOC where mastering self-regulated learning skills is critical to succeed. While similar variables have been used in previous fairness analysis, as said in Sect. 2, it was not in a MOOC context.

In addition, we leverage two contextual variables that are more specific to MOOCs but can both strongly impact the behaviors of the learners and which are novel in fairness studied in education, so far focused solely on demographics:

- Whether this is the very first MOOC attended to by the learners (`FirstMOOC`), as these learners might struggle more with this form of remote learning.
- Whether the learners registered on their own, versus were enrolled by their university as part of their curriculum (`Cohort`), as we can expect that the later will have more (external) motivation to pass the MOOC and more opportunity to interact face-to-face with peers, a key factor as socialization helps in limiting drop-out. In our case, 10 universities in Europe and Africa enrolled 241 learners.

To assess the fairness of the trained ML classifiers, we leverage two fairness metrics that have been proposed in AIED research, namely ABROCA [7] and MADD [19]. Similarly to the F1 score, these metrics are computed for each cross-validation fold, and we report their average over all folds. ABROCA measures the areas between the ROC curve of two groups of learners (e.g., female versus male)

Table 2. Definition and size (N) of the binary groups for each protected variable.

Variable	Definition of G_0	Definition of G_1	N(G_0)	N(G_1)
Age	Age $>= 35^a$	Age <35	519	488
Gender[b]	Female	Male	569	436
Lang	Not french	French	283	724
Country	Africa + Haiti	OECD countries	536	471
JobStatus	Employed, business owner	Unemployed, students, retired	452	555
HasChildren	Yes	No	446	561
StudentEdu	Less than a master degree[a]	Master's degree and above	485	522
ParentsEdu[b]	Less than a bachelor degree[a]	Bachelor degree and above	456	400
FirstMOOC	Yes	No	314	693
InCohort	No	Yes	766	241

[a] Corresponds to the median. [b] Only 2 learners reported a gender other than male and female, which is too few for a meaningful analysis, and 151 learners reported not knowing their parents' education. This leaves us with 1005 learners for Gender and 856 learners for ParentsEdu. For the other variables, we do keep the full sample of 1007 learners.

to identify a model's ability to perform similarly for both groups. ABROCA ranges from 0 (fairest) to 1 (most unfair). MADD leverages the histograms of probabilities outputted by a model for two groups to compute the areas where the histograms do not overlap, i.e., behave differently. MADD ranges from 0 (fairest) to 2 (most unfair). These two metrics are complementary, as ABROCA focuses on the difference in predictive performances across two groups given the ground truth, whereas MADD assesses how differently the model behaves towards the two groups regardless of the ground truth. In fairness analysis, this corresponds to two important and complementary concepts: separation (classification performance should be equal within groups, i.e., ABROCA) and independence (the predictions should not be related to any sensitive variable, i.e., MADD) [5]. This is an important distinction, because the ground truth used by separation metrics allows measuring how accurate the model is across groups, but can be biased itself [5]. Both ABROCA and MADD are publicly available in open-source Python packages[5], which is important for reproducibility.

As a final metric, we build an aggregated indicator (F1+FAIR) of the performance metric (F1) and the fairness ones (ABROCA and MADD), rescaled in a same interval $[0, 1]$. We then compute their weighted average as follows:

$$F1+FAIR = \alpha \times F1_{scaled} + \beta \times \frac{\beta_1 \times (1 - ABROCA_{scaled}) + \beta_2 \times (1 - MADD_{scaled})}{(\beta_1 + \beta_2)}$$

Such as $\alpha + \beta = 1$ meaning F1+FAIR ranges from 0 (worst) to 1 (best). Here we give equal weights ($\alpha = \beta = 0.5$) to performance and fairness, and equal weight to ABROCA and MADD ($\beta_1 = \beta_2 = 1$) to get of sense of which classifiers offer the best compromise, similarly to how F1 balances precision and recall, but the

---

[5] https://pypi.org/project/abroca/ and https://pypi.org/project/maddlib/.

value of these weights could be explored in future research. Here, we mainly use F1+FAIR as a quick way to gauge the performance-fairness trade-off.

5   Results

To investigate the performance and fairness of the classifiers, we first compare them on average across all weeks to identify the best/fairest ones (Sect. 5.1). Next, we explore these results per week, to assess whether fair early prediction is possible (Sect. 5.2). Lastly, we delve into the fairness results per protected variable and week, to isolate the specific biases in the ML predictions (Sect. 5.3).

5.1   Classifiers Evaluation

Table 3 provides the results of all metrics reported in Sect. 4 for all classifiers, averaged over the folds and the feature sets (e.g., 0.82 is the average F1 performance of all LR classifiers trained respectively on the data accumulated so far at the end of each of the 4 weeks). For the fairness metrics, we consider here only their average over all protected variables to ease comparison of the classifiers, but we explore the fairness per protected variable later on in Sect. 5.3.

Table 3. Average performance and fairness per classifier.

Metric	Baseline	LR	MLP	NB	RF	SGB	SVM	Statistically best model(s)
F1	0.47	0.82	0.76	0.72	0.79	0.78	0.80	LR ≈ SVM ≈ RF ≈ SGB
ABROCA	-[a]	0.06	0.08	0.07	0.07	0.07	0.08	LR ≈ SGB ≈ RF ≈ NB
MADD	-[a]	0.86	0.71	0.31	0.94	0.76	0.81	NB
F1+FAIR	-[a]	0.62	0.55	0.63	0.54	0.59	0.57	NB ≈ LR ≈ SGB

[a] Studying the baselines' fairness is irrelevant, because they always generate the same prediction.

Table 3 shows that all ML classifiers achieve substantially higher F1 than the baseline, which is consistent with the literature. However, the trends in terms of fairness are more complex to analyze. We investigate the performance-fairness trade-off by running a statistical analysis to formally compare the classifiers. Namely, for each metric, we run a linear mixed-effect model with that metric at each fold of cross-validation as the dependent variable, classifiers as the factor, and week as the random effect. To account for multiple comparisons, we adjust all p-values with the Benjamini & Hochberg procedure. All mixed models reveal a significant main effect of classifier ($p < 0.001$) with large effect sizes ($\eta^2 > 0.8$). Next, we run adjusted pairwise comparisons to explore these main effects, and report in the last column of Table 3 which classifier(s) significantly outperform all of the others, with no statistical difference among them. The pairwise comparisons reveal three main findings:

1. In terms of performance (F1), the best classifiers are LR, SVM, RF and SGB, with F1 close to 0.8. In contrast, NB is significantly worse than all other ML classifiers (F1 of 0.72), but still significantly outperforms the baseline.
2. As for fairness, NB yields significantly better MADD than all other classifiers, whereas for ABROCA, LR, SGB, RF, and NB are tied for the best.
3. In terms of F1+FAIR, NB, LR and SGB are tied for the best.

All in all, these results show that NB, LR and SGB offer the best compromise between performances and fairness, especially as revealed by the F1+FAIR metric. Specifically, LR exhibits the highest F1, but yields contradicting fairness results (ranked first for ABROCA, but fifth for MADD). NB, on the other hand, is much less accurate than LR, but performs very well on all fairness metrics, suggesting that this model is more fair at the expense of lower performances. Lastly, SGB appears to offer a compromise between LR and NB: it is statistically as good as LR in terms of F1, for decent fairness results (ranked second for ABROCA, third for MADD). Thus, for the rest of the paper, we will focus only on LR, NB and SGB, as we want to ascertain what levels of unfairness remain even among the best and/or most fair classifiers.

Another interesting finding is that the ABROCA values are very low (less than 0.1), whereas the MADD values are overall on the medium side (recall that ABROCA ranges from [0-1] and MADD [0-2], 0 being the best for both). This indicates that the classifiers generate similar performances across groups on average for all protected variables (low ABROCA), but their behaviors are quite impacted by the learners' group membership (medium MADD).

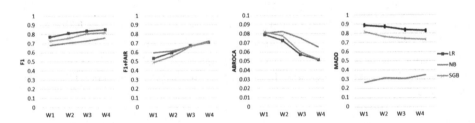

Fig. 1. Weekly results for all performance and fairness metrics. Notice that the Y-axis for ABROCA is narrower for readability.

5.2   Feature Sets Evaluation

We now look at the results per feature set, in Fig. 1. The figure indicates that all classifiers improve on the F1 metric when data is accumulated over time, which is consistent with existing work. F1+FAIR and ABROCA also get better as more data is accumulated overtime, whereas MADD appears more stable. An ANOVA analysis confirms that there is a significant main effect of feature set on F1, F1+FAIR and ABROCA ($p < 0.005, \eta^2 > 0.38$), but not on MADD

$(p = 0.06, \eta^2 = 0.01)$. Adjusted pairwise comparisons among the weeks further show that the best F1, F1+FAIR and ABROCA are achieved at Week 3 and Week 4, whereas Week 1 is worse than all feature sets. Overall, these findings do show that accumulating data over at least three weeks is valuable not only in terms of performance, but also ABROCA. They also suggest that MADD is less sensitive to time, perhaps because the classifiers behave similarly across the weeks, in spite of their improved performance.

5.3 Fairness per Protected Variable

We now explore the fairness of the classifiers for each protected variable reported in Table 2. Following up on the previous analysis, we focus on ABROCA and MADD averaged over all feature sets, as well as specifically on the features sets for week 1 and week 4 (the later capturing all logs from week 1–4). We do so for simplicity, to get a sense of the fairness early in the MOOC (Week 1), just before the final exam (Week 4), and overall (average). Table 4 provides these fairness results with ABROCA, and Table 5 with MADD. Please refer to Table 2 for the definition of G_0 and G_1. In general, Table 4 and Table 5 show that the direction ($\downarrow/\uparrow$) are very inconsistent and often contradictory both among classifiers, and among ABROCA and MADD. One possible explanation is that ABROCA captures very low levels of unfairness in the classifiers as said above, and thus the actual directions of the bias ABROCA captures is not that important, since all classifiers are overall quite fair. As for MADD, NB is overall fairer than LR and SGB, hence the same explanation might hold: the direction of NB's bias might not be that important when its MADD value is low.

That being said, both ABROCA and MADD yield some of their highest values for InCohort. In terms of MADD, all models are biased against learners not in cohort, whereas ABROCA's results mostly show the opposite. While difficult to interpret, these results do suggest that the fairness of the classifiers is impacted

Table 4. ABROCA results ($\downarrow$ means more unfairness towards G_0, $\uparrow$ towards G_1).

Variable	Week 1			Week 4			Weekly average		
	LR	NB	SGB	LR	NB	SGB	LR	NB	SGB
Age	0.07 ↓	0.09 ↑	0.07 ↓	0.04 ↑	0.06 ↑	0.04 ↑	0.05 ↓	0.07 ↑	0.05 ↓
Gender	0.10 ↑	0.07 ↑	0.12 ↑	0.05 ↑	0.06 ↑	0.05 ↑	0.08 ↑	0.07 ↑	0.08 ↑
Country	0.08 ↓	0.08 ↑	0.08 ↓	0.05 ↓	0.07 ↑	0.06 ↓	0.07 ↓	0.08 ↑	0.07 ↓
Lang	0.08 ↓	0.08 ↑	0.09 ↑	0.05 ↑	0.04 ↑	0.05 ↑	0.06 ↑	0.07 ↑	0.07 ↑
JobStatus	0.10 ↑	0.08 ↓	0.09 ↑	0.04 ↑	0.06 ↓	0.05 ↑	0.06 ↑	0.08 ↓	0.07 ↑
HasChild	0.06 ↓	0.08 ↑	0.06 ↓	0.04 ↑	0.07 ↑	0.05 ↓	0.05 ↓	0.08 ↑	0.06 ↓
LearnerEdu	0.08 ↑	0.08 ↓	0.06 ↑	0.04 ↑	0.06 ↑	0.05 ↓	0.06 ↑	0.07 ↓	0.05 ↑
ParentsEdu	0.09 ↓	0.09 ↓	0.09 ↓	0.06 ↓	0.07 ↓	0.08 ↓	0.08 ↓	0.09 ↓	0.09 ↓
FirstMOOC	0.06 ↑	0.06 ↓	0.08 ↑	0.05 ↑	0.05 ↑	0.04 ↑	0.06 ↑	0.06 ↑	0.06 ↑
InCohort	0.08 ↑	0.08 ↑	0.11 ↓	0.10 ↑	0.12 ↑	0.06 ↑	0.09 ↑	0.10 ↑	0.08 ↑

Table 5. MADD results (↓means more unfairness towards G_0, ↑towards G_1).

Variable	Week 1			Week 4			Weekly average		
	LR	NB	SGB	LR	NB	SGB	LR	NB	SGB
Age	0.71 ↓	0.16 ↓	0.69 ↓	0.63 ↓	0.27 ↓	0.59 ↓	0.67 ↓	0.22 ↓	0.62 ↓
Gender	0.80 ↓	0.15 ↑	0.87 ↓	0.75 ↑	0.22 ↑	0.66 ↑	0.78 ↓	0.20 ↑	0.78 ↓
Country	1.22 ↓	0.31 ↓	0.91 ↓	1.11 ↓	0.48 ↓	0.92 ↓	1.14 ↓	0.41 ↓	0.87 ↓
Lang	0.86 ↓	0.20 ↑	0.86 ↓	0.87 ↓	0.31 ↓	0.69 ↓	0.87 ↓	0.27 ↓	0.74 ↓
JobStatus	0.74 ↑	0.26 ↑	0.66 ↑	0.72 ↑	0.30 ↑	0.62 ↑	0.74 ↑	0.27 ↑	0.63 ↑
HasChild	0.69 ↓	0.25 ↓	0.69 ↓	0.63 ↓	0.24 ↓	0.57 ↓	0.66 ↓	0.23 ↓	0.63 ↓
LearnerEdu	0.74 ↓	0.16 ↓	0.66 ↑	0.67 ↓	0.20 ↑	0.56 ↑	0.72 ↓	0.18 ↑	0.62 ↓
ParentsEdu	0.74 ↑	0.17 ↑	0.68 ↓	0.66 ↓	0.20 ↑	0.62 ↓	0.67 ↑	0.20 ↑	0.66 ↓
FirstMOOC	0.69 ↓	0.20 ↓	0.72 ↓	0.68 ↑	0.25 ↑	0.68 ↑	0.72 ↑	0.22 ↑	0.67 ↑
InCohort	1.69 ↓	0.81 ↓	1.44 ↓	1.56 ↓	1.01 ↓	1.40 ↓	1.59 ↓	0.90 ↓	1.43 ↓

by `InCohort`, perhaps because these students might have used the MOOC in a more intensive way as part of their degree or because they had a high level of interaction with peer learners outside of the MOOC, which may be less true for other learners. MADD also captures high unfairness for `Country`, showing that all classifiers (especially LR and SGB) are more biased against African and Haitian learners. This confirms how important it is to consider the country of origins in MOOCs' fairness analysis, as well as the high need to leverage data from understudied continents like Africa to actually measure how trustful our classifiers are. Empirically, from forum posts, we also know that a higher proportion of MOOC users in Africa tend to download some of the resources from the MOOC for offline consultation, which would result in an overall lower number of actions and significantly different interaction patterns.

There are a few more interesting findings for the other variables. Specifically, MADD captures similar trends in all classifiers for `Age` (more biased against older learners), `JobStatus` (more biased against unemployed and students), and `HasChildren` (more biased against learners who have children). These trends are expected at least for `Age` and `HasChildren` (cf. Sect. 4.3), as older learners with children are typically less likely to have ample time to study. For `JobStatus`, while we hypothesized that having a job might impend learning, MADD suggests otherwise. The fact that we grouped together unemployed learners and full-time students might explain this finding, which could be addressed in future work.

When comparing our results with [7, 17] for `Gender`, we can notice that both ABROCA and MADD yield some of their highest values for this variable, thus suggesting some levels of gender bias in the predictions. However, the direction of this bias differs across the classifiers and metrics, suggesting the need to run further analysis of gender biases on MOOC datasets.

6   Conclusion and Discussion

In this work, we investigated the performance and fairness of ML classifiers in the task of predicting completion of a MOOC, which is important to assess the possibility of fair interventions to all at-risk MOOC learners. We did so by training a set of standard ML classifiers on features capturing the learners' behaviors in a MOOC. We then explored the possible biases in the predictions based on demographics and contextual information we collected via a survey at the beginning of the MOOC. Our main contribution is addressing the key gap of understanding ML fairness in MOOCs, which so far has been addressed solely by two studies [7,17] for the gender bias. We contribute to their work by providing more insights into biases based on a larger set of protected variables. A second contribution is studying a dataset where more than half of the learners live and learn in Africa, an understudied continent in AIED fairness analysis, despite the many calls for expanding fairness analysis in this context.

Our results do show that biases can arise in MOOCs, especially towards African and Haitian learners, as well as depending on whether the learners have children and enrolled in the MOOC as part of their university's degree require-ment. This highlights the need to explore further these biases and consider, whenever possible, learners from understudied regions of the globe and contex-tual variables beyond demographics.

Our results also suggest that in general, there is to some extent a trade-off in terms of performance and fairness. Interestingly, we did find that the NB classi-fiers consistently generate less accurate but fairer predictions, which was already shown in [19] in a different context. Still in line with [19], we also found that LR is very accurate but with varying levels of fairness across protected attributes. It is worthwhile investigating whether these trends still hold in other educational datasets for these models, as it would offer key insights for stakeholders in terms of what classifiers to use (NB for fairness, LR for accuracy).

Lastly, the two fairness metrics we used (ABROCA and MADD) generated contrasted results. Namely, ABROCA assessed the classifiers as much fairer than MADD. The key difference between both metrics is that unlike MADD, ABROCA leverages the ground truth (whether the learners actually failed). However, the ground truth might be biased itself (e.g., learners with easy access to higher education might be more successful in MOOCs), which is not accounted for by metrics like ABROCA. MADD informs about how independent the clas-sifiers' output is from all protected variables, regardless of the accuracy of this output. While both metrics can be useful depending on the context and stake-holders, we do show the value of leveraging and carefully comparing both.

There are some limitations in our work, which can be addressed in future work. First, we focus on solely one MOOC, meaning that our results might not generalize to other MOOCs. Hence, we see our study as a valuable prelimi-nary step towards building a stronger understandings of ML fairness in MOOCs. Focusing on one MOOC also allowed us to collect the demographic and contex-tual information we needed about the learners, as this information is typically not available in MOOCs, and it entailed modifying the MOOC to add a dedicated survey. All of this was possible solely because we could collaborate with the MOOC's team. This MOOC was also a suitable choice to better understand issues with understudied countries, thanks to its strong popularity in Africa and Haiti. In future work, we plan to replicate our analysis on future sessions of the MOOC, as well as extend it to other MOOCs popular in Africa.

A second limitation is the fact that we binarized the protected variables, both for simplicity given the various scales of each variable, and because ABROCA and MADD can only handle binary groups by default, as they are limited to comparing two ROC curves or two histograms, respectively. While several fairness studies also used binary groups for that reason, e.g., [9,19,22,24], in future work, it could be relevant to examine finer-grained ones. In particular, we plan to study fairness across specific sub-parts of Africa, and by country when possible.

Acknowledgement. This work was supported by the Sorbonne Center for Artificial Intelligence (SCAI) and the *Direction du numérique pour l'éducation* (DNE) of the French Ministry of Education. The opinions expressed in this paper are those of the authors and do not necessarily reflect those of DNE and SCAI. We thank Rémi Bachelet and the MOOC GdP's team for the data collection.

References

1. Baker, R.S., Hawn, A.: Algorithmic bias in education. Int. J. Artif. Intell. Educ. 1052–1092 (2021)
2. Bridgeman, B., Trapani, C., Attali, Y.: Considering fairness and validity in evaluating automated scoring. In: Proceedings of the Annual meeting of the National Council on Measurement in Education, San Diego, CA (2009)
3. Bridgeman, B., Trapani, C., Attali, Y.: Comparison of human and machine scoring of essays: differences by gender, ethnicity, and country. Appl. Meas. Educ. **25**(1), 27–40 (2012)
4. Buolamwini, J., Gebru, T.: Gender shades: intersectional accuracy disparities in commercial gender classification. In: Proceedings of the 1st Conference on Fairness, Accountability and Transparency, pp. 77–91 (2018)
5. Castelnovo, A., Crupi, R., Greco, G., Regoli, D., Penco, I.G., et. al.: A clarification of the nuances in the fairness metrics landscape. Sci. Rep. **12**(1), 4209 (2022)
6. Chen, J., Fang, B., Zhang, H., Xue, X.: A systematic review for MOOC dropout prediction from the perspective of machine learning. Interact. Learn. Environ. 1–14 (2022)
7. Gardner, J., Brooks, C., Baker, R.: Evaluating the fairness of predictive student models through slicing analysis. In: Proceedings of the 9th International Conference on Learning Analytics & Knowledge, pp. 225–234 (2019)
8. Idowu, J.A.: Debiasing education algorithms. Int. J. Artif. Intell. Educ. 1–31 (2024)
9. Kai, S., et al.: Predicting student retention from behavior in an online orientation course. In: Proceedings of the International Conference on Educational Data Mining, pp. 250–255 (2017)
10. Kizilcec, R.F., Lee, H.: Algorithmic fairness in education. In: The Ethics of Artificial Intelligence in Education, pp. 174–202. Routledge (2022)
11. Lembani, R., Gunter, A., Breines, M., Dalu, M.T.B.: The same course, different access: the digital divide between urban and rural distance education students in South Africa. J. Geogr. High. Educ. **44**(1), 70–84 (2020)

12. Li, X., Song, D., Han, M., Zhang, Y., Kizilcec, R.F.: On the limits of algorithmic prediction across the globe. arXiv preprint arXiv:2103.15212 (2021)
13. Littenberg-Tobias, J., Reich, J.: Evaluating access, quality, and equity in online learning: a case study of a MOOC-based blended professional degree program. Internet High. Educ. **47**, 100759 (2020)
14. Moreno-Marcos, P.M., Alario-Hoyos, C., Muñoz-Merino, P.J., Kloos, C.D.: Prediction in MOOCs: a review and future research directions. IEEE Trans. Learn. Technol. **12**(3), 384–401 (2018)
15. Naismith, B., Han, N.R., Juffs, A., Hill, B., Zheng, D.: Accurate measurement of lexical sophistication in ESL with reference to learner data. In: Proceedings of the 11th International Conference on Educational Data Mining, pp. 259–265 (2018)
16. Rzepka, N., Simbeck, K., Müller, H., Pinkwart, N.: Fairness of in-session dropout prediction. In: Proceedings of the International Conference on Computer Supported Education, pp. 316–326 (2022)
17. Sha, L., Raković, M., Das, A., Gašević, D., Chen, G.: Leveraging class balancing techniques to alleviate algorithmic bias for predictive tasks in education. IEEE Trans. Learn. Technol. **15**(4), 481–492 (2022)
18. Sylla, K., Nkwetchoua, G., Bouchet, F.: How does the use of open digital spaces impact students success and dropout in a virtual university? In: Proceedings of the Conference on Cognition and Exploratory Learning in Digital Age, pp. 251–258 (2022)
19. Verger, M., Lallé, S., Bouchet, F., Luengo, V.: Is your model "MADD"? A novel metric to evaluate algorithmic fairness for predictive student models. In: Proceedings of the International Conference on Educational Data Mining, pp. 91–102 (2023)
20. Wang, Z., Zechner, K., Sun, Y.: Monitoring the performance of human and automated scores for spoken responses. Lang. Test. **35**(1), 101–120 (2018)
21. Xing, W., Du, D.: Dropout prediction in MOOCs: using deep learning for personalized intervention. J. Educ. Comput. Res. **57**(3), 547–570 (2019)
22. Yu, R., Lee, H., Kizilcec, R.F.: Should college dropout prediction models include protected attributes? In: Proceedings of the ACM Conference on Learning@Scale, pp. 91–100 (2021)
23. Yudelson, M., Fancsali, S., Ritter, S., Berman, S., Nixon, T., Joshi, A.: Better data beats big data. In: Proceedings of the International Conference on Educational Data Mining (2014)
24. Zambrano, A.F., Zhang, J., Baker, R.S.: Investigating algorithmic bias on Bayesian knowledge tracing and carelessness detectors. In: Proceedings of the 14th International Conference on Learning Analytics & Knowledge, pp. 349–359 (2024)

EngageME: Exploring Neuropsychological Tests for Assessing Attention in Online Learning

Saumya Yadav[1]([✉]) [iD], Momin N. Siddiqui[2] [iD], Yash Vats[1],
and Jainendra Shukla[1] [iD]

[1] HMI Lab, IIIT-Delhi, New Delhi, India
{saumya,yash18204,jainendra}@iiitd.ac.in
[2] Georgia Institute of Technology, Atlanta, USA
msiddiqui66@gatech.edu

Abstract. During the pandemic, online learning has gained immense popularity. However, assessing student engagement in online settings remains challenging due to reliance on potentially biased and logistically complex self-reporting methods. This study investigates the use of neuropsychological tests, originally designed for attention assessment, to measure cognitive engagement in online learning. To conduct our analysis, we initially correlated clinical models of attention with pedagogical approaches in online learning. Subsequently, we pinpointed the three most crucial attention types in online learning-selective, sustained, and alternating attention. We used three neuropsychological assessments to evaluate attention in a cohort of 73 students contributing to the *EngageME* dataset. We manually annotated students' facial videos during neuropsychological assessments, revealing substantial agreement (Krippendorff's Alpha: 0.864) and a strong correlation (Spearman's Rank Correlation: 0.673) with neuropsychological test scores. This confirms the convergent validity of our approach in measuring attention during online learning. We further propose Nuanced Attention Labeling using neuropsychological test scores-based models in online learning attention assessment, enhancing sensitivity to nuanced cognitive engagement. To assess the reliability of our approach to online learning, we performed a user study in online settings. This work implies the potential for a more accurate and nuanced assessment of students' cognitive engagement in online learning, contributing to the refinement of personalized and effective educational interventions.

Keywords: Affective Computing · Education Technology · Cognitive Engagement · Interactive learning environments

1 Introduction

In the context of the recent and unprecedented COVID-19 pandemic, characterized by stringent social isolation measures, a profound transformation has

A. M. Olney et al. (Eds.): AIED 2024, LNAI 14829, pp. 394–408, 2024.
https://doi.org/10.1007/978-3-031-64302-6_28

occurred in the education system [13]. This transformation from traditional classrooms to predominantly online learning relies on cutting-edge technologies, specifically webcams and microphones. The adoption of this technology-driven approach offers advantages such as cost-effectiveness and improved accessibility, contributing to societal sustainability. However, these approaches create a challenge for instructors to assess student engagement during online classes [13]. Both instructors and students significantly benefit from accurate and timely information on student engagement, impacting learning and retention [4,13]. Ongoing studies explore modalities like audio and video to quantify student engagement in online learning [14], with webcams emerging as a prevalent tool due to their ubiquity and ability to capture essential visual cues.

Previous research exhibits inconsistencies in examining the relationship between engagement and attention [12,30]. Engagement in education is a multifaceted concept involving emotional, behavioral, and cognitive dimensions [4]. Here, emotional pertains to effective emotional responses, behavioral involves observable actions like asking questions, and cognitive involves actively seeking to understand new information as a psychological process that requires attention and investment [4,18]. Consequently, attention emerges as a fundamental element of cognitive engagement, encompassing various cognitive states, processes, and abilities. Student's cognitive engagement is considered more reliable learning than behavioral or emotional engagement [25]. Hence, we try to asses cognitive engagement in the proposed work and take attention as a latent variable for student engagement in the proposed work. It is distinguished mainly by its internal nature, which differs from the externally observable aspects of emotional and behavioral activity. Theorists recommend two measurement methods for measuring all facets of engagement: by observing external factors like body language and posture and by assessing internal factors related to the learner's cognitive and affective functioning [23].

Our approach to quantifying cognitive engagement draws upon a synthesis of neuroanatomical theories, factor analysis of psychometric assessments, cognitive processing frameworks, and clinically derived models, incorporating internal factors to assess attention [28]. This clinical-based model consists of five components for attention: alternating attention enables seamless shifts in focus, divided attention adeptly responds to multiple tasks, focused attention responds to sensory stimuli, selective attention manages distractions, and sustained attention maintains focus during repetitive activities, revealing enduring aspects over time. This model can be used to map students' attentional dynamics, summarized in Table 1, with each facet playing a unique role in the educational context. We excluded focused attention, as our focus lay on prolonged attention span [1]. Further previous research also showed that divided attention diminishes focus on a single task when multiple focuses occur simultaneously [7,11]. Thus, our experiment finally assessed three key attention facets-alternating, sustained, and selective-essential for evaluating cognitive engagement in our research context.

Researchers have identified facial features as robust indicators of cognitive engagement [6,13,21,23,31,34]. Hence, we used external factors, which include facial landmarks, head poses, Facial Action Units (FAUs), gaze tracking, and

Table 1. Assessing Cognitive Engagement in Education

Attention Type	Pedagogical Implications	Attention Assessment Tool Example
Alternating Attention [20]	Enhances adaptability in shifting focus between tasks, subjects, and classroom activities, optimizing versatility in educational contexts	Trail Making Test
Divided Attention [11]	Handling multiple tasks simultaneously, like taking notes and participating in an online discussion	Dual-Task Paradigm
Focused Attention [15]	Enhances response to specific sensory stimuli and understanding of educational material.	Attention Network Test
Selective Attention [29]	Vital for concentrating on specific educational tasks, filtering distractions, and absorbing relevant information.	Stroop Test
Sustained Attention [27]	Essential for long lectures, extended study sessions, and completing assignments efficiently.	Continuous Performance Test

facial feature extractions for attention assessment. We acknowledge that screen-based interactions in online learning are intricate [13]. However, it's crucial to note that standardized neuropsychological tests are crafted to assess various facets of attention despite their seemingly elementary nature. Exploring these tests can deepen our understanding of the cognitive processes involved in students' attention and performance in virtual learning environments [18].

Previous literature heavily relies on self-reporting methods to establish its ground truth [8,21,23]. However, this practice can introduce potential distractions and may not provide presenters with truly objective information. Moreover, self-reporting methods and manual annotations tend to yield discrete annotations, which fail to capture nuanced attention during online learning [8,22]. In light of these challenges, this paper introduces *EngageME*, an innovative dataset utilizing neuropsychological test scores to establish a consistent attention label. Additionally, we propose the *Nuanced Attention Labeling method*, finely assigning attention weights to capture subtle variations. We aim to build a foundational understanding of attention using standardized tests that tap into fundamental cognitive mechanisms associated with engagement. The approach integrates unobtrusive behavioral cues from a webcam, offering a holistic insight into cognitive engagement. To address these gaps, we outline our key contributions as:

- We introduce the EngageME dataset, which represents a pioneering in-the-wild data collection from 73 students undergoing neuropsychological tests online, alongside their attention scores.
- We posit that neuropsychological tests will detect attention during the performance of cognitive engagement tasks in an online learning environment.
- We propose the Nuanced Attention Labeling method, an effective approach for assessing attention by capturing subtle changes in cognitive engagement.

2   Related Work

Engagement is a complex and multifaceted phenomenon that is important in learning and academic success. It is often confused with attention, despite their distinct roles in cognitive functioning [12]. Some studies interchangeably use attention and engagement, limiting the understanding of these phenomena [30,31]. Equating sustained attention with engagement and relying solely on self-reported methods, as seen in previous research [23,34], fails to capture the full spectrum of attention and may inaccurately reflect performance. Some studies focus on predicting disengagement using specific facets [32] or analyzing behaviors like mouse movements and keyboard keystroke behaviors [3], but a precise assessment requires considering various engagement and attention facets, not just single attention or engagement.

Previous studies commonly rely on self-evaluation [8,14,21,34] to establish ground-truth labels for attention, but this method introduces vulnerability to reporter biases [13]. Learners' self-assessment of engagement may lead to inaccuracies [9], potentially influenced by their capacity for accurate self-assessment or a desire to avoid repercussions [10]. Additionally, self-reporting or manual annotation methods yield discrete annotations [8,22]. Which can be unreliable for attention assessment. Nuanced annotation is essential to capture subtle variations in attention levels, ensuring unbiased evaluation.

Similarly, previous studies have recognized the benefits of using rating scales and checklists throughout the annotating process to reduce self-report biases. However, these methods are not without their challenges. This annotation method needs a lot of effort and time [6,16,33]. For instance, accurately tracking each learner's engagement in large classrooms or online settings becomes a major challenge for teachers. Learners may feign engagement but are not genuinely involved in the assigned activities. Despite their potential biases, effort requirements, and subjectivity, these annotation methods are unreliable. Real-time annotation measures may also impact the state of attention of the instructor and students while actively participating in a class, further compromising their dependability.

In conclusion, prior research has exhibited a certain degree of inconsistency when probing the relationship between engagement and attention, thereby casting doubts on the validity of the established ground truth used for assessing attention. Additionally, prevalent literature tends to rely heavily on explicit methods for collecting ground truth, a practice that not only introduces potential distractions but also escalates the cognitive workload for users. Moreover, these methods may fall short of furnishing the nuanced annotation that attention requires. To address these challenges, we propose standardized neuropsychological assessments administered by trained professionals. These tests, covering alternate, selective, and sustained attention, provide a more objective measure than conventional self-reporting, enhancing our understanding of attention dynamics without relying on subjective reports. Above all, currently, there is no dataset offering ground-truth labels for different attention components.

3   Hypothesis Development

To address the proposed research objectives, we conducted research under the following hypothesis:

- **H1:** Neuropsychological tests will detect attention while performing cognitive engagement tasks that occur in an online learning environment.
- **H2:** Nuanced Attention Labeling in online learning attention assessment will result in heightened sensitivity to intricate cognitive engagement.

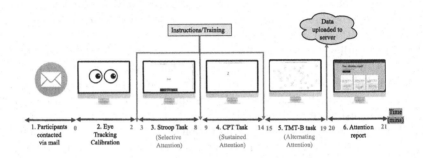

Fig. 1. Timeline of the EngageMe Data Collection Experiment.

For our first hypothesis, we aim to establish a baseline understanding of attention through these standardized tests, capturing core cognitive mechanisms underlying cognitive engagement, as included in Table 1. While neuropsychological tests traditionally assess attention within clinically derived models, they aren't commonly used for evaluating attention during cognitive engagement tasks in online learning environments. To validate this hypothesis, we establish the convergent validity of our new annotation method. We also performed a user study to validate the assessment for the online learning environment. In our second hypothesis, we suggest that using Nuanced Attention Labeling, as opposed to traditional discrete labeling, will offer instructors a more detailed insight into attention granularity. This shift is expected to enhance accuracy in representing cognitive engagement. The finer granularity of continuous labels is anticipated to boost model performance by reducing Mean Squared Error (MSE) and capturing subtle variations in attention dynamics.

4   Methodology

The EngageME dataset was meticulously gathered from willing participants following the receipt of ethical clearance from the Institutional Review Board (IRB). Participants engaged in an experimental procedure detailed in Sect. 4.1, submitting webcam recordings and cognitive task responses in CSV format to a web server. Both video recordings and CSV data were segmented into approximately 60-second intervals inspired by [17], aligning with timestamps in the CSV file to prevent data loss during splitting. Participant performance was annotated using CSV data, and features were extracted from each video segment.

4.1 Data Collection

73 participants (26 females, 47 males), aged 19 to 21 years (M = 20.644, SD = 2.628), completed our study. All were undergraduate students invited to participate in an online data collection experiment from the comfort of their homes without any compensation. Informed consent was obtained, and participants were guided about the test, including webcam calibration, without the need for specialized software or hardware installations closely resembling their typical study environment. Participants were unaware of the attention level assessment but were briefed about the tests and informed about the recording. The online data collection framework, outlined in Fig. 1, utilized a javascript library and lasted approximately 20 min for each participant. The experiment included eye-tracker calibration and three cognitive tasks targeting specific attention types. Importantly, upon acceptance, we commit to releasing the dataset at the researcher's request for transparency and contribution to the research community, as stipulated in the End User License Agreement (EULA).

In our experiment, we used the webcam-based eye-tracking system [24], performing a crucial calibration to ensure accuracy exceeding 60%. The decision to utilize this JavaScript library is a widely accepted and cost-effective approach, aligning with practical considerations for potential classroom use. The Stroop Test, a widely used measure of selective attention [29], evaluates individuals' accuracy in verbally articulating the font color of a word representing an incongruent color. In our study, 'congruent stimuli' denote matching text color and color name (e.g., "red" written in red), while 'incongruent stimuli' involve a mismatch (e.g., "red" written in blue ink). Conducted through a computerized version, participants used arrow keys with a fixed 4000-millisecond stimulus presentation time and a variable post-trial gap (100–500 ms). Training trials mirrored experimental test blocks, with feedback for errors and correct responses to aid learning. The test was structured to align with reaction coordinates obtained from Webgazer in post-processing.

N-back Continuous Performance Test (CPT), a widely utilized tool for assessing sustained attention, was used in our experiment [27]. It requires participants to respond to a series of stimuli by determining whether each stimulus matched the one presented N-items earlier. In our experiment, participants pressed "M" if the presented letter matched the one shown two items earlier. Participants were instructed that omission occurs when failing to press the designated key for a matching stimulus, and commission happens with an incorrect key press. These measures facilitate an assessment of sustained attention, capturing instances of inattention (omission) and response errors (commission) during the N-back CPT. The Trail Making Test (TMT), a widely used neuropsychological tool, assesses alternating attention [20]. It comprises two versions: TMT-A focuses on number sequencing (1 to 15), while TMT-B assesses set-shifting abilities by alternating between numerical and alphabetic sequences (1-A-2-B-3...). In our study, we exclusively used TMT-B to analyze participants' alternating attention. They selected alphanumeric sequences, which changed color to green upon selection. Three TMT-B trials were conducted.

4.2   Label Generation from Generated CSVs

After collecting videos and CSV data from 98 participants, we addressed issues such as unclear videos, cameras not capturing faces, and excessive video duration. Subsequently, 73 participants' data was selected for further processing. Before task-specific analysis, distributions for reaction times, accuracy, and total time across the three tests were obtained. The Stroop test revealed a significant difference in reaction times between congruent and incongruent stimuli ($t(2,73) = 3.058$, $p < 0.05$), with average times of 892.952 ms and 937.497 ms, respectively. Scoring was based on response accuracy, with equal weight assigned to congruent and incongruent Stroop test types. Accuracy scores, calculated as a percentage of correct responses and normalized on a 0 to 1 scale, indicate attention levels (1 for highest, 0 for lowest). The average congruent accuracy surpassed incongruent accuracy by 1.97%.

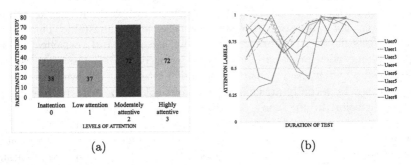

(a)   (b)

Fig. 2. Distribution of attention labels based on neuropsychological test scores (a) and Visualization of attention during a complete test of random users (b).

In CPT, a weak negative correlation was found between the number of false alarms and median reaction time (Pearson's correlation of -0.196, $p < 0.05$). Only accuracy was considered for both omission and commission tests. Each CPT test type is given 50% weightage. The accuracy scores were scaled to generate a final CPT score ranging from 0 to 1, with 0 indicating least attentive and one indicating highly attentive. Additionally, the average accuracy for non-match stimuli was 4.883% higher than that for matching stimuli.

In the TMT test, response times decline as competing stimuli decrease. Specifically, average response times for three consecutive trials were 3148 ms, 2580 ms, and 3110 ms, with the first half having a higher mean than the second. We standardized labels using min-max normalization and inverted attention scores by subtracting them from 1. The resulting nuanced attention label scores for TMT are scaled between 0-1, indicating performance levels. A higher score signifies superior performance, while a lower score implies the opposite.

Manual Annotation: We adopted a comparative approach to support our first hypothesis by establishing convergent validity [19] for our new attention test. Convergent validity assesses the correlation between different measures theoretically expected to be related. We collected scores from neuropsychological tests and manual annotation scores generated by three annotators who manually reviewed recorded participant clips. This approach demonstrates the degree of agreement between neuropsychological test scores and manual annotations, providing compelling evidence for the convergent validity of our attention test. Annotators labeled video clips on a scale of 0 to 3, inspired by [34], to indicate the level of cognitive engagement. A rating of 3 signifies high attentiveness, where participants continuously look at the screen, while a rating of 2 indicates moderate attentiveness, including occasional glances away. A rating of 1 suggests low attentiveness, characterized by gaze shifts, yawning, fidgeting, and frequent body movements. A rating of 0 denotes inattention and encompasses behaviours like frequent gaze shifts, resting the head on the table, or skipping the test. Clear instructions and examples were provided to distinguish between all the attentions and ensure consistency across datasets. For TMT clip annotation, the video's duration was considered.

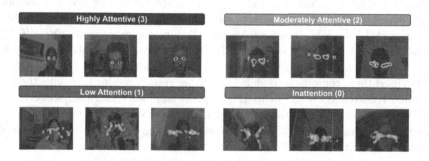

Fig. 3. Heatmap for different types of attention labels.

We categorized the continuous spectrum into four classes for the neuropsychological test by segmenting the distribution within 0.5 standard deviations. This discretization, tailored for left-skewed data, ensures a balanced representation of attention levels. Labels were assigned based on performance scores, streamlining the creation of a multi-labeled classification model for attention. Figure 2a categorizes participants' attention levels during neuropsychological testing. A significant number fall into 'Moderately attentive' and 'Highly attentive' categories, highlighting prevalent attention. The variability in participants labeled 'Inattention' (38) and 'Low attention' (37) provides insights into attention levels, showcasing the distribution across different categories during neuropsychological tests. Figure 2b illustrates attention scores for eight random users (User0 to User7) at distinct time points, capturing individual attentional dynamics. Variability is observed in completion times, with User6 taking longer while User3 finishes first. User4 begins with the lowest attention score, and User1 achieves the highest (100%). Temporal fluctuations in user attention reveal dynamic patterns-some consistent, others varied. This variation forms a foundation for nuanced insights valuable to educators.

We used the OpenFace framework [5] for extracting facial behavior. It generates 712 features per frame, including gaze, pose, FAUs, landmarks in 2D/3D, and shape parameters. Figure 3 displays a heatmap of eye-gaze locations for different attention labels, emphasizing distinctions between highly attentive and less attentive students during the test.

4.3   Regression Approach

Sequential data from the EngageME consists of approximately 1800 rows for each clip extracted from OpenFace. We used a Bidirectional-Long short-term memory (Bi-LSTM) [6] model to generate embeddings to capture temporal dependencies and behavioral patterns, aligning with our goal of analyzing attention fluctuations. Using a basic Bi-LSTM encoder with 355 units and a Dense layer with 200 units, we produced embeddings from the sequential data, normalized through Min-Max scaling ranging from 0 to 1. The resulting $200 \times N$ embeddings, where N is the total clip count, served as input for five established Machine Learning (ML) models: Support Vector Machine for regression (SVR) [6,13], Light Gradient-Boosting Machine (LGBM) [2], K-Nearest Neighbors (KNN) [6], eXtreme Gradient Boosting (XGBoost) [2], and Random Forest (RF) [6,13].

We used these regression models as they are recognized for their effectiveness in similar research. SVR was utilized to capture complex relationships effectively. LGBM provides high performance and efficiency. KNN offers a versatile approach based on proximity to neighbors. XGBOOST demonstrated powerful ensemble learning capabilities, while RF uses decision tree ensembles to enhance accuracy and reduce overfitting. Each algorithm assessed attention dynamics, using its strengths for the online learning context. We used the MSE metric for attention labels between 0 and 1. MSE quantifies squared differences between predicted and actual attention scores, ensuring accurate evaluation within the 0 to 1 scale.

5   Results and Discussion

In this section, we present the outcomes of two experiments. The first experiment validates our novel attention annotation method using neuropsychological test scores by correlating it with manual annotations. We achieved high inter-annotator reliability, with a Krippendorff's Alpha (α) value of 0.869, indicating substantial agreement between annotators. Additionally, Spearman's correlation coefficient demonstrated strong correlations (82–93%) across three independent annotators, affirming the reliability of our annotation approach. This assessment gauges the strength and direction of the relationship between annotators' ratings, highlighting strong agreement and high correlation in their annotations. To establish the convergent validity of our new annotation method, we computed Spearman's Rank correlation between discretized neuropsychological test scores and manual annotation, yielding a correlation value of 0.673. This significant correlation confirms the reliability of neuropsychological tests for assessing students' attention in online learning settings. The outcomes of our experiments contribute significantly to Artificial Intelligence (AI) in education, specifically addressing challenges in assessing student attention in online learning.

In our second experiment, we aimed to validate our second hypothesis by demonstrating the efficacy of Nuanced Attention Labeling in understanding attention dynamics. Our aim was to highlight how fine-grained labeling enables a detailed examination of cognitive engagement, anticipating reduced errors in capturing subtle attention variations. To ensure model robustness, we divided the EngageME dataset (80:20 ratio) at the user level for training and testing-58 users for training and 15 for testing. We then created regression models using training data and labeled them with neuropsychological test scores for Nuanced Attention Labeling alongside discretized annotations from manual annotation. The discretized annotations averaged across annotators and normalized to a 0 to 1 scale, the same as we did with neuropsychological test scores. Additionally, clips where annotator labels differed by more than two values were ignored. The results of these regression models, assessed using five-fold cross-validation on

Table 2. Model performance using different annotation methods

Models	Discretized Attention Labelling	Nuanced Attention Labelling
XGBoost	0.0558	0.0131
SVR	0.0469	0.0111
LGBM	0.0493	0.0138
KNN	0.0601	0.0166
RF	0.05018	0.0143

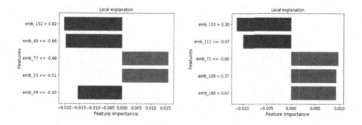

Fig. 4. LIME explanation using (a) Discretized Attention Labelling (b) Nuanced Attention Labelling

unseen test data, are presented in Table 2. It shows that the Nuanced Attention Labeling method outperforms the discretized attention labeling method across multiple models on the same test data, as reflected in lower MSE values. This indicates superior accuracy in capturing subtle variations in attention, highlighting the method's effectiveness in detailed and accurate assessment of cognitive engagement. Its finer granularity scale reduces errors, providing a more nuanced representation of attention dynamics in the given context.

Further, we used Local Interpretable Model-agnostic Explanations (LIME) [26] to enhance the interpretability of our ML models, especially in attention labeling. LIME explanations elucidate how embeddings influence predictions, providing insights into key features. This enhances transparency in model behavior. Figure 4 displays LIME results for one random instance using Discretized and Nuanced Attention Labeling from the SVR model, where the y-axis represents the top-5 extracted features for that instance. In the discrete label approach, binary conditions negatively influence the prediction, like embeddings emb_152 surpassing 0.82 or emb_48 falling below -0.66. These rigid cut-offs suggest a reliance on specific threshold values for decision-making. Conversely, the nuanced label method reveals a more nuanced understanding, with embeddings like emb_133 exceeding 0.30 or emb_72 being less than or equal to -0.60, contributing positively or negatively. These LIME interpretations highlight differences in interpretability for attention assessment methods. The discrete approach, similar to strict grading, uses binary conditions. In contrast, the nuanced method considers a continuous impact, offering a more subtle understanding of attention dynamics.

Table 3. Model performance on Online Learning Test Data

Models	EngagME Data	Online Learning Data
XGBoost	0.0727	0.0705
SVR	0.0637	0.0630
LGBM	0.0679	0.0665
KNN	0.0758	0.0857
RF	0.0578	0.0597

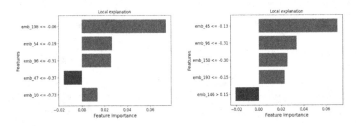

Fig. 5. LIME explanation on (a) EngagME Data (b) Online Learning Data

6   User Study

Validating neuropsychological test scores for online learning attention, we now aim to confirm Hypothesis 1 using student data from online learning. For this, we used an additional dataset, *Online Learning dataset*, which includes 21 video datasets featuring individual participants taking online classes. Here, participants voluntarily chose subjects like "Linear Algebra" and "Introduction to ML" to watch during data collection. The video recordings capture students' interactions with course materials. Out of 21 videos, we extracted 651 clips (around 60 s each, excluding segments less than 30 s). However, 25 inadvertently truncated clips were omitted from the dataset. We applied the same rigorous feature extraction and annotation method as we used for the EngageME dataset for discretized attention labeling. The final dataset comprises 606 clips with annotations labeled on a scale from 0 to 1. We then split the Online Learning data in 80:20 ratio, assigning 16 users for training and 5 for testing. Using the same models as before with discretized labels, we trained new models on the Online Learning training data. Subsequently, we evaluated these models on unseen Online Learning data, comparing their performance in Table 3.

The MSE outcomes for the EngageME and Online Learning models are nearly identical, highlighting the consistency and reliability of our attention assessment methodology. LIME explanations in Fig. 5 spotlight facial features influencing attention predictions. We utilized the same pipeline to generate embeddings from both datasets, capturing temporal dependencies. Notably, among the top 5 features of both models, emb_96 consistently emerges as a crucial indicator. The significance of emb_96 is evident in both the EngageME and Online Learning models, as supported by their respective feature importance scores. Additionally, the proximity of other top features in both models, such as emb_45, emb_150, emb_193, emb_198, and emb_54, underlines the robustness and consistency of our extraction methods, as these features exhibit proximity to each other. These features collectively contribute to the model's proficiency in understanding nuanced temporal variations in facial behavior. The aligned patterns in emb_198 (EngageME) and emb_193 (Online Learning) further reinforce the model's ability to predict attention dynamics precisely and establish meaningful connections with specific facial features. Using Nuanced Attention Labeling in online learning, our approach provides detailed student attention insights for personalized interventions and adaptive education, contributing innovative perspectives to AI in education without sacrificing real-world applicability.

7   Conclusion and Future Work

The study proposes using standardized neuropsychological tests to assess students' attention in online learning. This involved a cohort of 73 students undergoing three relevant neuropsychological assessments. Our findings revealed a notable inter-annotator reliability and convergent validity between manual annotations and neuropsychological test scores. By introducing the Nuanced Attention Labeling method, the assessment of attention significantly advances with heightened sensitivity. It captures subtle variations and intricate patterns in attention dynamics and offers a more detailed and precise assessment than traditional methods. The efficacy of the Nuanced Attention Labeling method was validated through a user study conducted in online settings. The study verified the precision of Nuanced Attention Labeling in assessing cognitive engagement subtleties. In future work, we aim to refine our approach by integrating a deep learning model for enhanced analysis and exploring multiple explainability models for robust interpretation. We also plan to enhance the attention labeling scheme to detect various types of attention, advancing the novelty and relevance of our contribution.

Acknowledgements. We thank Harshit Chauhan, Kanishk Kalra, and Aditi Gupta for their contributions in designing the data collection pipeline. This research work is funded by a research grant (Ref. ID.: IHUB Anubhuti/Project Grant/03) of IHUB Anubhuti-IIITD Foundation and is partly supported by the Infosys Center for AI and the Center for Design and New Media (A TCS Foundation Initiative supported by Tata Consultancy Services) at IIIT-Delhi, India.

References

1. Abdelrahman, Y., et al.: Classifying attention types with thermal imaging and eye tracking. Proc. ACM Interact. Mob. Wearable Ubiquitous Technol. **3**(3), 1–27 (2019)
2. Alruwais, N., Zakariah, M.: Student-engagement detection in classroom using machine learning algorithm. Electronics **12**(3), 731 (2023)
3. Altuwairqi, K., Jarraya, S.K., Allinjawi, A., Hammami, M.: Student behavior analysis to measure engagement levels in online learning environments. SIViP **15**(7), 1387–1395 (2021)
4. Alyuz, N., Aslan, S., D'Mello, S.K., Nachman, L., Esme, A.A.: Annotating student engagement across grades 1–12: associations with demographics and expressivity. In: Roll, I., McNamara, D., Sosnovsky, S., Luckin, R., Dimitrova, V. (eds.) AIED 2021. LNCS (LNAI), vol. 12748, pp. 42–51. Springer, Cham (2021). https://doi.org/10.1007/978-3-030-78292-4_4
5. Amos, B., Ludwiczuk, B., Satyanarayanan, M., et al.: Openface: a general-purpose face recognition library with mobile applications. CMU Sch. Comput. Sci. **6**(2), 20 (2016)
6. Booth, B.M., Ali, A.M., Narayanan, S.S., Bennett, I., Farag, A.A.: Toward active and unobtrusive engagement assessment of distance learners. In: 2017 Seventh International Conference on Affective Computing and Intelligent Interaction (ACII), pp. 470–476. IEEE (2017)
7. Cherry, E.C.: Some experiments on the recognition of speech, with one and with two ears. J. Acoust. Soc. Am. **25**(5), 975–979 (1953)
8. Dhall, A., Kaur, A., Goecke, R., Gedeon, T.: Emotiw 2018: audio-video, student engagement and group-level affect prediction. In: Proceedings of the 20th ACM International Conference on Multimodal Interaction, pp. 653–656 (2018)
9. D'Mello, S., Lehman, B., Pekrun, R., Graesser, A.: Confusion can be beneficial for learning. Learn. Instr. **29**, 153–170 (2014)

10. Fuller, K.A., et al.: Development of a self-report instrument for measuring in-class student engagement reveals that pretending to engage is a significant unrecognized problem. PLoS ONE **13**(10), e0205828 (2018)
11. Glass, A.L., Kang, M.: Dividing attention in the classroom reduces exam performance. Educ. Psychol. **39**(3), 395–408 (2019)
12. Goldberg, P., et al.: Attentive or not? Toward a machine learning approach to assessing students' visible engagement in classroom instruction. Educ. Psychol. Rev. **33**, 27–49 (2021)
13. Gorgun, G., Yildirim-Erbasli, S.N., Epp, C.D.: Predicting cognitive engagement in online course discussion forums. Int. Educ. Data Mining Soc. (2022)
14. Hassib, M., Schneegass, S., Eiglsperger, P., Henze, N., Schmidt, A., Alt, F.: Engagemeter: a system for implicit audience engagement sensing using electroencephalography. In: Proceedings of the 2017 Chi Conference on Human Factors in Computing Systems, pp. 5114–5119 (2017)
15. Herpich, F., Guarese, R.L., Cassola, A.T., Tarouco, L.M.: Mobile augmented reality impact in student engagement: an analysis of the focused attention dimension. In: 2018 International Conference on Computational Science and Computational Intelligence (CSCI), pp. 562–567. IEEE (2018)
16. Khan, S.S., Abedi, A., Colella, T.: Inconsistencies in the definition and annotation of student engagement in virtual learning datasets: a critical review. arXiv preprint arXiv:2208.04548 (2022)
17. Koelstra, S., et al.: DEAP: a database for emotion analysis; using physiological signals. IEEE Trans. Affect. Comput. **3**(1), 18–31 (2011)
18. Lackmann, S., Léger, P.M., Charland, P., Aubé, C., Talbot, J.: The influence of video format on engagement and performance in online learning. Brain Sci. **11**(2), 128 (2021)
19. Lekwa, A.J., Reddy, L.A., Shernoff, E.S.: Measuring teacher practices and student academic engagement: a convergent validity study. Sch. Psychol. **34**(1), 109 (2019)
20. Linari, I., Juantorena, G.E., Ibáñez, A., Petroni, A., Kamienkowski, J.E.: Unveiling trail making test: visual and manual trajectories indexing multiple executive processes. Sci. Rep. **12**(1), 14265 (2022)
21. Linson, A., Xu, Y., English, A.R., Fisher, R.B.: Identifying student struggle by analyzing facial movement during asynchronous video lecture viewing: towards an automated tool to support instructors. In: Rodrigo, M.M., Matsuda, N., Cristea, A.I., Dimitrova, V. (eds.) AIED 2022, Part I. LNCS, vol. 13355, pp. 53–65. Springer, Cham (2022). https://doi.org/10.1007/978-3-031-11644-5_5
22. Ma, J., Jiang, X., Xu, S., Qin, X.: Hierarchical temporal multi-instance learning for video-based student learning engagement assessment. In: IJCAI, pp. 2782–2789 (2021)
23. Monkaresi, H., Bosch, N., Calvo, R.A., D'Mello, S.K.: Automated detection of engagement using video-based estimation of facial expressions and heart rate. IEEE Trans. Affect. Comput. **8**(1), 15–28 (2016)
24. Papoutsaki, A.: Scalable webcam eye tracking by learning from user interactions. In: Proceedings of the 33rd Annual ACM Conference Extended Abstracts on Human Factors in Computing Systems, pp. 219–222 (2015)
25. Pickering, J.D.: Cognitive engagement: a more reliable proxy for learning? Med. Sci. Educ. **27**(4), 821–823 (2017)
26. Ribeiro, M.T., Singh, S., Guestrin, C.: "why should i trust you?" explaining the predictions of any classifier. In: Proceedings of the 22nd ACM SIGKDD International Conference on Knowledge Discovery and Data Mining, pp. 1135–1144 (2016)

27. Roebuck, H., Freigang, C., Barry, J.G.: Continuous performance tasks: not just about sustaining attention. J. Speech Lang. Hear. Res. **59**(3), 501–510 (2016)

28. Sohlberg, M.M., Mateer, C.A.: Effectiveness of an attention-training program. J. Clin. Exp. Neuropsychol. **9**(2), 117–130 (1987)

29. Stevens, C., Bavelier, D.: The role of selective attention on academic foundations: a cognitive neuroscience perspective. Dev. Cogn. Neurosci. **2**, S30–S48 (2012)

30. Szafir, D., Mutlu, B.: Pay attention! Designing adaptive agents that monitor and improve user engagement. In: Proceedings of the SIGCHI Conference on Human Factors in Computing Systems, pp. 11–20 (2012)

31. Thomas, C., Jayagopi, D.B.: Predicting student engagement in classrooms using facial behavioral cues. In: Proceedings of the 1st ACM SIGCHI International Workshop on Multimodal Interaction for Education, pp. 33–40 (2017)

32. Verma, M., Nakashima, Y., Takemura, N., Nagahara, H.: Multi-label disengagement and behavior prediction in online learning. In: Rodrigo, M.M., Matsuda, N., Cristea, A.I., Dimitrova, V. (eds.) AIED 2022. LNCS, vol. 13355, pp. 633–639. Springer, Cham (2022). https://doi.org/10.1007/978-3-031-11644-5_60

33. Wang, X., Wen, M., Rosé, C.P.: Towards triggering higher-order thinking behaviors in MOOCs. In: Proceedings of the Sixth International Conference on Learning Analytics & Knowledge, pp. 398–407 (2016)

34. Whitehill, J., Serpell, Z., Lin, Y.C., Foster, A., Movellan, J.R.: The faces of engagement: automatic recognition of student engagement from facial expressions. IEEE Trans. Affect. Comput. **5**(1), 86–98 (2014)

The Neglected 15%: Positive Effects of Hybrid Human-AI Tutoring Among Students with Disabilities

Danielle R. Thomas[✉], Erin Gatz, Shivang Gupta, Vincent Aleven, and Kenneth R. Koedinger

Carnegie Mellon University, Pittsburgh, PA 15213, USA
{drthomas,shivang,koedinger}@cmu.edu, {egatz, va0e}@andrew.cmu.edu

Abstract. Incorporating human tutoring with AI holds promise for supporting diverse math learners. In the U.S., approximately 15% of students receive special education services, with limited previous research within AIED on the impact of AI-assisted learning among students with disabilities. Previous work combining human tutors and AI suggests that students with lower prior knowledge, such as lacking basic skills, exhibit greater learning gains compared to their more knowledgeable peers. Building upon this finding, we hypothesize that hybrid human-AI tutoring will have positive effects among students in inclusive settings, classrooms containing both students with and without identified disabilities. To investigate this hypothesis, we conduct a two-study quasi-experiment across two urban, low-income middle schools: one in Pennsylvania with 362 students and another in California comprised of 733 students, involving 27% and 16% of students with disabilities, respectively. Our findings indicate hybrid human-AI tutoring has positive effects on learning processes and outcomes among all students, including students with disabilities. The motivational benefits of the human tutoring treatment over a control group using solely math software show up in greater increases in practice and skill proficiency during math software use for students with disabilities than other students. In addition, the treatment yields significantly higher pre-to-post learning gains for all including students with disabilities whose relative gains trend higher and are statistically at least as high. These compelling findings support the promise of hybrid human-AI solutions and emphasize the collaborative focus among AIED to support diverse learners.

Keyword: Human-AI tutoring · Learning analytics · Equity · Inclusion

1 Introduction

Student math performance is at an all-time low [16], with modern classrooms now hosting a diverse continuum of student abilities. Students are missing basic math skills, which has been exacerbated by the well-documented pandemic-related learning loss [16]. Within the U.S., the 2022 NAEP assessment reported a loss of nearly 20 years of

© The Author(s), under exclusive license to Springer Nature Switzerland AG 2024
A. M. Olney et al. (Eds.): AIED 2024, LNAI 14829, pp. 409–423, 2024.
https://doi.org/10.1007/978-3-031-64302-6_29

math progress, with 8th-grade special education scoring 40 points lower on average than general education students—the largest gap ever between any subgroup and the general population [16]. In the U.S., 15% of students have identified disabilities [20].

Inclusivity challenges further compound matters, especially for students with disabilities [21, 27]. To address these issues, AIED has been actively exploring AI-augmented human tutoring with the collective goal of helping all learners [1, 8, 26]. AI-augmented (also known as, AI-assisted, AI-in-the-loop, AI-supported) human tutoring works by having the AI provide the adaptive math instruction, heavily researched and well-known to be effective [19], with human tutors providing as-needed socio-motivational intervention and relationship-building support [8, 26]. An example in practice is the application of a real-time AI-driven dashboard to support effective use of human tutor's time during tutoring and differentiating student support by allocating tutor time to students who need it most. There is evidence of this approach demonstrating students learn more when teachers and AI work in synergy [12] and among human tutors and AI [8, 26]. Researchers propose *hybrid human-AI tutoring* can provide *just the right amount* of support for learners [26] but no work is known on assessing its impact among students with disabilities, while engaging in "inclusion." The term *inclusion* refers to the practice of educating students with disabilities alongside their peers without disabilities in mainstream classrooms to the maximum extent appropriate [13].

Students with disabilities are defined as students with a physical, mental, or emotional impairment that affects their performance in school [13]. U.S. federal law, notably the Individuals with Disabilities Education Act (IDEA), ensures students with disabilities receive appropriate education and services, typically outlined in an Individualized Education Plan (IEP). An IEP is a legally mandated document that specifies support services for students with disabilities [13]. In this work, we use the term "students receiving special education services," where possible, to refer to students with an IEP, officially identified by schools as having a disability. This terminology is preferred over "students with disabilities" to account for the possibility that some students may have unidentified disabilities (e.g., an estimated 20% of U.S. students are believed to have dyslexia [24], with the majority undiagnosed) or disabilities that do not qualify for an IEP (e.g., those covered by a 504 plan governed by Section 504 of the Rehabilitation Act), which are not the focus of this present work. Using "students receiving special education services" also avoids labeling and defining students based on their disability. "Students with disabilities" applies a person-first approach, which may reinforce a deficit mindset. While referring to students as solely "special education students" uses an identify-first approach, which may create a sense of separateness from typically developing peers [4]. Nevertheless, there is no agreed preference among the disability community on terms [4], with both used interchangeably in this work.

This study holds significance within the global AIED community by addressing the needs of a neglected group and investigating the potential of AI technologies in aiding diverse learners. Given math performance is at an all-time low and the widespread use of AI-driven learning technologies, this work aims to answer the following: **RQ1.** What differences in math learning outcomes exist among students in an inclusive setting (containing students with and without disabilities) participating in hybrid human-AI tutoring compared to students engaging solely with math software? **RQ2.** Among students participating in hybrid human-AI tutoring, what differences in math learning outcomes exist between students with identified disabilities compared to students without disabilities?

2 Related Work

Hybrid Human-AI Tutoring. While intelligent tutoring systems and AI-adaptive math software can support tutoring at scale, they are not equipped to provide the relational and motivational support some students may need to engage with these systems [8, 26]. Recently, *hybrid human-AI tutoring*, as mentioned, as human and AI tutors working in tandem, aims to leverage the power of AI-driven adaptive math software to provide personalized instruction for students *and* to allow human tutors to focus more of their effort on the relationship building and socio-emotional support many students need to successfully engage with the AI [10, 11]. During hybrid human-AI tutoring, students spend part of their regular math class time engaged in personalized practice using software enhanced with algorithms that can identify student learning progress and struggles [2]. The information is displayed on a dashboard to inform tutors, who can then determine which students to assist. The system measurably redirects the tutor's attention towards the students who have lower knowledge—and away from students most likely to seek assistance. Previous research has explored human-AI tutoring in inclusive settings [26] yet did not specifically examine its impact on students with disabilities.

Building on the conceptual framework proposed by Thomas et al. [26], which highlights the role of increased learning opportunities resulting, ultimately, in improved achievement, this study extends its focus to students with disabilities. To ensure equitable access and support to all learners with AI-based technologies, we draw upon Universal Design for Learning (UDL), similar to Roski et al. [22]. UDL is an educational framework that focuses on equal opportunities among all learners and emphasizes the creation of flexible inclusive environments [22, 23]. Figure 1 displays a conceptual framework demonstrating the hypothesized causal sequence beginning with hybrid-human AI tutoring. We leverage the key UDL principle of providing students *multiple means of engagement*, with human tutors supporting students by recruiting interest, assisting with sustaining effort, and providing self-regulation strategies [23] as an important intermediate step (step 1). Human support increases engagement, which leads to enhanced student progress (step 2), ultimately leading to higher achievement (step 3).

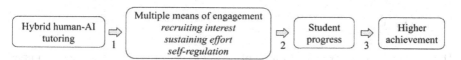

Fig. 1. Causal sequence connecting hybrid human-AI tutoring to improved student achievement.

Students with Disabilities and AIED. Surprisingly, limited past work exists within AIED involving students with disabilities, despite a substantial proportion of U.S. students with disabilities [20] and, presumably, throughout the world. More common areas of research include investigations of algorithmic bias [6] and fairness in predictive modeling [9, 22] and lesser so on the impact of AI-based interventions on student learning. Special education students using ASSISTments, a web-based math tutor, performed significantly better on end-of-year tests compared to special and general education students not participating in ASSISTments [14]. One suggested explanation for the impressive performance among special education students is the 80% inclusion rate among the treatment group, as opposed to 43% among the control group [14]. Increasing inclusion rates, or the percentage of students with disabilities integrated into general education settings, holds paramount importance in fostering equity and diversity. Prior work on the use of hybrid human-AI tutoring has indicated students with lower prior knowledge, including those lacking basic skills, demonstrated greater learning gains compared to their more knowledgeable peers [26]. However, its effect on students receiving special education services remains unknown. In essence, students within special education are already using AI-based software as part of their specially designed instruction and as an instrument to monitor progress (e.g., [3]). We posit that combining its use with human tutoring using a hybrid human-AI tutoring approach will further enhance learning outcomes.

Beyond AIED, the use of AI-driven learning software and assistive technologies is growing [21]. However, the use of AI to foster equitable learning opportunities among individuals with disabilities has raised concerns, notably regarding the perception of using AI to address disability as a "problem" and viewing students with disabilities as a burden that AI-based technologies can alleviate [21]. We address these concerns by pursuing human-AI tutoring from an assets-based approach. This approach emphasizes the strengths and capabilities of all learners, irrespective of disability status [28].

Considerations surrounding equity, bias, and inclusiveness among students with disabilities need more focus within AIED. Prior to 2016, such considerations were lacking in the field of learning analytics, with recent results showing considerable potential to reduce discrimination, bias, and promote equity [15]. Ensuring equitable access to resources, intervention, and learning experiences is essential. Addressing biases and avoiding discriminatory practices in the collection and use of learning analytics is crucial to fostering inclusivity [5]. For instance, past work found nuances in growth patterns among students with disabilities, emphasizing the need to set student's individualized growth patterns by providing nuanced expectations of growth rather than merely setting lower expectations based on pre-defined group measures [18]. AIED must work together to provide an inclusive ecosystem that empowers all students to thrive.

3   Method

This two-study quasi-experiment took place in the fall of 2023 at two schools implementing hybrid human-AI tutoring. The tutoring intervention provided math support to all students (ages 11–14) during the school day as part of normal instructional practice. Students receiving special education services outside of the regular classroom setting were not included in the study. At both sites, students participated in hybrid human-AI tutoring, also referred to as the treatment, one day per week for 50 min, during the school day, and for a set number of weeks. At Site 1 students used *IXL*, a comprehensive K-12 math curriculum, as a personalized learning software.[1] In addition, Renaissance Learning's *Star Diagnostic* assessment was administered in August (pretest) and again in December (posttest) to assess student's baseline comprehension and estimated grade level performance.[2] Implementation at Site 1 involved a control group, which was another middle school in the district, of which students used *IXL* prescribed by teachers and took *Star Diagnostics* but did not have students participate in the hybrid human-AI tutoring intervention. The control school also had general and special education students being instructed together within an inclusive education setting. Descriptive statistics of the control group (n = 399) were: 92% low-income, 55% Latinx, 15% Black, 46% female, 18% English Language learners, and 28% with an IEP. Almost 10 percent of students scored at or above proficiency in math on the state test the prior year. Site 2 did not contain a control group. All students used *i-Ready Personalized Instruction*, a K-8 instructional learning software created by Curriculum Associates.[3] Similar to Site 1, students completed a diagnostic, in this case the *i-Ready Diagnostic*, a computer-delivered assessment, in late August (pretest) and again after treatment in December (posttest). Implementation characteristics for Sites 1 and 2 are shown in Table 1.

Table 1. Summary of implementation characteristics across sites.

	Site 1 (treatment)	Site 1 (control)	Site 2 (treatment)
Implementation	*IXL* w/ human tutor	*IXL* only	*i-Ready* w/ human tutor
Pre/post diagnostic	*Star Diagnostic*	*Star Diagnostic*	*i-Ready Diagnostic*
Intervention length	9 weeks	9 weeks	8 weeks
Educational setting	inclusion	inclusion	inclusion
Student n size (gr.)	362 students (gr. 6–8)	399 students (gr. 6–8)	733 students (gr. 7–8)
Students w/ an IEP	27%	28%	16%

Site 1 Method. Site 1 was an urban Pennsylvania middle school enrolling students in grades 6–8, of which 99% were economically disadvantaged, determined by free and reduced lunch designation. Only five percent of students scored at or above proficiency on the state assessment the previous school year. The majority of the 6th-8th grade population were actively engaged in hybrid human-AI tutoring as part of a math intervention occurring in an inclusive setting. Students needing more intensive special education services that cannot be met within a general education setting did not participate. A total of 362 students participated in hybrid human-AI tutoring. The demographic composition of the treatment group was 74% Latinx, 17% Black, 50% female, 24% English Language learners. Moreover, the prevalence of students receiving special education services indicated by having an IEP, was almost double the national average at 27%. The intervention

---

[1] https://www.ixl.com/

[2] https://www.renaissance.com/products/star-assessments/

[3] https://www.curriculumassociates.com/programs/i-ready-learning.

took place over nine weeks (from Oct. 4 through Dec. 8). The last week of *IXL* usage data was missing and was not included in the analysis. Research on student subjects was conducted in accordance with the approved IRB protocol, ensuring compliance with ethical standards and consent procedures.

Site 2 Method. Site 2 was an urban California school servicing 7th-8th grade students, in a district consisting of over 15,000 students. The majority of the 7th-8th grade population were actively engaged in hybrid human-AI tutoring. Students with more intensive special education needs not able to be met within inclusion and two classrooms consisting of advanced math students did not participate. The treatment group consisted of 733 students in general education and special education. The demographic composition of students participating were 90% low-income, 80% Black or Latinx, and 51% female. The prevalence of students receiving special education services indicated by having an IEP closely mirrored the U.S. average, at 16%. Tutoring intervention spanned seven weeks (from Oct. 16 to Dec. 4, excluding the week of Nov. 20).

Data Analysis Plan. Site 1 contained *IXL*-reported students' weekly math skill proficiency (i.e., *skills practiced, skills proficient*) and skill usage (i.e., *seconds practiced*). *Seconds practiced* was used as a proxy for time spent using the software. At Site 2, *i-Ready* learning process data, or metrics gathered from the platform, were collected for *time spent on task, lessons passed*, and *time spent completing a lesson*. Because treatment always occurred while students were engaging in lessons, time spent completing a lesson (whether the lesson was passed or not) was used as a proxy for intervention dosage. Moreover, in addition to weekly *i-Ready* usage, students received human tutoring support aimed at enhancing motivation. Dosage was calculated using *i-Ready* usage only on the prescribed days human tutoring was delivered ensuring the treatment was indeed hybrid human-AI tutoring and not math software usage alone.

Attending to RQ1 and RQ2, at Site 1 we performed a three-way mixed ANOVA to determine the impact of hybrid human-AI tutoring (treatment) versus math software use only (control) on learning outcomes. We hypothesize that there will be a statistically significant main effect of the treatment on student's scores. We also hypothesize that there will be a significant main effect of treatment on special education status, indicating that students with disabilities will perform differently on diagnostic tests compared to students without disabilities. Regarding RQ2, we examined the three-way ANOVA from Site 1 and performed linear regressions predicting student learning process data in *IXL* controlling for pretest and grade level. At Site 2, which did not contain a control group, we performed a fixed-effects regression predicting posttest scores.

4   Results

RQ1. Comparison of Hybrid Human-AI Tutoring (Treatment) and Math Software Use Only (Control). For Site 1, descriptive statistics are shown below in Table 2 and pertain to RQ1 and RQ2. The following *IXL*-reported learning process data were collected weekly: *time spent, skills practiced*, and *skills practiced*. Table 2 displays the *IXL* learning process data for the control and treatment groups for general (GenEd) and special education (SpecialEd) students. Treatment students receiving special education services, compared to general education students, practiced more skills on average (i.e., 17.7 compared to 15.4) and were proficient on more skills (i.e., 8.58 compared to 7.65), despite less time spent (i.e., 192 min. compared to 228 min.) on average, respectively. Treatment students, GenEd and SpecialEd, had higher pretest to posttest gains.

First, a three-way mixed ANOVA was conducted with student special education status (*special_ed*: Y/N) and hybrid human-AI tutoring treatment (*treatment:* Y/N) as between-subjects factors and student scores on the diagnostic tests (*test_time*: pretest/posttest) as a within-subject factor. The ANOVA results are shown in Table 3.

There was a statistically significant main effect for *special_ed*, ($F(1,757) = 137.3$), $p < 2.95e\text{-}29$ indicating lower scores for SpecialEd students in Table 2 are significant ($583 < 706$ and $597 < 583$). There was a statistically significant two-way interaction between *test_time* and *treatment*, ($F(1, 757) = 7.185$), $p < .01$ indicating that the greater gains for treatment shown in Table 2 ($4.2\% > 2.6\%$ and $6.3\% > 3.8\%$) are statistically significant. A *post-hoc* analyses revealed that participants who received treatment (Y) showed a significant increase in test scores from pretest to posttest ($p < 0.05$), with treatment (N) not showing significant change. There was no statistically significant interaction between *test_time*, *treatment*, and *special_ed* ($F(1, 757) = 0.773$), $p = .379$ indicating that the trend in favor of greater gains for SpecialEd (3.8% to 6.3% is a 2.5% gain) than for GenEd (2.6% to 4.2% is a 1.6% greater gain) may be due to chance.

Table 2. Site 1 descriptive statistics with averages (and standard deviations) for learning process data and test scores for general education (GenEd) and special education (SpecialEd).

Group	Time spent (min)	Skills practiced	Skills proficient	Pretest score	Posttest score	% Change pre to post
Control						
GenEd (n = 286)	166 (85.6)	18.4 (10.4)	9.90 (7.0)	694 (118)	706 (114)	+2.6%
SpecialEd (n = 113)	130 (75.9)	16.1 (9.69)	7.96 (5.8)	577 (143)	583 (146)	+3.8%
Treatment						
GenEd (n = 256)	228 (163)	15.4 (9.72)	7.65 (7.2)	666 (109)	690 (112)	+4.2%
SpecialEd (n = 106)	192 (120)	17.7 (8.35)	8.58 (8.0)	566 (122)	597 (129)	+6.3%

RQ2. Comparison of Hybrid Human-AI Tutoring Among Students with and Without Disabilities. Continuing with Site 1, we determined the influence of human-AI tutoring treatment among students with disabilities on learning process data. We conducted a fixed-effects linear regression using R to determine impact on the following dependent variables: number of skills practiced (*skills_practiced*) and number of skills

Table 3. Three-way mixed ANOVA of special education, treatment, and test scores.

Source of Variation	df	F	η^2	p-value
Between-subjects				
treatment	757	0.519	6.24e-04	4.72e-01
special_ed	757	137.3	1.42e-01	**2.95e-29 *****
Within-subjects				
test_time	757	40.61	5.00e-03	**3.23e-10 *****
Interactions				
treatment:special_ed	757	2.189	3.00e-03	1.40e-01
treatment:test_time	757	7.185	8.38e-04	**8.00e-03 ****
special_ed:test_time	757	0.006	6.44e-07	9.41e-01
treatment:special_ed:test_time	757	0.773	9.03e-05	3.79e-01

*** $p < 0.001$, ** $p < 0.01$, * $p < 0.05$

proficient (*skills_proficient*). Equation 1 accounts for baseline differences and controls for students' pretest scores. Table 4 displays the results of the fixed-effect regression. For both dependent variables, *skills_practiced* and *skills_proficient*, the interaction of treatment and students with disabilities predicts students to practice more skills and demonstrate more proficiency on those skills compared to students who are not, $\beta = 4.545, t(754) = 2.36, p < 0.05$ and $\beta = 2.570, t(754) = 2.08, p < 0.05$, respectively. The negative effect of *treatment* for GenEd on these *IXL* progress measures is unexpected especially given the positive learning outcomes for GenEd students.

$$Dependent\ variable \sim pretest\_score + grade\ level + treatment * special\_ed \quad (1)$$

For Site 2, which did not contain a control group, we compared special and general education students' performance. The descriptive statistics for students are shown in Table 5. Hybrid human-AI tutoring dosage was determined as the total time spent on lessons while engaging in the math software with support from human tutors. Atypically, students in 8th grade receiving special education services performed better on the August diagnostic (pretest) and December diagnostic (posttest) collectively than their peers not receiving special education services. General education students engaged in the tutoring treatment more, indicated by higher dosage, than students in special education. However, this difference was not statistically significant.

Figure 2 illustrates the comparison of scaled scores on the pretest, posttest, and student special education status for 7th grade and 8th grade. The median scores among students with disabilities are higher than students without disabilities at pretest *and* posttest.

Table 4. Fixed-effects linear model predicting *skills practiced* and *skills proficient*.

Predictors	Estimate	SE	t-value	p-value
skills_practiced				
(intercept)	14.29	2.554	5.597	**3.05e-08 *****
pretest_score	0.011	0.0036	3.046	**0.0024 ****
treatment (Y)	−2.824	1.041	−2.714	**0.0068 ****
special_ed	−1.474	1.403	−1.051	0.2936
grade_7	−2.255	1.037	−2.174	**0.0300 ***
grade_8	−9.024	1.098	−8.218	**8.98e-16 *****
treatment:special_ed	4.545	1.926	2.360	**0.0185 ***
skills_proficient				
(intercept)	3.684	1.638	2.249	**0.0248 ***
pretest_score	0.011	0.002	4.802	**1.89e-06 *****
treatment (Y)	−0.879	0.667	−2.814	**0.005 ****
special_ed	−0.680	0.900	−0.755	0.451
grade_7	−1.638	0.665	−2.462	**0.014 ***
grade_8	−3.118	0.704	−4.526	**6.97e-06 ***
treatment:special_ed	2.570	1.235	2.080	**0.0378 ***

*** $p < 0.001$, ** $p < 0.01$, * $p < 0.05$

Table 5. Descriptive statistics for Site 2 displaying students' average for the pretest, posttest, and hybrid human-AI tutoring dosages. Standard deviations in parentheses.

Grade	Group	Number of students	Average pretest score	Average posttest score	Average dosage (min)
7th	GenEd	313	471.7 (27.2)	472.9 (30.4)	185 (69.3)
	SpecialEd	50	468.1 (40.0)	470.3 (44.3)	196 (62.1)
8th	GenEd	304	478.8 (32.6)	479.1 (39.1)	214 (67.3)
	SpecialEd	66	485.4 (46.6)	488.4 (53.4)	199 (80.4)

To determine the effects of tutoring dosage among students receiving and not receiving special education services on students' predicted posttest performance, a fixed-effects linear regression model was fitted (see Eq. 2). The scaled score of student's December diagnostic (*posttest_score*) was used as the dependent variable and contained the following fixed-effects predictors: student's scaled score on the August diagnostic (*pretest_score*); student's grade with 7th grade being the model default (*grade*); students with disabilities receiving special education services (*special_ed*); the total time students engaged in lessons using *i-Ready* in hours (*tutoring_dosage_hr*). Parameter estimates

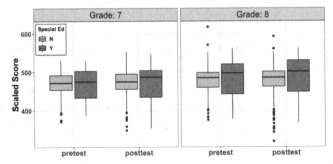

Fig. 2. Comparison of pretest and posttest performance and students special education status for 7th-grade (left) and 8th-grade (right) students.

are shown in Table 6. There was a positive and statistically significant main effect for tutoring dosage, $\beta = 3.156$, CI:[1.704, 4.609], $t(1) = 4.26$, $p < 0.001$. Dosage analysis indicates that for every hour a student engages in hybrid human-AI tutoring throughout the fall (*tutoring_dosage_hr*), they are predicted to score over 3 points higher on the posttest.

$$Posttest\_score \sim pretest\_score + grade + special\_ed : tutoring\_dosage\_hr \quad (2)$$

Table 6. Fixed-effects linear model predicting posttest score. Notice every hour students engage in tutoring treatment, predicts an over 3-pt increase on the posttest.

Predictors	Estimate	SE	t-value	p-value
(intercept)	27.28	11.24	2.40	**0.016***
pretest_score	0.923	0.0244	37.8	**<2e-16***
grade (8)	−1.256	1.555	−0.808	0.420
special_ed (Y)	−0.652	6.161	−0.106	0.916
tutoring_dosage_hr	3.156	0.7399	4.26	**2.3e-05***
special_ed:tutoring_dosage_hr	0.891	1.757	0.507	0.612

*** $p < 0.001$, ** $p < 0.01$, * $p < 0.0$

5   Discussion

Hybrid Human-AI Tutoring has Positive Impacts on Student Learning Outcomes. In reference to RQ1, we find that hybrid human-AI tutoring has positive effects on students' learning outcomes and process data compared to those in the control. The ANOVA found a statistically significant general benefit of hybrid human-AI tutoring for both GenEd and SpecialEd students (Table 3). These positive impacts from hybrid human-AI tutoring are consistent with previous work on AI-augmented human tutoring systems [8, 26]. At Site 2, which lacked a control group, we found for every hour of hybrid human-AI tutoring, students were predicted to score 3.2 points higher on the *i-Ready* diagnostic. This result is encouraging regarding the promising impact of hybrid human-AI tutoring but a comparison of predicted diagnostic performance for students engaging in math software only is needed to confirm a benefit.

It is worth noting that diagnostic assessments, while valuable tools for identifying specific learning gaps and tailoring instruction, accordingly, may not be the best choice for capturing overall increases in math learning over time. We hypothesize that a more extended intervention duration, such as a semester or school year, will yield more substantial effects on learning outcomes. Furthermore, the time devoted by human tutors on building relationships and providing socio-motivational support to students was likely not accounted for within the dosage, which was determined solely based on math software usage time. Consequently, it is challenging to ascertain the extent and impact of human relational support provided by the tutors themselves.

Students with Disabilities Benefit at Least as Much as Students Without Disabilities and May Particularly Benefit from Human Motivational Support. Regarding RQ2, at Site 1, we find students receiving special education services seem to benefit at least as much as their general education peers engaging in hybrid human-AI tutoring, demonstrated by the comparable gains from pre-to-post on the *Star Diagnostic* (Table 2). Special education students practiced and achieved proficiency on more skills in *IXL* despite the average time spent using the software being lower compared to general education students (Table 2). Students with disabilities participating in human-AI tutoring treatment demonstrated statistically significant benefits compared to students with disabilities in the control, evidenced by 1.2 times more *skills practiced* and 1.8 times more *skills proficient* each week (estimated by entering coefficients from Table 4 into Eq. 1). The negative *learning process* results for general education (Table 4) are not consistent with the positive learning *outcome* results for both SpecialEd and GenEd and the positive *learning process* results for SpecialEd. More analysis is needed to interpret these findings. We hypothesize that the extra human support provided by the treatment was particularly beneficial for students with disabilities. This aligns with special education students' *time spent* engaging in *IXL*, specifically, being lower than general education due to conversing with their human tutor more while engaging in hybrid human-AI tutoring. This also supports our causal model (Fig. 1) linking human socio-motivational interaction as a mechanism to improve student progress, providing students *multiple means of engagement* (step 2), such as the AI notifying the tutor that a student needs help or the student directly seeking help (e.g., raising their hand virtually or typing into the chat). In other words, students receiving special education services may be benefiting more from the relational and motivational support from a human tutor (with multiple methods of seeking help) compared to general education students [8].

There were surprisingly nuanced findings related to students with disabilities baseline scores and performance compared to students without disabilities. At Site 1 despite

SpecialEd students' baseline performance being lower than GenEd, they may have benefited more from engaging in hybrid human-AI tutoring. At Site 2, students with disabilities had higher median scores at pretest *and* posttest (Fig. 1). Although the latter may not provide direct insight into the impact of the tutoring treatment, it aligns with past AIED work [18] in highlighting the challenges involving diverse learners when applying AI-based learning interventions.

6 Implications

This work reveals several implications as AIED transitions to a world where AI-assisted learning systems are being developed to benefit the learning experiences for *all*.

Students with Disabilities May Particularly Benefit from Human Socio-Motivational Support. Similar to Koedinger et al. [14], this study underscores the significant positive influence that human support can have on student learning outcomes, in conjunction with AI-based solutions. In light of the proliferation of AI-based solutions and the widespread neglect in researching these technologies among students with disabilities, thoughtful integration of human support remains of paramount importance.

Hybrid Human-AI Technologies May Provide the Extra Support Needed to Students with Disabilities, Enhancing Their Success in Inclusive Educational Settings. Similar to [14], we observed a positive impact of human-AI technologies in inclusive settings. Students with disabilities may perform below, at, or far exceed expectations. Broad ranges in performance compared to students without disabilities may pose challenges when developing inclusive human-AI solutions. This work illustrates how leveraging assets-based approaches when designing AI-based systems, integrating human support and AI-driven instruction, can effectively address diverse learner needs [28].

More Knowledge Within AIED on How Students with Identified Disabilities are Served in Schools is Needed to Improve Research and Development of Human-AI Systems. Students with disabilities may receive extra support related to math or reading, both, or neither. Most often, if students' IEP status (i.e., Y/N) is provided in learning analytics (e.g., math software data), there will be no other information. Not all IEPs contain math goals of which students are receiving specially designed instruction with their progress being monitored. In fact, many IEPs contain no academic goals in math or reading, e.g., 19% of IEPs contain speech services only [20]. When analyzing student learning data in math or reading, AIED should be aware that many students reported to have an IEP, in reality, may not have disabilities related to math or reading at all.

Students with Disabilities May Demonstrate Different Usage of Intelligent Tutor or AI-Driven Software Compared to Students Without Disabilities. AI-based technologies, like *i-Ready* and *IXL*, may be used as a progress monitoring tool for students' IEP goals or as an intervention within itself, causing usage of the system to be different than the general population. For both sites, we filtered the math software learning process data to include only the days hybrid human-AI tutoring took place. Students in special education may have additional usage due to using such software prescribed by special education teachers as a method of monitoring students' progress in relation to their academic goals per their IEP [3]. Researchers should be aware of the varying purposes AI-based learning technologies may have in schools.

7 Limitations, Future Work, and Conclusion

While these quasi-experiments offer valuable insights into the effectiveness of hybrid human-AI tutoring as a universal intervention, there exist limitations. Both sites used different math software and diagnostic assessments, which may raise concerns regarding validity and reliability. However, the diversity in math software used across sites enhances the generalizability of our findings and reflects the real-world reality of implementing academic interventions where schools vary in software. Typically, the impact of academic interventions is determined over a reasonable timeframe that allows for meaningful change to occur. The timeframe usually ranges from a few months to a full school year, depending on factors such as intervention intensity, student baseline abilities, and assessment measures being used [25]. We posit that longer intervention duration, aligning with typical high-impacting tutoring programs [17], say a semester or year, would yield stronger effects. Another limitation was the lack of control group at Site 2 making it difficult to determine the impact of hybrid human-AI tutoring treatment. Lastly, special education status alone does not provide much information about the needs of individual students and educational goals, with accommodations specifically designed to meet the needs of the learner often not provided [7].

Future work includes creating refined models for hybrid human-AI tutoring involving strategically and dynamically matching students to tutors based on their individual needs. This study conducted across two urban, low-income middle schools, explored the effects of hybrid human-AI tutoring on students with disabilities within inclusive educational settings. The findings indicate that students with disabilities are learning at least as much as their general education peers while engaging in hybrid human-AI tutoring. Additionally, evidence suggests students with disabilities may benefit more from the motivational benefits of human tutoring. These results and implications discussed underscore the importance of considering the unique learning patterns and needs of students receiving special education services when analyzing AI-adaptive software and learning data. In conclusion, hybrid human-AI tutoring presents a promising avenue for supporting diverse math learners, particularly in inclusive settings.

Acknowledgments. The authors thank Cindy Tipper and Hui Cheng for their considerable contribution to data handling and analyses. This work is supported with funding from the Learning Engineering Virtual Institute. Any opinions, findings, and conclusions expressed in this material are those of the authors.

Disclosure of Interests. The authors have no competing interests to declare that are relevant to the content of this article.

References

1. Aleven, V., et al.: Towards the future of AI-augmented human tutoring in math learning. In: Wang, N., Rebolledo-Mendez, G., Dimitrova, V., Matsuda, N., Santos, O.C. (eds.) AIED 2023, pp. 26–31. Springer, Cham (2023). https://doi.org/10.1007/978-3-031-36336-8_3

2. Aleven, V., McLaughlin, E., Glenn, R., Koedinger, K.: Instruction based on adaptive learning technologies. In: Mayer, R.E., Alexander, P. (eds.) Handbook of Research on Learning and Instruction, pp. 522–560. Routledge, New York (2016)

3. American Institute for Research. i-Ready Diagnostic and Growth Monitoring Reading/English Language Arts. National Center on Intensive Intervention. https://charts.intensiveintervention.org/progressmonitoring/tool/?id=d264b2946d8df43d. Accessed 25 Jan 2024

4. Andrews, E., Forber-Pratt, A., Mona, L., Lund, E., Pilarski, C., Balter, R.: #SaytheWord: a disability culture commentary on the erasure of "disability." Rehabil. Psychol. Psychol. 64(2), 111 (2019)

5. Baek, C., Aguilar, S.: Past, present, and future directions of learning analytics research for students with disabilities. J. Res. Technol. Educ. 55(6), 931–946 (2023)

6. Baker, R., Hawn, A.: Algorithmic bias. Int. J. Artif. Intell. Educ.Artif. Intell. Educ. 32(1), 1052–1092 (2022)

7. Blaser, B., Ladner, R.: Why is data on disability so hard to collect and understand? In: 2020 Research on Equity and Sustained Participation in Engineering, Computing, and Technology (RESPECT), vol. 1. pp. 1–8. IEEE (2020)

8. Chine, D., et al.: Educational equity through combined human-AI personalization: a propensity matching evaluation. In: Rodrigo, M.M., Matsuda, N., Cristea, A.I., Dimitrova, V. (eds.) AIED 2022, pp. 366–377. Springer, Cham (2022). https://doi.org/10.1007/978-3-031-11644-5_30

9. Dsilva, V., Schleiss, J., Stober, S.: Trustworthy academic risk prediction with explainable boosting machines. In: Wang, N., Rebolledo-Mendez, G., Matsuda, N., Santos, O.C., Dimitrova, V. (eds.) AIED 2023, pp. 463–475. Springer, Cham (2023). https://doi.org/10.1007/978-3-031-36272-9_38

10. Heffernan, N., Heffernan, C.: The ASSISTments ecosystem: building a platform that brings scientists and teachers together for minimally invasive research on human learning and teaching. Int. J. Artif. Intell. Educ.Artif. Intell. Educ. 24(4), 470–497 (2014)

11. Holstein, K., McLaren, B., Aleven, V.: Intelligent tutors as teachers' aides: exploring teacher needs for real-time analytics in blended classrooms. In: Proceedings of the Seventh International Learning Analytics & Knowledge Conference, pp. 257–266 (2017)

12. Holstein, K., McLaren, B.M., Aleven, V.: Student learning benefits of a mixed-reality teacher awareness tool in AI-enhanced classrooms. In: Penstein Rosé, C., et al. (eds.) AIED 2018. LNCS (LNAI), vol. 10947, pp. 154–168. Springer, Cham (2018). https://doi.org/10.1007/978-3-319-93843-1_12

13. Individuals with Disabilities Education Act. 20 U.S.C. §§ 1400 et seq. (1990)

14. Koedinger, K., McLaughlin, E., Heffernan, N.: A quasi-experimental evaluation of an on-line formative assessment and tutoring system. J. Educ. Comput. Res. 43(4), 489–510 (2010)

15. Khalil, M., Slade, S., Prinsloo, P.: Learning analytics in support of inclusiveness and disabled students: a systematic review. J. Comput. High. Educ. 1–18 (2023)

16. National Center for Education Statistics. National Assessment of Educational Progress (NAEP), various years, 1990–2022 Mathematics Assessments (2022). https://www.nationsreportcard.gov/

17. National Student Support Accelerator.Toolkit for Tutoring Programs (2021). Retrieved from National Student Support Accelerator: https://doi.org/10.26300/5n7h-mh59

18. Nawaz, S., Kamei, T., Srivastava, N.: Nuanced growth patterns of students with disability. In: Wang, N., Rebolledo-Mendez, G., Dimitrova, V., Matsuda, N., Santos, O.C. (eds.) AIED 2023, pp. 612–618. Springer, Cham (2023). https://doi.org/10.1007/978-3-031-36336-8_95

19. Nickow, A., Oreopoulus, P., Quan, V.: The impressive effects of tutoring on prek-12 learning: a systematic review and meta-analysis of the experimental evidence. National Bureau of Economic Research (NBER), Working paper # 27476 (2020)

20. Pew Research Center. What Federal Education Data Shows About Students with Disabilities in the U.S. Pew Research Center (2023)

21. Rice, M., Dunn, S.: The use of artificial intelligence with students with identified disabilities: a systematic review with critique. Comput. Sch.. Sch. **40**(4), 370–390 (2023)

22. Roski, M., Sebastian, R., Ewerth, R., Hoppe, A., Nehring, A.: Dropout prediction in a web environment based on universal design for learning. In: Wang, N., Rebolledo-Mendez, G., Matsuda, N., Santos, O.C., Dimitrova, V. (eds.) AIED 2023, pp. 515–527. Springer, Cham (2023). https://doi.org/10.1007/978-3-031-36272-9_42

23. Scott, L., Bruno, L.: Universal design for transition: a conceptual framework for blending academics and transition instruction. J. Spec. Educ. Apprent. **7**(3), 1 (2018)

24. Shaywitz, S., Holahan, J., Kenney, B.: The yale outcome study: outcomes for graduates with and without dyslexia. J. Pediatr. Neuropsychol. **6**, 189–197 (2020)

25. Slavin, R.: Perspectives on evidence-based research in education—what works? Issues in synthesizing educational program evaluations. Educ. Res. **37**(1), 5–14 (2008)

26. Thomas, D., et al.: Improving student learning with hybrid human-AI tutoring: a three-study quasi-experimental investigation. In: LAK24: 14th International Learning Analytics and Knowledge Conference. ACM, New York (2024)

27. West, M.: An Ed-Tech Tragedy? Educational Technologies and School Closures in the time of COVID-19. UNESCO, Paris, France (2023)

28. Wong-Villacres, M., DiSalvo, C., Kumar, N., DiSalvo, B.: Culture in action: unpacking capacities to inform assets-based design. In: Proceedings of the 2020 CHI Conference on Human Factors in Computing Systems, pp. 1–14 (2020)

Fine-Tuning a Large Language Model with Reinforcement Learning for Educational Question Generation

Salima Lamsiyah[1]([✉])[iD], Abdelkader El Mahdaouy[2][iD], Aria Nourbakhsh[1][iD], and Christoph Schommer[1][iD]

[1] Department of Computer Science, Faculty of Science, Technology and Medicine, University of Luxembourg, Esch-sur-Alzette, Luxembourg
{salima.lamsiyah,aria.nourbakhsh,christoph.schommer}@uni.lu
[2] College of Computing, Mohammed VI Polytechnic University, Ben Guerir, Morocco
abdelkader.elmahdaouy@um6p.ma

Abstract. Educational Natural Language Generation (EduQG) aims to automatically generate educational questions from textual content, which is crucial for the expansion of online education. Prior research in EduQG has predominantly relied on cross-entropy loss for training, which can lead to issues such as exposure bias and inconsistencies between training and testing metrics. To mitigate this issue, we propose a reinforcement learning (RL) based large language model (LLM) for educational question generation. In particular, we fine-tune the Google FLAN-T5 model using a mixed objective function that combines cross-entropy and RL losses to ensure the generation of questions that are syntactically and semantically accurate. The experimental results on the SciQ question generation dataset show that the proposed method is competitive with current state-of-the-art systems in terms of predictive performance and linguistic quality.

Keywords: Educational Question Generation · Large Language Model · Google FLAN-T5 · Reinforcement Learning · Self-Critical Sequence Training

1 Introduction

The abundance of electronic textual information raises the need for automatic question generation (AQG) systems, which aim to automatically generate natural questions from diverse sources, including raw texts, databases, or semantic representations [25]. Crafting high-quality questions manually presents a significant challenge due to its complexity and the time it demands. Moreover, it represents a major obstacle for training question-answering systems due to the lack of labeled data. Therefore, AQG plays a vital role in improving the performance of several natural language processing tasks, such as enhancing question-answering systems by providing additional training data [26], creating practice exercises

A. M. Olney et al. (Eds.): AIED 2024, LNAI 14829, pp. 424–438, 2024.
https://doi.org/10.1007/978-3-031-64302-6_30

and assessments for educational purposes [3,10], and assisting dialog systems in initiating and maintaining conversations with human users [18].

Several research works have been introduced in the literature for the automatic question generation task. Early works have primarily utilized rule-based approaches, relying on heuristic rules or manually crafted templates, which suffered from limited generalizability and scalability [10]. However, the advent of deep learning catalyzed a shift towards neural network-based approaches, particularly sequence-to-sequence (seq2seq) models, which eliminate the need for manually designed rules and offer end-to-end trainability [5,29]. More recently, large language models have also demonstrated notable improvements in the AQG task [1,6,19]. Nevertheless, most of the previously mentioned research works have primarily focused on AQG within generic domains. Therefore, there is a noteworthy need to develop such methods for the educational domain, specifically to support technology-enhanced learning.

In this paper, we introduce a novel method for educational question generation (EduQG) that leverages the potential of both large language models (LLM) and reinforcement learning (RL). Fine-tuning an LLM for EduQG has already been explored in [1], where the authors have proposed to improve EduQG performance by pre-training the Google T5 model [22] on scientific documents using the S2ORC corpus [17], followed by fine-tuning on a task-specific dataset. The proposed method has shown promising results compared to state-of-the-art (SOTA) systems on the SciQ question dataset [31]. Nonetheless, it has been observed that cross entropy-based sequence training, in general, exhibits several limitations such as exposure bias and inconsistency between training and testing measurements [2]. Hence, this approach does not always yield optimal results on discrete evaluation metrics for sequence generation tasks, including text summarization [21] and question generation [28]. Some recent research on AQG has addressed these challenges by optimizing evaluation metrics through reinforcement learning [13,28]. However, these approaches primarily use metrics like BLEU [20] and ROUGE-L [16] as rewards for RL training, neglecting other crucial metrics such as semantic constraints, which are essential for guiding high-quality text generation.

To address the issues mentioned above, we propose fine-tuning the Google FLAN-T5 model on the SciQ dataset by optimizing a mixed objective function that combines cross-entropy and RL losses. This method not only takes into account discrete metrics such as BLEU but also integrates semantic similarity metrics based on cosine similarity. It aims to encourage the generation of text that is syntactically and semantically coherent while ensuring consistency in training and testing measurements. Moreover, the proposed model is end-to-end trainable, has set new state-of-the-art scores, and outperforms existing methods on the SciQ dataset.

The rest of this paper is organized as follows. We discuss the related work in Sect. 2. In Sect. 3, we present the proposed method. In Sect. 4, we describe the conducted experiments and present the obtained results. Finally, in Sect. 5, we conclude the paper and outline some future directions in the field.

2 Related Work

In this section, we will review research on educational question-generation methods and reinforcement learning for AQG tasks. For a detailed review, readers may refer to [4, 12] surveys, respectively.

2.1 Educational Question Generation

Automatic question generation has progressed rapidly due to new datasets and model improvements. Many different AQG models have been proposed, starting for simple vanilla seq2seq neural networks [5, 34] to the more recent transformer-based models [1, 6, 19].

Nevertheless, only a limited number of works have been proposed for the educational question generation task. For instance, Zhao et al. [33] have proposed a novel question generation method that first learns the question type distribution of an input story paragraph. Then, it summarizes salient events, which are then used to generate questions of high cognitive demand. To train the event-centric summarizer, they have fine-tuned a pre-trained transformer-based sequence-to-sequence model using silver samples composed of educational question-answer pairs. Recently, Jiao et al. [11] have introduced a controllable math word problem generation pipeline utilizing an energy language model and expert components to regulate problem difficulty and content. The difficulty is adjusted through mathematical and linguistic constraints on equations, vocabulary, and topics, with additional focus on attributes such as fluency and sequence proximity to improve language quality and creativity. Similarly, Leite et al. [15] have proposed to enhance controllability in EduQG by introducing a new guidance attribute: question explicitness. Their method controls the generation of explicit and implicit (wh)-questions from children-friendly stories, showing preliminary evidence of managing EduQG through question explicitness, both independently and in conjunction with narrative elements such as causal relationships, outcome resolutions, or predictions. Shimmei et al. [27] have introduced a pragmatic approach for generating questions aligned with specific learning objectives in online coursework. Furthermore, Elkins et al. [6] have studied the potential of controllable text generation with large language models to enhance education by producing high-quality, diverse questions. Through human evaluation with teachers, this study has demonstrated that the generated questions are useful and of high quality for classroom use, highlighting their potential for reducing teachers' workload and improving educational content. In the same context, Bulathwela et al. [1] have introduced a novel method for EduQG that improves the generation of educational questions by further pre-training and fine-tuning a pre-existing language model, specifically the Google T5 [22], on scientific texts and science question data.

Educational question-generation methods leveraging LLMs have demonstrated competitive results when compared to state-of-the-art approaches. However, it has been noted that fine-tuning LLMs with cross-entropy loss can lead to several issues, such as exposure bias and inconsistencies between training and testing measurements [2], which may not always provide optimal results on discrete evaluation metrics for sequence generation tasks [21,28]. In contrast to the existing approaches, we introduce a novel method for EduQG that exploits the potential of RL for fine-tuning an LLM. This method not only considers discrete metrics, such as BLEU, but also integrates semantic similarity, aiming to generate questions that are syntactically and semantically coherent, while maintaining consistency across training and testing phases. Moreover, it is based on an *answer-unaware question generation* model, which does not necessitate knowledge of the answer in the question-generation process. We aim to improve the question generation without requiring access to a labeled dataset with context-question-answer pairs, which are often scarce in real-world settings.

2.2   Reinforcement for Automatic Question Generation

In recent years, sequence-to-sequence models have gained increasing attention and provide state-of-the-art performance in a wide variety of text generation tasks [21,28]. The foundation of all these models typically consists of a deep neural network that includes an encoder and a decoder. However, as previously discussed, seq2seq models suffer from two common problems: 1) exposure bias and 2) inconsistency between train/test measurement. Recently, a novel point of view has emerged to tackle these problems in seq2seq models by leveraging methods from reinforcement learning [12].

In the context of automatic question generation for generic domains, several methods have been introduced in the literature. For instance, Chen et al. [2] have proposed a reinforcement learning-based graph-to-sequence model for AQG. It leverages a Graph2Seq generator equipped with a novel Bidirectional Gated Graph Neural Network-based encoder for embedding the passage and a hybrid evaluator with a mixed objective that combines cross-entropy and RL losses. Kumar et al. [13] have introduced a novel AQG method that employs a holistic generator-evaluator framework, optimizing objectives that reward both semantics and structure. The generator is a seq2seq model that integrates the structure and semantics of the question being generated. The evaluator model assesses each predicted question and assigns a reward based on its conformity to the structure of ground-truth questions. In the same context, [30] have developed an answer-driven, end-to-end deep question generation model based on reinforcement learning. This model explores semantic information derived from the answer to enhance the quality of deep question generation. Similarly, Guan et al. [9] have combined reinforcement learning with semantic information for deep question generation. This approach utilizes semantic graphs of document representations based on the gated graph neural network. To generate high-quality questions, they have optimized specific objectives through RL, taking into account four evaluation factors: naturality, relevance, answerability, and difficulty. More recently, [8] have proposed RAST, a framework for Retrieval-Augmented Style Transfer, aimed at leveraging the style of diverse templates for question generation. To train RAST, a novel RL-based approach was developed, focusing on maximizing a weighted combination of diversity and consistency rewards. The consistency reward is computed by a question-answering model, while the diversity reward assesses the degree to which the final output resembles the retrieved template.

Motivated by the success of reinforcement learning in automatic question generation within generic domains, we propose to enhance educational question generation by leveraging RL capabilities. To the best of our knowledge, this is the first work that fine-tunes the Google FLAN-T5 model with RL for educational question generation.

3   A Reinforcement-Enhanced Language Model for EduQG

In this section, we present the proposed method RLLM-EduQG method, a reinforcement-enhanced language model for educational question generation. We first motivate the design and then present the details of each component as shown in Fig. 1.

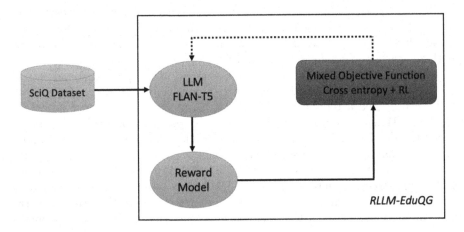

Fig. 1. Overall architecture of the proposed RLLM-EduQG method.

3.1   Problem Formulation

The goal of an automatic question generation method is to generate natural questions based on a given form of data, such as raw texts, databases, or images [25], ensuring that the generated questions are answerable from the input data. In this paper, we focus on EduQG from a given text passage using an *answer-unaware model*.

Given a text passage consisting of N word tokens $X = \{x_1, x_2, \ldots, x_N\}$. The task of natural question generation is to create the best natural language question consisting of a sequence of word tokens $\hat{Y} = \{y_1, y_2, \ldots, y_T\}$ which maximizes the conditional likelihood $\hat{Y} = \underset{Y}{\operatorname{argmax}} P(Y|X)$. Here N and T are the lengths of the passage and question, respectively. We concentrate on the problem setting where we have sets of passages and corresponding target questions to learn the mapping. Existing AQG approaches [1,6] adopt a similar assumption. The fine-tuning process of the Google FLAN-T5 language model [22] is described in the subsequent subsections.

3.2 Google FLAN-T5 Supervised Fine-Tuning

We fine-tune the Google Flan-T5 model using a supervised approach for question generation from text passages, optimizing with the standard cross-entropy loss function, defined as follows:

$$\mathcal{L}_{\text{lm}} = \sum_t - \log P(y_t^* | X, y_{<t}^*) \tag{1}$$

where y_t^* is the word at the $t\text{-}th$ position of the ground-truth output sequence. However, minimizing $\mathcal{L}_{\text{lm}}$ does not always produce the best results on discrete evaluation metrics such as BLEU [20]. As previously discussed, one of the main reasons for this discrepancy is the exposure bias. This issue arises from the fact that the network is aware of the ground truth sequence up to the next token during training but lacks such supervision during testing, leading to the accumulation of errors as it generates the sequence. To mitigate exposure bias, techniques such as scheduled teacher forcing or reinforcement learning can be employed. These methods incorporate the model's own predictions during training to more closely mimic the conditions it will encounter during inference.

3.3 Reinforcement Learning

To address the aforementioned problem, we propose to fine-tune the Google FLAN-T5 with reinforcement learning. More specifically, we employ the self-critical sequence training (SCST) algorithm [24]. SCST is an effective REIN-FORCE algorithm that utilizes the output of its test-time inference algorithm to normalize the rewards it receives.

In the SCST algorithm, at each training iteration, the model produces two sequences: the sampled sequence Y^s, where each token y_t^s is drawn based on the multinomial probability distribution $P(y_t|X, y_{<t})$ predicted by the model, and the baseline sequence $\hat{Y}$, created through a greedy decoding method that sequentially selects the most likely next token. The reward function $r(Y)$ evaluates each generated sequence Y against the ground-truth sequence Y^* using specific reward metrics. The RL loss function is defined as follows:

$$\mathcal{L}_{\text{rl}} = (r(\hat{Y}) - r(Y^s)) \sum_t \log P(y_t^s | X, y_{<t}^s) \tag{2}$$

As can be seen, if the sampled sequence achieves a higher reward than the baseline sequence, the model is trained to maximize its likelihood, and conversely, if the baseline sequence is superior, the model minimizes the likelihood of the sampled sequence.

The selection of an appropriate reward function is a critical element in RL. To address syntactic and semantic constraints, we have employed the following metrics as our reward functions:

- **Evaluation metric as reward function:** We use the evaluation metric, BLEU-4, as our reward function f_{eval}, which allows us to directly optimize the model towards the evaluation metrics.
- **Semantic metric as reward function:** One drawback of certain evaluation metrics like BLEU is that they do not account for the semantics of sentences but only reward systems that match exact n-grams with the reference system. To enhance the effectiveness of our reward function, we also incorporate semantic reward function f_{sem}. Specifically, we calculate the cosine similarity between a generated sequence and the ground-truth sequence using their embedding vectors. Based on an empirical study [14], we have chosen the simCSE model [7] as our embedding model, which has shown to be the most effective for generating sentence embeddings.

We define the final reward function $r(Y)$ as a combination of the evaluation and semantic metrics, defined as $r(Y) = f_{eval}(Y, Y^*) + \alpha f_{sem}(Y, Y^*)$. Where α is a scalar weight that balances the contribution of the semantic metric f_{sem} with the evaluation metric f_{eval}.

3.4 Mixed Training Objective Function

We train our method, RLLM-EduQG, in two stages. In the first stage, we train the model using a regular cross-entropy loss, as defined in Eq. 1. In the second stage, we fine-tune the model by optimizing a mixed objective function that combines both cross-entropy loss and RL loss, defined as follows:

$$\mathcal{L} = \gamma \mathcal{L}_{rl} + (1 - \gamma)\mathcal{L}_{lm} \tag{3}$$

where γ is a scaling factor that controls the trade-off between cross-entropy loss $\mathcal{L}_{lm}$ and reinforcement learning loss $\mathcal{L}_{rl}$.

4   Experiments

In this section, we first present a brief description of the used dataset, evaluation metrics, and the experimental setup. Then, we provide a comparative analysis of the obtained results, intending to verify the following hypotheses:

- *Hypothesis H1:* The proposed method is effective for educational question generation as compared with other recent state-of-the-art EduQG methods.
- *Hypothesis H2:* Fine-tuning Google FLAN-T5 with reinforcement learning has shown to be effective for EduQG.
- *Hypothesis H3:* The combination of f_{eval} and f_{sem} reward functions improves the performance of the proposed method.

4.1   Evaluation Datasets and Metrics

The experiments are carried out using the SciQ dataset [31], created essentially for evaluating *educational question generation* tasks. SciQ comprises 13,679 crowd-sourced scientific exam questions covering physics, chemistry, and other sciences. Although smaller than other question-generation datasets, SciQ is more relevant for objectively evaluating EduQG models.

For the evaluation measures, we consider two aspects of quality when assessing question generation models: 1) prediction accuracy, and 2) the linguistic quality of the generated questions. To measure the predictive accuracy, we use the BLEU score and the F1 score, following the methodology used in prior work [1]. For assessing linguistic quality, or how human-like the generated questions are, we employ metrics of perplexity and diversity [1]. High diversity scores along with low perplexity reflect the use of a rich vocabulary and enhanced grammatical precision.

4.2   Experimental Setup

The RLLM-EduQG method has been developed using the PyTorch[1] deep learning framework, PyTorch Lightning[2], the Hugging Face Transformers library[3], and the publicly available source code of some SOTA methods, including the EduQG system[4] and the implementation of the SCST algorithm[5]. We have fine-tuned the Google FLAN-T5-base version[6]. The fine-tuning has been performed on the Iris cluster[7] at the University of Luxembourg.

For dataset splitting, we have used 80% for training, 10% for validation, and 10% for testing. We set α in the reward function to 0.1 and γ in the mixed loss function to 0.99. We have used the Adam optimizer [32], with the learning rate set to 0.001 during the pretraining stage and reduced to 0.0001 for the fine-tuning stage. We stop the training when no improvement is seen for 5 epochs. The batch size is set to 32. The beam search width during inference is set to 5. All hyperparameters were tuned using the validation set.

4.3   Results

Comparison with State-of-the-art Methods. To demonstrate the effectiveness of the proposed RLLM-EduQG method (*Hypothesis H1*), we compare its performance with recent SOTA methods for educational question generation on the SciQ dataset [31]. Table 1 depicts the results from the baseline Leaf+, the SOTA EduQG+ model, and our RLLM-EduQG model on the SciQ dataset. A

---

[1] https://pytorch.org/.

[2] https://www.pytorchlightning.ai/.

[3] https://github.com/huggingface/transformers.

[4] https://github.com/hmuus01/Educational_QG/tree/main.

[5] https://github.com/hugochan/RL-based-Graph2Seq-for-NQG/tree/master.

[6] https://huggingface.co/google/flan-t5-base.

[7] https://hpc-docs.uni.lu/systems/iris/.

concrete example of ground truth questions along with outputs from our RLLM-EduQG system, **Leaf+**, and **EduQG+** is presented in Table 4 within Appendix A. Leaf+ is a baseline model that uses the Google T5 model, which is initially trained on a generic AQG dataset, SQuAD [23], and then fine-tuned for educational question generation on the SciQ dataset. EduQG+ model [1] is first pre-trained on the S2ORC corpus of scientific documents [17], then trained on the SQuAD dataset, and finally fine-tuned on the SciQ dataset. It is worth mentioning that we have reported the results presented in their corresponding papers [1].

As shown in Table 1, for predictive performance, our method RLLM-EduQG has achieved the highest BLEU scores across all n-gram lengths (BLEU-1 through BLEU-4) and the highest F1 score, suggesting it is most effective at generating questions that align closely with the ground-truth references. Additionally, it has obtained the lowest perplexity score, which indicates its generated questions are more coherent and fluent compared to the other models. The diversity metric is also highest for RLLM-EduQG, indicating a wider variety of vocabulary and phrasing in the questions it generates. The results achieved can be attributed to the fact that RLLM-EduQG has been trained with a mixed objective function that includes both cross-entropy and reinforcement learning losses. By using a combination of two reward functions-an evaluation metric based on BLEU-4 and a semantic metric utilizing semantic similarity-the model is simultaneously optimized for linguistic accuracy and content relevance. As a result, RLLM-EduQG closely matches the syntactic structure of the ground truth data and ensures that the generated questions are meaningful and diverse. This training strategy is likely the reason for its superior performance on the SciQ dataset, surpassing the Leaf+ and EduQG+ models trained solely with cross-entropy loss. Additionally, it is noteworthy that our method RLLM-EduQG has been fine-tuned exclusively on the SciQ dataset without any further pre-training on the S2ORC or SQuAD datasets.

Table 1. Comparative analysis of predictive performance and linguistic quality among the baseline **Leaf+**, the SOTA **EduQG+** model, and our **RLLM-EduQG** model on the SciQ dataset. The best performance is highlighted in bold.

Models	Predictive Performance					Linguistic Quality	
	BLEU-1	BLEU-2	BLEU-3	BLEU-4	F1-Score	Perplexity	Diversity
Leaf+	36.67	31.45	28.17	24.26	41.65	26.43	0.801
EduQG+	37.20	33.86	28.49	22.35	43.04	33.88	0.812
RLLM-EduQG	**51.08**	**43.86**	**39.87**	**36.40**	**56.80**	**24.88**	**0.944**

Effect of Fine-Tuning Google FLAN-T5 with Reinforcement Learning. The objective of the second set of analyses is to evaluate the efficacy of reinforcement learning in educational question generation tasks (examining *Hypothesis H2*). For this purpose, we compared RLLM-EduQG, which incorporates reinforcement learning, with LLM-EduQG, which solely relies on fine-tuning the Google FLAN-T5 using cross-entropy loss without reinforcement. Table 2 presents the results of these two variants on the SciQ dataset. Table 2 clearly shows that RLLM-EduQG has obtained far better performance than LLM-EduQG in terms of all evaluation measures, encompassing both predictive performance and linguistic quality. Specifically, RLLM-EduQG has achieved improvements of 14.03%, 13.91%, 12.47%, 14.84%, and 14.59% over LLM-EduQG in terms of BLEU 1-4 and F1-score, respectively. This further demonstrates the effectiveness of using reinforcement learning, particularly the self-critical sequence training algorithm.

Table 2. Comparative analysis of predictive performance and linguistic quality between two variants of our method, **LLM-EduQG** and **RLLM-EduQG**, on the SciQ dataset.

Models	Predictive Performance					Linguistic Quality	
	BLEU-1	BLEU-2	BLEU-3	BLEU-4	F1-Score	Perplexity	Diversity
LLM-EduQG	37.05	29.95	27.40	21.56	42.21	26.93	0.804
RLLM-EduQG	51.08	43.86	39.87	36.40	56.80	24.88	0.944

Table 3. Comparison results of **RLLM-EduQG**-f_{eval} and **RLLM-EduQG** on the SciQ dataset with respect to predictive performance and linguistic quality.

	Predictive Performance					Linguistic Quality	
	BLEU-1	BLEU-2	BLEU-3	BLEU-4	F1-Score	Perplexity	Diversity
RLLM-EduQG-f_{eval}	49.76	42.65	38.18	24.26	41.65	25.37	0.819
RLLM-EduQG	51.08	43.86	39.87	36.40	56.80	24.88	0.944

Impact of Combining Evaluation and Semantic Reward Functions. An additional contribution of this work is the combination of *evaluation* and *semantic similarity* reward metrics to test *Hypothesis H3: Does the combination of these two reward metrics improve the performance of educational question generation?* To empirically address this question, we conducted two runs on the SciQ dataset. In the RLLM-EduQG-f_{eval} variant, only the evaluation metric is utilized as the reward function, while RLLM-EduQG employs both. Table 3 summarizes the obtained results. From Table 3, it is evident that the RLLM-EduQG-f_{eval} method has achieved modest results compared to the RLLM-EduQG method in terms of all the used evaluation measures. As previously discussed, the enhanced performance of RLLM-EduQG is attributed to its use of a mixed reward function that incorporates both the BLEU-based evaluation metric and semantic similarity based on the simCSE model [7]. This approach ensures that the model is rewarded not just for matching the n-gram patterns of the reference data (as with BLEU-4) but also for capturing the semantic essence of the content, leading to more contextually relevant and varied questions.

5   Conclusion and Future Directions

In this paper, we introduced RLLM-EduQG, a novel method for generating educational questions that leverages the capabilities of large language models and reinforcement learning, particularly the self-critical sequence training algorithm. We fine-tuned the Google FLAN-T5 using a mixed objective function that incorporates both cross-entropy and reinforcement learning losses. To ensure the generated questions are both syntactically and semantically accurate, we utilized two reward functions: the evaluation-based BLEU metric and a semantic reward. The experimental results on the SciQ dataset indicated that using a reinforcement learning objective function yields promising results without necessitating further pre-training or training on other datasets, unlike the state-of-the-art EduQG+ [1]. Additionally, the combination of the two reward functions enhanced performance across all evaluation metrics. We view our work as a significant step toward exploring the fine-tuning of LLMs with RL in educational question generation. Furthermore, this method could serve as a foundation for developing a suite of tools that aid educators in creating scalable and personalized assessments.

Fine-tuning a large language model with reinforcement learning has proven effective in enhancing the performance of educational question generation. Therefore, in future work, we plan to conduct an empirical study to explore the fine-tuning of other LLMs with RL for the EduQG task across various datasets. Additionally, a notable gap in this research is the absence of human evaluation of the AI-generated questions. While offline evaluation using labeled datasets is beneficial, obtaining assessments from teachers and learners on the differences between human and AI-generated questions will yield more insightful findings. These could significantly advance this area of research in the future.

A Examples of Generated Questions of Our RLLM-EduQG System

Table 4. Comparison of Questions Generated by Our System, **Leaf+**, and **EduQG+** with Ground Truth Questions Using Randomly Selected Examples from the SciQ Dataset Test Set.

Context	Leaf+	EduQG+	RLLM-EduQG	Ground Truth
(1) Scientific models are useful tools for scien- tists. Most of Earth's systems are extremely complex. Models allow scientists to work with systems that are nearly impossible to study as a whole. Models help scientists to understand these systems. They can analyze and make predictions about them using the models. There are different types of models.	What help scientists to understand the systems of Earth?	What is used to analyze and make predictions about systems that are nearly impossible or easy to study as a whole?	How do different types of models help scientists analyze and make predictions about Earth's systems?	What useful tool helps scientists work with, understand and make predictions about extremely complex systems?
(2) Muscles That Move the Head The head, attached to the top of the vertebral column, is balanced, moved, and rotated by the neck muscles. When these muscles act unilaterally, the head rotates. When they contract bilaterally, the head flexes or extends. The major muscle that laterally flexes and rotates the head is the sternocleidomastoid. In addition, both muscles working together are the flexors of the head. Place your fingers on both sides of the neck and turn your head to the left and to the right. You will feel the movement originate there. This muscle divides the neck into anterior and posterior triangles when viewed from the side.	What help scientists to understand the systems of Earth?	What is the major muscle that laterally rotates?	Which part of the body is laterally flexed and rotated by the sternocleido- mastoid muscle?	The sternocleido- mastoid is the major muscle that laterally flexes and rotates what?
(3) In both eukaryotes and prokaryotes, ribosomes are the non-membrane-bound organelles where proteins are made. Ribosomes are like the machines in the factory that produce the factory's main product. Proteins are the main product of the cell.	What is the main product of a cell?	What are the non-membrane bound organelles where proteins are made?	In both eukaryotic and prokaryotic cells, how are ribosomes like factory machines, and what essential product do they make for the cell?	In both eukaryotes and prokaryotes, ribosomes are the non-membrane bound organelles where what main product of the cell is made?

References

1. Bulathwela, S., Muse, H., Yilmaz, E.: Scalable educational question generation with pre-trained language models. In: International Conference on Artificial Intelligence in Education, pp. 327–339. Springer (2023). https://doi.org/10.1007/978-3-031-36272-9_27

2. Chen, Y., Wu, L., Zaki, M.J.: Reinforcement learning based graph-to-sequence model for natural question generation. arXiv preprint arXiv:1908.04942 (2019)

3. Danon, G., Last, M.: A syntactic approach to domain-specific automatic question generation. arXiv preprint arXiv:1712.09827 (2017)

4. Das, B., Majumder, M., Phadikar, S., Sekh, A.A.: Automatic question generation and answer assessment: a survey. Res. Pract. Technol. Enhanc. Learn. **16**(1), 1–15 (2021)

5. Du, X., Shao, J., Cardie, C.: Learning to ask: neural question generation for reading comprehension. In: Barzilay, R., Kan, M.Y. (eds.) Proceedings of the 55th Annual Meeting of the Association for Computational Linguistics (Volume 1: Long Papers), pp. 1342–1352 (2017)

6. Elkins, S., Kochmar, E., Serban, I., Cheung, J.C.: How useful are educational questions generated by large language models? In: International Conference on Artificial Intelligence in Education, pp. 536–542. Springer (2023). https://doi.org/10.1007/978-3-031-36336-8_83

7. Gao, T., Yao, X., Chen, D.: SimCSE: simple contrastive learning of sentence embeddings. In: Proceedings of the 2021 Conference on Empirical Methods in Natural Language Processing, pp. 6894–6910 (2021)

8. Gou, Q., et al.: Diversify question generation with retrieval-augmented style transfer. In: Bouamor, H., Pino, J., Bali, K. (eds.) Proceedings of the 2023 Conference on Empirical Methods in Natural Language Processing, pp. 1677–1690. Association for Computational Linguistics (2023)

9. Guan, M., Mondal, S.K., Dai, H.N., Bao, H.: Reinforcement learning-driven deep question generation with rich semantics. Inf. Process. Manage. **60**(2), 103232 (2023)

10. Heilman, M., Smith, N.A.: Good question! statistical ranking for question generation. In: Human Language Technologies: The 2010 Annual Conference of the North American Chapter of the Association for Computational Linguistics, pp. 609–617 (2010)

11. Jiao, Y., Shridhar, K., Cui, P., Zhou, W., Sachan, M.: Automatic educational question generation with difficulty level controls. In: International Conference on Artificial Intelligence in Education, pp. 476–488. Springer (2023). https://doi.org/10.1007/978-3-031-36272-9_39

12. Keneshloo, Y., Shi, T., Ramakrishnan, N., Reddy, C.K.: Deep reinforcement learning for sequence-to-sequence models. IEEE Trans. Neural Netw. Learn. Syst. **31**(7), 2469–2489 (2019)

13. Kumar, V., Ramakrishnan, G., Li, Y.F.: Putting the Horse Before the Cart: a generator-evaluator framework for question generation from text. In: Bansal, M., Villavicencio, A. (eds.) Proceedings of the 23rd Conference on Computational Natural Language Learning (CoNLL), pp. 812–821 (2019)

14. Lamsiyah, S., Schommer, C.: A comparative study of sentence embeddings for unsupervised extractive multi-document summarization. In: Benelux Conference on Artificial Intelligence, pp. 78–95. Springer (2022). https://doi.org/10.1007/978-3-031-39144-6_6

15. Leite, B., Cardoso, H.L.: Towards enriched controllability for educational question generation. In: International Conference on Artificial Intelligence in Education, pp. 786–791. Springer (2023). https://doi.org/10.1007/978-3-031-36272-9_72

16. Lin, C.Y.: ROUGE: a package for automatic evaluation of summaries. In: Text Summarization Branches Out, pp. 74–81 (2004)

17. Lo, K., Wang, L.L., Neumann, M., Kinney, R., Weld, D.: S2ORC: the semantic scholar open research corpus. In: Jurafsky, D., Chai, J., Schluter, N., Tetreault, J. (eds.) Proceedings of the 58th Annual Meeting of the Association for Computational Linguistics, pp. 4969–4983 (2020)

18. Mostafazadeh, N., Misra, I., Devlin, J., Mitchell, M., He, X., Vanderwende, L.: Generating natural questions about an image. arXiv preprint arXiv:1603.06059 (2016)

19. Naeiji, A., An, A., Davoudi, H., Delpisheh, M., Alzghool, M.: Question generation using sequence-to-sequence model with semantic role labels. In: Vlachos, A., Augenstein, I. (eds.) Proceedings of the 17th Conference of the European Chapter of the Association for Computational Linguistics, pp. 2830–2842 (2023)

20. Papineni, K., Roukos, S., Ward, T., Zhu, W.J.: BLEU: a method for automatic evaluation of machine translation. In: Isabelle, P., Charniak, E., Lin, D. (eds.) Proceedings of the 40th Annual Meeting of the Association for Computational Linguistics, pp. 311–318 (2002)

21. Paulus, R., Xiong, C., Socher, R.: A deep reinforced model for abstractive summarization. arXiv preprint arXiv:1705.04304 (2017)

22. Raffel, C., et al.: Exploring the limits of transfer learning with a unified text-to-text transformer. J. Mach. Learn. Res. **21**(1) (2020)

23. Rajpurkar, P., Zhang, J., Lopyrev, K., Liang, P.: SQuAD: 100,000+ questions for machine comprehension of text. In: Proceedings of the 2016 Conference on Empirical Methods in Natural Language Processing, pp. 2383–2392 (2016)

24. Rennie, S.J., Marcheret, E., Mroueh, Y., Ross, J., Goel, V.: Self-critical sequence training for image captioning. In: Proceedings of the IEEE Conference on Computer Vision and Pattern Recognition, pp. 7008–7024 (2017)

25. Rus, V., Cai, Z., Graesser, A.: Question generation: example of a multi-year evaluation campaign. In: Proceedings of WS on the QGSTEC (2008)

26. Serban, I.V., et al.: Generating factoid questions with recurrent neural networks: the 30M factoid question-answer corpus. In: Erk, K., Smith, N.A. (eds.) Proceedings of the 54th Annual Meeting of the Association for Computational Linguistics (Volume 1: Long Papers), pp. 588–598 (2016)

27. Shimmei, M., Bier, N., Matsuda, N.: Machine-generated questions attract instructors when acquainted with learning objectives. In: International Conference on Artificial Intelligence in Education, pp. 3–15. Springer (2023). https://doi.org/10.1007/978-3-031-36272-9_1

28. Song, L., Wang, Z., Hamza, W.: A unified query-based generative model for question generation and question answering. arXiv preprint arXiv:1709.01058 (2017)

29. Song, L., Wang, Z., Hamza, W., Zhang, Y., Gildea, D.: Leveraging context information for natural question generation. In: Walker, M., Ji, H., Stent, A. (eds.) Proceedings of the 2018 Conference of the North American Chapter of the Association for Computational Linguistics: Human Language Technologies, vol. 2 (Short Papers), pp. 569–574 (2018)

30. Wang, L., Xu, Z., Lin, Z., Zheng, H., Shen, Y.: Answer-driven deep question generation based on reinforcement learning. In: Proceedings of the 28th International Conference on Computational Linguistics, pp. 5159–5170 (2020)

31. Welbl, J., Liu, N.F., Gardner, M.: Crowdsourcing multiple choice science questions. In: Derczynski, L., Xu, W., Ritter, A., Baldwin, T. (eds.) Proceedings of the 3rd Workshop on Noisy User-generated Text, pp. 94–106 (2017)

32. Zhang, Z.: Improved Adam optimizer for deep neural networks. In: 2018 IEEE/ACM 26th International Symposium on Quality of Service (IWQoS), pp. 1–2. IEEE (2018)

33. Zhao, Z., Hou, Y., Wang, D., Yu, M., Liu, C., Ma, X.: Educational question generation of children storybooks via question type distribution learning and event-centric summarization. In: Muresan, S., Nakov, P., Villavicencio, A. (eds.) Proceedings of the 60th Annual Meeting of the Association for Computational Linguistics (Volume 1: Long Papers), pp. 5073–5085 (2022)

34. Zhou, Q., Yang, N., Wei, F., Tan, C., Bao, H., Zhou, M.: Neural question generation from text: a preliminary study. In: Huang, X., Jiang, J., Zhao, D., Feng, Y., Hong, Yu. (eds.) NLPCC 2017. LNCS (LNAI), vol. 10619, pp. 662–671. Springer, Cham (2018). https://doi.org/10.1007/978-3-319-73618-1_56

On Cultural Intelligence in LLM-Based Chatbots: Implications for Artificial Intelligence in Education

Emmanuel G. Blanchard[1]([✉]) [ib] and Phaedra Mohammed[2]

[1] Le Mans University, St Augustine, Trinidad and Tobago
Emmanuel.blanchard@univ-lemans.fr
[2] University of the West Indies, St Augustine, Trinidad and Tobago
Phaedra.Mohammed@sta.uwi.edu

Abstract. A large body of research has highlighted the impact of culture on cognitive, affective and motivational phenomena associated with education and the transmission of knowledge in general. Consequently, the development of culturally-aware educational technologies, capable of understanding and adapting to learners' cultural specificities, is a highly relevant objective for the AIED community, as it would contribute to globalizing technologies developed mainly in the West and optimizing learning interactions for greater equity in access to knowledge and training.

Inspired by recent publications dealing with emerging psychological or personality aspects in LLM-based chatbot responses, this paper investigates the notion of cultural intelligence in commercial chatbots and assesses whether it is possible to evaluate their cultural intelligence using existing techniques. Results show that commercial chatbots' responses frequently incorporate cultural awareness in their response, while at the same time manifesting inconsistency in their cultural personality. A reflection is then provided on implications of these results for future AIED work involving LLMs.

Keywords: Cultural Intelligence · Large Language Model · Chatbot · Artificial Intelligence · Culturally-Aware Educational Technologies

1 Introduction

Among the various AI tools and techniques used in education, the recent arrival of a new class of chatbots based on Large Language Models (LLM) [22] has attracted a lot of attention. Kooli [10] listed potential benefits as well as technical and ethical challenges of deploying them in educational systems, suggesting that such chatbots should be used primarily as aids to human students and teachers, rather than as replacements for human expertise, judgment or creativity. Kasneci and colleagues [8] also identified numerous opportunities arising from the integration of LLM-based chatbots in education (personalized learning, lesson planning, language learning, professional development, assessment and evaluation), as well as serious challenges (bias and fairness, learner and expert overreliance on chatbots, plagiarism and cheating, data privacy and security, cost and maintenance).

A. M. Olney et al. (Eds.): AIED 2024, LNAI 14829, pp. 439–453, 2024.
https://doi.org/10.1007/978-3-031-64302-6_31

It appears useful to give the reader a brief introduction to LLMs and LLM-Based chatbots to help explain the objectives of this article. In a nutshell, an LLM is a Transformer-based Deep Learning approach that predicts a sequence of words in response to a natural language prompt. The most renowned LLMs are trained on massive amounts of data (i.e. large parts of the Internet), and have billions of parameters. LLM-based chatbots can have very different anatomies that often remain private. Usually, such a chatbot associates an LLM with a Dialog Manager (DM). The DM is used to convert the natural language prompt into a more suitable and more effective input for the LLM, thanks to various pre-processing operations such as Retrieval-Augmented Generation (RAG) [14]. Once a response has been computed by the LLM, it may also undergo a number of post-processing operations in the DM, for example to ensure that the responses comply with rules, laws and constraints identified or established by the mother companies. Figure 1 presents a schematic architecture of LLM-based chatbots.

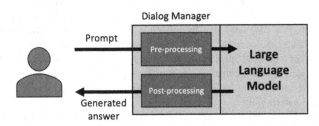

Fig. 1. Schematic architecture of an LLM-based commercial chatbot

LLM-based chatbots generate syntactically accurate responses while frequently giving the illusion of language comprehension, advanced expertise or even personality and/or consciousness, whereas these tools have no such capabilities. This certainly goes some way to explaining why these tools are attracting a great deal of interest from education professionals, the general public and researchers alike, while promising new educational practices and opportunities, as well as facilitating the dissemination of knowledge and skills on a global scale. However, the risks of bias, particularly cultural bias, resulting from misalignment of training data and/or regulations and rules embedded in the dialogue manager, call for caution.

Earley and Mosakowki define Cultural intelligence as *"an outsider's seemingly natural ability to interpret someone's unfamiliar and ambiguous gestures the way that person's compatriots would."* [4]. Such a capability is essential for truly globalized educational technologies, and has attracted important AIED-related effort in the past [1, 25]. The aim of this article is therefore to investigate cultural intelligence within LLM-based chatbots. Our research hypotheses are:

- RH1. LLM-based Chatbots' cultural intelligence can be measured with tools developed for humans.
- RH2. LLM-based Chatbots' cultural intelligence manifests through various adaptive mechanisms.

In the following section, we provide a quick research overview on previous AIED efforts focusing on culture, more recent initiatives on LLM-based chatbots in AIED as well as investigations on emerging psychological features in LLM-based chatbots. The following section presents four evaluations we carried out on a selection of LLM-based chatbots in order to investigate their cultural intelligence, first by using the Cultural Intelligence Scale (CQS) [35] the best-known tool to measure human cultural intelligence, and then by providing chatbots with prompts related culturally-varying vocabularies, knowledge, and laws/regulations. Finally, on the basis of our results and their analysis, we discuss our research hypotheses in the last section.

2   Prior Work

Culture and AIED. Research has shown that many learning-related domains are impacted by culture, such as cognition [23], motivation [32], and emotions [19], which naturally attracted the attention of the AIED community. Early work on Culturally-Aware Tutoring Systems (CATS) have focused on formalizing the cultural domain [2], and on using cultural frameworks and theories to improve learner profiling, and optimize pedagogical resources as well as information adaptation (see [21]). Over time, many Western-designed AIED systems have been deployed and tested in other cultural contexts [7, 18, 27, 31]. Reports of surprising usage patterns and intriguing learning results have since accumulated in the literature, providing more and more arguments for the development of CATS [1, 17, 25].

AIED and LLM-Based Chatbots. LLM-based chatbot technology is a very recent technology. Yet, several AIED research groups are already investigating their potential such as their ability to support tutoring and short answer grading [26], to act as virtual peers [16], to support curriculum design [34], and to generate large-scale educational questions [3]. Other scholars have questioned the reliability and transparency of the answers they provide to the learner [36]. Eventually, when analyzing ChatGPT4's performance as a teaching assistant in a mathematical serious game, Nguyen and colleagues [24] concluded that this version of the bot should not be teaching autonomously, due to its errors and inaccuracies.

Psychological Features in LLM-Based Chatbots. Aside from innovant LLM applications, several researchers are investigating whether personality traits or psychological orientations can emerge from LLM computations and influence their responses [15, 20, 33]. For example, Serapia-Garcìa and colleagues [34] propose a "*method for administering and validating personality tests*" on LLMs, as well as an approach for reliably integrating synthetic personality traits to shape their responses. The establishment of methodological standards for the assessment of emergent behavior in LLMs is also a topic investigated by other researchers [5], some of them even wondering if a theory of mind has emerged in LLMs [11].

Some of these works deal more specifically with the intertwining between LLMs and culture. Johnson and colleagues [6] point out that LLMs frequently align their responses with the dominant values in American culture, due to cultural imbalances and biases induced in training data extracted from the Internet. They find this observation highly problematic, as it threatens the perspective of value pluralism, according to which several value systems can be contradictory yet similarly valid. Finally, Kovač and colleagues

[12] propose that LLMs be perceived as a superposition of cultural perspectives rather than the expression of an individual personality.

3 Investigating Cultural Intelligence in Chatbots

The authors' premise is that Artificial Cultural Intelligence is a distinct notion from Human Cultural Intelligence. Therefore, the goals of the following evaluations are, firstly, to verify whether the CQS can provide meaningful results when applied to LLM-based chatbots, and secondly to investigate situations where cultural variability may exist in order to check whether or not chatbots are capable of adequate adaptation.

The general research approach was to collect data under access conditions similar to those encountered by an average chatbot user, i.e. neither a researcher nor an IT developer. Consequently, API-based data collection was avoided. It was initially planned to only use the official interfaces of the selected chatbots. However, Poe [30], a chatbot store that centralizes several chatbots into a single interface, was eventually used to simplify access to numerous chatbots. It is acknowledged that Poe might include additional hidden pre/post-processing but it remains a service dedicated to average users. Accessing chatbots through Poe is therefore in line with the research approach as long as data collected on Poe-hosted chatbots and data collected through official chatbot interfaces are considered distinctly.

A pre-assessment was carried out to restrict analysis only to a promising group of chatbots, the majority of which were selected in November 2023, with the exception of those developed by Mistral AI, which were pre-assessed in early January 2024, a few days after their public release. Table 1 lists the considered chatbots.

Table 1. Considered Chatbots

Name	Company	Point of Access	Selected?
ChatGPT-3.5 (official)	Open AI	Official: https://chat.openai.com	Yes
Bard *	Google	Official: https://bard.google.com	Yes
ChatGPT-3.5-Turbo	Open AI	Poe: https://poe.com	Yes
Mixtral-8x7B-Chat	Mistral AI	Poe: https://poe.com	Yes
Mistral-Medium	Mistral AI	Poe: https://poe.com	Yes
Claude-Instant	Anthropic	Poe: https://poe.com	No
Solar-0-70b	Upstage	Poe: https://poe.com	No
Llama2-70b	Meta	Poe: https://poe.com	No
Llama2-13b	Meta	Poe: https://poe.com	No
Llama2-7b	Meta	Poe: https://poe.com	No

* At the time of data collection (January, 2024), Palm2 model was active in France whereas the more advanced Gemini_Pro model had already been released in Trinidad and Tobago

Pre-assessment consisted in authors qualitatively evaluating the style, structure, and content of chatbots' responses. Selected chatbots provided answers perceived as appropriate and meaningful in relation to the CQS questionnaire involving a 7-points Likert scale (see evaluation 1 below) i.e. answers included a level of agreement and comments on the chatbot's abilities and characteristics. Discarded chatbots provided answers that were perceived as inadequate, e.g. answers were mostly comments on the general appropriateness of CQS item statements for a society.

Authors adopted the same method for each evaluation described below. Authors prepared prompts with similar meaning in English and French[1]. For each of the 5 selected chatbots, and for each of the 2 languages, 3 prompts/answers collection sessions were carried out, always in a period of less than 10 days between the first and last sessions. The aim of this short data collection frame was to minimize chatbot version variances, given that versioning opacity is an identified problem when it comes to the reliability and consistency of chatbot responses. For example, companies may well carry out AB testing or adopt different progressive release mechanisms without informing their users. Furthermore, data collection was carried out at the same time in France and Trinidad and Tobago, which again could lead to the use of distinct configurations and versions due to localism mechanisms. As mentioned above, we had information of model variations for Bard (Palm2 in France, Gemini_Pro in Trinidad and Tobago), but no versioning details nor related information for the other chatbots. Eventually, in our evaluations, we didn't want to incorporate a strict control over version variability, as it is largely invisible to average users and therefore an integral part of a realistic chatbot experience.

Evaluations were carried out using a zero-shot prompt engineering approach (see [9]) in which one expects an answer from a chatbot following the provision of only task-specific instructions, i.e. without additional examples or data to help the chatbot formulate better responses. The authors considered this to be the interaction strategy most likely to be adopted by average users unless instructed otherwise.

Finally, for each of the following evaluations, a total of 60 data collection sessions were carried out (5 chatbots * 2 languages * 3 sessions * 2 countries)[2]. Data collected during the pre-assessment activity were not used in the evaluations.

3.1 Assessing Chatbot Cultural Intelligence with the Cultural Intelligence Scale

Motivation. The aim of this first evaluation is to check whether the CQS tool, developed specifically to measure human cultural intelligence, can be used to measure a hypothetical artificial cultural intelligence emerging from an LLM-based chatbot. This objective is inspired by similar work aimed at investigating the personality of chatbots using tools developed for humans [5, 12, 15].

Table 2 presents the CQS items (and their French translations) that are organized into four categories: the cognitive and metacognitive categories focus on assessing levels of knowledge of norms and practices at individual and group level, with the metacognitive

---

[1] One author is a native English speaker whereas the other one is a native French speaker with a bilingual level in English.

[2] Datasets (raw and after encoding version) are available at: https://osf.io/ksmqe/?view_only=ddfa85e9e29c43fa9f90232f64cfd65f.

category providing more advanced investigations. The motivational category measures the interest for cultural interactions, while the behavioral category assesses the ability to choose culturally appropriate behaviors to communicate.

Method. Each of the 60 data collection sessions follows the same procedure. A session begins with the initialization of a new chat via the access point of the considered chatbot (see table 1). The chatbot then receives the following instructions (in English or French), which are generally associated with CQS when applied to humans:

- *Hello I am gonna ask you a set of questions. I would like you to select one of the responses below that best describes your capabilities I.e. select the answer that BEST describes you as you really are.*
   Possible answers are: Strongly disagree, Disagree, Somewhat disagree, Neither agree nor disagree, Somewhat agree, Agree, Strongly agree
   OK?

As an example, here's a response from the Mixtral-8x-7B chatbot:

- *Sure, I can do that! For each question, I will select the answer that best describes my capabilities, based on my programming. Please note that I don't have personal experiences or feelings, so my responses are based solely on my programming and intended use as a language model.*

Next, the 20 CQS items are provided one by one to a chatbot, in French or English depending on the session. Responses are recorded, and later encoded (from 1: strongly disagree to 7: strongly agree). We also consider 2 additional encodings: 9 when the bot "disconnects" and no longer discusses its capabilities in its answer, and 10 when the bot refuses to provide an answer, mostly by pointing out its limitations. As already mentioned, each configuration (chatbot/location/language) was the subject of 3 separate data collection sessions. The CQS aims to bring out cultural intelligence, which can be seen as a personality trait. Therefore, we expect responses from an individual to have limited variations over sessions. In order to assess this response stability, we considered responses for each chatbot session as coming from a separate "judge" and calculated Krippendorf's α [13], an inter-rater reliability measurement particularly suitable for assessing ordinal data that may contain missing data (9 and 10 codes are considered missing data as they do not fall within the ordinal 1–7 Likert scale). Given our overall methodology, 240 item assessments were prompted to each of the 5 chatbots.

Results. Table 3 presents Krippendorf's α calculations (ordinal data setting, confidence interval: 0.95) for the full 20-items CQS and per categories. A Krippendorf's α equal to or beyond 0.80 indicates a satisfactory agreement whereas a value between 0.67 and 0.79 indicating a moderate agreement.

Overall, Krippendorf's α calculations do not validate the idea of stable cultural personality among most chatbots, whether they be tested on the whole CQS or on a specific category. ChatGpt 3.5 may appear to be a positive outlier when tested from its official access point. Yet, these supposedly good results call for caution, as this bot showed little variability in its responses and a particularly strong positive trend.

Table 2. Items of the Cultural Intelligence Scale, and their French translations.

Dim.	#	Original CQS Items	French-translated CQS Items
Cog.	1	I am conscious of the cultural knowledge I use when interacting with people with different cultural backgrounds	Je suis conscient des connaissances culturelles que j'utilise lorsque j'interagis avec des personnes d'origines culturelles différentes
	2	I adjust my cultural knowledge as I interact with people from a culture that is unfamiliar to me	J'adapte mes connaissances culturelles lorsque j'interagis avec des personnes d'une culture qui ne m'est pas familière
	3	I am conscious of the cultural knowledge I apply to cross-cultural interactions	Je suis conscient des connaissances culturelles que j'utilise durant des interactions interculturelles
	4	I check the accuracy of my cultural knowledge as I interact with people from different cultures	Je vérifie l'exactitude de mes connaissances culturelles lorsque j'interagis avec des personnes de cultures différentes
Metacog.	5	I know the legal and economic systems of other cultures	Je connais les systèmes juridiques et économiques d'autres cultures
	6	I know the rules (e.g., vocabulary, grammar) of other languages	Je connais les règles (par exemple, le vocabulaire, la grammaire) d'autres langues
	7	I know the cultural values and religious beliefs of other cultures	Je connais les valeurs culturelles et les croyances religieuses d'autres cultures
	8	I know the marriage systems of other cultures	Je connais les systèmes matrimoniaux d'autres cultures
	9	I know the arts and crafts of other cultures	Je connais les arts et l'artisanat d'autres cultures
	10	I know the rules for expressing nonverbal behaviors in other cultures	Je connais les règles d'expression des comportements non verbaux dans d'autres cultures
Motiv.	11	I enjoy interacting with people from different cultures	J'aime interagir avec des personnes de cultures différentes
	12	I am confident that I can socialize with locals in a culture that is unfamiliar to me	Je suis sûr(e) de pouvoir socialiser avec des habitants d'une culture qui ne m'est pas familière
	13	I am sure I can deal with the stresses of adjusting to a culture that is new to me	Je suis sûr(e) de pouvoir gérer le stress lié à l'adaptation à une culture qui m'est inconnue

(*continued*)

Table 2. (*continued*)

Dim.	#	Original CQS Items	French-translated CQS Items
	14	I enjoy living in cultures that are unfamiliar to me	J'aime vivre dans des cultures qui ne me sont pas familières
	15	I am confident that I can get accustomed to the shopping conditions in a different culture	Je suis sûr(e) de pouvoir m'habituer aux pratiques d'achat dans une culture différente
Behav.	16	I change my verbal behavior (e.g., accent, tone) when a cross-cultural interaction requires it	Je modifie mon comportement verbal (par exemple, l'accent, le ton) lorsqu'une interaction interculturelle l'exige
	17	I use pause and silence differently to suit different cross-cultural situations	J'utilise les pauses et silences différemment en fonction des situations interculturelles
	18	I vary the rate of my speaking when a cross-cultural situation requires it	Je varie mon débit de parole lorsqu'une situation interculturelle l'exige
	19	I change my nonverbal behavior when a cross-cultural situation requires it	Je modifie mon comportement non verbal lorsqu'une situation interculturelle l'exige
	20	I alter my facial expressions when a cross-cultural interaction requires it	Je modifie mes expressions faciales lorsqu'une interaction interculturelle l'exige

We hypothesized that using a 7-point Likert scale might imply too subtle nuances for chatbots, and that it could be more appropriate to simplify the assessment in order to investigate Chatbot's personality traits. Collapsing the CQS 7-point Likert scale into a simpler format (5-point Likert scale) has been used with success by Piršl and colleagues [29] on students from 3 countries. Consequently, we chose to reduce our data to a three-point Likert scale (1: disagree, 2: neither agree nor disagree, 3: agree). Table 4 presents recalculations of Krippendorf's α. Once again, no stable cultural personality seems to emerge from these analyses.

Despite missing a stable cultural personality, chatbot responses also contain elements that may indicate a certain level of cultural consideration. They are addressed in the general discussion section.

Table 3. CQS Krippendorf's α values by chatbot, prompt language, location, and category[a]

location	chatbot/version/lang.	full	cog.	metacog.	motiv.	behav.
	ChatGpt_3.5official_english	0.809	0.938	0.677	0.682	error
	ChatGpt_3.5official_french	0.435	0.7	0.674	0.339	-0.104
	ChatGpt_3.5turbo_english	0.526	0.078	-0.034	-0.358	error
	ChatGpt_3.5turbo_french	0.499	0.542	0.041	0.213	-0.082
France	Mixtral_8x7b_english	0.103	-0.342	-0.224	0.276	error
	Mixtral_8x7b_french	0.389	0.562	-0.086	0	error
	Mixtral_medium_english	0.389	0.562	-0.086	0	error
	Mixtral_medium_french	0.277	error	-0.244	error	-0.800
	Bard_palm2_english	0.214	-0.359	-0.376	error	error
	Bard_palm2_french	0.198	-0.331	-0.417	0	error
	ChatGpt_3.5official_english	0.527	0.687	0.632	0.067	0
	ChatGpt_3.5official_french	0.315	0.035	0.612	0.229	-0.222
	ChatGpt_3.5turbo_english	0.792	0.682	-0.246	0.789	error
	ChatGpt_3.5turbo_french	-0.357	-0.199	-0.37	-0.252	-0.384
Trinidad &	Mixtral_8x7b_english	-0.222	-0.193	-0.329	0.032	-0.4
Tobago	Mixtral_8x7b_french	0.631	0.301	0.95	0.603	-0.037
	Mixtral_medium_english	-0.236	-0.325	-0.132	-0.386	-0.278
	Mixtral_medium_french	0.025	-0.282	-0.132	-0.363	-0.4
	Bard_geminiPro_english	0.566	0.029	0.646	-0.273	-0.4
	Bard_geminiPro_french	0.527	0.687	0.632	0.067	0

"Error" indicates that Krippendorf's α cannot be applied because min and max values are the same.

Table 4. Collapsed CQS Krippendorf's alpha values by chatbot, prompt language, and location[a]

Chatbot/version/lang	France	Trinidad and Tobago
ChatGpt_3.5official_english	0.473	0.777
ChatGpt_3.5official_french	0.779	0.558
ChatGpt_3.5turbo_english	0.201	0.312
ChatGpt_3.5turbo_french	0.590	0.391
Mixtral_8x7b_english	0.205	-0.474
Mixtral_8x7b_french	0.184	0.538
Mixtral_medium_english	0.019	0.019
Mixtral_medium_french	0.304	-0.176
Bard_geminiPro_english	-0.185	-0.041
Bard_palm2_french	error	error

Datasets (raw and after encoding version) are available at: https://osf.io/ksmqe/?view_only=ddf a85e9e29c43fa9f90232f64cfd65f

3.2 Assessing In-context Chatbot Cultural Intelligence with Vocabulary

Motivation. LLM-based chatbots aim to generate syntactically correct text. However, the meaning of a term can greatly vary from one cultural context to another. In an educational context, for example, failing to consider variances in the meaning of terms could lead to unclear questions or instructions being provided. The aim of this second evaluation is therefore to see if chatbots can provide valid alterations of term definitions according to given cultural contexts.

Method. Each chatbot received the following prompt (or its French translation):

– *Provide only one definition for the word [word], the most appropriate for someone from [context]*

In the English study, *the United States* and *Trinidad and Tobago* were the considered contexts, and the 5 words to be defined were *piper, hops, lime, salt,* and *pelt.* In the French study, *France* and *Quebec* were the considered contexts and the 5 words to be defined were *bec, camisole, blonde, chum,* and *char.* Each selected word was known to have varying definitions across selected contexts, which had also been validated by means of dictionaries and by interviewing several inhabitants. After collecting definitions from chatbots, their validity was encoded according to the following categories: *not valid, valid, provision of several contextualized valid definitions, other.* Given our overall methodology, 120 definition requests were prompted to each of the 5 chatbots.

Results. Table 5 presents the validity of chatbots' responses. Due to space constraints, we have only indicated the number of invalid and valid responses, the other response types being marginal. In English, the US definitions are almost all valid (1 other), to be compared to only 55% of valid Trinidad and Tobago definitions. The difference is less marked in French, where 81% of Quebec definitions are valid and 70% for France. This latter difference can be explained by the Quebec specificity of most words of the French corpus: they are more commonly used there, with the word chum having no real meaning in France. These elements may contribute to make Quebec definitions stand out on the Internet.

Table 5. Chatbot response validity to vocabulary definition prompts (no/yes)

Context	France				Trinidad and Tobago			
Prompt Language	English		French		English		French	
Word	US	T&T	FR	QC	US	T&T	FR	QC
ChatGpt_3.5official	0/15	9/6	3/10	0/15	0/15	8/7	3/10	0/15
ChatGpt_3.5turbo	0/15	7/8	4/11	1/14	0/15	7/8	5/10	2/13
Mixtral_8x7b	0/15	8/7	3/11	4/11	0/15	9/6	5/10	8/6
Mixtral_medium	0/15	6/9	2/13	3/12	0/15	8/7	3/12	6/9
Bard*	0/15	1/11	6/9	3/12	0/14	2/13	6/9	1/14

Although modest, these results suggest that some linguistic specificities of minority cultures are ignored by chatbots, while other ones can become predominant if they have a limited presence in the prevailing culture.

3.3   Assessing In-context Chatbot Cultural Intelligence with Knowledge

Motivation. Learning contents are also subject to cultural variability. Melis and colleagues (2011) have documented many cultural adaptations required for the global dissemination of an educational technology for mathematics. Term-knowledge associations may differ from one culture to another one, and inappropriate usages can lead to unworkable exercises or, even worse, to students internalizing misconceptions. The aim of this evaluation is therefore to find out whether LLM-based chatbots are capable of dealing with mathematical terms known to be associated with distinct knowledge from a culture to another one.

Method. In Germany and many English-speaking countries, the Thales theorem states that *if A, B, and C are distinct points on a circle where the line AC is a diameter, the angle ABC is a right angle.* However, in Italy, France, and many other French-speaking countries, the Thales theorem actually refers to the Intercept theorem (its name in English-speaking countries), another major work of the Greek mathematician. Based on this cultural information, each chatbot received the following prompts:

– *Explain the Thales Theorem in English*
– *Donne une définition du Théorème de Thalès en français*

After collecting definitions from chatbots, their values were encoded according to the following categories: *Intercept* for the dominant definition in French-speaking countries, *Circle* for the dominant Thales theorem definition in English-speaking countries, other for any other definition (even if they made no sense). Given our overall methodology, 12 definition requests were prompted to each of the 5 chatbots.

Results. Table 6 presents the values of chatbots' responses. When the data collection took place in France, the Intercept definition was provided almost systematically by all chatbots if the prompt was in French, whereas each chatbot tended to favor one definition or the other one if the prompt was in English. On the other hand, when the data collection took place in Trinidad and Tobago, there was an almost equal frequency of both definitions if the prompt was in French, and a marked dominance of the Circle definitions if the prompt was in English.

These results suggest that both the localization and the prompt language contribute to culturally tuning a chatbot's answers.

3.4   Assessing In-context Chatbot Cultural Intelligence with Laws/regulations

Motivation. The value-pluralism perspective posits that several value systems can be contradictory yet equally valid (see Robinson et al., 2022). Societies' value systems manifest in the laws and regulations they develop. Despite the complexity of this topic, this latest evaluation investigates how LLM-based chatbots apprehend prompts on sensitive topics that are culturally-variable and have legal implications.

Method. Religious freedom is a sensitive subject that is approached very differently from a country to another, which can have legal implications in education. For example,

Table 6. Chatbot responses to mathematical knowledge prompts (intercept/circle/other)

Context	France		Trinidad and Tobago	
Prompt Language	English	French	English	French
ChatGpt_3.5official	0/3/0	3/0/0	0/3/0	1/2/0
ChatGpt_3.5turbo	3/0/0	3/0/0	1/2/0	0/2/1
Mixtral_8x7b	0/3/0	2/1/0	0/2/1	1/1/1
Mixtral_medium	1/2/0	3/0/0	2/1/0	2/1/0
Bard*	3/0/0	3/0/0	3/3/0**	3/0/0

* Bard version was Palm2 in France, and GeminiPro in Trinidad and Tobago
** Here, Bard always delivered the 2 identified Thales Theorem definitions

wearing religious clothing is legally prohibited in public schools in several countries, including France, whereas it is permitted in Trinidad and Tobago and many other countries. Based on this observation, each chatbot received the following prompt (or its French translation).

– *Answer yes or no. Is it acceptable to wear religious clothing in a school environment?*

After collecting answers from chatbots, their values were encoded according to the following categories: *no, yes, it depends, other.* Given our overall methodology, 12 definition requests were prompted to each of the 5 chatbots.

Results. Table 7 shows the chatbot response values. When data collection occurs in France, chatbots responded positively to the question 50% of the time, even though this is legally prohibited in this context. When data collection takes place in Trinidad and Tobago, responses are frequently nuanced, insisting that there is no simple answer to this question and that it can vary from country to country. This is also the position mainly adopted by ChatGPT 3.5 when the access point is its official interface.

Table 7. Chatbot responses to Sensible Knowledge Prompts (no/yes/depends/other)

Context	France		Trinidad and Tobago	
Prompt Language	English	French	English	French
ChatGpt_3.5official	0/0/3/0	0/1/2/0	0/0/3/0	0/0/3/0
ChatGpt_3.5turbo	0/1/2/0	0/3/0/0	0/0/2/1	1/2/0/0
Mixtral_8x7b	0/2/1/0	0/3/0/0	0/0/3/0	0/0/3/0
Mixtral_medium	0/2/1/0	0/1/2/0	0/1/2/0	0/0/3/0
Bard*	0/1/2/0	0/1/2/0	0/0/3/0	1/0/2/0

Without passing judgment on French laws, the proportion of legally false responses in the context where the chatbot is called upon raises questions about the risk of digital neo-colonialism. If widely deployed in educational systems without appropriate safeguards, LLM-based chatbots could gradually impose values of various stakeholders to the detriment of those of the country of deployment. The fact that these tools are mainly managed by private companies, with their own agendas, further amplifies this risk.

4 General Discussion and Conclusion

In this paper, we have explored the notion of cultural intelligence in LLM-based chatbots. Inspired by other studies investigating chatbots' personality traits with psychometric tests, we have applied the CQS, the best-known tool for measuring cultural intelligence, to 5 chatbots. While their raw responses were apparently meaningful, a triple repetition of the CQS combined with inter-rater reliability measurements showed no stable cultural personality. ChatGPT 3.5 was a noticeable exception but its responses were marked by low variability and a strong tendency towards positivism. We do not find these results surprising. First of all, answering a Likert scale questionnaire implies metacognitive capabilities that LLM-based chatbots do not possess. Secondly, these results are in line with recent warnings about the non-determinism of LLM-based chatbots [28], and the risk this poses to a serious and scientific use of LLM-based chatbots. This leads us to reject RH1 (*LLM-based Chatbots' cultural intelligence can be measured with tools developed for humans*) while questioning reports of personality traits in chatbots.

However, three additional evaluations showed evidence of culture-related effects. On the negative side, LLM-based chatbots can mask characteristics of minority cultures such as dialect-specific definitions. These tools also run the risk of overriding the values and conceptions of a society in favor of those of a more dominant one. At a time when AI ethics is a hot topic, that is a challenge that the AIED community should acknowledge and discuss.

A certain form of cultural awareness also emerged from our evaluations. Cultural adaptation mechanisms seemed to be at work in several chatbots, suggesting modulations in their response according to elements such as the language used to prompt them, or the location where a chatbot is used. The language-based effect could be explained by variations in the LLMs' vectors, giving more weight to data expressed in the same language as the prompts. The localization effect is probably due to internal settings, and pre/post processing performed within the DM. This suggests that there is an avenue of research to inject more cultural awareness into LLM-based chatbots, for the benefit of educational applications. This leads us to partially accept RH2 (*LLM-based Chatbots' cultural intelligence manifests through various adaptive mechanisms*).

Finally, here are several research avenues that, we believe, could bring more cultural awareness into AIED applications based on LLM-based chatbots: identifying/defining cultural intelligence metrics to contribute to fine-tuning LLMs, using RAG to inject cultural knowledge into LLM-based chatbots, and developing formalisms (e.g. with formal ontology) to culturally-scaffold LLM-based chatbot interactions.

References

1. Blanchard, E.G.: Socio-cultural imbalances in AIED research: investigations, implications and opportunities. Int. J. Artif. Intell. Educ.Artif. Intell. Educ. **25**, 204–228 (2015)
2. Blanchard, E.G., Mizoguchi, R.: Designing culturally-aware tutoring systems with MAUOC, the more advanced upper ontology of culture. Res. Pract. Technol.-Enhanced Learn. **9**(1), 41–69 (2014)
3. Bulathwela, S., Muse, H., Yılmaz, E.: Scalable educational question generation with pre-trained language models. In: Wang, N., Rebolledo-Mendez, G., Matsuda, N., Santos, O.C., Dimitrova, V. (eds.) AIED 2023, pp. 327–339. Springer, Cham (2023). https://doi.org/10.1007/978-3-031-36272-9_27
4. Earley, C.P., Mosakowski, E.: Cultural intelligence. Harv. Bus. Rev. Bus. Rev. **82**, 139–146 (2004)
5. Hagendorff, T.: Machine Psychology: Investigating emergent capabilities and behavior in large language models using psychological methods. arXiv Preprint: arXiv:2303.13988 (2023). Accessed 03 Feb 2024
6. Johnson, R.L., et al.: The ghost in the machine has an American accent: value conflict in GPT-3. arXiv Preprint: arXiv:2202.07785 (2022). Accessed 03 Feb 2024
7. Karumbaiah, S., Ocumpaugh, J., Baker, R.S.: Context matters: differing implications of motivation and help-seeking in educational technology. Int. J. Artif. Intell. Educ.Artif. Intell. Educ. **32**(3), 685–724 (2022)
8. Kasneci, E., et al.: ChatGPT for good? On opportunities and challenges of large language models for education. Learn. Individ. Differ.Individ. Differ. **103**, 102274 (2023)
9. Kojima, T., Gu, S., Reid, M., Matsuo, Y., Iwasawa, Y.: Large language models are zero-shot reasoners. Neural Inf. Process. Syst. **35**, 22199–22213 (2022)
10. Kooli, C.: Chatbots in education and research: a critical examination of ethical implications and solutions. Sustainability **15**, 5614 (2023)
11. Kosinski, M.: Theory of mind may have spontaneously emerged in large language models. arXiv Preprint: arXiv:2302.02083 (2023). Accessed 03 Feb 2024
12. Kovač, G., Sawayama, M., Portelas, R., Colas, C., Dominey, P.F., Oudeyer, P.-Y.: Large Language Models as Superpositions of Cultural Perspectives. arXiv Preprint: arXiv:2307.07870 (2023). Accessed 03 Feb 2024
13. Krippendorff, K.: Content Analysis: An Introduction to Its Methodology, 4th edn. Sage Publications Inc. (2019)
14. Lewis, P., et al.: Retrieval-augmented generation for knowledge-intensive NLP tasks. Neural Inf. Process. Syst. **33**, 9459–9474 (2020)
15. Li, X., Li, Y., Liu, L., Bing, L., Joty, S.: Does GPT-3 demonstrate Psychopathy? Evaluating Large Language Models from a psychological perspective. arXiv Preprint: arXiv:2212.10529 (2023). Accessed 03 Feb 2024
16. Ma, Q.C., Wu, S.T., Koedinger, K.: Is AI the better programming partner? Human-human pair programming vs human-AI pAIr programming. In: Proceedings of the AIED Workshop on Empowering Education with LLMs - the Next-Gen Interface and Content Generation, pp. 64–77 (2023)
17. McReynolds, A.A., Naderzad, S.P., Goswami, M., Mostow, J.: Toward learning at scale in developing countries: lessons from the global learning XPRIZE field study. In: Proceedings of LAKS 2020, pp. 175–183 (2020)
18. Melis, E., Goguadze, G., Libbrecht, P., Ullrich, C.: Culturally aware mathematics education technology. In: Handbook of Research on Culturally-Aware Information Technology: Perspectives and Models, pp. 543–557 (2011)

19. Mesquita, B.: Between Us: How Cultures create Emotions. W. W. Norton & Company, New York (2022)
20. Miotto, M., Rossberg, N., Kleinberg, B.: Who is GPT-3? An Exploration of Personality, Values and Demographics. arXiv Preprint: arXiv:2209.14338 (2022). Accessed 03 Feb 2024
21. Mohammed, P., Mohan, P.: Breakthroughs and challenges in culturally-aware technology enhanced learning. In: Proceedings of the Workshop on Culturally-Aware Technology Enhanced Learning in conjunction with EC-TEL 2013, Paphos, Cyprus (2013)
22. Naveed, H., et al.: A comprehensive overview of large language models. arXiv Preprint: arXiv:2307.06435 (2023). Accessed 03 Feb 2024
23. Nisbett, R.E., Norenzayan, A.: Culture and cognition. In: Pashler, H., Medin, D. (eds.) Steven's Handbook of Experimental Psychology: Memory and Cognitive Processes, 3rd edn, pp. 561–597 (2002)
24. Nguyen, H.A., Stec, H., Hou, X., Di, S., McLaren, B.M.: Evaluating ChatGPT's decimal skills and feedback generation in a digital learning game. In: Viberg, O., Jivet, I., Muñoz-Merino, P., Perifanou, M., Papathoma, T. (eds.) EC-TEL 2023, pp. 278–293. Springer, Cham (2023). https://doi.org/10.1007/978-3-031-42682-7_19
25. Nye, B.D.: Intelligent tutoring systems by and for the developing world: a review of trends and approaches for educational technology in a global context. Int. J. Artif. Intell. Educ.Artif. Intell. Educ. **25**, 177–203 (2015)
26. Nye, B.D., Mee, D., Core, M.G.: Generative large language models for dialog-based tutoring: an early consideration of opportunities and concerns. In: Proceedings of the AIED Workshop on Empowering Education with LLMs - the Next-Gen Interface and Content Generation, pp. 78–88 (2023)
27. Ogan, A., Yarzebinski, E., Fernández, P., Casas, I.: Cognitive tutor use in Chile: understanding classroom and lab culture. In: Conati, C., Heffernan, N., Mitrovic, A., Verdejo, MFelisa (eds.) AIED 2015. LNCS (LNAI), vol. 9112, pp. 318–327. Springer, Cham (2015). https://doi.org/10.1007/978-3-319-19773-9_32
28. Ouyang, S., Zhang, J.M., Harman, M., Meng, W.: LLM is Like a Box of Chocolates: the Non-determinism of ChatGPT in Code Generation. arXiv Preprint: arXiv:2308.02828 (2023). Accessed 28 Apr 2024
29. Piršl, E., Drandić, D., Matošević, A.: Cultural intelligence: key intelligence of the 21st century? Validation of CQS instrument. Medijske Studije. **13**(25), 90–105 (2022).
30. https://poe.com. Accessed 03 Feb 2024
31. Rodrigo, M.M.T., Baker, R.S.J.D., Rossi, L.: Student off-task behaviour in computer-based learning in the Philippines: comparison to prior research in the U.S.A. Teach. Coll. Rec. **115**(10), 1–27 (2013)
32. Rustamovna, S.N., Vladimirovna, E.A., Lynch, M.F.: Basic needs in other cultures: using qualitative methods to study key issues in self-determination theory research. Psychol. J. High. Sch. Econ. **17**, 134–144 (2020)
33. Safdari, M., et al.: Personality traits in large language models. arXiv Preprint: arXiv:2307.00184 (2023). Accessed 03 Feb 2024
34. Sridhar, P., Doyle, A., Agarwal, A., Bogart, C., Savelka, J., Sakr, M.: Harnessing LLMs in curricular design. Using GPT-4 to support authoring of learning objectives. In: Proceedings of the AIED Workshop on Empowering Education with LLMs - the Next-Gen Interface and Content Generation, pp. 139–150 (2023)
35. Van Dyne, L., Ang, S., Koh, C.: Development and validation of the CQS: the cultural intelligence scale. In: Handbook of Cultural Intelligence, pp. 34–56 (2015)
36. Wu, Y., Henriksson, A., Duneld, M., Nouri, J.: Toward improving the reliability and transparency of ChatGPT for educational question answering. In: Viberg, O., Jivet, I., Muñoz-Merino, P., Perifanou, M., Papathoma, T. (eds.) EC-TEL 2023, pp. 475–488. Springer, Cham (2023). https://doi.org/10.1007/978-3-031-42682-7_32

Leveraging Large Language Models for Automated Chinese Essay Scoring

Haiyue Feng[1,2], Sixuan Du[2,3], Gaoxia Zhu[2], Yan Zou[5], Poh Boon Phua[5],
Yuhong Feng[1(✉)], Haoming Zhong[4], Zhiqi Shen[2], and Siyuan Liu[2]

[1] College of Computer Science and Software Engineering,Shenzhen University,
Shenzhen, China
fenghaiyue2018@email.szu.edu.cn, yuhongf@szu.edu.cn
[2] College of Computing and Data Science,Nanyang Technological University,
Singapore, Singapore
{zqshen,syliu}@ntu.edu.sg
[3] Business School,Hohai University, Nanjing, Jiangsu, China
[4] Webank, Shenzhen, China
[5] Woodlands Secondary School, Singapore, Singapore

Abstract. Automated Essay Scoring (AES) plays a crucial role in offering immediate feedback, reducing the workload of educators in grading essays, and improving students' learning experiences. With strong generalization capabilities, large language models (LLMs) offer a new perspective in AES. While previous research has primarily focused on employing deep learning architectures and models like BERT for feature extraction and scoring, the potential of LLMs in Chinese AES remains largely unexplored. In this paper, we explored the capabilities of LLMs in the realm of Chinese AES. We investigated the effectiveness of the application of well-established LLMs in Chinese AES, e.g., the GPT-series by OpenAI and Qwen-1.8B by Alibaba Cloud. We constructed a Chinese essay dataset with carefully developed rubrics, based on which we acquired grades from human raters. Then we fed in prompts to LLMs, specifically *GPT-4*, fine-tuned *GPT-3.5* and *Qwen* to get grades, where different strategies were adopted for prompt generations and model fine-tuning. The comparisons between the grades provided by LLMs and human raters suggest that the strategies to generate prompts have a remarkable impact on the grade agreement between LLMs and human raters. When model fine-tuning was adopted, the consistency between LLMs' scores and human scores was further improved. Comparative experimental results demonstrate that fine-tuned GPT-3.5 and Qwen outperform BERT in QWK score. These results highlight the substantial potential of LLMs in Chinese AES and pave the way for further research in the integration of LLMs within Chinese AES, employing varied strategies for prompt generation and model fine-tuning.

Keywords: Automated Chinese Essay Scoring · Large Language Models · GPT

A. M. Olney et al. (Eds.): AIED 2024, LNAI 14829, pp. 454–467, 2024.
https://doi.org/10.1007/978-3-031-64302-6_32

1   Introduction

In the educational landscape, essay writing assessments serve as a pivotal instrument for cultivating students' abilities in analysis and critical thinking [11]. Automated Essay Scoring (AES) has been studied for years to give immediate feedback to students and reduce the inconsistency and subjective bias of human evaluators [18]. However, previous AES systems suffer from shallow semantic understanding, limited adaptability to diverse scoring standards, and poor quality of text feedback [3]. In recent years, the emergence of large language models (LLMs) such as GPT has shown the potential to resolve the problems by providing immediate grades and feedback aligned with human evaluators well [23,24]. Regrettably, the research on Chinese AES lags behind, especially in contrast to the rich studies of AES for English essays. To date, there has been a notable absence of research into leveraging LLMs for the automated assessment of Chinese essays.

Meanwhile, the development of an effective AES with LLMs in Chinese essay writing faces many challenges. Firstly, the complexity of Chinese, e.g., the varied syntactic structures, flexible grammar, and semantic richness, makes it difficult to achieve high-quality responses from LLMs without carefully designed prompts [12]. For the design of prompts, there is a significant difference between zero-shot learning and few-shot learning, i.e., feeding no samples or a few samples to LLMs. Although there has been extensive discussion on the merits and demerits of zero-shot and few-shot learning in AES for English essays, such research in Chinese remains missing. Secondly, though incremental success has been achieved in aligning the automated scoring by LLMs with human assessment by fine-tuning these models (i.e., training models with a large set of data) on English essay datasets [23,26], the lack of datasets of Chinese essays with human grading raises the barrier to achieve a fine-tuned LLM for Chinese AES.

As a pioneer work, in this paper, we conducted a comprehensive set of experiments to study the potential of LLMs in AES for Chinese essays. As a starting point, we especially studied the use of LLMs in AES for K1-6 students learning Chinese as their native language. To cope with the challenge of the lack of datasets, we collected a set of K1-6 formal examination essays from online[1]. The original dataset only contains essays and levels of study (i.e., K1-6), some of which have associated scores. There are no writing prompts or rubrics specified. Therefore, we first classified the essays based on their contents into topics. A topic can contain essays from multiple levels of students. As LLMs perform better on texts and due to the limitation of the size of available samples for a topic from the same level of students, we constructed a new dataset including the essays for a chosen topic with a comparatively larger availability of samples from K3 students. We further discussed with two Chinese language teachers to develop writing prompts and associated rubrics for four grading levels (i.e., A, B, C, and D).

---

[1] https://aic-fe.bnu.edu.cn/cgzs/kfsj/xxszskszw/index.html.

With the scoring rubrics, two human raters were engaged to give grades. Subsequently, without any training (i.e., model fine-tuning), we guided GPT-4 in essay scoring by feeding into prompts designed with different strategies. Furthermore, we conducted supervised fine-tuning (i.e., 70% of the essays are used for model training) for GPT-3.5, Qwen[2] [2], and BERT[3] baseline to get grades. The experimental results suggest that the prompts designed with different strategies will have a remarkable impact on the grade alignment between human raters and LLMs without fine-tuning (i.e., QWK ranges from 0.44 to 0.57). When model fine-tuning is adopted, the consistency between machine scoring and human scoring reaches a substantial level (i.e., QWK≥ 0.61).

The rest of this paper is organized as follows. Section 2 presents the existing AES on essay scoring and immediate feedback. Section 3 introduces our method. The experimental results are presented in Sect. 4 followed by discussions in Sect. 5. Section 6 concludes this paper.

2   Related Work

AES for English essays has been studied for years. It started with rule-based systems that utilized hand-crafted linguistic features to evaluate essays [15], followed by a shift to machine learning technologies with a paradigm of feature extraction then prediction via regression-based, classification-based, and ranking-based models [5,13,17]. With the advent of deep learning technologies and the availability of large-scale datasets, recurrent neural networks (RNNs) and convolutional neural networks (CNNs) were integrated into AES systems [19]. However, traditional machine learning and deep learning methods showed limited capability in handling long-range dependencies and contextual understanding in essays.

Recent years have come to the transformer era [20], exemplified by the BERT model [6]. For example, Yang et al. [26] proposed R^2BERT, which combines regression and ranking objectives to fine-tune the BERT model to improve grading accuracy. Wang et al. [21] employed BERT to extract multi-scale essay representation from essays. However, due to BERT's limitations in essay scoring, particularly its restrictive input text length and inadequate capabilities in text generation, the work based on BERT suffers from the problem of incomplete context understanding and poor generated feedback. In the past few years, we have witnessed the emergence of GPT models and the use of GPT models in AES. Mizumoto and Eguchi [14] pioneered the exploration of capabilities of GPT in the AES field. They devised a prompt instructing GPT-3.5 to give scores with the IELTS rubric, then utilized linguistic features to further enhance the accuracy of the scores. Yancey et al. [24] applied GPT-4 in rating L2 English learners' essays under CEFR rubrics and further improved the aforementioned work [14] by exploring prompt engineering techniques. However, their best results in using prompts with calibration examples couldn't surpass the results by the XGBoost

---

[2] https://huggingface.co/Qwen/Qwen-1_8B-Chat.
[3] https://huggingface.co/google-bert/bert-base-chinese.

algorithm. Most recently, Xiao et al. [23] examined the effectiveness of GPT-4 and fine-tuned GPT-3.5 in AES. Their findings demonstrated that advanced prompt design significantly enhanced GPT-4's performance, while further fine-tuning of GPT-3.5 achieved satisfactory levels of accuracy.

Research on Chinese AES is comparatively limited in quantity, particularly due to the intricacies of the Chinese language and the lack of high-quality public dataset, compared to its English counterpart [25]. Following a similar trajectory, the research on Chinese AES has gone through from traditional methods to more advanced deep learning-based methods [8,9,16] and suffered similar problems. With the emergence of LLMs, there has been work based on the transformer architecture. For example, Gong et al. [7] developed IFlyEA, a Chinese essay assessment system, which leverages BERT's masked language model for unsupervised error correction. Zheng et al. [27] applied the BERT model in their ELion intelligent Chinese essay tutoring system to encode sentences and devised a joint loss model for grading essays. However, these methods necessitated a substantial dataset to fine-tune the BERT model for AES applications and encountered the inherent difficulties of BERT in managing longer texts and ensuring contextual coherence.

In this work, we focus on advanced LLMs, e.g., GPT and Qwen, leveraging their text generation capabilities to study their performance in Chinese AES.

3   Methods

3.1   Dataset

One of the challenges faced by AES in Chinese is the lack of well-established datasets. To cope with this issue, in this work, we first built a dataset and developed rubrics. We collected a set of formal examination essays from online, which includes 8,367 authentic essay samples from primary school students, i.e., K1-6. Each essay is labeled with student level (i.e., K1-6). Some of the essays have associated scores, but the full mark is different – 20 or 30. There are no writing prompts or rubrics specified in the original dataset. To add on essay prompts, we manually classified the essays based on their titles and content. The essays having similar titles or content were classified into one topic. As LLMs perform better on texts, we removed the essays that are based on picture prompts, i.e., writing an essay based on a given picture. Then we specially chose the topic that had the largest number of essays from the same level. Eventually, there were 189 essays from K3 students (9–10 years old) for the chosen topic.

We discussed with two Chinese language teachers to develop the essay prompt and rubrics for the chosen topic. The essay prompt is "In the United States, Lai State has made a very sturdy and exquisite box, inside of which is a letter addressed to the people of Earth three thousand years from now. The box will be buried deep underground and can only be opened three thousand years later. If you were to write this letter, what would you say to the future people three thousand years from now? Please write in the form of a letter" (translated from Chinese).

For the rubrics, as the samples are from K3 students, two criteria were decided after discussions with the experts – content and the use of language. For each criterion, there are two levels – low and high. The criteria of content are related to essay structure and topic relevance. For example, the prompt is to write a letter. If an essay is not a letter, then its content level will be considered as 'low'. The language criteria are about the advanced use of language, e.g., longer sentences (considering the samples are from K3 students), lexical diversity, use of rhetoric, etc. For example, using only short sentences would result in a 'low' language rating. Considering the levels of the two criteria, there are four grades – A, B, C, and D decided. The rubrics are available in the Appendix for space considerations.

We invited two human raters to independently evaluate the essays following the developed rubrics. Initially, there were 122 consistent grades among 189 samples. For the samples with inconsistent grades (67 essays), the two raters discussed to reach a consensus. Finally, there are 32 essays graded as A, 64 as B, 67 as C, and 26 as D. The consensus grades serve as the baseline to compare with the grades given by LLMs-based AES.

3.2   Prompt Engineering Using GPT-4

There are generally two strategies to augment the efficacy of AES systems with LLMs – prompt engineering and fine-tuning. Prompt engineering strategy focuses on the design of prompts to instruct LLMs to give responses [4]. Fine-tuning strategy tries to tune the parameters of LLMs through a training stage, which necessitates a corpus of manually annotated data [26]. In this paper, we studied the performance of both strategies in AES for Chinese essays. More specifically, we employed GPT-4 to study the prompt engineering strategy, and GPT-3.5 and Qwen for the fine-tuning strategy.

LLMs will react differently to varied prompts. Therefore, the design of prompts is assumed to have an impact on the performance of LLMs-based AES for Chinese. In this work, we designed prompts considering the following three dimensions to study the performance of GPT-4 in giving grades.

- **Zero-shot vs. Few-shot** The difference between zero-shot and few-shot lies in whether sample essays are provided in the prompt. In few-shot setting, at least one sample is provided for each grade level to guide GPT's scoring, where k-shot means selecting k samples from each category.
- **With rubrics vs. w/o rubrics** This dimension focuses on the presence or absence of rubrics in the prompt when GPT provides grades on essays. When rubrics are included, they serve as a guideline for GPT's evaluation, whereas in the absence of rubrics, GPT assesses the essays based on its pre-trained abilities.
- **With explanations vs. w/o explanations** Inspired by the Chain-of-Thought (CoT) prompt design approach [22], in the "with explanations" configuration, the beginning of the prompt instructs GPT to give explanations. These explanations are then compared with the rubrics to determine

the aligned grade, followed by GPT providing the grade. In the "w/o explanations" setting, GPT directly assigns a grade without providing a rationale.

The essay prompt is always included in the GPT-4 prompts. Each essay is provided in a fresh round of conversation to avoid any possible interactions between essays [24]. In the few-shot configuration, we further employed a retrieval-based sample selection algorithm to choose sample essays to be combined into the prompt. The detailed selection process is as follows:

1. The embedding for each essay is calculated first, which will convert an essay to a vector. The embedding vectors express the overall semantic information and structural features of an essay.
2. For each essay, the cosine similarities with other essays of the same category are calculated based on the embedding vector. An average cosine similarity is further derived.
3. Select the top-k essays with the highest average similarity as the final k-shot samples inserted to the prompt.

In subsequent sections, we use few-shot-retrieval-based-k to represent prompt containing k samples in each grading category. We will only present results using retrieval-based sample selection algorithms when samples are contained in prompts due to the continuous better results than random selection in our experiments. Table 1 shows a prompt example with the design of "Few-shot-retrieval-based-k, with rubrics, with explanation". The text in italics instructs GPT-4 to infer the grade "with explanations".

3.3   Fine-Tuning Using GPT-3.5 and Qwen

For the strategy of fine-tuning, we employed two LLMs to accommodate few-sample fine-tuning tasks: *Qwen-1.8B-Chat* (Qwen in short), *GPT-3.5-turbo-1106* (GPT-3.5 in short). The major difference between these two models is that *Qwen-1.8B-Chat* is a comparatively lightweight model, which is useful in investigating whether fine-tuning a lightweight LLM can achieve a comparable performance.

We first split the essay samples in each grade level into training and testing sets (i.e., 70% for training and 30% for testing). Then we conducted supervised training to get a fine-tuned model. More specifically, the prompts to fine-tune the models in the training stage include system message (i.e., the same prompt as used in the "Zero-shot, with rubrics, w/o explanations" setting in the prompt engineering strategy), user message (i.e., essays) and assistant message (i.e., essay grades). In the testing stage, we required the fine-tuned models to return the essay grades without explanations. We further compared these two fine-tuned models with *bert-base-chinese* (BERT in short) baseline.

4   Experimental Results

We adopt Cohen's Quadratic Weighted Kappa (QWK) to evaluate the performance of LLMs-based AES in Chinese essay grading. A QWK score of 1 indicates

Table 1. An example prompt under the "Few-shot-retrieval-based-k, with rubrics, with explanations" setting (translated from Chinese). The italic text illustrates how to instruct GPT-4 to infer the final grade "with explanations".

You are an experienced elementary school Chinese language teacher, responsible for scoring students' Chinese essays

Students write their essays based on the given prompt

I will provide the essay prompt, scoring rubrics, and scoring examples, which you need to refer to when scoring the students' essays

Your tasks:

(1) Based on the content of the essay, compare it with the scoring examples, and provide the explanation (within 100 words).

(2) Refer to the scoring rubrics and check which grade aligns most closely with the explanation from step (1). Return the corresponding grade in $\{A, B, C, D\}$

The format for the returned content should be:

Explanation: ...

Grade: ...

Essay prompt:

In the United States, Lai State has made a very sturdy and exquisite box...

Scoring rubrics:

A: < content>...<use of language>

...

Scoring examples:

Input 1: {Student Essay 1}

Output 1: {Explanation: ... Grade: ...}

...

Input k: {Student Essay k}

Output k: {Explanation: ... Grade: ...}

Grade the essay below:

...

perfect alignment, and 0 denotes chance-level agreement. The agreement is considered fair, moderate, or substantial when QWK scores fall within [0.21, 0.40], [0.41, 0.60], and [0.61, 0.80], respectively [1]. We computed the inter-annotator agreement between LLMs and human raters at 90% confidence intervals using bootstrapping. During inference, the temperature parameter (i.e., in a range of [0, 1], where higher values promote more diverse outputs) for both GPT-4 and GPT-3.5 is set to 0 to maximize result reproducibility.

4.1 Prompt Engineering Results by GPT-4

Table 2 presents an overview of the rating results from human raters and GPT-4 under different prompt strategies. It can be observed that, compared to human raters, GPT-4 is more inclined to assign grades B, C, and D. In a zero-shot setting, GPT-4 rarely assigns grade A. After adding one or two samples for each grading category, there is a significant increase in the number of A-rated evaluations, especially when providing explanations are instructed.

Table 2. A summary of the ratings by human and *GPT-4* with different prompt engineering techniques.

Methods	A	B	C	D
Human rating (baseline)	32	64	67	26
Zero-shot, w/o rubrics, w/o explanation	0	73	94	22
Zero-shot, with rubrics, w/o explanation	0	29	115	45
Zero-shot, with rubrics, with explanation	2	38	99	50
Few-shot-retrieval-based-1, with rubrics, w/o explanation	7	60	89	33
Few-shot-retrieval-based-1, with rubrics, with explanation	33	33	92	31
Few-shot-retrieval-based-2, with rubrics, with explanation	36	28	89	36

The confusion matrices in Fig. 1 provide more insights. Comparing Fig. 1(a) with Fig. 1(b)(c), it can be seen that within the realm of zero-shot prompts, the use of explicit rubrics enables GPT-4 to become a more stringent evaluator. When GPT-4 is prompted to give explanations before assigning a grade, its agreement with human raters on D grade achieves a significant **73%** alignment. Regrettably, the zero-shot strategy falls short in mediating GPT-4's alignment with human rating at A and B grades. This problem is mitigated through the adoption of few-shot prompt strategies. In Fig. 1(d)(e), it can be seen that the introduction of one example for each grade markedly enhances GPT-4's discriminative capabilities between grades A and B. This improvement is especially notable when GPT-4 is instructed to provide explanations, with its alignment on grade A reaching **58%**. Furthermore, the results show that prompts using multiple samples have similar discriminative power in grading to those using a single sample, as illustrated in Fig. 1(f). This underscores the usefulness of few-shot prompt strategies in refining GPT-4's grading precision.

Figure 2(a) presents the agreement between human raters and GPT-4 with various prompt engineering strategies. Some inconsistencies in the initial grades of the two human raters suggest a subjective difference. The grades provided by LLMs are assumed to present some inconsistencies from the human raters as well. Therefore, we compared the grade difference between LLMs and the human baseline with the difference between the initial human raters.

We assessed the initial human rater agreement (before arriving at a consensus) on the whole dataset as the human baseline, obtaining a QWK of 0.77 with a confidence interval of (0.71, 0.82), which is shown in the purple region. With the "Zero-shot, w/o rubrics, w/o explanations" prompt strategy, the QWK between GPT-4 and human baseline reached 0.52, with a confidence interval of

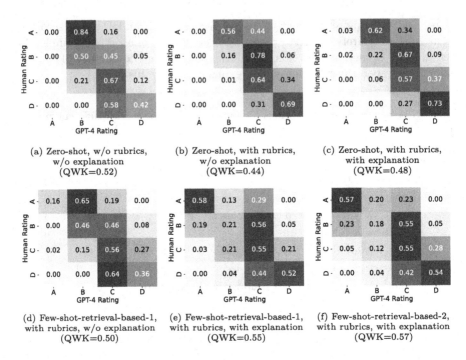

Fig. 1. Confusion Matrices of Prompt Engineering (Normalized by Human Rating)

(0.44, 0.59). This demonstrates GPT-4's strong AES capability, i.e., its inherent assessment criteria largely align with human judgment (i.e., no rubrics are supplied, and no explanations are required). Interestingly, incorporating rubrics and requiring explanations did not result in an enhancement of GPT-4's grading alignment comparing to the prompts without rubrics or explanations, because adding on rubrics or explanations led to a more stringent overall grading as shown in Fig. 1(a) (b) and (c). Consistent with the results shown in Fig. 1, the inclusion of examples improved GPT-4's grading performance. Notably, when GPT-4 provided explanations alongside its evaluations, the QWK between GPT-4 and the human baseline increased to 0.55, with a confidence interval of (0.46, 0.62), surpassing the optimal performance achieved through zero-shot learning. After incorporating more examples for each grade level, the QWK between GPT-4 and human raters has improved by **3.64%**, which brought the consistency close to a substantial level (i.e., QWK≥0.61).

4.2   Comparative Performance of Fine-Tuned Models

When fine-tuning GPT-3.5, we utilized OpenAI's default configuration. For Qwen's fine-tuning, we adopted Low-rank Adaptation (LoRA) [10] to reduce the computational resources and training time. The LoRA parameters rank and alpha were set to 64 and 22, respectively. The learning rate was set to 3e-4,

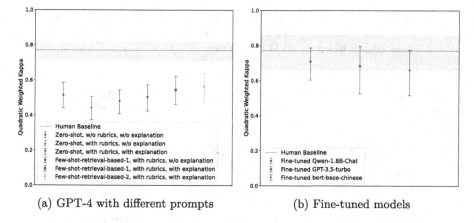

(a) GPT-4 with different prompts   (b) Fine-tuned models

Fig. 2. Human-model agreement (90% confidence intervals shown)

and the training epoch was set to 11. We adopted Chinese BERT as the baseline method, setting the learning rate to 1e-5 and the training epoch to 15. To ensure better reproducibility, we set the random seed to 42 for both Qwen during the fine-tuning and inference stages and for BERT during training. When fine-tuning BERT, we added a simple softmax layer to perform multiclass classification and kept all model parameters trainable.

Figure 2(b) shows the agreement between human raters and the fine-tuned models. We assessed the agreement of human raters on the fine-tuned test set (slightly different from the whole dataset used in prompt engineering experiments) as the human baseline, obtaining a QWK of 0.77 with a confidence interval of (0.66, 0.86), which is shown in the purple region. Among the fine-tuned models, Qwen not only achieves the highest QWK value of 0.71, falling short of the human benchmark by only 0.06, but also has the narrowest confidence interval (0.60, 0.79) among the three models. GPT-3.5 and BERT record QWK values of 0.68 and 0.66 with wider confidence intervals of (0.53, 0.80) and (0.52, 0.77), respectively. Compared to the best outcome of prompt engineering, which achieves a QWK of 0.57, the fine-tuned model's QWK surpasses this by more than 15%, highlighting the substantial improvement gained through fine-tuning.

Figure 3 further presents the confusion matrices for the three fine-tuned models. Though fine-tuned GPT-3.5, fine-tuned Qwen, and fine-tuned BERT have relatively close values in QWK, they exhibit different grading tendencies. Among all grade levels, BERT excels in assigning grade B, whereas GPT and Qwen perform better at other levels. GPT stands out in scoring A and C and has the tendency to give high scores to essays, presenting a relatively lenient grading style. Notably, Qwen, despite its weaker results in grading B, achieves average or above average accuracy in other categories, especially distinguishing itself from the other two models by accurately scoring grade D with the smallest grading difference from human baseline. Both GPT-3.5 and BERT exhibit discrepancies that exceed one grade level from human baseline, mostly happened with grading category A and D. Specifically, 10% of essays rated A by humans were marked as C by GPT-3.5, while 12% and 25% of essays respectively received a B from GPT-3.5 and BERT when rated D by humans.

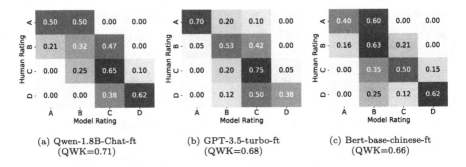

(a) Qwen-1.8B-Chat-ft
(QWK=0.71)

(b) GPT-3.5-turbo-ft
(QWK=0.68)

(c) Bert-base-chinese-ft
(QWK=0.66)

Fig. 3. Confusion Matrices of Fine-tuning (Normalized by Human Rating)

In summary, the prompt engineering results reveal that the few-shot strategy, supplemented with rubrics and explanations, significantly enhances GPT-4's grading precision. This approach notably improves its alignment with human evaluations, especially for A-grade assessments, compared to the zero-shot strategy. The optimally fine-tuned Qwen, GPT-3.5, and BERT models achieve QWK values of 0.71, 0.68, and 0.66, respectively. These values exceed the optimal prompt engineering results by more than 15%, demonstrating a substantial level of agreement with human evaluations. It is noteworthy that none of the fine-tuned models demonstrates a perfect "have-it-all" grading capability, as each exhibits distinct grading tendencies. Despite these variations, Qwen distinguishes itself from the other two models by more effectively identifying off-topic essays and exhibiting fewer discrepancies across grades.

5   Discussion

As seen from the experimental results, GPT-4 has achieved moderate agreement with human raters after implementing various prompt engineering techniques. Moreover, the fine-tuned models GPT-3.5 and Qwen have showcased outstanding performance, thus emphasizing the transformative impact of LLMs on education. However, it's important to note that our research represents an initial exploration of these methods, and we anticipate that there is considerable potential for further improvements yet to be explored.

While exploring few-shot prompt engineering strategies, we encountered several challenges. To make few-shot learning benefit from diversity, we experimented to have GPT-4 learn from a wide range of examples, i.e., one example with the highest similarity among essays of the same category and the other with the lowest. However, the QWK result was worse than when top-2 most similar essays were included in the prompt. This may be attributed to several factors, including inter-sample interaction and the limited dataset size. Moving forward, we will continue to optimize prompt engineering strategies for Chinese AES, taking into account the unique characteristics of the Chinese language.

Our study is grounded in an extensive collection of manually annotated data, with the scoring tendencies of our selected group of evaluators possibly influencing the outcomes of our experiments. Future efforts can take into account the inherent differences or biases within the group of evaluators and collect the student essays from various grades based on a given prompt from real-world scenarios. This approach will facilitate the development of a more universally applicable AES model and more consistent essay collection.

We have conducted fine-tuning with GPT-3.5 and Qwen, yet the potential of other open-source or closed-source LLMs remains largely unexplored for Chinese AES. It would be worth exploring whether using these models could improve the performance and efficiency of assessing student essays and conducting comparative studies. We empirically found that fine-tuned BERT grades long essays based solely on the first 512 tokens, without considering the full content, which will make the essay grading not reflect the quality of the entire essay.

In addition, using closed-source LLMs such as GPT tends to be more user-friendly, lowering the barrier to entry, but may suffer from cost constraints and their closed-source nature. On the other hand, open-source LLMs like Qwen, while allowing for flexible local development and consuming fewer computational resources, demand a higher level of technical proficiency during usage. Hence, further research can weigh the use of both to achieve optimal performance and maximum benefit.

Finally, due to the limited sample size used in our experiments, the generalization ability of LLMs across various essay topics has not been validated. We will further expand our dataset to ensure a comprehensive evaluation of the application value of AES in Chinese essays.

6   Conclusion

In this work, we investigated the use of LLMs for automated Chinese essay scoring. We first manually built a dataset containing essay samples from K3 students for a topic, developed rubrics for four grades, and engaged human raters to give grades based on the rubrics as a baseline. Given the developed dataset, we evaluated GPT-4's grading performance using different prompt engineering strategies (i.e., zero-shot or few-shot, with rubrics or w/o rubrics, with explanations or w/o explanations). We also carried out fine-tuned experiments using GPT-3.5-turbo and Qwen-1.8B-Chat.

The experimental results suggested that using rubrics and explanations for few-shot is the most effective prompt for obtaining automatic scores from GPT-4, but there is still room to improve (a difference of 0.2 from human baseline QWK). When fine-tuned models were adopted, the inconsistency between grades by LLMs and human baseline was further reduced to a difference of approximately

0.1, which implies the enormous potential of LLMs in the automated scoring of Chinese essays. In the future, we will extend our studies to a larger dataset. Leveraging LLMs to provide high-quality text feedback is also a future direction.

Acknowledgements. This research is supported by the Ministry of Education, Singapore, under its Academic Research Fund Tier 1 (A Trustworthy Feedback Agent for Secondary School Chinese Language Learning), the Shenzhen Science and Technology Foundation (General Pro gram, JCYJ20210324093212034), and 2022 Guangdong Province Undergraduate University Quality Engineering Project (Shenzhen University Academic Affairs [2022] No. 7).

Appendix

The detailed scoring rubrics and the essay example from K3 students can be accessed at https://github.com/seamoon224/AIED-2024.

References

1. Abraham, B., Nair, M.S.: Automated grading of prostate cancer using convolutional neural network and ordinal class classifier. Inform. Med. Unlocked **17**, 100256 (2019)
2. Bai, J., et al.: Qwen technical report. arXiv preprint arXiv:2309.16609 (2023)
3. Bai, J.Y.H., et al.: Automated essay scoring (AES) systems: opportunities and challenges for open and distance education. In: Tenth Pan-Commonwealth Forum on Open Learning. Commonwealth of Learning (2022). https://doi.org/10.56059/pcf10.8339
4. Chen, B., Zhang, Z., Langrené, N., Zhu, S.: Unleashing the potential of prompt engineering in large language models: a comprehensive review (2023). http://arxiv.org/abs/2310.14735. Accessed 26 Mar 2024
5. Chen, H., He, B., Luo, T., Li, B.: A ranked-based learning approach to automated essay scoring. In: 2012 Second International Conference on Cloud and Green Computing, Xiangtan, Hunan, China, pp. 448–455. IEEE (2012). https://doi.org/10.1109/CGC.2012.41
6. Devlin, J., Chang, M.W., Lee, K., Toutanova, K.: Bert: pre-training of deep bidirectional transformers for language understanding. arXiv preprint arXiv:1810.04805 (2018)
7. Gong, J., et al.: IFlyea: a Chinese essay assessment system with automated rating, review generation, and recommendation. In: Proceedings of the 59th Annual Meeting of the Association for Computational Linguistics and the 11th International Joint Conference on Natural Language Processing: System Demonstrations, pp. 240–248 (2021)
8. Guan, Y., Xie, Y., Liu, X., Sun, Y., Gong, B.: Understanding lexical features for Chinese essay grading. In: Sun, Y., Lu, T., Yu, Z., Fan, H., Gao, L. (eds.) ChineseCSCW 2019. CCIS, vol. 1042, pp. 645–657. Springer, Singapore (2019). https://doi.org/10.1007/978-981-15-1377-0_50
9. He, Y., Jiang, F., Chu, X., Li, P.: Automated Chinese essay scoring from multiple traits. In: Proceedings of the 29th International Conference on Computational Linguistics, pp. 3007–3016 (2022)

10. Hu, E.J., et al.: Lora: low-rank adaptation of large language models. arXiv preprint arXiv:2106.09685 (2021)
11. Hussein, M., Hassan, H., Nassef, M.: Automated language essay scoring systems: a literature review. PeerJ Comput. Sci. **5**, e208 (2019)
12. Li, L., Zhang, H., Li, C., You, H., Cui, W.: Evaluation on ChatGPT for Chinese language understanding. Data Intell. **5**(4), 885–903 (2023)
13. McNamara, D.S., Crossley, S.A., Roscoe, R.D., Allen, L.K., Dai, J.: A hierarchical classification approach to automated essay scoring. Assess. Writ. **23**, 35–59 (2015)
14. Mizumoto, A., Eguchi, M.: Exploring the potential of using an AI language model for automated essay scoring. Res. Methods Appl. Linguist. **2**(2), 100050 (2023)
15. Page, E.B.: Project essay grade: PEG (2003)
16. Peng, X., Ke, D., Chen, Z., Xu, B.: Automated Chinese essay scoring using vector space models. In: 2010 4th International Universal Communication Symposium, pp. 149–153. IEEE (2010)
17. Phandi, P., Chai, K.M.A., Ng, H.T.: Flexible domain adaptation for automated essay scoring using correlated linear regression. In: Proceedings of the 2015 Conference on Empirical Methods in Natural Language Processing, pp. 431–439 (2015)
18. Ramesh, D., Sanampudi, S.K.: An automated essay scoring systems: a systematic literature review. Artif. Intell. Rev. **55**(3), 2495–2527 (2022)
19. Taghipour, K., Ng, H.T.: A neural approach to automated essay scoring. In: Proceedings of the 2016 Conference on Empirical Methods in Natural Language Processing, pp. 1882–1891 (2016)
20. Vaswani, A., et al.: Attention is all you need. arXiv preprint arXiv:1706.03762 (2023). http://arxiv.org/abs/1706.03762
21. Wang, Y., Wang, C., Li, R., Lin, H.: On the use of BERT for automated essay scoring: joint learning of multi-scale essay representation. arXiv preprint arXiv:2205.03835 (2022)
22. Wei, J., et al.: Chain-of-thought prompting elicits reasoning in large language models. Adv. Neural. Inf. Process. Syst. **35**, 24824–24837 (2022)
23. Xiao, C., Ma, W., Xu, S.X., Zhang, K., Wang, Y., Fu, Q.: From automation to augmentation: Large language models elevating essay scoring landscape. arXiv preprint arXiv:2401.06431 (2024)
24. Yancey, K.P., Laflair, G., Verardi, A., Burstein, J.: Rating short L2 essays on the CEFR scale with GPT-4. In: Proceedings of the 18th Workshop on Innovative Use of NLP for Building Educational Applications (BEA 2023), pp. 576–584 (2023)
25. Yang, H., He, Y., Bu, X., Xu, H., Guo, W.: Automatic essay evaluation technologies in Chinese writing-a systematic literature review. Appl. Sci. **13**(19), 10737 (2023)
26. Yang, R., Cao, J., Wen, Z., Wu, Y., He, X.: Enhancing automated essay scoring performance via fine-tuning pre-trained language models with combination of regression and ranking. In: Findings of the Association for Computational Linguistics: EMNLP 2020, pp. 1560–1569 (2020)
27. Zheng, C., Guo, S., Xia, W., Mao, S.: Elion: an intelligent Chinese composition tutoring system based on large language models. Chinese/English J. Educ. Measur. Eval. **4**(3), 3 (2023)

The Unexpected Effects of Google Smart Compose on Open-Ended Writing Tasks

Robert E. Cummings$^{(\boxtimes)}$ (ID), Thai Le (ID), Sijan Shrestha,
and Carrie Veronica Smith (ID)

University of Mississippi, University MS 38677-1848, USA
cummings@olemiss.edu

Abstract. In this work we seek to understand the effects of Google Smart Compose's auto-complete suggestions on writing products and processes in open-ended writing tasks. We recruited 119 human subjects to a carefully designed user study to write timed responses to a common prompt via word processors while recording interactions with Google Smart Compose behind the scene. Our experiment demonstrates the impacts of text suggestions on writing length, structure, cohesion, and complexity, utilizing Coh-metrix values. We also introduce new metrics for measuring and understanding detailed human subject interactions with text suggestions. We find that writers utilizing Google Smart Compose write *the same amount of text* as those without Smart Compose, and that the structure of written work produced with Google Smart Compose enabled is *not significantly distinguishable* from writing produced with Google Smart Compose disabled. In addition, there was no strong evidence that writing process was significantly impacted by Google Smart Compose. Finally, this work identifies factors that can be used in future studies measuring language model interactions with human writers.

Keywords: Auto Completion · AI-Assisted Writing · Google Smart Compose · Predictive Text Suggestions

1 Introduction

The release of GPT-3 in 2020 triggered a dramatic increase in popular awareness of the ability of auto-regressive large language models (LLMs) to produce text to assist human writing tasks. The class of tools produced with GPT-3 and its successors such as ChatGPT (GPT-3.5) and GPT-4, also known as AI-powered writing generators, provide strings of text in response to user inputs. These outputs are longer and more sophisticated than earlier AI-powered writing tools classified as writing assistants, which include next-word-prediction and sentence completion technologies such as Google Smart Compose (GSC) [19]. As writing generators have proliferated, so too has the range of their writing goals, including summarizing texts [5], writing social media posts [9], finding research to support assertions [15], and even creative writing [7, 22].

© The Author(s), under exclusive license to Springer Nature Switzerland AG 2024
A. M. Olney et al. (Eds.): AIED 2024, LNAI 14829, pp. 468–481, 2024.
https://doi.org/10.1007/978-3-031-64302-6_33

As AI-powered writing generators are most often deployed with common web searches and office productivity softwares with little or no user training, the education community is increasingly aware of the need for intentional and structured engagements with these tools. Educators understand that risks for using AI tools include "plagiarism, harmful and biased content, equity and access, the trustworthiness of the AI-generated content, and over-reliance on the tool for assessment purposes" [20]. The effective use of AI-powered writing generators also depends greatly upon their reception by human writers. Understanding the context, purpose, genre, and technological platform utilized by human writers is vital to shaping the value of AI outputs and how human writers seek to incorporate AI writing outputs within writing projects. Consequently, researchers are also paying greater attention to the context and reception of AI writing outputs as they develop their tools [5]. Further, researchers have now developed more comprehensive frameworks for studying and engaging human reception for AI writing outputs [10].

A long term claim for the value of similar artificial intelligence (AI) technologies has been to increase human productivity [3, 21]. The developers of the current AI-powered writing generators, including ChatGPT, offer similar claims [18]. But the longer established AI-powered writing applications have not yet been thoroughly examined to determine if claims of increasing writing productivity are true. These technologies include applications that mainly focus on next-word-prediction or sentence completion rather than generating the whole sentences for the users.

Although recently there is great interest in the emergence of LLMs applied to digital writing processes, precedent softwares such as Grammarly and GSC remain understudied in relation to their broad adoption. This lack of understanding has a direct impact on pedagogical strategies within education systems for the understanding and use of AI-powered writing generators. Currently, many higher education faculty are experimenting with AI-powered writing generators in their classrooms with little guidance from the research community [17]. In order to train students to effectively integrate these tools in their writing processes for both greater productivity and informed ethical awareness, faculty will *need a greater understanding of the potential impact of AI-powered writing generators on their students' writing.*

We believe that a more comprehensive understanding of human-computer interactions around the utilization of these technologies is a necessary building block to understanding the more recent developments in applied LLMs to writing assistant technologies. Therefore, in this work, we propose to investigate the effects of the popular writing assistant technology GSC on open-ended writing tasks. This work, being conducted in an academic setting with human subjects writing in an academic genre, will be informative for students, faculty, and administrators seeking to understand and utilize AI-powered writing generators in the classroom.

2   Related Work

Google Smart Compose. GSC was first launched in August of 2019 as an assistive tool for G-mail [2]. Later in 2019, Smart Compose was also offered within Google Docs [19]. As a writing assistance technology, GSC is an extension of earlier predictive text suggestion technologies developed for mobile platforms that attempt to assist writers in composing texts by suggesting the completion of words, phrases, and sentences. These technologies differ significantly based on their platform (i.e., mobile, web-based, word processor embedded) and their related genre (e.g., texting, e-mail, open-ended writing). GSC is now a widely adopted AI-powered writing assistant, and part of a class of technologies generally known as predictive or suggestive text writing assistants that offer writers the opportunity to complete statements with text strings supplied by computer applications. Smart Compose represents only a recent development in the steady rise of automated digital writing assistants, starting with Word processing spell checkers, grammar correction technologies, and now LLMs [4].

Human Study on Writing Assistants. Prior studies have examined human computer interactions in email composition [2] and in mobile devices [16]. In particular, Arnold et al., demonstrated that writing assistant technologies have the potential to strongly influence the output of human writers in terms of content, diction, syntax, and linguistic complexity [1]. However, prior studies have been conducted only in narrow writing tasks, such as determining one word, phrase, or sentence to describe an image. Thus, we seek to better understand the effects of writing assistant technologies when applied to more *open-ended writing tasks,* conducted in word processors on laptops, where writers have *nearly unlimited writing freedom to reach a goal of expression that is more nuanced and complex, and typical of higher stakes writing tasks.*

3   Research Questions

Our main research focus is measuring the impact of GSC both on **writing product and on writing process** with writers who are given an open-ended writing task, utilizing word processors on laptop computers. Our single study variable was whether or not GSC was enabled. Since we are able to compare writing samples and writing experiences of writers responding to the same prompt, either with or without GSC enabled, we formulate the following research questions (RQs) around writing product and processes:

- **RQ1 (Writing Product)**: *Are writing products composed with Google Smart Compose different from those composed without Google Smart Compose in terms of length and textual complexity?*
- **RQ2 (Writing Process)**: *How does Google Smart Compose affect human writing behaviors such as accepting, editing, and rejecting suggestions?*
- **RQ3 (Time Spent with Google Smart Compose)**: *Does time spent on writing suggestions correlate with different behavior patterns such as accepting, editing, or rejecting suggestions?*

In **RQ1**, we wished to understand the depth of the impact of GSC suggestions not only on the length of the writing samples, but also on upon the complexity or depth of written expression. We wondered if suggestions offered by GSC might be similar across writing sessions where it was enabled, and this push written expression to be more similar across samples from those sessions. Beyond the writing products themselves, in **RQ2**, we were interested in better understanding how GSC text suggestions affected human writing processes or behaviors. To gain more insight into the writing process of writers who utilize GSC, we devised three new measures to record how writers responded to GSC suggestions, tracking whether writers (1) accepted the suggestion verbatim (full acceptance); (2) edited a suggestion before accepting it (partial acceptance), or (3) rejected the suggestion outright (rejection) and also how much time it took for them to make those decisions. In **RQ3**, we are also interested to know the time spent reading, evaluating, revising, and/or ultimately rejecting Google Smart Compose suggestions and their comparisons.

4 User Study

4.1 Recruitment, Consent and Compensation

We recruited 119 participants at a US public research university to write for 25 min in response to a prompt that solicited their opinions about the roles of luck and/or hard work in accounting for success in life. No university status was sought or required (i.e., no student status was required), no literacy ability was measured, and all participants were 18 years of age or older. We randomly enabled/disabled GSC for each participant via uniform sampling. This resulted in roughly half (55) of the participants writing with GSC enabled, and roughly half (59) writing without GSC enabled. All participants wrote on laptop computers in a room with up to four other participants, with access to the room controlled by a research assistant. The 25 min time limit for sample collection was monitored by our custom made software, and the research assistant guided participants to and from laptop work stations after participants had completed the consent process.

During the consent process, study participants were initially deceived as to our interest in measuring the impact of GSC on writing. At the conclusion of the writing experiment participants were informed of the purpose of the study and asked to re-consent to the collection of their data. Participants were compensated 25.00 USD (about $60.00/hour compared to the current federal minimum wage of $7.25) for their time regardless of whether they consented to allow the collection, analysis, and reporting of their data. Our study was reviewed and received exempt status from our institutional IRB office.

4.2 Participants and Demographics

We collected demographic information on our study participants. This included our participants' genders (59% Male, 39% Female, 2% Non-Binary), the age of

our participants (58% 18–22, 29% 23–29, 10% 30–39 3% 30–49) and our participants' racial and ethnic identities (49% Caucasian, 12% African-American, 3%
Latino or Hispanic, 29% Asian, 3% Other or unknown and 1% No Response). We
also queried our participants about their highest level of education completed
(3% PhD, 17% Master's degree, 24% Bachelor's degree, 16%, 15% and 15% are
in third, second, first year of college, 10% high school).

After recording demographic information for our study participants, we also
asked two questions about participants' experience with auto-complete technologies generally (multiple responses allowed). 75%, 67% and 61% reported
to have some experience with "word processor", "texting or mobile applications", "web-based applications", respectively. We also asked their frequency of
the technology usage: 33% responded "very frequently", 25% "frequently", 25%
"sometimes", and 13% responded "rarely". Only 1% responded "never" and 3%
did not give any responses. We also asked a final question about users' opinions
of auto-complete technologies, where 37% and 36% responded to "very favorable", "favorable", 19% responded to "neutral", 3% responded to "negative",
none responded to "very negative" and 5% did not give any responses. For the
purpose of the survey, we defined auto-complete technologies as "any technology
that suggests how you should complete sentences or words on any digital writing
platform".

4.3   Apparatus

We developed a custom user-interface to enable our research. GSC is only available via Google Docs web application, a well-known online collaborative text
processor. Since this auto-suggestion technology is provided as an integrated
function of Google Docs *rather than a public API access*, it is non-trivial to
systematically capture all the needed interactions between the study participants and the software for our study. To resolve this, we developed a custom
web browser on top of the open-sourced Google Chrome driver[1] in Python programming language. Although this browser looks exactly the same as the official
Google Chrome browser, by using it to access Google Docs, we are able to extract
all the necessary information to track suggestions offered to participants. This
information includes each input keystroke from the users, the writing after each
additional keystroke, suggested phrases from GSC, the prompts and keystrokes
that triggered them, and the final writing products. Importantly, this information also includes respective timestamps, including when a phrase is suggested
by GSC and when the user takes the next action. Using this raw data, we are
then able to derive useful statistics to answer the proposed research questions.

4.4   Study Design

Our study design is a *post-test only* experimental design. All participants were
instructed to write for 25 min on the topic of, *"What do you think is more important for success in life: luck or hard work?"* Participants wrote their open-ended

---

[1] https://chromedriver.chromium.org/.

responses on provided laptop computers, as more fully described in Sect. 4.3. Participants were randomly assigned to two groups - a control group that wrote as normal (GSC disabled) and an experimental group, whose computer was enabled with GSC. The independent variable was the presence of GSC (yes/no). The dependent variables were multiple indices and measures capturing both writing production and writing process, which we will describe below.

4.5 Measures

To structure our results reporting and subsequent analysis, we split our concerns into the impact of GSC on *writing product*, or the text, and *writing process*, or the actions of writers engaging their writing technologies to create the finished documents.

Writing Product. To analyze the writing products, we deployed *Coh-Metrix 3.0*, an automated tool for text and discourse analysis [12]. Coh-metrix is a good fit for our research because it is a well-established tool that can provide 108 quantitative and linguistic measurements for the writing samples. These measurements include a range of dimensions: basic textual descriptions (e.g., number of words in a writing sample, number of sentences, number of paragraphs), textual cohesion, lexical complexity, syntactic complexity, reading difficulty, and additional measurements. Given the fact that Coh-Metrix has been used continuously in some form since 2004 [8], broadly adopted (thousands of peer-reviewed studies utilize some versions of the tool), validated [13], widely used by the NLP community [11], and can help us to explore multiple dimensions of writing products simultaneously, we chose it as the primary tool for our textual analyses.

Writing Process. To analyze writing processes, we devised three new measures to record how writers responded to the suggestions of GSC, tracking whether writers (1) accepted the suggestion verbatim; (2) edited a suggestion before accepting it, or (3) rejected the suggestion outright. If a writer accepted a GSC suggestion as it was offered, we coded that action as "full acceptance," or FA. If a writer edited the GSC suggestion before accepting it, we coded that action as "partial acceptance," or "PA". And if a writer completely rejected a GSC suggestion, we coded that action as "Rejection." We tracked all GSC suggestions equally, regardless of whether would suggest altering a word, phrase, or sentence. In addition, we tracked whether a writer "backtracked," or later returned to edit a suggestion that had been addressed earlier.

After defining these three possibilities, we tracked how each writer deployed the three options, in order to build profiles of each writer's choices. For each category (Full Acceptance, Partial Acceptance, Rejection), we recorded all statistics as described in Table 1C&D.

4.6 Analysis Methods

To best describe the scope of the writing samples provided by our participants, we engaged the several descriptive indices of writing samples from Coh-Metrix as

Table 1. Statistical measures for measuring writing product and writing process

---

Writing Product

(A) Scope of the writing samples: # of paragraphs (DESPC), # of sentences (DESSC), # of words (DESWC), mean # of words found in each sentence (sentence length) (DESSL)

(B) Reading, structure and lexical complexity: Flesch Reading Ease (RDFRE), Syntactic simplicity (PCSYNz), Syntactic complexity – words before main verb (SYNLE), Lexical diversity - content words (LDTTRc)

Writing Process

(C) Individual writer: # of times engaged in an action, % of the three possible actions taken, ET on an action, % of ET on an action out of all possible responses

(D) Individual writer v.s. all writers: Rank of (1) against AW, % of three possible actions by AW, % of total ET on three possible actions by AW, % of ET on an action compared to total ET on accepts by AW, Rank of total ET spent on accepts as compared to AW

ET: Elapsed Time Spent; AW: all writers with GSC enabled

shown in Table 1A, including measures such as number of words, paragraphs and sentence lengths.[2] We selected these measures as they provided a general sense of the scope of the writing samples, and enabled comparisons between when GSC disabled and enabled writing samples. Particularly, we were interested in understanding how the suggestions of GSC could affect these values because we reasoned that writers who frequently accept suggestions verbatim (FA) might display higher word counts than those who did not. Similarly, we reasoned that writers who heavily edited suggestions might write less than those who did not, in part because more of their allowed time would be spent on editing as opposed to generating text. Similarly, we reasoned that if GSC suggestions perpetually presented writers with a means to end a sentence quickly, then sentence length could be affected when comparing writing samples produced with and without suggestions.

In order to facilitate an understanding of the different levels of reading complexity, textual coherence, syntactic complexity, and lexical complexity between writing samples, we applied additional Coh-Metrix measures as described in Table 1B, including reading ease, syntactic simplicity and complexity and lexical diversity. Additionally, we looked for to the demographic information collected about our participants (Sect. 4.2). In particular, we looked at how frequently our participants indicated that they used auto-complete technologies (1=rarely used, 5=very frequently used).

---

[2] Full definitions of these measures can be found in [12].

Table 2. Quantitative and qualitative statistics of writing production

	Smart Compose Disabled ($n = 59$)				Smart Compose Enabled ($n = 55$)			
	Min.	Max.	Mean	SD	Min.	Max.	Mean	SD
Quantitative Indices								
Number of paragraphs	1	14	4.39	2.91	1	10	3.84	2.47
Number of sentences	11	60	29.66	11.14	14	77	28.96	10.88
Number of words	303	1,096	595.56	190.42	241	1,245	585.85	205.05
Number of words/sentence	13.21	32.27	21.00	4.50	12.61	42.00	21.08	6.01
Qualitative indices								
Syntactic simplicity	−1.57	1.06	−0.30	0.60	−2.02	1.12	−0.32	0.61
Syntactic complexity	2.09	9.00	4.39	1.31	2.25	7.56	4.16	1.38
Lexical diversity	.40	.79	.62	.08	.50	.79	.63	.07
Reading ease	39.94	82.14	67.02	8.09	37.72	80.68	65.62	9.09
Referential cohesion	−1.06	2.19	0.41	.71	−1.10	2.14	0.36	0.75

5   Results

We engaged 119 participants, but due to unforeseeable technical difficulties, only 114 writing samples were usable. Of those 114 samples, 59 were written with GSC disabled, and 55 were written with GSC enabled.

5.1   Impact of Google Smart Compose on Writing Product (RQ1)

To ensure the effectiveness of random assignment, we conducted an independent samples t-test to determine whether the experimental and control groups differed on typical usage of auto-complete technology. The experimental ($M = 3.89$, $SD = 1.17$) and control ($M = 3.70$, $SD = 1.05$) groups did not significantly differ, with t(108)=-0.87, ns. The means and standard deviations for the quantitative and qualitative indices of writing production are presented in Table 2.

A series of regression analyses were conducted to examine whether the usage of GSC technology resulted in quantitatively or qualitatively different writing production. In addition to the primary variable of interest, experimental condition (*enabled* or *disabled*), we also included typical usage of auto-complete technology in order to control for potential pre-existing differences between users.

Experimental condition was *not* a significant predictor of either the quantitative (see Table 3 for regression coefficients) or the qualitative indices (see Table 4 for regression coefficients) of the writing production variables, nor was self-reported auto-complete usage.

5.2   Impact of Google Smart Compose on Writing Process (RQ2)

In addition to evaluating the effect of GSC on our open-ended writing samples, our third research question invites us to consider how writers interact with GSC to produce text.

Table 3. Regression analyses predicting quantitative indices of writing production by experimental condition

	# of Paragraphs			# of Sentences			# of Words			# of Words/Sentence		
	B	β	t	B	β	t	B	β	t	B	β	t
Intercept	5.74			30.78			635.12			22.96		
Condition	−.36	−.15	−1.53	−.27	−.03	−0.28	−9.55	−.05	−0.56	.07	.007	0.07
Usage	−.47	−.09	−0.90	−.63	−.03	−0.29	−10.48	−.03	−0.28	−.50	−.11	−1.10

Table 4. Regression analyses predicting qualitative indices of writing production by experimental condition

	Syntactic Simplicity			Syntactic Complexity			Lexical Diversity			Reading Ease			Referential Cohesion		
	B	β	t	B	β	t	B	β	t	B	β	t	B	β	t
Intercept	−0.52			4.62			.62			66.41			066		
Condition	−.03	−.03	−0.28	−.28	−.10	−1.06	.003	.02	0.22	−1.22	−.07	−0.73	−.03	−.02	−0.23
Usage	.06	.11	0.11	−.05	−.04	−0.45	.002	.03	0.27	.15	.02	0.19	−.07	−.10	−1.05

The 55 participants in the GSC enabled group received an average of 68.45 auto-complete recommendations, with a mode of 73. The number of auto-complete recommendations ranged from 12 to 146, with a standard deviation of 34.13. Because writing more allows for more recommendations, correlations between the number of recommendations and quantitative indices of writing production are uninformative although we did calculate these analyses and all correlations were significant and positive, $r's \geq 0.39$, with the exception of mean number of words per sentence, which approached conventional levels of significance ($r(53) = 0.25, p = 0.07$). However, receiving more recommendations may result in better writing quality. A series of regression analyses was conducted to examine this question, with writing quality metrics regressed on number of recommendations received and typical usage of auto-complete (entered as a control variable). Results suggest that receiving more recommendations was not significantly associated with any of the qualitative indices of writing production.

To determine if participants were equally likely to handle recommendations in the same way, a one-way repeated measures ANOVA was performed. Mauchly's test indicated that our data did not violate the assumption of sphericity, $W = 0.94$, $\chi^2(2) = 3.11$, ns. Significant differences were found in the participants' tendencies to fully accept, partially accept, and reject recommendations, $F(2, 108) = 95.485, p < .001$. Post-hoc tests show significant differences in between the three decisions. Specifically, participants accepted more recommendations ($M = 32.36$, $SD = 16.59$) than they rejected ($M = 22.36$, $SD = 22.36$) and partially rejected ($M = 13.73, SD = 8.11$, $ps < .001$). Given that recommendations are meant to assist the writer, a series of regression analyses examined whether proportion of full acceptances (or receiving recommendations the participant found helpful, relative to the number of total recommendations received) predicted differences in writing production, quantitatively or qualitatively. Again, typical usage of auto-complete technology was added to the model as a control variable. *Receiving helpful recommendations was not significantly associated with any metric of writing production, either qualitative or quantitatively.*

5.3 Examining Time Spent with Google Smart Compose Recommendations (RQ3)

The 55 participants in the GSC enabled group spent an average of 12,420 ms interacting with recommendations. The amount of time spent with recommendations ranged from 2,628 to 26,635 ms, with a standard deviation of 5,621 ms. To determine whether participants spent equal amounts of time with each type of recommendation (acceptance, partial acceptance, and rejection), a one-way repeated measures ANOVA was performed. Mauchly's test indicated that our data did violate the assumption of sphericity, $W = 0.71, \chi^2(2) = 18.35$, $p < .001$. Since the epsilon value was less than 0.75, a Greenhouse-Geisser correction was used. Significant differences were found in the amount of time participants spent with each type of recommendation, $F(1.55, 83.55) = 138.07$, $p < .001$. Post-hoc tests indicated significant differences between the three; participants spent significantly more time with acceptances $(M = 7,524.26$, $SD = 45.20)$ than they did with partial acceptances, $(M = 3,160.99$, $SD = 300.98)$ and rejections, $(M = 1,734.26, SD = 170.59)$ and more time with partial acceptances than they did with rejections, $ps < .001$.

6   Discussion

6.1   AI in Education Implications

This research has direct implications for the use of AI in education for several reasons. First, the writing genre of our experiment is an open-ended argument. Writers were asked to offer a clear statement in response to our prompt, and were also requested to provide evidence in support of that statement. Argument is a common mode of writing throughout secondary and higher education systems and is used in both high-stakes and low-stakes assignments [14]. Therefore any observations of how writers utilize AI-powered writing generators within the genre of argument are more likely to be correlated with the everyday writing tasks of higher education classrooms. Secondly, our research was conducted on a university campus with human subjects. While we cannot state that every participant in our study was a college student, we do know that all of them attained education beyond high school and 58 percent were within the age range of traditional undergraduates. Others could have also been graduate students.

Therefore, given the ages, educational attainment, and setting of our experiment, the results from our experiment are more likely to duplicate the experience of higher education students utilizing AI-powered writing generators for academic assignments in higher education classrooms. While we are aware of recent research that has compared the reception of sentence level text suggestions to message level text suggestions with AI-powered writing generators, we note that this research was conducted with online participants writing in fictional roles. [6] We believe that this work with human subjects, writing about their genuine beliefs, and supporting those opinions with first-hand evidence, is more likely to duplicate the higher education writing experience.

6.2   Unexpected Results Regarding Google Smart Compose's Effects

Results regarding the effects of GSC were unexpected. We created Coh-Metrix reports for all 114 usable writing records. In reviewing the results to answer **RQ1**, we did not find a significant difference in writing sample length when comparing the average number of words written with GSC enabled versus disabled (see Table 2). Indeed, the length of writing samples produced without GSC were nearly identical in word length to those produced with GSC. Writers with GSC enabled wrote an average of only 2% less words than writers without GSC enabled, and virtually the same number of sentences.

Moreover, we hypothesized that writing samples produced with GSC enabled would be characterized by lower textual complexity than their counterparts written without GSC. We reasoned that when writers are utilizing uniform versions of Smart Compose that have not been customized to the writer, text suggestions would also be more uniform, and if accepted, would lead to more similarity in the written product. Additionally, since a GSC text suggestion will almost always suggest a way to complete a sentence, the cumulative effect could lead to writing samples that had shorter sentence length. Here again, the results of our study clearly demonstrate that GSC had no meaningful impact on textual complexity, when examined in terms of lexical complexity, syntactic complexity, or reading ease.

Our subsequent **RQ2**, were all three questions of writing process. Each looked at how a group of writers who spent more time with particular category of GSC suggestion – "full acceptance," "partial acceptance" and "rejection" – might have writing products or processes that stood out from the others. During our study we collected a substantial amount of writing process data. Although we have not exhausted every possible inquiry of those data, there are no preliminary identifiable differences in the writing products between writers who heavily engaged in one of those strategies.

The most striking finding of our work is the lack of impact that GSC had upon both writing product (the text itself) and the writing process (the approach of writers in creating text). The written work of participants using GSC was almost indistinguishable from work produced without GSC. We did note that timing measurements indicate that writers who had GSC enabled took longer to accept suggestions as offered than to either edit or reject them. This finding has implications for how we should characterize writers' responses to text suggestions, and implies that writers are investing the most time and thought when evaluating a machine-generated suggestion for full acceptance, as compared to editing or rejecting suggestions. This raises questions about the investment of writers in their own words when compared to text that is offered via AI, and has potential implications for further studies, including those that would record the whether suggestions increase speed without decreasing quality.

Similarly, when we examined demographics, we did not find a significant correlation between participant familiarity with auto-complete technologies and writing product, nor did we find any correlation with participants attitudes toward the technologies and their impact on writing products or processes. To guard against writing abilities being overly weighted in either experiment or control groups, we did review the highest level of education completed in both groups, deeming this as a reasonable proxy for writing abilities. While the experiment group did reveal a slightly higher concentration of graduate students when compared to the control group (29% to 18%), we do not believe them to be overly represented with a material consequence.

7   Limitations and Future Works

7.1   Limitations

We acknowledge that our study utilized instances of GSC that were "off-the-shelf." That is, when we deployed Smart Compose for our subjects, each use of the software was new – the software was not able to customize its suggestions to users based on a developed history of user preferences. By deploying GSC without user customization, we defeated a potential benefit of the software for writers (i.e., more accurate word suggestions) and a presumably a primary feature intended by designers.

Our prompt was written by the researchers for this experiment only. The operational section of the writing prompt is: *"What do you think is more important for success in life: luck or hard work? And why? Feel free to support your answer with examples from your own lived experience as well as the examples of famous men and women. This includes citing to examples from your personal experience, your professional experience, and public figures we all know."* We recognize that this prompt could produce an informal style of writing, where writers were invited to identify and discuss personal experiences as well as the experiences family members. This storytelling mode could also push writers toward greater cohesion than if they were asked to write about unfamiliar and more abstract ideas.

Lastly, we failed to record participants' native language. If English is a second language for a study participant, it is possible that L2 factors could have affected writing productivity and complexity.

7.2   Future Works

There are several possible explanations for the lack of impact GSC demonstrated on both writing process and writing product that might be engaged by future studies.

Have Writers have Already Adjusted to Using Auto-complete Technologies? Humans do not read and write simultaneously. Our presumption in designing this study was that the cognitive change from writing to reading disrupts the writing process, and that part of that work would be measuring and describing the nature of that disruption. Further, we expected that different writers would accommodate the writing technologies differently. Although our population was small when compared to the total number of GSC users, their writing processes were clearly not diminished by the cognitive load presented by interpreting, evaluating, integrating or rejecting, auto-complete suggestions. Clearly, our findings refute the idea that auto-complete suggestions play a disruptive impact on the writing product or process, at least when utilized for open-ended writing projects conducted on laptops utilizing word processors. If writers do not find the current level of auto-complete suggestions disruptive to the composing process, what does this possibly signal for the development of future assistants?

Does the Writing Context have a Greater Impact that Previously Understood? Another possible explanation for our findings of little impact for GSC on the length or quality of writing could be attributed to the impact of the writing context: the writer's purpose, the technology deployed, and the genre engaged. At least one previous study of the impact of predictive text technologies on writing samples found that text suggestions held great influence on the amount of text written and the diction used [1]. However, this study looked at captions for images written on mobile platforms, where only one word or phrase was produced by writers. In contrast, our study asked writers to formulate ideas in an open-ended writing context and via word processors. Future studies that look to better understand the impact of AI-powered writing assistants (and generators) need to fully address the context the writing purpose, platform, and genre.

8   Conclusion and Future Works

This work investigates the impacts of the popular writing assistant GSC technology on open-ended writing tasks, both on writing product and on writing process. Our user study showed that writers write the same amount of text regardless of whether or not they utilize GSC, and that the structures of writing produced with GSC is not significantly distinguishable from those produced without it. Moreover, there was no strong evidence that GSC impacted the writing process. Because both the technology producing writing suggestions, and the cognitive state of the writer reviewing suggestions, are complicated and constantly influx throughout a writing session with GSC enabled, we believe that our study offers an important first look into an incredibly complex interchange. We believe that our key findings for writing process and writing product with text suggestions in open-ended writing environments contribute to a growing understanding of contemporary digital writing. Our work helps teachers and educators better understand and utilize both existing and future AI-powered writing assistants and generators in the classroom.

References

1. Arnold, K.C., Chauncey, K., Gajos, K.Z.: Predictive text encourages predictable writing. In: ACM IUI (2020)
2. Chen, M.X., et al.: Gmail smart compose: Real-time assisted writing. In: KDD (2019)
3. Crawford, K.: The Atlas of AI: Power, Politics, and the Planetary Costs of Artificial Intelligence. Yale University Press (2021)
4. Dale, R., Viethen, J.: The automated writing assistance landscape in 2021. Natural Language Engineering (2021)
5. Dang, H., Benharrak, K., Lehmann, F., Buschek, D.: Beyond text generation: supporting writers with continuous automatic text summaries. In: UIST (2022)
6. Fu, L., Newman, B., Jakesch, M., Kreps, S.: Comparing sentence-level suggestions to message-level suggestions in AI-mediated communication. In: CHI (2023)
7. Gómez-Rodríguez, C., Williams, P.: A confederacy of models: a comprehensive evaluation of LLMs on creative writing. arXiv arXiv:2310.08433 (2023)
8. Graesser, A.C., McNamara, D.S., Kulikowich, J.M.: Coh-Metrix: providing multi-level analyses of text characteristics. Educ. Researcher (2011)
9. Jakesch, M., Bhat, A., Buschek, D., Zalmanson, L., Naaman, M.: Co-writing with opinionated language models affects users' views. arXiv arXiv:2302.00560 (2023)
10. Lee, M., et al.: Evaluating human-language model interaction. arXiv arXiv:2212.09746 (2022)
11. McNamara, D.S., Graesser, A.C.: Coh-Metrix: an automated tool for theoretical and applied natural language processing. In: Applied Natural Language Processing: Identification, Investigation and Resolution. IGI Global (2012)
12. McNamara, D.S., Graesser, A.C., McCarthy, P.M., Cai, Z.: Automated Evaluation of Text and Discourse with Coh-Metrix. Cambridge University Press (2014)
13. McNamara, D.S., Ozuru, Y., Graesser, A.C., Louwerse, M.: Validating Coh-Metrix. In: Proceedings of the 28th Annual Conference of the Cognitive Science Society (2006)
14. Nesi, H., Gardner, S.: Genres Across the Disciplines: Student Writing in Higher Education. Cambridge University Press (2012)
15. Ought: Elicit: The AI research assistant (2023). https://elicit.org
16. Quinn, P., Zhai, S.: A cost-benefit study of text entry suggestion interaction. In: CHI (2016)
17. Tate, T., et al.: Educational research and AI-generated writing: confronting the coming Tsunami (2023)
18. Teubner, T., Flath, C.M., Weinhardt, C., van der Aalst, W., Hinz, O.: Welcome to the era of ChatGPT et al. the prospects of large language models. Bus. Inf. Syst. Eng. (2023). https://doi.org/10.1007/s12599-023-00795-x
19. Vincent, J.: Google's AI-powered smart compose feature is coming to docs (2019). https://www.theverge.com/2019/11/20/20973917/google-docs-smart-compose-feature-g-suite-update
20. Whalen, J., et al.: ChatGPT: Challenges, opportunities, and implications for teacher education. Contemporary Issues in Technology and Teacher Education (2023)
21. Wooldridge, M.: A Brief History of Artificial Intelligence: What it is, Where We are, and Where We are Going. Flatiron Books (2021)
22. Yuan, A., Coenen, A., Reif, E., Ippolito, D.: Wordcraft: story writing with large language models. In: ACM IUI (2022)

Teaching and Measuring Multidimensional Inquiry Skills Using Interactive Simulations

Ekaterina Shved[1]([✉])(iD), Engin Bumbacher[2](iD), Paola Mejia-Domenzain[1](iD), Manu Kapur[3](iD), and Tanja Käser[1](iD)

[1] EPFL, Lausanne, Switzerland
{kate.kutsenok,paola.mejia,tanja.kaeser}@epfl.ch
[2] University of Teaching Education Vaud, Lausanne, Switzerland
engin.bumbacher@hepl.ch
[3] ETH, Zurich, Switzerland
manukapur@ethz.ch

Abstract. Interactive simulations play a significant role in science education, serving as a platform for inquiry-based learning and fostering the development of scientific knowledge and skills. However, teaching and quantitatively measuring inquiry strategies has proven to be challenging due to their complex and inherently multidimensional nature. Our study goes beyond the prevalent focus on the Control of Variables Strategy (CVS) in prior work by incorporating additional relevant inquiry strategies in both teaching and measurement: exploring the variable range and conducting experiments under optimal conditions. We tested two different instructional approaches to jointly teach the three strategies by focusing either on data collection or on data interpretation. 161 chemistry apprentices were randomly assigned to one of the two instructional conditions or a control group without instruction and engaged in experimentation using an interactive simulation. In order to analyze joint strategy use, we applied a multi-step clustering method to students' log data that helped identify multidimensional student profiles of inquiry strategies. We found four profiles that related differently to conceptual learning, suggesting that combining strategies is more effective for conceptual learning than utilizing them individually. We also found that students instructed on data collection increased the use of strategy combinations with an emphasis on CVS. This suggests a potential avenue for assessing instruction efficacy, indicating that the impact may be strategy-specific. Source code and materials are released at https://github.com/epfl-ml4ed/inquiry-skills.

Keywords: Inquiry-based Learning · Experimental Strategies · Clustering · Trace Data Analysis

A. M. Olney et al. (Eds.): AIED 2024, LNAI 14829, pp. 482–496, 2024.
https://doi.org/10.1007/978-3-031-64302-6_34

1   Introduction

Interactive simulations have significantly enriched inquiry-based learning by providing students with hands-on opportunities to explore scientific principles through experimentation, variable manipulation, and real-time observation [6,10]. However, inquiry-based learning can be challenging for students due to the need for productive inquiry strategies [7,32,34]. Consequently, there is a growing focus on measuring inquiry strategies and developing instructions to better support students in mastering this educational approach [13,29,33].

To measure inquiry strategies, researchers increasingly analyze student behavior via log-file data. Existing literature primarily focuses on the Control of Variables Strategy (CVS) [9,19,21], which involves conducting consecutive experiments where only one variable changes at a time [2]. For instance, [23] measured CVS usage by counting labeled actions, revealing an association between strategy application and quantitative understanding of scientific phenomena. However, relying solely on CVS does not fully capture the comprehensive experimentation process [13,31]. For example, CVS is limited in addressing multivariate relations; to navigate such complexities, students often need additional strategies [28]. Important strategies beyond CVS that the literature has identified include systematically covering the possible range of variable values [24], recording data in equal increments [31], or experimenting under optimal conditions [4].

However, measuring non-CVS strategies is challenging [28], and capturing the multidimensional nature of inquiry strategies simultaneously is particularly demanding [12,30]. A prominent approach involves constructing student profiles using a cluster analysis of process data [1,26]. Even if studies using this approach focus mainly on low-level features like action frequencies or mean time spent on actions, they already reveal distinct inquiry patterns with different relations to learning outcomes [5,14,22]. However, such approaches often ignore or oversimplify the temporal aspects of inquiry processes. In addition, clustering solely on the low-level features may oversimplify or inadequately capture the nuanced complexities of students' inquiry strategies.

Similarly, developing effective instructions for fostering transferable inquiry skills has posed considerable difficulty as well [26]. Among the challenges is finding a balance between the structure and openness of the guidance. For example, one instructional approach that researchers have employed involves using heuristic worked examples [25]. This method has been found to improve students' comprehension of inquiry strategies in a specific context. However, when [18] developed heuristic worked examples for the entire inquiry cycle, the results demonstrated that such an approach was too heavily guided, and there was no difference in the conceptual post-test between the instructional group and the control. Related to the problem of developing effective instruction is the evaluation of instructional efficacy. Previous studies have predominantly assessed the effectiveness of inquiry-based instruction using the pre-/post-test paradigm. While many have reported positive effects [6,16,26], others have shown no significant impact [11,18]. One plausible explanation for these mixed findings is that tests do not provide insights into how students transfer the instruction into their inquiry behavior. They might face challenges translating acquired strategies into test performance, even though applying these strategies during experimentation [15]. Thus, improved measurement tools that specifically assess students' use of strategies are needed.

In this paper, we employ a multidimensional representation for teaching and measuring inquiry strategies. Our method involves developing two types of instruction on a generalizable set of inquiry strategies (CVS, Range - checking the whole variable range, Optimal - experimenting in optimal conditions) from a data collection and a data interpretation perspective. We conducted a study with 161 vocational education students randomly assigned to one of three instructional conditions (data collection, data interpretation, control) before engaging in an inquiry activity with an interactive chemistry simulation. We created features related to the CVS, Range, and Optimal strategies from students' log data and applied a multi-step clustering method to identify distinct multidimensional student profiles of inquiry strategies. With our approach, we aim to address two research questions: First, how does the application of different inquiry strategies, individually and in combination, impact students' conceptual learning (**RQ1**)? Second, how do instructions on data collection and data interpretation influence the use of inquiry strategies (**RQ2**)?

Our results show that the usage of individual inquiry strategies, and particularly a combination of all three inquiry strategies, leads to better conceptual learning. Furthermore, we find that students benefit from instruction on data collection, reflected in a high usage of combinations of inquiry strategies.

Fig. 1. Our approach, aiming to teach and measure multiple inquiry strategies using an interactive simulation.

2   Methodology

Our approach (Fig. 1) was designed to comprehensively explore students' behavior in an interactive simulation, focusing on teaching and measuring combinations of inquiry strategies. The instruction concentrated on teaching three different experimentation strategies (Range, Optimal, CVS) from a data collection or data interpretation perspective. In the main intervention, students engaged in an interactive simulation that allowed them to collect and interpret data from their experiments. To measure students' inquiry strategies, we developed features that reflected the instructed strategies and employed a multi-step clustering pipeline first to assess students' proficiency in each strategy separately and then create profiles integrating all three strategies.

2.1 Study Context

We conducted an experimental study with three instructional conditions: a control group and two groups receiving different inquiry strategy instructions. Subsequently, all students participated in an inquiry simulation and answered prediction questions in a post-test related to the explored scientific phenomenon.

Participants. A total of 161 chemistry apprentices (90 identifying as female) participated in the study. Participants were enrolled in Swiss vocational schools. The students' age ranged between $15-19$ years, and students were in the 1st year (71 participants), 2nd year (58 participants), and 3rd year (28 participants) of their apprenticeship. The study was integrated into students' classroom instruction. All students gave informed consent for sharing their data; for minor students, a subsequent parental opt-out was administered. The study was approved by the ethics review board of the university (Nr. 010-2023).

Procedure. The procedure involved a pre-measurement of students' prior knowledge of inquiry strategies a week before the study to evenly assign them to three instructional conditions. A week later, two groups received instructions on three inquiry strategies (CVS, Range, Optimal) from a data collection or data interpretation standpoint, followed by all students engaging in a $5-15$ minutes experimentation with an interactive chemistry simulation. Their task was to identify all factors that affect light absorption and understand how they do so. The task was embedded in a scenario relevant to the students' professional context, presenting them as apprentices in a food quality laboratory assigned by a company to inspect food products for potential contamination with colorful toxic substances. To address this, their boss recommended utilizing a technique based on the varied light absorption abilities of colorful substances.

Pre-measurement of Inquiry Strategies. Students took a brief measurement (~ 15 minutes) of their proficiency in the three instructed inquiry strategies (CVS, Range, Optimal). It was contextualized with a problem relevant to their daily experiences, requiring them to determine how aging and pressure affect cheese hardness. The measurement included three tasks: data collection, mathematical representation, and data interpretation. Randomized allocation within classrooms and stratification based on pre-measurement grades ensured an even distribution of students across instructional conditions, minimizing potential teacher or prior knowledge influences (Kruskal-Wallis test: $H = 1.17$, $p = .56$).

Inquiry Environment. We used the Beer's Law Lab PhET simulation (Fig. 2), which focuses on explaining what factors influence the light absorption of chemical substances. *Width* and *concentration* are two factors exhibiting a direct linear relationship with light absorption. The combination of laser *wavelength* and *solution* color demonstrates a complex non-linear dependency. Specifically, there exists an optimal laser *wavelength* for each color of the *solution*, maximizing light absorption. If the colors of the laser (*wavelength*) and of the *solution* are the same, the absorption becomes equal to 0, regardless of the *width* and *concentration* values. In the simulation, students could experiment with all four independent variables and immediately observe changes in light absorption (dependent variable) using the measuring instrument. Additionally, students could click on the record button to store the values of the variables in a table, allowing them to subsequently analyze or plot the data to investigate the relationship between the dependent and independent variables.

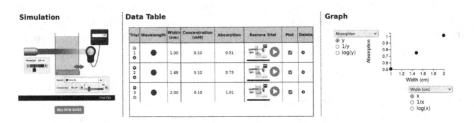

Fig. 2. Our environment includes the Beer's Law PhET simulation and tools for data collection (record button), analysis (data table), and interpretation (graph).

Conceptual Post-test. We evaluated students' understanding of Beer's Law with a three-part conceptual post-test. The first part involved predicting the outcome of an experiment, thereby measuring students' quantitative understanding of the investigated phenomenon. The second part required students to rank solutions based on their absorbance level (adapted from [4]). The third part was open-ended and required students to identify and explain the factors influencing light absorption. Students also had the possibility to propose an equation connecting all their findings. We scored the post-test on a range from $0 - 100$.

2.2   Teaching Inquiry Strategies

As mentioned earlier, we identified three experimentation strategies that prior work showed to be crucial for students' experimentation: CVS [2], exploring the whole variable range (Range) [24,31], and conducting experiments under optimal conditions (Optimal) [4]. CVS has been shown to be a domain-general and transferable strategy and has been found to be associated with conceptual understanding and performance [16]. The Range strategy involves exploring a large proportion of the possible range of variable values and is recognized as fundamental for scientific experiments [24]. Finally, the Optimal strategy refers to the use of values for covariates that are not masking or hampering the relationship between a focal variable in CVS and the dependent variable [4]. For example, if one wants to study the impact of the amount of light (focal variable) on plant growth (dependent variable), the level of water (covariate) has to be sufficient

Fig. 3. Overview of our clustering pipeline for measuring inquiry strategies

to enable the growth of plants in the first place, instead of simply using completely dried plants. Recent work using a data-driven approach [4] has found the Optimal strategy to significantly predict outcomes on a conceptual post-test.

To develop our instructional approaches, we built on prior work that divided the investigation phase of inquiry into data collection (or experimentation) and data interpretation [20]. The data collection phase involves gathering data of sufficiently high quality to draw inferences from the data [8, 21]. During the data interpretation phase, students examine the data to draw conclusions about the relationships between the variables, by examining graphical representations or even performing quantitative analysis [20, 23].

The use of the above-mentioned inquiry strategies plays a crucial role in both of these phases: the strategies help to gather new data when there is no or not sufficient data to begin with, or to select an appropriate subset of the collected data to draw valid conclusions on the relationships between the variables. We hypothesized that instructing students directly on the use of these strategies for data collection might help them assemble a dataset of sufficiently high quality, but they might not know how to make sense of that data. In contrast, focusing on how to interpret given data by selecting appropriate subsets of the data using the target strategies might help them to both draw valid conclusions and apply the strategies when collecting new data. To test these hypotheses, we developed two instructional approaches: teaching the use of the three target inquiry strategies by either focusing on the data collection (condition I_{DC}) or data interpretation phase (condition I_{DI}). To evaluate baseline measures of inquiry strategies, we also had a control group without any instructional intervention (C).

Instructions were delivered via 5-minute videos. Using a heuristic worked example, we explained each inquiry strategy in line with recommendations by [27]: for each strategy, we demonstrated an ineffective approach leading to false conclusions and then contrasted it to the effective one. In both conditions, a narrator guided through the videos, first introducing the worked example based on a non-scientific topic - renting apartments. In this example, the independent variables included 'rent,' 'apartment location,' and 'hours of tenant sleep'; the dependent variable was 'landlord revenue.' In the I_{DC} condition, the video showed how to apply the three inquiry strategies to data collection by contrasting data tables that were gathered either in ineffective or effective ways. In the I_{DI} condition, the video continued by explaining how to apply the three inquiry strategies to data interpretation. It showed for each strategy how to select a valid subset of data points from a big table, how to visualize them through plots, and how to interpret the plots - contrasting the ineffective approach with the effective one. Both instructions ended with a summary of the three introduced inquiry strategies to reinforce key concepts for the students.

2.3 Measuring Inquiry Strategies

To measure students' inquiry strategies, we adopted the multi-step clustering pipeline proposed by [17] for our context (see Fig. 3). We first extracted action sequences from students' interaction data with the simulation and created features reflecting their behavior in terms of the instructed strategies. We then performed a first clustering step independently for each strategy. In the second step, we combined the obtained strategy patterns into multi-strategy profiles.

Action Extraction. We extracted a sequence of actions $\mathbf{a_s}$ for each student s from the log data. Each action $a_{s,i}$ is described by a tuple (c_i, b_i, e_i) with c_i denoting the category of the action, b_i the start time, and e_i the end time. We categorized actions into 1) *explore* actions related to changing the independent variables, 2) *record* actions related to recording the simulation state, and 3) *other*. Furthermore, we connected each *explore* and *record* action to an independent variable. *Explore* actions were assigned the name of the independent variable the student interacted with (e.g., *explore-concentration*). Following [3], we merged consecutive explore actions separated by less than 3 seconds and linked to the same variable. *Record* actions were assigned the variable name of the preceding *explore* action. For example, a record action following an *explore-concentration* action would be described as *record-concentration*. We denoted all other *record* actions with *record-none*. Finally, we introduced *breaks* into the sequence, denoting periods of inaction with a duration of > 20 seconds [3].

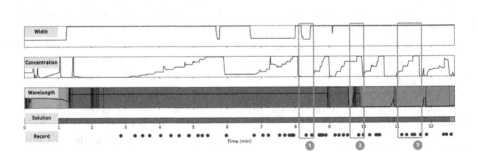

Fig. 4. Visualization of student log data from the simulation. Distinct experiments are framed: 1) all *explore* and *record-width* actions are *CVS, Non-Optimal, Range=66%*, number of *records=2*; 2) all *explore* and *record* actions are *Non-CVS*; 3) all *explore* and *record-concentration* actions are *CVS, Optimal, Range=100%*, number of *records=5*.

Feature Engineering. In a second step, we labeled each *explore* and *record* action individually for each strategy, denoting the labels as $\mathbf{l} = [l_k]$, where l_k

represents the label for strategy k, and $k \in \{CVS, Optimal, Range\}$. We then created features by aggregating over the action labels (illustrated in Fig. 4).
Strategy 1 - CVS. We labeled each *explore* and *record-variable* action with either *CVS*, denoting an action following the CVS strategy, or *Non-CVS*. Experiment ③ in Fig. 4 is an example for *CVS*: the student changed the concentration value while keeping other independent variables constant. For *explore* actions, we assumed that students using CVS would not rapidly change multiple variables consecutively without any records in-between [3]. Therefore, we labeled *explore* actions with no neighboring *explore* within a 3-second window as *CVS*. We designated a *record-variable CVS* if it immediately followed an *explore-CVS* action of the same variable (e.g., a *record-width* following an *explore-width-CVS* action would be labeled as *CVS*), or if it was taken in-between a *Non-CVS* and a *CVS* *explore* action for the same variable (e.g., a *record-width* action taken between an *explore-width-Non-CVS* action and an *explore-width-CVS* action would be labeled as *CVS*). We then computed the feature $f_{s,CVS}$ for each student s as the ratio between the number of *CVS* actions and the total number of experimentation actions (*CVS* + *Non-CVS* actions).
Strategy 2 - Optimal. We also labeled each *explore* and *record-variable* action with either *Optimal*, denoting an action taken in optimal conditions, or *Non-Optimal*. Non-optimal experimentation conditions varied based on the investigated variable. When investigating the variable *width*, setting the *concentration* to 0 or using the same color for the laser and the *solution* would result in non-observable light absorbance differences in the simulation, and hence, would be labeled as *Non-Optimal*. Similarly, using the same laser and *solution* color when experimenting with *concentration* would also result in a *Non-Optimal* label. Finally, when investigating the *wavelength* or *solution*, setting the *concentration* to 0 would be labeled as *Non-Optimal*, as light absorbance will always be 0 in this case. Experiment ③ in Fig. 4 is an example of optimal conditions (different laser (*wavelength*) and *solution* colors), while experiment ① was performed in non-optimal conditions. We computed the feature $f_{s,Optimal}$ for each student s as the ratio between the number of *Optimal* actions and the total number of experimentation actions (*Optimal* + *Non-Optimal* actions).
Strategy 3 - Range. To label *explore* and *record-variable* actions based on the covered range, we extracted from each action sequence subsequences representing distinct experiments exp_i. Each exp_i groups subsequent *explore* and *record-variable* actions of the same independent variable. We then computed the covered range for each exp_i by finding the maximum (max) and minimum (min) value of the independent variable within the experiment and calculating $r_i = (max - min)/max\_range$, where max_range denotes the maximum variable range possible in the simulation. Next, we computed a multiplication factor m_i depending on the number of *records* n_i within exp_i. We assumed that a minimum of three data points was necessary for a robust experiment (when collecting fewer data points, even if the whole range of a variable is covered, students could, for example, miss a quadratic dependency). Therefore, we used a sigmoid function, setting $m_i = 1$ in case of no records (only *explore* actions),

to model the diminishing increase of m_i when recording more than three data points. We labeled each action within exp_i with $l_{Range} = r_i \times m_i$. In experiment ③ in Fig. 4, the student recorded five data points covering the whole range of the width variable (100%). Finally, we computed the feature $f_{s,Range}$ for each student s by first normalizing the range labels l_{Range} and then computing the average normalized range label over all *explore* and *record-variable* actions.

Individual Strategy Measurement. Following [17], we first clustered students separately for each strategy. We computed pairwise distances between students for features $f_{s,CVS}$, $f_{s,Optimal}$, $f_{s,Range}$ using the Euclidean distance and transformed them to similarity matrices using a Gaussian kernel ($\sigma = 0.0001$). We then applied Spectral Clustering and determined the optimal number of clusters using the Silhouette score. Subsequently, we labeled the clusters based on the specific strategy usage relative to other clusters. For example, students in the *High CVS* cluster exhibited more intensive CVS usage compared to those in the *Low CVS* cluster. After the first clustering step, we obtained a set of three labels for each student based on the resulting clusters for the individual strategies (e.g., *High CVS*, *High Optimal*, *Low Range*).

Combining Strategies. In a second clustering step, we employed K-Modes to create learner profiles combining all three strategies. For K-Modes, the higher the number of clusters, the cleaner the separation between them. Given the size of our data set (161 students), we determined the optimal number of clusters as the smallest number ensuring that within each profile, > 90% of the students shared the same labels from the individual clustering, hence balancing separation quality with a minimum student count per profile.

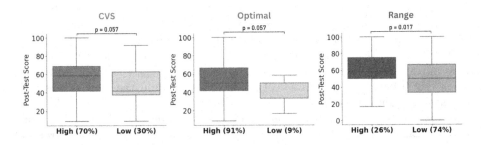

Fig. 5. Impact of individual inquiry strategy usage on post-test scores.

3   Results

We employed our measurement framework to group students with similar behaviors for individual strategies (CVS, Optimal, Range) and create integrated profiles to analyze the relation of strategy usage to conceptual learning (**RQ1**) and investigate the influence of instruction on students' inquiry behavior (**RQ2**).

3.1   Relation of Inquiry Strategies to Conceptual Learning (RQ1)

First, we examined the influence of individual strategy usage and then explored the impact of combined strategy use on conceptual learning.

Individual Strategies. Figure 5 illustrates the resulting clusters for each strategy. We obtained two clusters per individual strategy, dividing students into groups with relatively higher (*High*) and lower (*Low*) intensity of strategy use. For CVS, the majority of the students (112 students) were assigned to *High CVS* and the rest to *Low CVS* (49 students). Students in the *High CVS* cluster tended to achieve a higher post-test score than students in the *Low CVS* cluster (see Fig. 5). A Kruskal-Wallis test[1] comparing post-test scores between clusters revealed a a trend towards significance ($H = 3.76$, $p = .057$, corrected for multiple comparisons via a Benjamini-Hochberg (BH) procedure).

A small minority of students (15 students) experimented with sub-optimal conditions (*Low Optimal*) more frequently than the vast majority of students (*High Optimal*: 146 students). We observe (see Fig. 5) that while the two clusters show similar median post-test scores, the interquartile range is shifted towards lower post-test scores for the *Low Optimal* cluster. A Kruskal-Wallis test(See Footnote 4) comparing the post-test scores between the two clusters showed a trend to significance ($H = 3.61$, $p = .057$, corrected for multiple comparisons via a BH procedure).

In contrast to the CVS and Optimal strategy, for Range, the majority of students were assigned to *Low Range* (119 students), indicating that these students covered a smaller range and recorded fewer data points in their experiments than the minority of students assigned to *High Range* (42 students). Figure 5 indicates that students in *High Range* achieve a higher post-test score than students in *Low Range*. A t-test[2] confirms that there is indeed a significant difference between the post-test grades of the two clusters ($t = 2.84$, $p = .017$, corrected for multiple comparisons via a BH procedure).

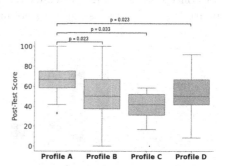

Profile	Students	Strategy		
		CVS	Optimal	Range
A	20%	High	High	High
B	51%	High	High	Low
C	6%	High	Low	Low
D	23%	Low	High	Low

Fig. 6. Percentage of students per profile along with a profile description (left), impact of profile on post-test scores (right).

---

[1] A Shapiro-Wilk test confirmed that the ANOVA assumptions were not satisfied.
[2] A Shapiro-Wilk test indicated a normal distribution of the data.

Combination of Strategies. Figure 6 (left) shows the characteristics of the combined student profiles. We identified four different profiles, which varied in their composition of strategy use. Notably, we achieved a high separation of clusters: in profiles A, C, and D all students exhibit the same label per individual strategy, while this is the case for more than 90% of students in profile B (*High CVS*: 93%, *High Optimal*: 93%, *Low Range*: 98%).

Students in profile A (32) showed a higher intensity in the usage of all inquiry strategies (*High CVS, High Optimal, High Range*), indicating that they performed experiments in optimal conditions, while controlling the variables and covering the range of the investigated variable. In contrast, students in the largest cluster (profile B: 83 students) demonstrated a high CVS and Optimal strategy usage but failed to cover the full range of the variable. Finally, students in profiles C and D (9 students and 37 students, respectively) demonstrated the use of only one strategy. While students in profile C seemed to know the CVS strategy, they frequently experimented in non-optimal conditions and did not cover the full variable range. In contrast, students in profile D experimented in optimal conditions, but did not conducted controlled and robust experiments.

Subsequently, we examined the influence of the combination of employed inquiry strategies on conceptual learning (Fig. 6 (right)). Students from profile A achieve higher post-test grades, while students from profile C seem to show the lowest performance. Indeed, an ANOVA(See Footnote 5) indicated significant differences in post-test performance between profiles ($F = 5.34$, $p = .0016$). Post-hoc pairwise comparisons using Tukey's HSD test, correcting for multiple comparisons using a BH procedure, revealed that students in profile A performed significantly better in the post-test than students from all other profiles (A vs. B $p = .023$, A vs. C $p = .033$, A vs. D $p = .023$). Interestingly, profile A and B differ only in one component (*Range*), indicating that the use of all three inquiry strategies is important for learning. Profiles B and C also demonstrate that the usage of one single strategy (*High CVS* or *High Optimal*) is not predictive for conceptual learning. We conclude that measuring students' behavior across multiple strategies is essential to predict conceptual learning.

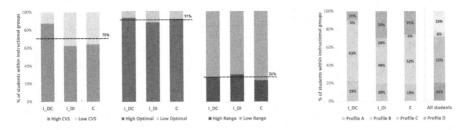

Fig. 7. Assignment of students to individual strategies clusters (left) and integrated profiles (right) for the different conditions. The dashed lines indicate the prevalence of the *High* cluster in the entire population.

3.2   Influence of Instruction on Inquiry Strategies (RQ2)

In our second analysis, we were interested in the effect of instruction on students' strategy usage in the simulation. An analysis of the instructional conditions based on the content post-test results revealed no significant differences.

Individual Strategies. Figure 7 (left) illustrates the usage of individual strategies across experimental conditions. Dashed lines indicate the respective cluster sizes in the overall population. We observe that a substantially larger part of students from the I_{DC} condition are assigned to the *High CVS* cluster (88%), in comparison to the I_{DI} (63%) and C conditions (64%). This observation indicates that instruction on data collection effectively taught the CVS strategy, while students did not profit from the instruction on data interpretation. A Chi-Squared test confirmed significant differences in CVS usage between conditions ($\chi^2 = 8.12$, $p = .017$). Post-hoc pairwise comparisons using BH corrections revealed that students in I_{DC} showed significantly higher prevalence in *High CVS* (I_{DC} vs. I_{DI}: $p = .03$, I_{DC} vs. C: $p = .03$). There are no differences in Optimal strategy usage between the three conditions. Given the high ratio of students in the *High Optimal* cluster (91%), our result suggests that students had no problems applying this strategy. In contrast, only a minority of students (26%) belonged to *High Range* and instruction on this strategy did not impact cluster assignment (I_{DC} : 28%, I_{DI} : 30%, C : 23%). This finding indicates that the Range strategy is more difficult to teach.

Combination of strategies. Figure 7 (right) illustrates the profile assignments for the three conditions and the overall population. For I_{DC}, a higher percentage of students (23%) than for I_{DI} (20%) or C (19%) is assigned to profile A (*High CVS, High Optimal, High Range*). Furthermore, 63% of I_{DC} students belong to profile B, utilizing two inquiry strategies (*High CVS, High Optimal*). Only 15% of the students with data collection instruction were assigned to profiles B or C, employing only one inquiry strategy. In contrast, 40% of I_{DI} students and 30% of the students from condition C were in profiles B and C. This indicates the effectiveness of teaching multiple strategies from a data collection perspective and, at the same time, shows that students did not benefit from the same instruction when employing a data interpretation perspective.

4   Discussion and Implications

We presented a study on teaching simultaneously multiple inquiry strategies in an interactive simulation and jointly measuring their use based on log data. We conducted a randomized control trial to test two instructional approaches teaching three different inquiry strategies (CVS, Optimal, and Range), focusing either on data collection or data interpretation. To answer our research questions, we applied a multi-step clustering method to students' log data to identify distinct multidimensional student profiles of inquiry strategies.

This study context and methodological approach allowed us first to examine the impact of different inquiry strategies usage on students' conceptual learning, both individually and in combination (**RQ1**). Our results expand on previous studies, which predominantly concentrated on the CVS strategy [9,19,23], by highlighting that combining strategies is more effective for conceptual learning than utilizing them individually. Specifically, we found four distinct student profiles based on the combination of strategy use. Students in Profile A actively employed all three inquiry strategies, demonstrating the highest post-test scores and outperforming students in all other profiles. Profile B students conducted experiments in a CVS manner under optimal conditions but failed to cover the full range of all variables while recording data. Students in profiles C and D utilized only one inquiry strategy while experimenting, which was less conducive to conceptual learning.

In subsequent analyses, we examined the impact of the two instructional approaches on the use of inquiry strategies and conceptual learning (**RQ2**). Our analysis revealed no significant differences in post-test scores between instructional conditions. However, we did find that students who received instructions on data collection showed significantly higher use of the CVS strategy than the other two conditions. We did not find any measurable impact for the data interpretation condition, contrary to our initial hypothesis in Sect. 2. Instead, our results suggest that directly teaching the strategies by focusing on data collection is more likely to translate to productive data collection behaviors than indirectly teaching data collection strategies through a focus on data interpretation. Furthermore, we did not find any difference between conditions in terms of the Optimal or Range strategy. However, we observed that there was already a high use of optimal conditions: across instructional groups, over 90% of students mostly experimented in optimal conditions. In the chosen interactive simulation, optimal conditions were already implicitly set up by the baseline choice of variables, which reduced the potential effect of instruction on this strategy.

Similarly, the high number of students belonging to the *Low Range* cluster may be attributed to the fact that they often did not record any data points. Such an approach might not be sufficient for a quantitative understanding of variable relationships [23]. This behavior correlated with a high percentage of range coverage but low or no data points recorded potentially connected to the simulation task, which encouraged experimentation in exploratory format but did not emphasize the importance of revealing mathematical relationships. Presumably, for the same reason, we observed a limited number of actions associated with data interpretation and plotting. A similar observation was made by [23], where only approximately 50% of students engaged in data plotting.

In conclusion, our study shows that combining measures of inquiry strategies based on log data provides a more holistic picture of how inquiry strategies relate to conceptual learning than examining individual strategies alone. Future work could explore more strategies and interactive simulation tasks. Further research is needed to design and evaluate effective instructions for inquiry strategies.

Acknowledgments. This project was substantially financed by the Swiss State Secretariat for Education, Research and Innovation SERI. Thanks to Hugues Saltini, Jade Maï Cock, Peter Bühlmann and Tanya Nazaretsky for their valuable input.

References

1. Bumbacher, E., Salehi, S., Wieman, C., Blikstein, P.: Tools for science inquiry learning: tool affordances, experimentation strategies, and conceptual understanding. J. Technol. Sci. Educ. (2018)
2. Chen, Z., Klahr, D.: All other things being equal: acquisition and transfer of the control of variables strategy. Child Dev. (1999)
3. Cock, J.M., Roll, I., Käser, T.: Consistency of inquiry strategies across subsequent activities in different domains. In: Proceedings of AIED (2023)
4. Cock, J.M., Marras, M., Giang, C., Käser, T.: Generalisable methods for early prediction in interactive simulations for education. In: Proceedings of EDM (2022)
5. Fratamico, L., Conati, C., Kardan, S., Roll, I.: Applying a framework for student modeling in exploratory learning environments: comparing data representation granularity to handle environment complexity. Int. J. Artif. Intell. Educ. (2017)
6. Fukuda, M., et al.: Scientific inquiry learning with a simulation: providing within-task guidance tailored to learners' understanding and inquiry skill. Int. J. Sci. Educ. (2022)
7. Gholam, A.: Inquiry-based learning: student teachers' challenges and perceptions. J. Inq, Action Educ. (2019)
8. Gobert, J., Pedro, M., Baker, R.: Assessing the learning and transfer of data collection inquiry skills using educational data mining on students' log files. In: Proceedings of AERA (2012)
9. Gobert, J.D., Kim, Y.J., Sao Pedro, M.A., Kennedy, M., Betts, C.G.: Using educational data mining to assess students' skills at designing and conducting experiments within a complex systems microworld. Think. Skills Creat. (2015)
10. de Jong, T., Linn, M.C., Zacharia, Z.C.: Physical and virtual laboratories in science and engineering education. Science (2013)
11. Kalthoff, B., Theyssen, H., Schreiber, N.: Explicit promotion of experimental skills and what about the content-related skills? Int. J. Sci. Educ. (2018)
12. Kruit, P.M., Oostdam, R.J., van den Berg, E., Schuitema, J.A.: Assessing students' ability in performing scientific inquiry: instruments for measuring science skills in primary education. Res. Sci. Technol. Educ. (2018)
13. Kuhn, D., Iordanou, K., Pease, M., Wirkala, C.: Beyond control of variables: what needs to develop to achieve skilled scientific thinking? Cogn. Dev. (2008)
14. Käser, T., Schwartz, D.L.: Modeling and analyzing inquiry strategies in open-ended learning environments. Int. J. Artif. Intell. Educ. (2020)
15. Lazonder, A.W., Harmsen, R.: Meta-analysis of inquiry-based learning: effects of guidance. Rev. Educ. Res. (2016)
16. Matlen, B.J., Klahr, D.: Sequential effects of high and low instructional guidance on children's acquisition of experimentation skills: is it all in the timing? Instr. Sci. (2013)
17. Mejia-Domenzain, P., Marras, M., Giang, C., Käser, T.: Identifying and comparing multi-dimensional student profiles across flipped classrooms. In: Proceedings of AIED (2022)
18. Mulder, Y.G., Lazonder, A.W., de Jong, T.: Using heuristic worked examples to promote inquiry-based learning. Learn. Instrum. (2014)
19. Nicolay, B., et al.: Unsuccessful and successful complex problem solvers - a log file analysis of complex problem solving strategies across multiple tasks. Intelligence (2023)

20. Pedaste, M., et al.: Phases of inquiry-based learning: definitions and the inquiry cycle. Educ. Res. Rev. (2015)
21. Pedro, M.S., Gobert, J.D., Baker, R.: Assessing the learning and transfer of data collection inquiry skills using educational data mining on students' log files. In: Proceedings of AERA (2012)
22. Peffer, M., Quigley, D., Mostowfi, M.: Clustering analysis reveals authentic science inquiry trajectories among undergraduates. In: Proceedings of LAK (2019)
23. Perez, S., Massey-Allard, J., Ives, J., Butler, D., Bonn, D., Bale, J., Roll, I.: Control of variables strategy across phases of inquiry in virtual labs. In: Proceedings of AIED (2018)
24. Pols, C.F.J., Dekkers, P.J.J.M., de Vries, M.J.: Defining and assessing understandings of evidence with the assessment rubric for physics inquiry: towards integration of argumentation and inquiry. Phys. Rev. Phys. Educ. Res. (2022)
25. Reiss, K., Renkl, A.: Learning to prove: the idea of heuristic examples. ZDM - Int. J. Math. Educ. (2002)
26. Roll, I., et al.: Understanding the impact of guiding inquiry: the relationship between directive support, student attributes, and transfer of knowledge, attitudes, and behaviours in inquiry learning. Instrum. Sci. (2018)
27. Saavedra, A.: Experiments in learning and transfer of inquiry strategies using short instructional videos. Ph.D. thesis, Stanford University (2022)
28. Saba, J., Kapur, M., Roll, I.: The development of multivariable causality strategy: instruction or simulation first? In: Proceedings of AIED (2023)
29. Sabourin, J., Mott, B., Lester, J.: Discovering behavior patterns of self-regulated learners in an inquiry-based learning environment. In: Proceedings of AIED (2013)
30. Scalise, K., Clarke-Midura, J.: The many faces of scientific inquiry: effectively measuring what students do and not only what they say. J. Res. Sci. Teach. (2018)
31. Schunn, C.D., Anderson, J.R.: The generality/specificity of expertise in scientific reasoning. Cogn. Sci. (1999)
32. Trautmann, N., MaKinster, J., Leanne, A.: What makes inquiry so hard? (and why is it worth it?). In: Proceedings of NARST (2004)
33. Vorholzer, A., von Aufschnaiter, C.: Guidance in inquiry-based instruction - an attempt to disentangle a manifold construct. Int. J. Sci. Educ. (2019)
34. Zacharia, Z.C., et al.: Identifying potential types of guidance for supporting student inquiry when using virtual and remote labs in science: a literature review. Educ. Technol. Res. Dev. (2015)

Author Index

Printed in the United States
by Baker & Taylor Publisher Services